Nanoplasmonics

Chemistry Centre, London,
United Kingdom
16–18 February 2015

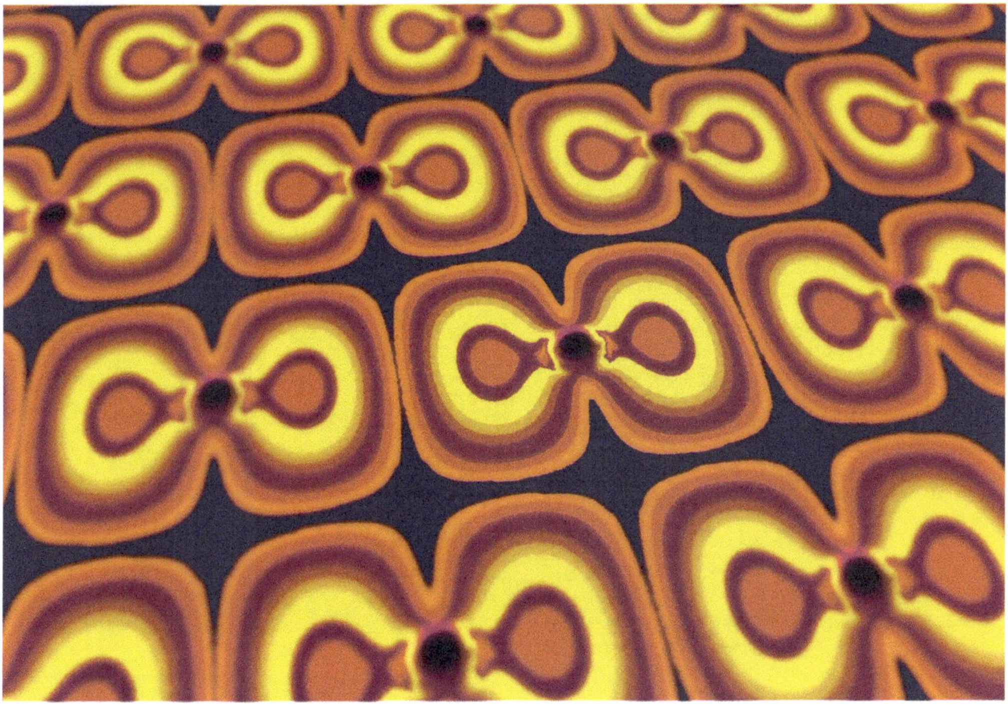

FARADAY DISCUSSIONS

Volume 178, 2015

ROYAL SOCIETY OF CHEMISTRY

The Faraday Division of the Royal Society of Chemistry, previously the Faraday Society, founded in 1903 to promote the study of sciences lying between Chemistry, Physics and Biology.

Faraday Discussions (Print ISSN 1359-6640, Electronic ISSN 1364-5498) is published 9 times a year by the Royal Society of Chemistry, Thomas Graham House, Science Park, Milton Road, Cambridge, UK CB4 0WF.

Volume 178 ISBN-13: 978-1-78262-178-2

2015 annual subscription price: print+electronic £877, US $1635; electronic only £833, US $1553. Customers in Canada will be subject to a surcharge to cover GST. Customers in the EU subscribing to the electronic version only will be charged VAT. All orders, with cheques made payable to the Royal Society of Chemistry, should be sent to RSC Order Department, Royal Society of Chemistry, Thomas Graham House, Science Park, Milton Road, Cambridge, CB4 0WF, UK.
Tel +44 (0) 1223 432398; E-mail www.orders@rsc.org

If you take an institutional subscription to any RSC journal you are entitled to free, site-wide web access to that journal. You can arrange access via Internet Protocol (IP) address at www.rsc.org/ip. Customers should make payments by cheque in sterling payable on a UK clearing bank or in US dollars payable on a US clearing bank.

Printed in the UK

MIX
Paper from responsible sources
FSC www.fsc.org FSC® C013604

Faraday Discussions are unique international discussion meetings that focus on rapidly developing areas of chemistry and its interfaces with other scientific disciplines.

Nanoplasmonics

Faraday Discussions

www.rsc.org/faraday_d

A General Discussion on Nanoplasmonics was held in London, UK on the 16th, 17th and 18th of February 2015.

RSC Publishing is a not-for-profit publisher and a division of the Royal Society of Chemistry. Any surplus made is used to support charitable activities aimed at advancing the chemical sciences. Full details are available from www.rsc.org

CONTENTS

ISSN 1359-6640; ISBN 978-1-78262-178-2

Cover
The electric field of surface plasmon polariton Bloch modes simulated over the plasmonic crystal with a hexagonal lattice formed by the array of apertures in a thin gold film. Image courtesy of Dr Ryan McCarron, Dr Daniel O'Connor and Prof. Anatoly Zayats, King's College London, UK.

INTRODUCTORY LECTURE

PAPERS AND DISCUSSIONS

CONCLUDING REMARKS

ADDITIONAL INFORMATION

Faraday Discussions

PAPER

Introductory lecture: nanoplasmonics

Mark L. Brongersma*

Received 31st March 2015, Accepted 31st March 2015
DOI: 10.1039/c5fd90020d

Nanoplasmonics or nanoscale metal-based optics is a field of science and technology with a tremendously rich and colourful history. Starting with the early works of Michael Faraday on gold nanocolloids and optically-thin gold leaf, researchers have been fascinated by the unusual optical properties displayed by metallic nanostructures. We now can enjoy selecting from over 10 000 publications every year on the topic of plasmonics and the number of publications has been doubling about every three years since 1990. This impressive productivity can be attributed to the significant growth of the scientific community as plasmonics has spread into a myriad of new directions. With 2015 being the International Year of Light, it seems like a perfect moment to review some of the most notable accomplishments in plasmonics to date and to project where the field may be moving next. After discussing some of the major historical developments in the field, this article will analyse how the most successful plasmonics applications are capitalizing on five key strengths of metallic nanostructures. This Introductory Lecture will conclude with a brief look into the future.

1 A brief history of the field of nanoplasmonics

The field of nanoplasmonics, or plasmonics for short, is concerned with the study and application of a surprisingly rich set of optical phenomena that can be elicited from metallic nanostructures. At the origin of these phenomena are easily accessible surface plasmon excitations, which are collective charge oscillations in a metal. They can take various forms, ranging from freely propagating electron density waves along metal surfaces to localized charge oscillations in metal nanoparticles. Their unique properties have motivated many scientific studies and a wealth of practical applications in physics, chemistry, biology and medicine. This Introductory Lecture of the Faraday Discussion 178 analyses how the most successful applications of plasmonics have effectively capitalized on five key strengths of plasmonic structures. These strengths have become evident over time as researchers learned more about the many unusual properties of nanoscale metallic objects. To understand the successes in plasmonics, it is thus important to understand the historical development of the field, and this is the topic of this first section.

Geballe Laboratory for Advanced Materials, Stanford University, Stanford, CA 94305, USA. E-mail: brongersma@stanford.edu; Fax: +1 650 736 1984; Tel: +1 650 736 2154

Michael Faraday's research marked the birth of the plasmonics field. Back in the 1850s he performed pioneering, systematic studies on the optical properties of gold (Au) leaf that was beaten so thin it was rendered translucent, transmitting in the green and reflecting the yellow part of incident sunlight.[1] He also synthesized and investigated Au colloidal particles and examined their distinct ruby red colours. Fig. 1 shows a glass flask with Au nanoparticles produced by Faraday himself.

In 1904, Maxwell Garnett provided a detailed explanation for the origin of the vibrant colours of the suspensions produced by Faraday when he described the optical properties of transparent media doped with a high density of metallic nanoparticles.[3] In his mathematical treatment, he used the newly developed Drude theory of metals and the understanding of the electromagnetic properties of small spheres developed by Lord Rayleigh. In his analysis, the optical properties of the small particles were derived in terms of bulk optical constants (real and imaginary part of the refractive index). A more physically intuitive description in terms of surface plasmon excitations in the metal nanoparticles only came much later.

Another early set of optical experiments involving metals were those by Heinrich Hertz in 1887, who discovered the photoelectric effect when he studied the impact of ultraviolet light on an electrical discharge between metallic electrodes.[4] The key explanation of this phenomenon was provided in 1905 by Albert Einstein.[5] More detailed photoemission studies on metallic films in the 1960s revealed the possible use of plasmon resonances to enhance light absorption in the near-surface region of the metal and thereby to enhance the photoemission

Fig. 1 A glass flask containing a suspension of nanoscale Au colloidal particles in water produced by Michael Faraday. The synthesis and optical properties of such colloidal suspensions and of beaten Au leaf were discussed in the famous Bakerian Lecture written by Michael Faraday in 1857.[1] Au colloids were typically prepared using aqueous solutions of Au-containing compounds, such as sodium chloroaurate ($NaAuCl_4$). The addition of a reducing solution, *e.g.* phosphorus in carbon disulfide in a two-phase system, would produce the ruby red liquid seen in this photograph. Although not visible in any optical microscope at the time, it was Turkevich who used electron microscopy to show that such a synthesis results in Au nanoparticles of a few to a few tens of nanometers.[2] Reproduced with permission from the Royal Institution of Great Britain (© Paul Wilkinson).

rate.[6,7] In metallic nanostructures, light absorption can be increased even further through the excitation of localized surface plasmon resonances and this topic is currently receiving renewed attention with an eye on its use for photocatalysis and solar energy harvesting.[8]

The first observation of propagating surface plasmons on metal films dates back to 1902 when Robert W. Wood perceived some unexplained features in optical reflection measurements he performed on metallic gratings.[9] The origins of these features, now termed Wood anomalies, were elucidated in 1941 in physically intuitive terms by Ugo Fano,[10] who suggested the important role of surface wave excitation. Using detailed experiments of the dispersive behaviour of the surface waves, these excitations were unequivocally linked to surface plasmons by Rufus Ritchie in 1968.[11]

Meanwhile, in 1956 David Pines theoretically described the characteristic energy losses experienced by fast electrons traveling through metals and attributed these losses to collective oscillations of free electrons inside the metal.[12] In analogy to earlier work on plasma oscillations in gas discharges, he named them "plasmons". This original research kicked off an ever-increasing stream of publications on the topic of plasmon science and technology (see Fig. 2).

In 1957 a theoretical study was published by Rufus Ritchie showing that electrons crossing the boundary of a metal film can experience a lowered electron energy loss compared to electrons moving through the bulk of a metal. He

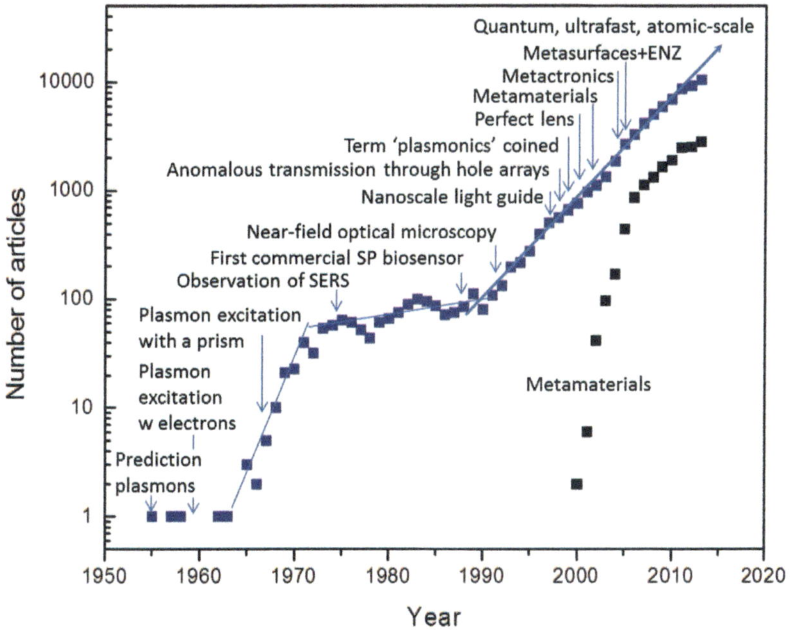

Fig. 2 The rapid growth of the field of plasmonics is illustrated by the number of scientific articles published annually containing the word "plasmon" or "surface plasmon" in either the title or abstract. This graph is updated from the original graph shown in the book "*Surface Plasmon Nanophotonics*," ed. M. L. Brongersma and P. G. Kik, Springer Series in Optical Sciences, 2007, vol. 131,[13] reprinted with kind permission from Springer Science and Business Media.

attributed this to the excitation of surface collective oscillations.[14] Powell and Swan experimentally demonstrated such excitations[15] for which the quanta were termed "surface plasmons" by Stern and Ferrell.[16]

A major advance in the study of surface plasmons was made in 1968 when Andreas Otto,[17] Erich Kretschmann[18] and Heinz Raether[19] presented methods for optically exciting surface plasmons on metal films, making experiments on plasmon phenomena easily accessible to many researchers. This sparked a series of fundamental studies on surface plasmon propagation along metallic films, starting with the work of Eleftherios Economou in 1969.[20] As a result, knowledge quickly built on how various experimental geometries can be optimized to excite highly confined short-range surface plasmons and more poorly confined long-range surface plasmons.[21,22] As the field continued to develop and the importance of the coupling between the oscillating electrons of a surface plasmon and the electromagnetic field was more appreciated, Stephen Cunningham *et al.* introduced the term surface-plasmon-polariton (SPP) in 1974.[23]

In the 1970s a completely new approach was explored to excite SPPs, which later played a key role in the development of active plasmonic sources. It did not rely on the gratings used by Wood or on optical prisms, which were rapidly gaining popularity at the time. Instead, it was based on the use of quantum emitters. This approach found its origin in the early experiments of Hans Kuhn[24] and Karl H. Drexhage,[25] who studied the changes in the fluorescent decay rate of molecules when they were spaced at different heights above a silver mirror (Fig. 3). They argued that most of their data could be explained using an intuitive classical model, in which the constructive/destructive interference of light that was directly emitted to free space and the reflected light from the mirror could lead to an enhancement/reduction in the decay lifetime. They did not give a detailed explanation for the faster-than-expected decay rates close to the metal beyond noting that this must be linked to energy losses into the metal. Ronald R. Chance, Alfred Prock and Robert J. Silbey provided an electromagnetic theory that included far-field as

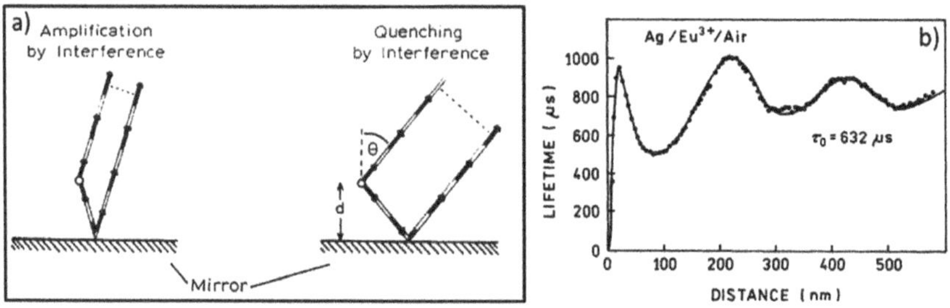

Fig. 3 (a) Schematics used by Drexhage to explain how the emission of an excited quantum emitter is influenced by the presence of a metallic mirror. He argued that the interference of the light that is directly emitted to free space and reflected light from the mirror could cause enhancements or reductions in the radiative decay rate of the emitter.[25] Reprinted from ref. 25 with permission from Elsevier. (b) Normalized photoluminescence lifetimes for an Eu^{3+} complex placed at different distances from a silver mirror. The solid line is the best fit using the model proposed by Chance, Prock, and Silbey,[27] which includes the impact of free space reflection of propagating light waves as well as the excitation of surface plasmon polaritons and electron hole pair excitation (*i.e.* 'lossy waves') in the metal.[28] Figure reprinted from ref. 27 with permission from John Wiley & Sons.

well as near-field effects and correctly predicted the decay rate at emitter spacings much smaller than the dominant emission wavelength λ_e.[26]

However, the physical nature of the decay process only became clear when the analogy between the radiative decay of molecules and the coupling of radio antennas to radio-wave propagation near the surface of the earth became apparent.[29] It was then argued that, as an emitter approaches a metal surface to a distance that is just a small fraction of λ_e, the excitation rate of propagating surface plasmon waves can become very effective.[30–33] Bo N. J. Persson then also argued that very close to the surface ($d < 10$ nm) electron–hole pair excitation can become dominant.[34] The first experimental demonstration of the coupling of emitters to surface plasmon waves was performed by Yu M. Gerbshtein and co-workers in 1975.[35] He showed that excited surface plasmon waves again could be decoupled to free-space photons with the help of a high-index prism. These works also stimulated later research in manipulating radiative decay and a range of other photophysical processes near finite-sized metallic nanoparticles.[36]

At this point the basic properties of surface plasmons were quite well-established. However the connection to the optical properties of metal nanoparticles had not yet been made. In 1970, more than sixty years after Maxwell Garnett's work on the colours of metal-doped glasses, Uwe Kreibig and Peter Zacharias initiated a rich set of studies in which they analysed the electronic and optical response of gold and silver nanoparticles.[37] In this work, they, for the first time, described the optical properties of metal nanoparticles in the context of localized surface plasmon excitations. Excellent reviews capture the great diversity and depth of the research on the optical properties of metallic nanoparticles.[38,39] Research in this time period discussed details of the importance of the size, shape, spatial arrangement and environment,[40] and even temperature,[41] quantum[42–44] and non-local[45] effects, in the optical response of nanometer-sized noble metal particles. With the increased control over nanostructure synthesis and arrangement as well as improved characterization tools, these topics remain of significant interest to the present day.

Another major discovery in the area of metal optics occurred in that same year, when Martin Fleischmann and co-workers observed strong Raman scattering from pyridine molecules in the vicinity of roughened silver surfaces.[46] Richard van Duyne[47] and J. Alan Creighton[48] independently came to the conclusion that the increase in surface area over a smooth surface could not explain the significant (>factor of 10) enhancements seen in the measured Raman signals. Although it was not realized at the time, the Raman scattering process, which involves an exchange of energy between photons and molecular vibrations, was enhanced by the concentrated electromagnetic fields near the rough silver surface due to the presence of surface plasmons. The link to plasmons was made by Martin Moskovits[49] and this facilitated the development of detailed electromagnetic theories that quantify the observed enhancements in Raman spectra by Moskovits,[50] George Schatz, Matthew Young and Richard Van Duyne.[51] The link to plasmons also inspired Mark Stockman, Vladimir Shalaev and co-workers to design more complex structures, such as fractal clusters, that could provide very significant field and Raman enhancements.[52] Ultimately, it became possible to take Raman spectra of individual molecules, as shown in the work by Katrin Kneipp.[53] These collective works led to the now well-established field of Surface Enhanced Raman Scattering (SERS).[54,55]

The 1980s saw the birth of the scanning near-field optical microscope (SNOM) through the work of Dieter Pohl and Aaron Lewis,[56–58] a tool that can both capitalize on surface plasmon excitations to concentrate light below the diffraction limit and also facilitate imaging of plasmon fields with high spatial resolution. Over time, tremendous progress has been made in improving the spatial resolution, reducing data acquisition time, and adding new functionality such as nanoscale spectroscopy[59,60] and the measurement of amplitude, phase and vectorial properties of light at the nanoscale.[61]

Around 1990, the growth in the annual number of papers on surface plasmons showed a sudden increase, and since this time the number of publications has doubled approximately every three years. This is linked to the development of the first commercial surface plasmon resonance (SPR)-based sensor in 1986 (ref. 62) based on the work by Bo Liedberg, Claes Nylander and Ingemar Lundström.[63] At present, still a very substantial fraction of all publications on surface plasmons involves the use of plasmons for sensing and biodetection.

Around the turn of the century, a range of new applications for metallic nanostructures were considered that are fundamentally different in character. In 1997 Juinichi Takahara and co-workers suggested that metallic nanowires could enable the guiding of optical beams below the free-space diffraction limit (see Fig. 4a),[64] in 1998 Thomas Ebbesen *et al.* reported on the extraordinary optical transmission through subwavelength metal apertures,[65] and in 2001 Sir John Pendry suggested that a thin metallic film may serve as a "perfect lens"[66] (see Fig. 4). These applications were aimed at manipulating the flow of light with metallic nanostructures below the diffraction limit to realize new optical functionalities. With the realization that surface plasmons, like electrons, and photons, could now form the basis for an entirely new device technology, Mark Brongersma and Harry Atwater coined the term "plasmonics" for the field in 1999.[67]

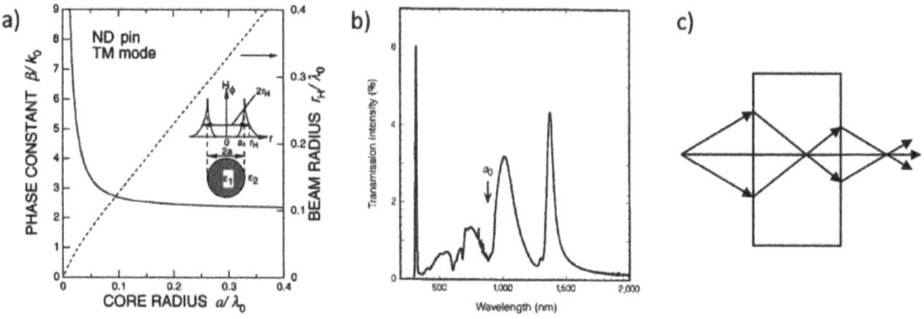

Fig. 4 (a) Plots of the phase constant and beam radius for a metallic nanowire waveguide used by Takahara and co-workers to illustrate how such waveguides can guide light with a deep subwavelength mode diameter.[64] Reprinted from ref. 64 with permission from OSA Publishing. (b) Transmission spectrum of a Ag film with a square array of holes that Ebbesen and co-workers took to prove that properly designed arrays can exhibit a light transmission which was larger than expected from knowledge of the total area of the holes.[65] Reprinted by permission from Macmillan Publishers Ltd from ref. 65, copyright 1998. (c) Schematic used by Sir John Pendry to illustrate how negative refraction can make a perfect lens.[66] Figure reprinted with permission from ref. 66. Copyright 2000 by the American Physical Society.

From the turn of the century, the development of plasmonics saw a tremendous diversification into many new areas of science and technology. Some of the most promising and fast-growing areas included plasmon-assisted catalysis,[8,36,68–73] nanoscale thermal engineering for nanostructure synthesis and phase transformations,[74–79] heat-assisted magnetic recording[80,81] and treatment of cancer,[82,83] optical trapping[84–86] and sensing,[87–89] optical imaging[90–94] and lithography,[95,96] engineering emission from quantum emitters,[97–107] thermal radiation engineering,[108–112] ultrafast and nonlinear plasmonics,[113–116] plasmonics for solar energy harvesting,[8,69,70,117–128] and active chipscale plasmonic devices including most notably nanoscale sources,[129–139] modulators[140–145] and detectors.[146–152] Interestingly, the first applications mentioned in this list capitalized on light-induced heating, which was originally considered to be a major weakness of plasmonics. Many of the advances in these areas resulted from a deeper understanding of the light concentrating properties of resonant antennas and cavities[153–164] as well as the guiding properties of plasmonic waveguides[165–173] and nanocircuits.[137,174–178]

The rapid growth of the plasmonics field in the early part of the 21st century was in part facilitated by the development and commercialization of powerful electromagnetics simulation codes, the introduction of new plasmonic materials and the introduction of powerful nanofabrication and optical/physical analysis techniques. System-level simulation tools and design rules have also become available to analyse plasmonic structures of increased complexity, such as nanoparticle clusters, metamaterials, and non-periodic plasmonic devices. I briefly analyse the different areas of progress below.

Firstly, this time period saw an increase in the general availability of excellent commercial computational design tools for nanophotonic structures. These included now popular simulation techniques such as Finite-Difference Time-Domain (FDTD), Finite-Difference Frequency-Domain (FDFD), the Discrete Dipole Approximation (DDA) and the Boundary Element Method (BEM).[179] As a result, the computation of optical properties of nanostructures was no longer just accessible to a few electromagnetics experts.

Plasmonics also saw a large influx of new, high-performance materials. Of course, there was an ongoing quest to produce high-quality single crystal noble metal structures with ultrasmooth surfaces.[180–184] At the same time, Complementary Metal Oxide Semiconductor (CMOS) technology moved into the sub-100 nm size range and now facilitates the scalable and reproducible fabrication of plasmonic elements making it the truly the next chip-scale technology.[138,185–188] This technology allowed for the incorporation of an increasingly large materials set that is of great practical value to the plasmonics community. Alexandra Boltasseva, Vladimir Shalaev, Harry Atwater and others have strongly promoted the use of a wide variety of new Si-compatible materials for plasmonics and characterized their performance for different applications.[189–191]

Most recently, a set of new, more exotic materials have also been entering into the plasmonics research space. Unique 2-dimensional materials such as graphene both support plasmon propagation[192–194] and can benefit from the highly confined plasmon fields to realize new types of modulators and detectors.[195–200] The 2-dimensional transition metal dichalcogenides also exhibit a range of very unique optical properties that will no doubt unlock new plasmoptoelectronic device applications.[201–203]

New materials are also developed for active plasmonic devices. The down-scaling of active elements to a subwavelength footprint constitutes a particularly formidable challenge, as one cannot benefit from the long interaction lengths that conventional active devices can rely on to achieve a desired active functionality. This is particularly true for a range of important electro-optical switching, modulation and cavity-tuning devices, which tend to rely on weak electro-absorption or electro-refraction. To address this point, new metals have been introduced that feature lower carrier densities than the noble metals, such as transparent oxides, graphene and titanium nitride. Their carrier density can be tuned by chemical doping and various electrical gating strategies. As such, these materials can serve as actively-tunable metals, offering new pathways to modify the resonant optical properties beyond passive size and shape tuning.[145,204–209] New materials showing extreme changes in their optical properties in *e.g.* metal-to-insulator transitions, or changes in structural phases, are also receiving an increased interest for tuning of the optical properties of plasmonic elements by placing such materials in their environment.[210,211]

New materials that offer strong field concentration or surface wave guiding in other frequency regimes are also gaining more interest. For example, mid-infrared plasmonics has recently received more attention as researchers hope to manipulate infrared light and control near- and far-field thermal emission.[108–110,212–217] In the development of infrared and THz plasmonics, the concept of spoof surface plasmons also played an important role. Whereas regular surface plasmon waves are very poorly confined at long wavelengths where the conductivity of most metals is very high, it was shown that corrugation of the metal surfaces opens up the possibility to support more confined spoof surface plasmon waves that in many ways resemble conventional surface plasmons.

Finally, recent history has seen a reignited passion for quantum plasmonic materials and systems. Optical antennas are now routinely capable of concentrating light to physical dimensions where quantum size effects naturally occur in various nanomaterials, and it thus seems logical that links to various subdisciplines of solid state physics are currently being made. With the ability to realize increasingly small cavities, the Purcell effect, the flow of light from emitters near metal nanostructures and strong coupling effects are now investigated to new levels of detail.[98,100,218,219] The recent work by Isabel Romero, Javier Aizpurua, Garnett W. Bryant and F. Javier García de Abajo has also opened new lines of research into quantum plasmonic phenomena by showing that nearly touching metallic nanoparticles feature a singular optical response in the limit where they become touching. Here, the possibility to physically transfer charge between the adjacent particles has a dramatic impact on the plasmonic response/the supported plasmonic modes. This has spurred an intense debate and a flurry of experiments that aim to show the impact of quantum tunnelling and non-local effects between closely-spaced metal nanostructures on their optical response.[220–224]

New and powerful fabrication methods including printing tools, nanosphere lithography, soft lithography, focussed ion beam milling, two-photon lithography, self-assembly, and novel materials chemistry approaches have further accelerated progress.[225–233] Using these techniques, nanostructures can now be created with extreme control and over increasingly large areas. A similar substantive impact on the field comes from the advances in near-field and far-field optical characterization tools for plasmonic structures. A variety of commercial scanning near-field microscopes are now available and this has made this quite advanced route of characterizing nanostructures accessible to a large user base. In recent research it has also become eminently clear that the successful realization of future, ultra-compact photonic devices will require the development of new techniques capable of correlating the nanostructural properties of materials and devices to their optical performance. In this regard, one cannot underestimate the recent development of electron-beam-based techniques that can simultaneously map the morphology, atomic structure and optical modes of plasmonic devices at very high spatial resolution (see Fig. 5a).[234–238] Several leading groups, including those of Christian Colliex, Albert Polman, F. Javier García de Abajo, Nikolay Zheludev and Paul Midgley have shown that techniques such as cathodoluminescence and electron energy loss spectroscopy (EELS) can provide valuable information on the spatial distribution of the local density of optical states and the plasmonic modes that can be excited in nanometallic structures, at an astounding ~10 nm spatial resolution and most recently in 3 dimensions.[239]

With all of these technological advances, it has become possible to realize and understand the optical properties of plasmonic structures of increasing complexity. By placing metallic nanostructures first in pairs[240] and later in more intricate multi-particle arrangements, one could realize nanoscale waveguides,[166] complex antenna structures capable of directing light emission[99,134,241] and engineered particle clusters that exhibit Fano-like resonances[242] and structural-induced transparency.[243,244] Nordlander and Halas provided a key tool to intuitively understand the plasmon response of complex metal particle arrays based on

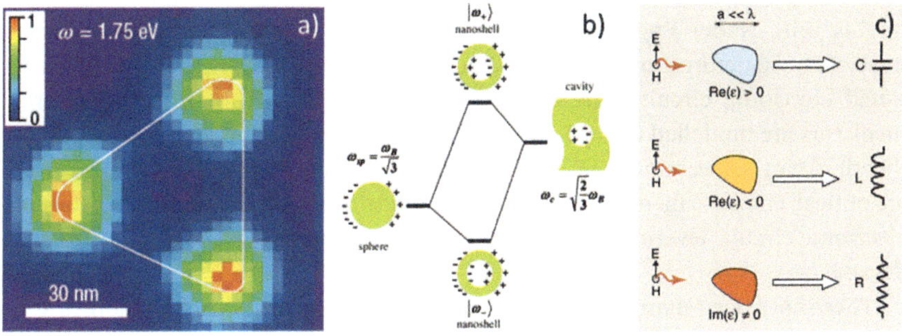

Fig. 5 (a) Experimental EELS map of a plasmonic mode of an Ag nanoprism. It was created by raster-scanning the electron beam from a scanning transmission electron microscope over the structure and by plotting the spatial distribution of the electrons that lost 1.75 eV through the excitation of this particular plasmonic mode.[234] Reprinted by permission from Macmillan Publishers Ltd from ref. 234, copyright 2007. (b) An energy-level diagram used to describe the concept of plasmon hybridization. Here, the optical properties of a metal nanoshell can be thought of as resulting from the interaction between sphere and cavity plasmons.[245] Figure from ref. 245, reprinted with permission from AAAS. (c) The illustrations show how nanoparticles with different dielectric properties can be viewed as lumped nanocircuit elements for light. Such modularization facilitates the rapid design of complex nanophotonic architectures consisting of a plurality of nano-elements. This forms the basis of metactronics.[274] Figure from ref. 274, reprinted with permission from AAAS.

the interaction or "hybridization" of elementary plasmons supported by constituent building blocks, in analogy to molecular orbital theory (see Fig. 5b).[245]

A natural extension of this has also been the development of metamaterials constructed from a dense array of subwavelength elements. These building blocks are typically metallic/plasmonic and can support both electrical and magnetic resonances. As a result new materials could be realized that exhibit exotic functionalities that go beyond those found in nature, including a negative index,[246–248] a zero index,[249] cloaking,[250] imaging below the diffraction limit,[90–92,94,251,252] lasing,[253] as well as unique anisotropic behavior,[254] chiral properties and optical activity.[255,256] Transformation optics has provided a theoretical framework for the design of new exotic metamaterial-based optical components, and suggests how one can achieve ultimate control over the flow of light by manipulating the properties of optical materials at a subwavelength scale.[257,258] Much of the work in this area was stimulated by the early works of Victor Veselago and Sir John Pendry that pointed to some of the wonderful opportunities for the metamaterials field.[66,257,259,260] Currently, new developments are emerging in the realization of flat optical elements[261–264] and essentially two-dimensional metamaterials, termed metasurfaces.[265–267] There is also a natural evolution to active metadevices.[252] These developments will ultimately enable completely flat and active optical components for optical beam/guided wave manipulation. Such planar elements afford facile integration with electronic, mechanical, magnetic, chemical and biotechnologies.

As the complexity of plasmonic systems increased, the development of simple design rules for components became absolutely essential. The power of good design rules lies in their ability to hide much of the complexity within an individual device. Instead, the aim is to capture the essence of the device function and focus on its interactions with other devices. Such simplifications then enable the construction of system-level theories and simulators that can predict the behaviour of larger plasmonic circuits and advanced multicomponent metamaterials. In this vein, Nader Engheta developed an elegant theoretical framework that treats nanostructured optical or "metactronic" circuits, much akin to conventional electronic circuitry, in 2005 (see Fig. 5c).[268,269] Within this framework, insulators are modelled as capacitors, metals as inductors, and energy dissipation (heating) can be accounted for by introducing resistors. The desired response of an optical circuit can now be realized simply through the optimization of an electronic circuit, levering knowledge built up by electrical engineers over many decades.

There have been many sceptical voices about the usefulness of plasmonics as a device technology. However, after roughly a decade and a half of research and development of plasmonic functionalities, researchers in the field have managed to figure out how to deal with the biggest challenge: optical losses in metallic systems.[270] By capitalizing on the strengths of plasmonic structures, new useful functionalities are now being designed that can either live with some degree of optical loss or capitalize on this loss to facilitate local heating. The state of the art is discussed in a number of excellent reviews.[113,271–273] In all successful plasmonics applications the metal structures perform a unique set of functions that are very hard or impossible to accomplish with dielectrics and semiconductors. In the following discussion, we will analyse the real strengths of nanometallic elements and their physical properties in more detail.

2 Plasmonics applications: capitalizing on key strengths

The previous section described the colourful and impressive history of the field of nanoplasmonics and argued how recent developments in nanotechnology gave rise to a significant acceleration in the developments in this area. In order to deeply appreciate the success and growth of this field, it is also important to understand the key strengths and weaknesses of plasmonic structures and devices. Below, I will argue how the most successful applications in the field make use of at least one of five unique properties of metallic nanostructures. These strengths are illustrated with examples in Fig. 6 and can be summarized as follows: (1) extreme light concentration and light manipulation below the free space diffraction limit; (2) straightforward tunability of the optically-resonant response of metallic nanostructures by engineering their size, shape or dielectric environment; (3) simple building blocks offer a tremendous design flexibility to create an almost unimaginably large number of optical functions; (4) efficient light-to-heat conversion with high spatial and temporal control; (5) multi-functionality of nanometallic elements in a single physical space.

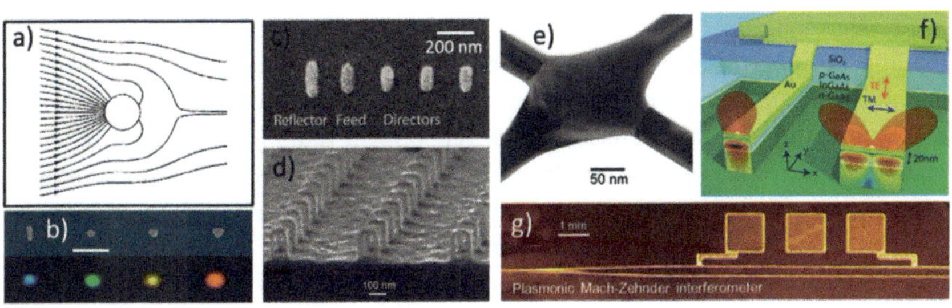

Fig. 6 (a) Image showing how the excitation of a localized surface plasmon redirects the flow of light (Poynting vector) towards and into a 20 nm-diameter Al nanoparticle.[275] Reproduced with permission from ref. 275. Copyright 1983, American Association of Physics Teachers. (b) Scanning electron microscopy (SEM) and corresponding dark-field optical microcroscopy images of Ag nanoparticles of different shapes (the scale bar is 300 nm).[226] Figure reprinted from ref. 226 with permission from John Wiley & Sons. (c) SEM image of an optical Yagi–Uda antenna driven by quantum dots. The image shows metallic reflector, feed, and director elements.[99] Figure from ref. 99, reprinted with permission from AAAS. (d) SEM image of an array of staple-shaped Au nanostructures forming a metamaterial with a negative permeability.[276] Figure reprinted with permission from ref. 276. Copyright 2005 by the American Physical Society. (e) Transmission electron microscopy (TEM) image of two Ag nanowires that were optically welded together through the excitation of a surface plasmon resonance at the crossing-point of the two wires.[277] Figure reprinted from ref. 277 with permission from Nature Publishing Group. (f) Schematic showing two Au antenna electrodes of different widths used to simultaneously inject current into an underlying quantum well and directing the emission from this quantum well in either a dipolar or quadrupolar angular emission pattern.[134] Figure reprinted from ref. 134 with permission from Nature Publishing Group. (g) Optical microscope image of a thermo-optic Mach–Zender interferometric modulator, in which Au strips simultaneously serve as long-range surface plasmon waveguides and electrical wires that can be heated to modify the index of a thermo-optic polymer surrounding the stripes and to achieve switching.[140] Reprinted with permission from ref. 140. Copyright 2004, AIP Publishing LLC.

Arguably the most well-recognized strength of metallic nanostructures is their ability to concentrate light to the nanoscale. Craig F. Bohren beautifully illustrated this point by showing that even a simple spherical metallic nanoparticle can concentrate light to the nanoscale (see Fig. 6a).[275] It essentially can operate like a miniature version of a radio antenna that can effectively capture and concentrate radio waves. The time-varying electric field associated with a light wave can exert a force on the gas of negatively charged electrons in a metallic nanostructure and drive them into a collective oscillation. The driven motion of the oscillating electrons can be described in analogy to a simple mass–spring system, in which the relevant mass is the electron mass and the restoring force is produced by the displaced electron charge. As such, it is expected that the system can feature a strong resonance in which all of the electrons are effectively set into oscillation by converting electromagnetic energy into kinetic energy of the electrons.

The ability to concentrate light and/or to produce high local field intensities has traditionally been the domain of dielectric lenses and resonators. These objects provide a convenient way to manipulate light and enhance a range of linear and non-linear phenomena with virtually no optical material loss. However, the fundamental laws of diffraction dictate that dielectric lenses cannot focus light to spot sizes less than about half a wavelength of light ($\sim\lambda/2$) and dielectric resonators exhibit electromagnetic mode volumes limited to $V_m \approx (\lambda/2)^3$, where λ is the wavelength of light inside the dielectric medium. The nature of nanometallic light concentrators and resonators is quite distinct from their dielectric counterparts, and nanometallic structures do not exhibit these limitations. Plasmonic antennas and lenses can convert optical radiation into intense, engineered, localized near-field distributions or deep sub-wavelength guided modes. Thus, wherever sub-wavelength control over light is required, nanometallic structures are likely to play an important role.

By squeezing light into subwavelength volumes, plasmonic structures can achieve a number of very valuable functions. In their role as optical antennas they can efficiently mediate interactions between propagating radiation and nanoscale objects and devices. This of course comes at the cost of some unavoidable optical losses that are intrinsically linked to the way metals manipulate light by means of plasmon excitations. However, in many applications one can live with some optical losses if given a dramatic increase in another, desired performance parameter. For example, in chip-scale devices, a reduced footprint can directly lead to an increased speed (which typically scales with a linear device dimension) and a reduced power consumption (which typically scales with the capacitance or area of a device). Smaller device structures tend to also produce steeper thermal gradients when heated and this affords more rapid cooling. In addition, the subwavelength dimensions of plasmonic structures facilitate increased light–matter interaction. Despite modest values of the optical quality factor Q (typically between 10–100), metallic cavities with sufficiently small mode volumes can often outperform higher-Q, dielectric cavities ($Q \sim 10^6$) and benefit from having a more broadband response and easy electrical access *via* the highly-conductive metals forming the cavity.[133,134,137,193] The strong confinement can ultimately lead to the realization of new types of useful quantum plasmonic devices and circuits for quantum information processing, including quantum computing, cryptography and metrology.[176,278–282]

The second key strength of plasmonic structures lies in the possibility to engineer their optically-resonant response in very straightforward ways through a modification of their size, geometry and dielectric environment.[39] An example is shown in the scanning electron microscopy (SEM) micrographs and associated dark field optical microscopy images of several differently-shaped Ag nanoparticles (see Fig. 6b). The observable changes in the colour of the scattered light from these structures can be understood by realizing that the different particle shapes cause incident light to drive different plasmon-currents and displaced-charge distributions. As a result, the restoring force produced by the displaced charges and the associated resonance frequency will also be different. The fact that very simple changes in the shape and size can result in dramatically different responses constitutes a tremendously powerful notion in the design of plasmonic functions. It affords a tremendous design flexibility and opens the door to the creation of a large diversity of possible devices.

The third key strength of plasmonics lies in the fact that simple nanometallic building blocks offer a tremendous design flexibility to create an almost unimaginably large number of optical functions. Here, one can take advantage of the fact that metallic nanostructures can feature useful near- and far-field optical interactions.[39,245,283] For example, Fig. 6c shows an SEM image of a five element Yagi–Uda antenna capable of directing the emission from a quantum dot.[99] Complex multi-nanoparticle assemblies also offer the opportunity to create sharp optical resonances to more effectively concentrate light and gain increased control over a range of physical processes, including light absorption, fluorescence and Raman emission, hot electron ejection, and more.[8,242,284,285] The sensitive dependence of plasmonic scattering signals on the distance and angle between metallic nanostructures has also facilitated the development of new nanoscale measurement, sensing and tomographic imaging techniques for chemistry and biology.[286–288] In further increasing complexity, the assembly of many plasmonic building blocks has led to the realization of bulk metamaterials and metasurfaces (see Fig. 6d).[276] Most recently, there has also been a growing interest in realizing non-periodic plasmonic structures and devices that can that outperform their periodic counterparts. To identify and realize such devices is a nontrivial task due to the need to solve Maxwell's equations repeatedly over an enormous design space. Here, brute-force optimization is typically impractical and therefore these type of optimizations rely on a variety of techniques to reduce the computation time. The reward can however be significant, and recently, new types of aperiodic plasmonic couplers, wavelength splitters, mode transformers, as well as designer solar light trapping and sensing schemes have been explored.[118,289–291]

Whereas in many plasmonics applications the generation of heat is undesirable, the ability to rapidly raise and lower the temperature in nanoscale volumes of material has also found new applications. Metal nanostructures are arguably the most efficient objects at converting incident light to heat.[292,293] As such, local heat generation can be viewed as the fourth key strength of metallic nanostructures. Early experiments on the propagation of surface plasmon polaritons along metal films already highlighted the efficient light-to-heat conversion inside the metal. In careful measurements, the heat produced by the radiationless decay of the propagating plasmons could be quantified in a photoacoustic study.[294] Subwavelength metallic nanostructures, in particular, can be heated very effectively by capitalizing on their large optical absorption cross section.[295,296] By engineering their size, shape, and dielectric environment, one can control the way in which they absorb and scatter light as well as the relative importance of these processes.[297–300] Several techniques have been developed to measure the local temperature near metallic nanostructures.[301–303] The ability to effectively heat and measure temperature has been exploited in many practical plasmonics applications. Recent active areas of research include the selective identification and killing of cancer cells,[304] photothermal nanotherapeutics,[305] modification of polymer surfaces,[306] local control over phase transitions,[307,308] growth of individual metallic nanoparticles,[79] semiconductor nanowires and carbon nanotubes,[309,310] nanofluidics and chemical separation,[311] heterogeneous catalysis,[312,313] drug delivery,[314] reversible photothermal melting of DNA,[315,316] heat assisted magnetic recording[81] and steam generation.[317,318] Fig. 6e shows another photothermal application of plasmonics, in which two metallic nanowires were welded together with the help of local heat generation by surface plasmon excitations.[277] The large change in the plasmonic resonance properties upon the realization of a solid (electrical) connection between the two wires caused a desirable reduction in the heating efficiency and a natural self-limitation of the welding process. This process was found to be very effective in the creation of large-area transparent metallic electrodes. In many of the above-mentioned applications, the need to only heat locally with plasmons rather than globally in a furnace has resulted in major increases in control, speed, and energy efficiency with an accompanying reduction in cost. Complex assemblies of metallic nanostructures may be designed to thermally engineer nanoscale environments in which local heating occurs in a desired, prescribed fashion. This opens up a wide range of opportunities for the creation of new devices in which thermally-stimulated processes can be manipulated and controlled.

One final unique feature of metallic nanostructures is that they can perform multiple functions at the same time. This is rooted in the fact that metals can display a number of desirable physical properties, including a high electrical conductivity, an excellent thermal conductivity, a good mechanical stability, a high catalytic activity, a high temperature stability, *etc.* Of course, different metals perform better in terms of some of these properties than others, but many metals exhibit several of these desirable traits. Examples of the use of multifunctionality can be found in chip-scale optoelectronic devices and systems. Here, the metal wiring is traditionally only used to inject/extract current or to apply electric fields. However, the metallic leads in photodetectors and sources can simultaneously be used as electrodes and as optical antennas capable of enhancing the way light is received[193,319,320] or emitted.[134,137,321] Fig. 6f shows an example where metallic antenna-electrodes are used to inject current into a light-emitting quantum well and to control the angular and polarization properties of the emitted light. In a different approach, the metals in plasmonic waveguides can be used to transmit optical signals, but also serve as electrical heaters or as contacts to apply an electrical field to modulate the guided signal.[140,142] The pioneering work by Sergey Bozhevolnyi on plasmonic modulators elegantly illustrates this concept (see Fig. 6g). Here, long-range surface plasmons are modulated in a thermo-optic Mach–Zender interferometric device constructed from 15 nm-thin and 8 μm-wide gold stripes embedded in a thermo-optic polymer. One of the metallic stripes can be heated by running an electrical current through it. The resulting temperature increase induces a change in the local refractive index in the polymer surrounding the heated stripe. This in turn causes the interference of the plasmon signals propagating along the two parallel stripes of the interferometer to be changed from constructive to destructive. In this way, the device can modulate incoming signals with great efficiency and extinction ratios as high as 30 dB were achieved. By cleverly combining metals and semiconductors, entirely new functions can also be realized. For example, it was recently shown that the metallic leads to a Si nanowire photodetector could render the device 'invisible' (*i.e.* significantly reduce the scattering cross section of the structure).[319] In this example, the metal again performs multiple functions at the same time and in the same physical space. These types of multifunctional devices can only be achieved by moving beyond traditional design principles in optoelectronics that prescribe an initial optimization of the performance of the individual electronic and optical building blocks, followed by an assembly of the pieces. In the design of multifunctional devices, the boundaries between electronic, photonic and other functional elements are less clear (or completely gone) and these distinct functions need to be optimized in unison. The benefit may be an increased functionality of chip-scale devices and new unexplored opportunities for ultra-dense integration.

3 The future of nanoplasmonics

The field of nanoplasmonics has come a long way since the early studies by Michael Faraday. He would have been pleased to learn about the tremendous wealth of pure science and exciting practical applications that developed over the next century and a half. Whereas the early days of plasmonics focused primarily

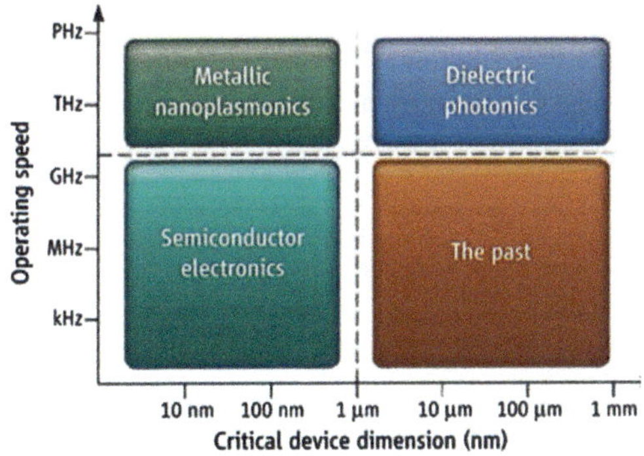

Fig. 7 Schematic elucidating the role of plasmonics with respect to other chip-scale device technologies. It shows how semiconductor electronics, dielectric photonics and metallic nanoplasmonics occupy well-defined domains on a graph of the operating speed and device dimensions. The dashed lines indicate physical/practical limitations of different technologies; semiconductor electronics in electronic processors tend to be limited in speed by thermal and interconnect delay time issues to about 10 GHz. Dielectric photonics is limited in its critical dimensions by the fundamental laws of diffraction. Plasmonics can serve as an effective bridge between similar-speed dielectric photonics and similar-size nano-electronics. Figure adapted from ref. 185 (M. L. Brongersma and V. M. Shalaev, *Science*, 2010, **328**, 440–441). Reprinted with permission from AAAS.

on fundamental science, we have seen a gradual transition from these fundamental studies to more application-driven research. In particular, the last two decades have seen a tremendous diversification of the application areas, and the reach of plasmonics has expanded into many different disciplines. Everywhere that light is used, it seems of value to think whether plasmonics can help to accomplish something new, better, faster, lower-power or smaller. Over time, researchers in the field have gained a better appreciation for the weaknesses of plasmonics and learned to capitalize on its key strengths to achieve many valuable results. It has also become more obvious as to how plasmonics can augment other technologies. As an example, Fig. 7 shows how plasmonics complements semiconductor electronics and dielectric photonics.[138,185]

From the graph it is clear that each of the presented device technologies can perform unique functions that play to the strength of the key materials that these technologies are based on. The electrical properties of semiconductors enable the realization of truly nanoscale elements for computation and information storage. The high transparency of dielectrics (*e.g.* glass) facilitates information transport over long distances and at incredible data rates. Unfortunately, semiconductor electronics is limited in speed by thermal and interconnect resistor–capacitor–inductor (RCL) delay-time issues and photonics is limited in size by the fundamental laws of diffraction. Plasmonics offers precisely what electronics and photonics do not have: the size of electronics and the speed of photonics. Plasmonic devices might therefore naturally interface with similar-speed photonic devices and with similar-size electronic components, increasing the synergy between these technologies.

Similar analyses can be applied to the linking of plasmonics to other fields of science and technology. As the control over the size, shape and placement of plasmonic structures approaches the atomic scale, plasmonics will more effectively and naturally interface with nanomechanics, nanomagnetics, nanobiology and ultimately chemistry. At these very small length scales, quantum, non-local, non-linear and ultrafast phenomena will naturally play a very big role in the design of new plasmonic functions. With the many recent developments and great future potential, it is clear that plasmonics is here to stay. It will be exciting to watch what the next 150 years will bring.

4 Concluding remarks

This Introductory Lecture has provided a brief overview of some of the key developments of the field of nanoplasmonics. I have also tried to provide my personal assessment of the limitations, key strengths and most exciting future opportunities for the field. In this relatively short Lecture, it is obviously impossible to capture the breath of this still very rapidly expanding field. I apologize to those researchers whose great scientific accomplishments were not covered. Fortunately, the many excellent lectures that were presented as part of this Faraday Discussion nicely captured the state of the art of the field and some of the frontier topics not highlighted here. I hope that this Faraday Discussion will inspire the next generation of young plasmonics scientists to take the field into exciting new directions and that they join us in defining and shaping the future of nanoplasmonics.

Acknowledgements

The author acknowledges support from a Multidisciplinary University Research Initiative grant from the Air Force Office of Scientific Research (grant no. FA9550-10-1-0264).

References

1 M. Faraday, *Philos. Trans. R. Soc. London*, 1857, **147**, 145–181.
2 J. Turkevich, P. C. Stevenson and J. Hillier, *Discuss. Faraday Soc.*, 1951, **11**, 55–75.
3 J. C. Maxwell Garnett, *Philos. Trans. R. Soc., A*, 1904, **203**, 385–420.
4 H. Hertz, *Ann. Phys. Chem.*, 1887, **267**, 983–1000.
5 A. Einstein, *Ann. Phys.*, 1905, **322**, 132–148.
6 J. Hofmann and W. Steinmann, *Phys. Status Solidi*, 1968, **30**, K53–K56.
7 J. E. Sipe and J. Becher, *J. Opt. Soc. Am.*, 1981, **71**, 1286.
8 M. L. Brongersma, N. J. Halas and P. Nordlander, *Nat. Nanotechnol.*, 2015, **10**, 25–34.
9 R. W. Wood, *Philosophical Magazine Series 6*, 1902, **4**, 396–402.
10 U. Fano, *J. Opt. Soc. Am.*, 1941, **31**, 213–222.
11 R. H. Ritchie, E. T. Arakawa, J. J. Cowan and R. N. Hamm, *Phys. Rev. Lett.*, 1968, **21**, 1530–1533.
12 D. Pines, *Rev. Mod. Phys.*, 1956, **28**, 184–198.
13 *Surface Plasmon Nanophotonics*, ed. M. L. Brongersma and P. G. Kik, Springer Series in Optical Sciences, 2007.
14 R. H. Ritchie, *Phys. Rev.*, 1957, **106**, 874–881.
15 C. J. Powell and J. B. Swan, *Phys. Rev.*, 1959, **115**, 869–875.
16 E. A. Stern and R. A. Ferrell, *Phys. Rev.*, 1960, **120**, 130–136.
17 O. Schwelb, *J. Lightwave Technol.*, 2004, **22**, 1380–1394.
18 E. Kretschmann, *Z. Phys.*, 1971, **241**, 313–324.
19 E. Kretschmann and H. Raether, *Z. Naturforsch.*, 1968, **23**, 2135–2136.
20 E. N. Economou, *Phys. Rev.*, 1969, **182**, 539–554.
21 J. J. Burke, G. I. Stegeman and T. Tamir, *Phys. Rev. B: Condens. Matter Mater. Phys.*, 1986, **33**, 5186–5201.
22 D. Sarid, *Phys. Rev. Lett.*, 1981, **47**, 1927–1930.
23 S. L. Cunningham, A. A. M. Maradudin and R. F. Wallis, *Phys. Rev. B: Condens. Matter Mater. Phys.*, 1974, **10**, 3342–3355.
24 H. Kuhn and H. Kuhn, *J. Chem. Phys.*, 1970, **53**, 101–108.
25 K. H. Drexhage, *J. Lumin.*, 1970, **1–2**, 693–701.
26 R. R. Chance, A. Prock and R. Silbey, *J. Chem. Phys.*, 1974, **60**, 2744–2748.
27 R. R. Chance, A. Prock and R. Silbey, in *Advances in Chemical Physics*, ed. I. Prigogine and S. A. Rice, 1978, vol. 37.
28 R. R. Chance, A. H. Miller, A. Prock and R. Silbey, *J. Chem. Phys.*, 1975, **63**, 1589–1595.
29 A. Sommerfeld, *Ann. Phys.*, 1909, **28**, 665–736.
30 H. Morawitz and M. Philpott, *Phys. Rev. B: Condens. Matter Mater. Phys.*, 1974, **10**, 4863–4868.
31 M. R. Philpott, *J. Chem. Phys.*, 1975, **62**, 1812–1817.

32 C. F. Eagen, W. H. Weber, S. L. McCarthy and R. W. Terhune, *Chem. Phys. Lett.*, 1980, **75**, 274–277.
33 G. W. Ford and W. H. Weber, *Phys. Rep.*, 1984, **113**, 195–287.
34 B. N. J. Persson and N. D. Lang, *Phys. Rev. B: Condens. Matter Mater. Phys.*, 1982, **26**, 5409–5415.
35 Y. Gerbshtein and I. Merkulov, *JETP Lett.*, 1975, **22**, 35–36.
36 J. I. Gersten and A. Nitzan, *Surf. Sci.*, 1985, **158**, 165–189.
37 U. Kreibig and P. Zacharias, *Z. Phys. B: Condens. Matter*, 1970, **231**, 128–143.
38 U. Kreibig and L. Genzel, *Surf. Sci.*, 1985, **156**, 678–700.
39 K. L. Kelly, E. Coronado, L. L. Zhao and G. C. Schatz, *J. Phys. Chem. B*, 1992, **107**, 668–777.
40 S. Person, M. Jain, Z. Lapin, J. J. Sáenz, G. Wicks and L. Novotny, *Nano Lett.*, 2013, **13**, 1806–1809.
41 U. Kreibig, *J. Phys. F: Met. Phys.*, 2001, **4**, 999–1014.
42 F. Ouyang, P. Batson and M. Isaacson, *Phys. Rev. B: Condens. Matter Mater. Phys.*, 1992, **46**, 15421–15425.
43 O. Keller, M. Xiao and S. Bozhevolnyi, *Opt. Commun.*, 1993, **102**, 238–244.
44 J. Tiggesbaumker, L. Koller and A. Liebsch, *Phys. Rev. A*, 1993, **48**, R1749–R1752.
45 A. Ljungbert and S. Lundqvist, *Surf. Sci.*, 1985, **156**, 839–844.
46 M. Fleischmann, P. J. Hendra and A. J. McQuillan, *Chem. Phys. Lett.*, 1974, **26**, 163–166.
47 D. L. Jeanmaire and R. P. Van Duyne, *J. Electroanal. Chem. Interfacial Electrochem.*, 1977, **84**, 1–20.
48 M. G. Albrecht and J. A. Creighton, *J. Am. Chem. Soc.*, 1977, **99**, 5215–5217.
49 M. Moskovits, *J. Chem. Phys.*, 1978, **69**, 4159.
50 M. Moskovits, *J. Chem. Phys.*, 1982, **77**, 4408–4416.
51 G. C. Schatz, M. A. Young and R. P. Van Duyne, *Top. Appl. Phys.*, 2006, **103**, 19–46.
52 M. I. Stockman, V. M. Shalaev, M. Moskovits, R. Botet and T. F. George, *Phys. Rev. B: Condens. Matter Mater. Phys.*, 1992, **46**, 2821–2830.
53 K. Kneipp, Y. Wang, H. Kneipp, L. T. Perelman, I. Itzkan, R. R. Dasari and M. S. Feld, *Phys. Rev. Lett.*, 1997, **78**, 1667–1670.
54 M. Moskovits, *Rev. Mod. Phys.*, 1985, **57**, 783–826.
55 E. C. Le Ru and P. G. Etchegoin, *MRS Bulletin*, 2013, **38**, 631–640.
56 D. W. Pohl, W. Denk and M. Lanz, *Appl. Phys. Lett.*, 1984, **44**, 651–653.
57 A. Lewis, M. Isaacson, A. Harootunian and A. Muray, *Ultramicroscopy*, 1984, **13**, 227–231.
58 E. Betzig, J. K. Trautman, T. D. Harris, J. S. Weiner and R. L. Kostelak, *Science*, 1991, **251**, 1468–1470.
59 A. Hartschuh, E. J. Sánchez, X. S. Xie and L. Novotny, *Phys. Rev. Lett.*, 2003, **90**, 095503.
60 T. Ichimura, N. Hayazawa, M. Hashimoto, Y. Inouye and S. Kawata, *Phys. Rev. Lett.*, 2004, **92**, 220801.
61 N. Rotenberg and L. Kuipers, *Nat. Photonics*, 2014, **8**, 919–926.
62 B. Liedberg, C. Nylander and I. Lunström, *Sens. Actuators*, 1983, **4**, 299–304.
63 B. Liedberg, C. Nylander and I. Lundström, *Biosens. Bioelectron.*, 1995, **10**, i–ix.

64 J. Takahara, S. Yamagishi, H. Taki, A. Morimoto and T. Kobayashi, *Opt. Lett.*, 1997, **22**, 475–477.
65 T. Ebbesen, H. Lezec, H. Ghaemi, T. Thio and P. Wolff, *Nature*, 1998, **391**, 667–669.
66 J. B. Pendry, *Phys. Rev. Lett.*, 2000, **85**, 3966.
67 M. L. Brongersma, J. W. Hartman and H. A. Atwater, *Mater. Res. Soc. Symp. Proc.*, 1999, **502**, H10.5.
68 K. Watanabe, D. Menzel, N. Nilius and H.-J. Freund, *Chem. Rev.*, 2006, **106**, 4301–4320.
69 I. Thomann, B. A. Pinaud, Z. Chen, B. M. Clemens, T. F. Jaramillo and M. L. Brongersma, *Nano Lett.*, 2011, **11**, 3340–3446.
70 S. Linic, P. Christopher and D. B. Ingram, *Nat. Mater.*, 2011, **10**, 911–921.
71 M. Maillard, P. Huang and L. Brus, *Nano Lett.*, 2003, **3**, 1611–1615.
72 S. Mukherjee, F. Libisch, N. Large, O. Neumann, L. V. Brown, J. Cheng, J. B. Lassiter, E. A. Carter, P. Nordlander and N. J. Halas, *Nano Lett.*, 2013, **13**, 240–247.
73 W. Hou and S. B. Cronin, *Adv. Funct. Mater.*, 2013, **23**, 1612–1619.
74 L. Cao, D. N. Barsic, A. R. Guichard and M. L. Brongersma, *Nano Lett.*, 2007, **7**, 3523–3527.
75 J. R. Adleman, D. A. Boyd, D. G. Goodwin and D. Psaltis, *Nano Lett.*, 2009, **9**, 4417–4423.
76 H. H. Wei, H. I-Kai, A. Bushmaker, R. Kumar, J. Theiss and S. B. Cronin, *Nano Lett.*, 2008, **8**, 3278–3282.
77 A. O. Govorov and H. H. Richardson, *Nano Today*, 2007, **2**, 30–38.
78 H. H. Richardson, A. C. Thomas, M. T. Carlson, M. E. Kordesch and A. O. Govorov, *J. Electron. Mater.*, 2007, **36**, 1587–1593.
79 K. F. MacDonald, V. A. Fedotov, S. Pochon, K. J. Ross, G. C. Stevens, N. I. Zheludev, W. S. Brocklesby and V. I. Emel'Yanov, *Appl. Phys. Lett.*, 2002, **80**, 1643–1645.
80 H. Chalabi, D. Schoen and M. L. Brongersma, *Nano Lett.*, 2014, **14**, 1374–1380.
81 W. A. Challener, C. Peng, A. V. Itagi, D. Karns, W. Peng, Y. Peng, X. Yang, X. Zhu, N. J. Gokemeijer, Y.-T. Hsia, G. Ju, R. E. Rottmayer, M. A. Seigler and E. C. Gage, *Nat. Photonics*, 2009, **3**, 303.
82 L. R. Hirsch, R. J. Stafford, J. A. Bankson, S. R. Sershen, B. Rivera, R. E. Price, J. D. Hazle, N. J. Halas and J. L. West, *Proc. Natl. Acad. Sci. U. S. A.*, 2003, **100**, 13549–13554.
83 P. K. Jain, I. H. El-Sayed and M. A. El-Sayed, *Nano Today*, 2007, **2**, 18–29.
84 L. Novotny, R. Bian and X. Xie, *Phys. Rev. Lett.*, 1997, **79**, 645–648.
85 Y. Pang and R. Gordon, *Nano Lett.*, 2012, **12**, 402–406.
86 J.-H. Kang, K. Kim, H.-S. Ee, Y.-H. Lee, T.-Y. Yoon, M.-K. Seo and H.-G. Park, *Nat. Commun.*, 2011, **2**, 582.
87 J. Homola, S. S. Yee and G. Gauglitz, *Sens. Actuators, B*, 1999, **54**, 3–15.
88 J. N. Anker, W. P. Hall, O. Lyandres, N. C. Shah, J. Zhao and R. P. Van Duyne, *Nat. Mater.*, 2008, **7**, 442–453.
89 N. Liu, M. L. Tang, M. Hentschel, H. Giessen and A. P. Alivisatos, *Nat. Mater.*, 2011, **10**, 631–636.
90 N. Fang, H. Lee, C. Sun and X. Zhang, *Science*, 2005, **308**, 534–537.
91 R. J. Blaikie and D. O. S. Melville, *J. Opt. A: Pure Appl. Opt.*, 2005, **7**, S176–S183.
92 Z. Jacob, L. V. Alekseyev and E. Narimanov, *Opt. Express*, 2006, **14**, 8247–8256.

93 A. Salandrino and N. Engheta, *Phys. Rev. B: Condens. Matter Mater. Phys.*, 2006, **74**, 1–5.
94 L. Zhaowei, H. Lee, Y. Xiong, C. Sun and X. Zhang, *Science*, 2012, **315**, 1686.
95 P. G. Kik, A. L. Martin, S. A. Maier and H. A. Atwater, *Proc. SPIE*, 2002, **4810**, 7–13.
96 Z. Ye, S. Zhang, Y. Wang, Y.-S. Park, T. Zentgraf, G. Bartal, X. Yin and X. Zhang, *Phys. Rev. B: Condens. Matter Mater. Phys.*, 2012, **86**, 155148.
97 P. Anger, P. Bharadwaj and L. Novotny, *Phys. Rev. Lett.*, 2006, **96**, 3–6.
98 A. Kinkhabwala, Z. Yu, S. Fan, Y. Avlasevich, K. Müllen and W. E. Moerner, *Nat. Photonics*, 2009, **3**, 654–657.
99 A. G. Curto, G. Volpe, T. H. Taminiau, M. P. Kreuzer, R. Quidant and N. F. van Hulst, *Science*, 2010, **329**, 930–933.
100 K. C. Y. Huang, Y. C. Jun, M. Seo and M. L. Brongersma, *Opt. Express*, 2011, **19**, 19084.
101 E. C. Garnett and P. Yang, *J. Am. Chem. Soc.*, 2008, **130**, 9224–9225.
102 A. F. Koenderink, *Opt. Lett.*, 2010, **35**, 4208–4210.
103 D. Gérard, J. Wenger, N. Bonod, E. Popov, H. Rigneault, F. Mahdavi, S. Blair, J. Dintinger and T. Ebbesen, *Phys. Rev. B: Condens. Matter Mater. Phys.*, 2008, **77**, 045413.
104 B. C. Buchler, T. Kalkbrenner, C. Hettich and V. Sandoghdar, *Phys. Rev. Lett.*, 2005, **95**, 063003.
105 S. Kühn, U. Håkanson, L. Rogobete and V. Sandoghdar, *Phys. Rev. Lett.*, 2006, **97**, 017402.
106 L. Rogobete, V. Sandoghdar and S. Ku, *Phys. Rev. Lett.*, 2006, **017402**, 1–4.
107 T. H. Taminiau, S. Karaveli, N. F. van Hulst and R. Zia, *Nat. Commun.*, 2012, **3**, 979.
108 J.-J. Greffet, R. Carminati, K. Joulain, J.-P. Mulet, S. Mainguy and Y. Chen, *Nature*, 2002, **416**, 61–64.
109 J. A. Schuller, T. Taubner and M. L. Brongersma, *Nat. Photonics*, 2009, **3**, 658–661.
110 H. Chalabi, E. Hasman and M. L. Brongersma, *Phys. Rev. B: Condens. Matter Mater. Phys.*, 2015, **91**, 014302.
111 A. P. Raman, M. A. Anoma, L. Zhu, E. Rephaeli and S. Fan, *Nature*, 2014, **515**, 540–544.
112 H. T. Miyazaki, K. Ikeda, T. Kasaya, K. Yamamoto, Y. Inoue, K. Fujimura, T. Kanakugi, M. Okada, K. Hatade and S. Kitagawa, *Appl. Phys. Lett.*, 2008, **92**, 141114.
113 M. I. Stockman, *Opt. Express*, 2011, **19**, 22029–22106.
114 K. F. MacDonald, Z. L. Samson, M. I. Stockman and N. I. Zheludev, *Nat. Photonics*, 2008, **3**, 55–58.
115 L. Cao and M. L. Brongersma, *Nat. Photonics*, 2009, **3**, 12–13.
116 M. Kauranen and A. V. Zayats, *Nat. Photonics*, 2009, **6**, 737–748.
117 M. Westphalen, U. Kreibig, J. Rostalski, H. Lu and D. Meissner, *Sol. Energy Mater.*, 2000, **61**, 97–105.
118 R. A. Pala, J. S. Q. Liu, E. S. Barnard, D. Askarov, E. C. Garnett, S. Fan and M. L. Brongersma, *Nat. Commun.*, 2013, **4**, 2095.
119 R. A. Pala, J. White, E. Barnard, J. Liu and M. L. Brongersma, *Adv. Mater.*, 2009, **21**, 3504–3509.
120 B. P. Rand, P. Peumans and S. R. Forrest, *J. Appl. Phys.*, 2004, **96**, 7519–7526.

121 D. Derkacs, S. H. Lim, P. Matheu, W. Mar and E. T. Yu, *Appl. Phys. Lett.*, 2006, **89**, 20–23.
122 S. Pillai, K. R. Catchpole, T. Trupke and M. A. Green, *J. Appl. Phys.*, 2007, **101**, 0–8.
123 H. A. Atwater and A. Polman, *Nat. Mater.*, 2010, **9**, 865–865.
124 M. Kahl, T. Thomay, V. Kohnle, K. Beha, J. Merlein, M. Hagner, A. Halm, J. Ziegler, T. Nann, Y. Fedutik, U. Woggon, M. Artemyev, F. Pérez-Willard, A. Leitenstorfer and R. Bratschitsch, *Nano Lett.*, 2007, **7**, 2897–2900.
125 O. Neumann, A. S. Urban, J. Day, S. Lal, P. Nordlander and N. J. Halas, *ACS Nano*, 2013, **7**, 42–49.
126 S. Mubeen, J. Lee, N. Singh, S. Krämer, G. D. Stucky and M. Moskovits, *Nat. Nanotechnol.*, 2013, **8**, 247–251.
127 S. V. Boriskina, H. Ghasemi and G. Chen, *Mater. Today*, 2013, **16**, 375–386.
128 S. Hayashi, K. Kozaru and K. Yamamoto, *Solid State Commun.*, 1991, **79**, 763–767.
129 D. J. Bergman and M. I. Stockman, *Phys. Rev. Lett.*, 2003, **90**, 027402.
130 M. T. Hill, Y.-S. Oei, B. Smalbrugge, Y. Zhu, T. de Vries, P. J. van Veldhoven, F. W. M. van Otten, T. J. Eijkemans, J. P. Turkiewicz, H. de Waardt, E. J. Geluk, S.-H. Kwon, Y.-H. Lee, R. Nötzel and M. K. Smit, *Nat. Photonics*, 2007, **1**, 589–594.
131 D. M. Koller, A. Hohenau, H. Ditlbacher, N. Galler, F. Reil, F. R. Aussenegg, A. Leitner, E. J. W. List and J. R. Krenn, *Nat. Photonics*, 2008, **2**, 684–687.
132 R. F. Oulton, V. J. Sorger, T. Zentgraf, R.-M. Ma, C. Gladden, L. Dai, G. Bartal and X. Zhang, *Nature*, 2009, **461**, 629–632.
133 A. C. Hryciw, Y. C. Jun and M. L. Brongersma, *Opt. Express*, 2009, **17**, 185–192.
134 Y. C. Jun, K. C. Y. Huang and M. L. Brongersma, *Nat. Commun.*, 2011, **2**, 283.
135 M. P. Nezhad, A. Simic, O. Bondarenko, B. Slutsky, A. Mizrahi, L. Feng, V. Lomakin and Y. Fainman, *Nat. Photonics*, 2010, **4**, 395–399.
136 Y.-J. Lu, J. Kim, H.-Y. Chen, C. Wu, N. Dabidian, C. E. Sanders, C.-Y. Wang, M.-Y. Lu, B.-H. Li, X. Qiu, W.-H. Chang, L.-J. Chen, G. Shvets, C.-K. Shih and S. Gwo, *Science*, 2012, **337**, 450–453.
137 K. C. Y. Huang, M.-K. Seo, T. Sarmiento, Y. Huo, J. S. Harris and M. L. Brongersma, *Nat. Photonics*, 2014, **8**, 244–249.
138 R. Zia, J. Schuller, A. Chandran and M. Brongersma, *Mater. Today*, 2006, **9**, 20–27.
139 P. Neutens, L. Lagae, G. Borghs and P. Van Dorpe, *Nano Lett.*, 2010, **10**, 1429–1432.
140 T. Nikolajsen, K. Leosson and S. I. Bozhevolnyi, *Appl. Phys. Lett.*, 2004, **85**, 5833–5835.
141 T. Nikolajsen, K. Leosson and S. I. Bozhevolnyi, *Opt. Commun.*, 2005, **244**, 455–459.
142 W. Cai, J. S. White and M. L. Brongersma, *Nano Lett.*, 2009, **9**, 4403–4411.
143 J. Dionne, K. Diest, L. Sweatlock and H. Atwater, *Nano Lett.*, 2009, **9**, 897.
144 A. Melikyan, L. Alloatti, A. Muslija, D. Hillerkuss, P. C. Schindler, J. Li, R. Palmer, D. Korn, S. Muehlbrandt, D. Van Thourhout, B. Chen, R. Dinu, M. Sommer, C. Koos, M. Kohl, W. Freude and J. Leuthold, *Nat. Photonics*, 2014, **8**, 229–233.
145 H. W. Lee, G. Papadakis, S. P. Burgos, K. Chander, A. Kriesch, R. Pala, U. Peschel and H. A. Atwater, *Nano Lett.*, 2014, **14**, 6468.

146 T. Ishi, J. Fujikata, K. Marita, T. Baba and K. Ohashi, *Jpn. J. Appl. Phys.*, 2005, **44**(part 2), L364–L366.
147 J. S. White, G. Veronis, Z. Yu, E. S. Barnard, A. Chandran, S. Fan and M. L. Brongersma, *Opt. Lett.*, 2009, **34**, 686–688.
148 L. Tang, S. E. Kocabas, S. Latif, A. K. Okyay, D.-S. Ly-Gagnon, K. C. Saraswat and D. A. B. Miller, *Nat. Photonics*, 2008, **2**, 226–229.
149 P. Neutens, P. Van Dorpe, I. De Vlaminck, L. Lagae and G. Borghs, *Nat. Photonics*, 2009, **3**, 283–286.
150 I. Goykhman, B. Desiatov, J. Khurgin, J. Shappir and U. Levy, *Nano Lett.*, 2011, **11**, 2219–2224.
151 C. Scales and P. Berini, *IEEE J. Quantum Electron.*, 2010, **46**, 633–643.
152 D.-S. Ly-Gagnon, K. C. Balram, J. S. White, P. Wahl, M. L. Brongersma and D. A. B. Miller, *Nanophotonics*, 2012, **1**, 9–16.
153 P. Mühlschlegel, H.-J. Eisler, O. J. F. Martin, B. Hecht and D. W. Pohl, *Science*, 2005, **308**, 1607–1609.
154 P. J. Schuck, D. P. Fromm, A. Sundaramurthy, G. S. Kino and W. E. Moerner, *Phys. Rev. Lett.*, 2005, **94**, 017402.
155 S. I. Bozhevolnyi and T. Søndergaard, *Opt. Express*, 2007, **15**, 10869–10877.
156 P. Bharadwaj, B. Deutsch and L. Novotny, *Adv. Opt. Photonics*, 2009, **1**, 438–483.
157 Y. Kurokawa and H. Miyazaki, *Phys. Rev. B: Condens. Matter Mater. Phys.*, 2007, **75**, 035411.
158 V. J. Sorger, R. F. Oulton, J. Yao and G. Bartal, *Nano Lett.*, 2009, **9**, 3489–3493.
159 H. Ditlbacher, A. Hohenau, D. Wagner, U. Kreibig, M. Rogers, F. Hofer, F. Aussenegg and J. Krenn, *Phys. Rev. Lett.*, 2005, **95**, 257403.
160 T. H. Taminiau, F. D. Stefani and N. F. van Hulst, *New J. Phys.*, 2008, **10**, 105005.
161 T. H. Taminiau, F. D. Stefani, F. B. Segerink and N. F. van Hulst, *Nat. Photonics*, 2008, **2**, 234–237.
162 D. T. Schoen, A. C. Atre, A. García-Etxarri, J. A. Dionne and M. L. Brongersma, *Nano Lett.*, 2015, **15**, 120–126.
163 T. H. Taminiau, F. D. Stefani and N. F. Van Hulst, *Nano Lett.*, 2011, **11**, 1020–1024.
164 M. L. Brongersma, *Nat. Photonics*, 2008, **2**, 270–272.
165 M. Quinten, A. Leitner, J. R. Krenn and F. R. Aussenegg, *Opt. Lett.*, 1998, **23**, 1331–1333.
166 M. L. Brongersma, J. W. Hartman and H. A. Atwater, *Phys. Rev. B: Condens. Matter Mater. Phys.*, 2000, **62**, R16356–R16359.
167 J.-C. Weeber, Y. Lacroute and A. Dereux, *Phys. Rev. B: Condens. Matter Mater. Phys.*, 2003, **68**, 1–10.
168 R. Zia, M. D. Selker, P. B. Catrysse and M. L. Brongersma, *J. Opt. Soc. Am. A. Opt. Image Sci. Vis.*, 2004, **21**, 2442–2446.
169 R. Zia, M. Selker and M. Brongersma, *Phys. Rev. B: Condens. Matter Mater. Phys.*, 2005, **71**, 1–9.
170 R. Zia and M. L. Brongersma, *Nat. Nanotechnol.*, 2007, **2**, 426–429.
171 D. F. P. Pile, T. Ogawa, D. K. Gramotnev, Y. Matsuzaki, K. C. Vernon, K. Yamaguchi, T. Okamoto, M. Haraguchi and M. Fukui, *Appl. Phys. Lett.*, 2005, **87**, 1–3.

172 G. Veronis, Z. Yu, S. Kocabas, D. Miller, M. Brongersma and S. Fan, *Chin. Opt. Lett.*, 2009, **7**, 302–308.
173 S. I. Bozhevolnyi, *Opt. Express*, 2006, **14**, 9467–9476.
174 S. I. Bozhevolnyi, V. S. Volkov, E. Devaux, J. Laluet and T. W. Ebbesen, *Nature*, 2006, **440**, 508–511.
175 D. K. Gramotnev and S. I. Bozhevolnyi, *Nat. Photonics*, 2010, **4**, 83–91.
176 N. P. De Leon, M. D. Lukin and H. Park, *IEEE J. Sel. Top. Quantum Electron.*, 2012, **18**, 1781–1791.
177 W. Cai, W. Shin, S. Fan and M. L. Brongersma, *Adv. Mater.*, 2010, **22**, 5120–5124.
178 W. Shin, W. Cai, P. B. Catrysse, G. Veronis, M. L. Brongersma and S. Fan, *Nano Lett.*, 2013, **13**, 4653–4758.
179 G. A. E. Vandenbosch, in *Plasmonics – Principles and Applications*, ed. K. Y. Kim, InTech, 2012, pp. 23–48, ISBN: 978-953-51-0797-2.
180 P. Nagpal, N. C. Lindquist, S.-H. Oh and D. J. Norris, *Science*, 2009, **325**, 594–597.
181 Y. Xia, Y. Xiong, B. Lim and S. E. Skrabalak, *Angew. Chem., Int. Ed.*, 2009, **48**, 60–103.
182 Z. Yan, Y. Bao, U. Manna, R. A. Shah and N. F. Scherer, *Nano Lett.*, 2014, **14**, 2436–2442.
183 X. Zhu, Y. Zhang, J. Zhang, J. Xu, Y. Ma, Z. Li and D. Yu, *Adv. Mater.*, 2010, **22**, 4345–4349.
184 H. Gao, Y. Hu, Y. Xuan, J. Li, Y. Yang, R. V Martinez, C. Li, J. Luo, M. Qi and G. J. Cheng, *Science*, 2014, **346**, 1352–1356.
185 M. L. Brongersma and V. M. Shalaev, *Science*, 2010, **328**, 440–441.
186 S. Papaioannou, D. Kalavrouziotis, K. Vyrsokinos, J.-C. Weeber, K. Hassan, L. Markey, A. Dereux, A. Kumar, S. I. Bozhevolnyi, M. Baus, T. Tekin, D. Apostolopoulos, H. Avramopoulos and N. Pleros, *Sci. Rep.*, 2012, **2**, 1–9.
187 S. Zhu, G. Q. Lo and D. L. Kwong, *Int. J. Electr. Comput. Electron. Commun. Eng.*, 2013, **7**, 712–715.
188 J. T. Kim, J. J. Ju, S. Park, M. Kim, S. K. Park and M.-H. Lee, *Opt. Express*, 2008, **16**, 13133–13138.
189 A. Boltasseva and H. A. Atwater, *Science*, 2011, **331**, 290–291.
190 P. R. West, S. Ishii, G. V. Naik, N. K. Emani, V. M. Shalaev and A. Boltasseva, *Laser Photon. Rev.*, 2010, **4**, 795–808.
191 W. Li, U. Guler, N. Kinsey, G. V. Naik, A. Boltasseva, J. Guan, V. M. Shalaev and A. V. Kildishev, *Adv. Mater.*, 2014, **26**, 7959–7965.
192 Z. Fei, A. S. Rodin, G. O. Andreev, W. Bao, A. S. McLeod, M. Wagner, L. M. Zhang, Z. Zhao, M. Thiemens, G. Dominguez, M. M. Fogler, A. H. Castro Neto, C. N. Lau, F. Keilmann and D. N. Basov, *Nature*, 2012, **487**, 82–85.
193 A. L. Falk, F. H. L. Koppens, C. L. Yu, K. Kang, N. de Leon Snapp, A. V. Akimov, M.-H. Jo, M. D. Lukin and H. Park, *Nat. Phys.*, 2009, **5**, 475–479.
194 J. Chen, M. Badioli, P. Alonso-González, S. Thongrattanasiri, F. Huth, J. Osmond, M. Spasenović, A. Centeno, A. Pesquera, P. Godignon, A. Z. Elorza, N. Camara, F. J. García de Abajo, R. Hillenbrand and F. H. L. Koppens, *Nature*, 2012, **487**, 77–81.

195 T. J. Echtermeyer, L. Britnell, P. K. Jasnos, A. Lombardo, R. V. Gorbachev, A. N. Grigorenko, A. K. Geim, A. C. Ferrari and K. S. Novoselov, *Nat. Commun.*, 2011, **2**, 458.
196 Z. Fang, Z. Liu, Y. Wang, P. M. Ajayan, P. Nordlander and N. J. Halas, *Nano Lett.*, 2012, **12**, 3808–3813.
197 Z. Fang, Y. Wang, Z. Liu, A. Schlather, P. M. Ajayan, F. H. L. Koppens, P. Nordlander and N. J. Halas, *ACS Nano*, 2012, **6**, 10222–10228.
198 J. M. Woo, M. S. Kim, H. W. Kim and J. H. Jang, *Appl. Phys. Lett.*, 2014, **104**, 081106.
199 V. Thareja, J. Kang, H. Yuan, K. Milaninia, H. Y. Hwang, Y. Cui, P. G. Kik and M. Brongersma, *Nano Lett.*, 2015, **15**, 1570–1576.
200 J. Niu, Y. Jun Shin, Y. Lee, J.-H. Ahn and H. Yang, *Appl. Phys. Lett.*, 2012, **100**, 061116.
201 J. Mann, Q. Ma, P. M. Odenthal, M. Isarraraz, D. Le, E. Preciado, D. Barroso, K. Yamaguchi, G. Von Son Palacio, A. Nguyen, T. Tran, M. Wurch, A. Nguyen, V. Klee, S. Bobek, D. Sun, T. F. Heinz, T. S. Rahman, R. Kawakami and L. Bartels, *Adv. Mater.*, 2014, **26**, 1399–1404.
202 L. Britnell, R. M. Ribeiro, A. Eckmann, R. Jalil, B. D. Belle, A. Mishchenko, Y.-J. Kim, R. V. Gorbachev, T. Georgiou, S. V. Morozov, A. N. Grigorenko, A. K. Geim, C. Casiraghi, A. H. Castro Neto and K. S. Novoselov, *Science*, 2013, **340**, 1311–1314.
203 A. Splendiani, L. Sun, Y. Zhang, T. Li, J. Kim, C. Y. Chim, G. Galli and F. Wang, *Nano Lett.*, 2010, **10**, 1271–1275.
204 E. Feigenbaum, K. Diest and H. A. Atwater, *Nano Lett.*, 2010, **10**, 2111–2116.
205 A. P. Vasudev, J.-H. Kang, J. Park, X. Liu and M. L. Brongersma, *Opt. Express*, 2013, **21**, 26387.
206 M. Abb, P. Albella, J. Aizpurua and O. L. Muskens, *Nano Lett.*, 2011, **11**, 2457–2463.
207 A. V. Krasavin and A. V. Zayats, *Phys. Rev. Lett.*, 2012, **109**, 053901.
208 V. J. Sorger, N. D. Lanzillotti-Kimura, R.-M. Ma and X. Zhang, *Nanophotonics*, 2012, **1**, 17–22.
209 P. R. West, S. Ishii, G. V. Naik, N. K. Emani, V. M. Shalaev and A. Boltasseva, *Laser Photonics Rev.*, 2010, **4**, 795–808.
210 K. Appavoo, B. Wang, N. F. Brady, M. Seo, J. Nag, R. P. Prasankumar, D. J. Hilton, S. T. Pantelides and R. F. Haglund, *Nano Lett.*, 2014, **14**, 1127–1133.
211 A.-K. U. Michel, D. N. Chigrin, T. W. W. Maß, K. Schönauer, M. Salinga, M. Wuttig and T. Taubner, *Nano Lett.*, 2013, **13**, 3470–3475.
212 J. Schuller, R. Zia, T. Taubner and M. Brongersma, *Phys. Rev. Lett.*, 2007, **99**, 107401.
213 S. Vassant, A. Archambault, F. Marquier, F. Pardo, U. Gennser, A. Cavanna, J. L. Pelouard and J. J. Greffet, *Phys. Rev. Lett.*, 2012, **109**, 237401.
214 S. Law, D. C. Adams, A. M. Taylor and D. Wasserman, *Opt. Express*, 2012, **20**, 12155–12165.
215 Y. Zhong, D. Wasserman, Y. Zhong, S. D. Malagari, T. Hamilton and D. Wasserman, *J. Nanophotonics*, 2015, **9**, 093791.
216 M. Wagner, A. S. Mcleod, S. J. Maddox, Z. Fei, M. Liu, R. D. Averitt, M. M. Fogler, S. R. Bank, F. Keilmann and D. N. Basov, *Nano Lett.*, 2014.
217 R. Hillenbrand, T. Taubner and F. Keilmann, *Nature*, 2002, **418**, 159–162.

218 T. Schwartz, J. A. Hutchison, C. Genet and T. W. Ebbesen, *Phys. Rev. Lett.*, 2011, **106**, 196405.
219 C. Sauvan, J. P. Hugonin, I. S. Maksymov and P. Lalanne, *Phys. Rev. Lett.*, 2013, **110**, 237401.
220 J. A. Scholl, A. García-Etxarri, A. L. Koh and J. A. Dionne, *Nano Lett.*, 2013, **13**, 564–569.
221 K. J. Savage, M. M. Hawkeye, R. Esteban, A. G. Borisov, J. Aizpurua and J. J. Baumberg, *Nature*, 2012, **491**, 574–577.
222 J. Zuloaga, E. Prodan and P. Nordlander, *Nano Lett.*, 2009, **9**, 887–891.
223 C. Marinica, A. K. Kazansky, P. Nordlander, J. Aizpurua and A. G. Borisov, *Nano Lett.*, 2012, **12**, 1333–1339.
224 G. Toscano, S. Raza, W. Yan, C. Jeppesen, S. Xiao, M. Wubs, A.-P. Jauho, S. I. Bozhevolnyi and N. A. Mortensen, *Nanophotonics*, 2013, **2**, 161–166.
225 B. Kobrin, E. S. Barnard, M. L. Brongersma, M. K. Kwak and L. J. Guo, in *Proc. SPIE 8249, Advanced Fabrication Technologies for Micro/Nano Optics and Photonics V*, ed. W. V. Schoenfeld, R. C. Rumpf and G. von Freymann, 2012, vol. 8249, p. 82490.
226 W. A. Murray and W. L. Barnes, *Adv. Mater.*, 2007, **19**, 3771–3782.
227 V. E. Ferry, M. A. Verschuuren, H. B. T. Li, R. E. I. Schropp, H. A. Atwater and A. Polman, *Appl. Phys. Lett.*, 2009, **95**, 183503.
228 Y. Xia and G. Whitesides, *Annu. Rev. Mater. Sci.*, 1998, **28**, 153–184.
229 S. Kawata, H. B. Sun, T. Tanaka and K. Takada, *Nature*, 2001, **412**, 697–698.
230 J. Fischer and M. Wegener, *Laser Photonics Rev.*, 2013, **7**, 22–44.
231 A. I. Kuznetsov, R. Kiyan and B. N. Chichkov, *Opt. Express*, 2010, **18**, 21198–21203.
232 J. Serbin, A. Egbert, A. Ostendorf, B. N. Chichkov, R. Houbertz, G. Domann, J. Schulz, C. Cronauer, L. Fröhlich and M. Popall, *Opt. Lett.*, 2003, **28**, 301–303.
233 J. Henzie, M. H. Lee and T. W. Odom, *Nat. Nanotechnol.*, 2007, **2**, 549–554.
234 J. Nelayah, M. Kociak, O. Stéphan, F. J. García de Abajo, M. Tencé, L. Henrard, D. Taverna, I. Pastoriza-Santos, L. M. Liz-Marzán and C. Colliex, *Nat. Phys.*, 2007, **3**, 348–353.
235 E. J. R. Vesseur, R. de Waele, M. Kuttge and A. Polman, *Nano Lett.*, 2007, **7**, 2843–2846.
236 F. J. García de Abajo, *Rev. Mod. Phys.*, 2010, **82**, 209–275.
237 M. V. Bashevoy, F. Jonsson, A. V. Krasavin, N. I. Zheludev, Y. Chen and M. I. Stockman, *Nano Lett.*, 2006, **6**, 1113–1115.
238 D. T. Schoen, A. C. Atre, A. García-Etxarri, J. A. Dionne and M. L. Brongersma, *Nano Lett.*, 2015, **15**, 120–126.
239 O. Nicoletti, F. de la Peña, R. K. Leary, D. J. Holland, C. Ducati and P. A. Midgley, *Nature*, 2013, **502**, 80–84.
240 W. Rechberger, A. Hohenau, A. Leitner, J. R. Krenn, B. Lamprecht and F. R. Aussenegg, *Opt. Commun.*, 2003, **220**, 137–141.
241 I. M. Hancu, A. G. Curto, M. Castro-López, M. Kuttge and N. F. Van Hulst, *Nano Lett.*, 2014, **14**, 166–171.
242 J. A. Fan, C. Wu, K. Bao, J. Bao, R. Bardhan, N. J. Halas, V. N. Manoharan, P. Nordlander, G. Shvets and F. Capasso, *Science*, 2010, **328**, 1135–1138.
243 N. Liu, H. Guo, L. Fu, S. Kaiser, H. Schweizer and H. Giessen, *Nat. Mater.*, 2008, **7**, 31–37.

244 R. D. Kekatpure, E. S. Barnard, W. Cai and M. L. Brongersma, *Phys. Rev. Lett.*, 2010, **104**, 243902.
245 E. Prodan, C. Radloff, N. J. Halas and P. Nordlander, *Science*, 2003, **302**, 419–422.
246 D. R. Smith, J. B. Pendry and M. C. K. Wiltshire, *Science*, 2004, **305**, 788–792.
247 V. M. Shalaev, *Nat. Photonics*, 2007, **1**, 41–48.
248 C. M. Soukoulis and M. Wegener, *Nat. Photonics*, 2011, 1–8.
249 E. J. R. Vesseur, T. Coenen, H. Caglayan, N. Engheta and A. Polman, *Phys. Rev. Lett.*, 2013, **110**, 013902.
250 A. V. Kildishev, W. Cai, U. K. Chettiar and V. M. Shalaev, *New J. Phys.*, 2008, **10**, 115029.
251 A. Salandrino and N. Engheta, *Phys. Rev. B: Condens. Matter Mater. Phys.*, 2006, **74**, 075103.
252 N. I. Zheludev and Y. S. Kivshar, *Nat. Mater.*, 2012, **11**, 917–924.
253 N. I. Zheludev, S. L. Prosvirnin, N. Papasimakis and V. A. Fedotov, *Nat. Photonics*, 2008, **2**, 351–354.
254 A. Poddubny, I. Iorsh, P. Belov and Y. Kivshar, *Nat. Photonics*, 2013, **7**, 948–957.
255 V. Fedotov, P. Mladyonov, S. Prosvirnin and N. Zheludev, *Phys. Rev. E: Stat., Nonlinear, Soft Matter Phys.*, 2005, **72**, 056613.
256 E. Plum, X. X. Liu, V. A. Fedotov, Y. Chen, D. P. Tsai and N. I. Zheludev, *Phys. Rev. Lett.*, 2009, **102**, 113902.
257 J. B. Pendry, D. Schurig and D. R. Smith, *Science*, 2006, **312**, 1780–1782.
258 U. Leonhardt, *Science.*, 2006, **312**, 1777–1780.
259 V. G. Veselago, *Phys.-Usp.*, 1968, **10**, 509–514.
260 A. Boardman, *Journal of Optometry*, 2010, **13**, 020401.
261 P. B. Catrysse, L. Verslegers, Z. Yu, J. S. White, E. S. Barnard, M. L. Brongersma and S. Fan, *Nano Lett.*, 2009, **9**, 235–238.
262 E. Hasman, Z. Bomzon, A. Niv, G. Biener and V. Kleiner, *Opt. Commun.*, 2002, **209**, 45–54.
263 L. Martín-Moreno, F. García-Vidal, H. Lezec, A. Degiron and T. Ebbesen, *Phys. Rev. Lett.*, 2003, **90**, 167401.
264 F. Afshinmanesh, J. S. White, W. Cai and M. L. Brongersma, *Nanophotonics*, 2012, **1**, 125–129.
265 N. Yu, P. Genevet, M. A. Kats, F. Aieta, J.-P. Tetienne, F. Capasso and Z. Gaburro, *Science*, 2011, **334**, 333–337.
266 X. Ni, S. Ishii, A. V. Kildishev and V. M. Shalaev, *Light: Sci. Appl.*, 2013, **2**, e72.
267 A. Kildishev, A. V. Boltasseva and V. M. Shalaev, *Science*, 2013, **339**, 1232009.
268 N. Engheta, A. Salandrino and A. Alù, *Phys. Rev. Lett.*, 2005, **95**, 095504.
269 N. Engheta, *Science*, 2007, **317**, 1698–1702.
270 J. B. Khurgin, *Nat. Photonics*, 2015, **10**, 2–6.
271 W. L. Barnes, A. Dereux and T. W. Ebbesen, *Nature*, 2003, **424**, 824–830.
272 J. A. Schuller, E. S. Barnard, W. Cai, Y. C. Jun, J. S. White and M. L. Brongersma, *Nat. Mater.*, 2010, **9**, 193–204.
273 S. Maier, M. Brongersma and P. Kik, *Adv. Mater.*, 2001, **13**, 1501–1505.
274 N. Engheta, *Science*, 2007, **317**, 1698–1702.
275 C. F. Bohren, *Am. J. Phys.*, 1983, **51**, 323.
276 S. Zhang, W. Fan, B. K. Minhas, A. Frauenglass, K. J. Malloy and S. R. J. Brueck, *Phys. Rev. Lett.*, 2005, **94**, 037402.

277 E. C. Garnett, W. Cai, J. J. Cha, F. Mahmood, S. T. Connor, M. Greyson Christoforo, Y. Cui, M. D. McGehee and M. L. Brongersma, *Nat. Mater.*, 2012, **11**, 241–249.
278 D. Chang, A. Sørensen, P. Hemmer and M. Lukin, *Phys. Rev. Lett.*, 2006, **97**, 053002.
279 D. E. Chang and M. D. Lukin, *Proc. SPIE*, 2008, **6904**, 69040I.
280 M. S. Tame, K. R. McEnery, S. K. Ozdemir, J. Lee, S. A. Maier and M. S. Kim, *Nat. Phys.*, 2013, **9**, 329–340.
281 M. Kuttge, E. J. R. Vesseur, J. Verhoeven, H. J. Lezec, H. A. Atwater and A. Polman, *Appl. Phys. Lett.*, 2008, **93**, 113110.
282 J. S. Fakonas, H. Lee, Y. A. Kelaita and H. A. Atwater, *Nat. Photonics*, 2014, **8**, 317–320.
283 W. Rechberger, A. Hohenau, A. Leitner, J. R. Krenn, B. Lamprecht and F. R. Aussenegg, *Opt. Commun.*, 2003, **220**, 137–141.
284 N. I. Zheludev, S. A. Maier, N. J. Halas, P. Nordlander, H. Giessen and C. T. Chong, *Nat. Mater.*, 2010, **9**, 707–715.
285 A. Sundaramurthy, K. Crozier, G. Kino, D. Fromm, P. Schuck and W. Moerner, *Phys. Rev. B: Condens. Matter Mater. Phys.*, 2005, **72**, 165409.
286 N. Liu, M. Hentschel, T. Weiss, A. P. Alivisatos and H. Giessen, *Science*, 2011, **332**, 1407–1410.
287 C. Sönnichsen, B. M. Reinhard, J. Liphardt and A. P. Alivisatos, *Nat. Biotechnol.*, 2005, **23**, 741–745.
288 F. Shafiei, C. Wu, Y. Wu, A. B. Khanikaev, P. Putzke, A. Singh, X. Li and G. Shvets, *Nat. Photonics*, 2013, **7**, 367–372.
289 A. Baron, E. Devaux, J. C. Rodier, J. P. Hugonin, E. Rousseau, C. Genet, T. W. Ebbesen and P. Lalanne, *Nano Lett.*, 2011, **11**, 4207–4212.
290 A. Gopinath, S. V Boriskina, B. M. Reinhard and L. Dal Negro, *Opt. Express*, 2009, **17**, 3741–3753.
291 T. Tanemura, K. C. Balram, D. S. Ly-Gagnon, P. Wahl, J. S. White, M. L. Brongersma and D. A. B. Miller, *Nano Lett.*, 2011, **11**, 2693–2698.
292 M. I. Tribelsky, A. E. Miroshnichenko, Y. S. Kivshar, B. S. Luk'yanchuk and A. R. Khokhlov, *Phys. Rev. X*, 2011, **1**, 1–9.
293 G. Baffou and R. Quidant, *Laser Photonics Rev.*, 2013, **7**, 171–187.
294 B. Rothenhäusler, J. Rabe, P. Korpiun and W. Knoll, *Surf. Sci. Lett.*, 1984, **137**, 373–383.
295 C. F. Bohren, *Am. J. Phys.*, 1983, **51**, 323.
296 S. Toroghi and P. G. Kik, *Phys. Rev. B*, 2014, **90**, 205414.
297 G. Baffou, R. Quidant and C. Girard, *Appl. Phys. Lett.*, 2009, **94**, 153109.
298 G. Baffou and R. Quidant, *Laser Photonics Rev.*, 2013, **7**, 171–187.
299 H. H. Richardson, M. T. Carlson, P. J. Tandler, P. Hernandez and A. O. Govorov, *Nano Lett.*, 2009, **9**, 1139–1146.
300 G. Baffou, R. Quidant and F. J. García de Abajo, *ACS Nano*, 2010, **4**, 709–716.
301 G. Baffou, M. P. Kreuzer, F. Kulzer and R. Quidant, *Opt. Express*, 2009, **17**, 3291–3298.
302 D. Boyer, P. Tamarat, A. Maali, B. Lounis and M. Orrit, *Science*, 2002, **297**, 1160–1163.
303 G. Baffou, P. Bon, J. Savatier, J. Polleux, M. Zhu, M. Merlin, H. Rignault and S. Monneret, *ACS Nano*, 2012, **6**, 2452–2458.

304 L. R. Hirsch, R. J. Stafford, J. A. Bankson, S. R. Sershen, B. Rivera, R. E. Price, J. D. Hazle, N. J. Halas and J. L. West, *Proc. Natl. Acad. Sci. U. S. A.*, 2003, **100**, 13549–13554.
305 V. P. Zharov, K. E. Mercer, E. N. Galitovskaya and M. S. Smeltzer, *Biophys. J.*, 2006, **90**, 619–627.
306 L. Röntzsch, K.-H. Heinig, J. A. Schuller and M. L. Brongersma, *Appl. Phys. Lett.*, 2007, **90**, 044105.
307 H. H. Richardson, A. C. Thomas, M. T. Carlson, M. E. Kordesch and A. O. Govorov, *J. Electron. Mater.*, 2007, **36**, 1587–1593.
308 B. Soares, F. Jonsson and N. Zheludev, *Phys. Rev. Lett.*, 2007, **98**, 153905.
309 D. A. Boyd, L. Greengard, M. Brongersma, M. Y. El-Naggar and D. G. Goodwin, *Nano Lett.*, 2006, **6**, 2592–2597.
310 L. Cao, D. N. Barsic, A. R. Guichard and M. L. Brongersma, *Nano Lett.*, 2007, **7**, 3523–3527.
311 D. A. Boyd, J. R. Adleman, D. G. Goodwin and D. Psaltis, *Anal. Chem.*, 2008, **80**, 2452–2456.
312 J. Adleman, D. Boyd, D. Goodwin and D. Psaltis, *Nano Lett.*, 2009, **9**, 4417–4423.
313 L. Greengard, M. Brongersma and D. Boyd, *US Pat.* 7,998,538, 2004.
314 S. R. Sershen, S. L. Westcott, N. J. Halas and J. L. West, *J. Biomed. Mater. Res.*, 2000, **51**, 293–298.
315 J. Stehr, C. Hrelescu, R. A. Sperling, G. Raschke, M. Wunderlich, A. Nichtl, D. Heindl, K. Kürzinger, W. J. Parak, T. A. Klar and J. Feldmann, *Nano Lett.*, 2008, **8**, 619–623.
316 M. Reismann, J. C. Bretschneider, G. von Plessen and U. Simon, *Small*, 2008, **4**, 607–610.
317 O. Neumann, A. S. Urban, J. Day, S. Lal, P. Nordlander and N. J. Halas, *ACS Nano*, 2013, **7**, 42–49.
318 O. Neumann, C. Feronti, A. D. Neumann, A. Dong, K. Schell, B. Lu, E. Kim, M. Quinn, S. Thompson, N. Grady, P. Nordlander, M. Oden and N. J. Halas, *Proc. Natl. Acad. Sci. U. S. A.*, 2013, **110**, 11677–11681.
319 P. Fan, U. K. Chettiar, L. Cao, F. Afshinmanesh, N. Engheta and M. L. Brongersma, *Nat. Photonics*, 2012, **6**, 380–385.
320 P. Fan, K. C. Y. Huang, L. Cao and M. L. Brongersma, *Nano Lett.*, 2013, **13**, 392–396.
321 K. C. Y. Huang, M.-K. Seo, Y. Huo, T. Sarmiento, J. S. Harris and M. L. Brongersma, *Nat. Commun.*, 2012, **3**, 1005.

Faraday Discussions

PAPER

Peculiar terminals in light at the extreme

A. M. Mahmoud and N. Engheta*

Received 15th October 2014, Accepted 24th November 2014
DOI: 10.1039/c4fd00205a

Materials with both relative permeability and permittivity near zero offer unprecedented features in light-matter interactions. Here, we explore how terminal endings involving bound regions of such materials may exhibit unusual electromagnetic behaviour. Physical intuitions and potential implications of such peculiar effects are also discussed.

Introduction

Tailoring materials parameters in the context of metamaterials has provided exciting possibilities in science and the engineering of light interactions with matter. In the past several years, numerous studies have been conducted in the exploration of the electromagnetic properties of metamaterials.[1–3] Materials with parameters attaining near-zero values have particularly attracted a lot of attention in recent years.[4–28] In such platforms, the effective refractive indices exhibit low values, and therefore, the waves experience little variations of phase throughout the structures filled with such media. A variety of novel phenomena have been investigated in epsilon-near-zero (ENZ),[4–16] mu-near-zero (MNZ),[17,18] epsilon-and-mu-near-zero (EMNZ) and zero-index structures.[19–28]

In the present work, we explore another exciting feature of EMNZ media, namely, how the terminals or "end points" of bound structures filled with an EMNZ medium behave electromagnetically. First, we assume a hypothetical EMNZ structure in which $\varepsilon(\omega_{\mathrm{op}})/\varepsilon_{\mathrm{o}} \approx 0$ and $\mu(\omega_{\mathrm{op}})/\mu_{\mathrm{o}} \approx 0$ at the operating frequency ω_{op}, and explore some of the unconventional characteristics of terminals in such bound EMNZ. Later, we examine our findings about the terminals in one of the proposed structures,[28] which effectively imitate the EMNZ electromagnetic behaviour at this operating frequency, as it interacts with the "outside".

Results and discussion

Extended short-circuited volume

In the EMNZ media, the electricity and magnetism become decoupled even though the fields are temporally dynamic.[28] This implies that some of the

University of Pennsylvania, Department of Electrical and Systems Engineering, 200 South 33rd Street, Philadelphia, PA 19104, USA. E-mail: engheta@ee.upenn.edu

phenomena such as the "short-circuited" terminal, "open-circuited" terminal, "end-fire" apertures, *etc.* may behave unconventionally. Let us start with the conventional notion of short circuit. Imagine we have a parallel-plate waveguide (PPW) made of a perfectly electric conducting (PEC) ceiling and floor (Fig. 1a) and filled with air. We assume a transverse electromagnetic (TEM) wave propagating inside this PEC PPW from left to right. At the right end of this PPW, we close the waveguide with a PEC terminal. As it is very well known, the incident TEM wave is completely reflected from this PEC terminal and, consequently, a complete standing wave is formed in this waveguide (Fig. 1b). As we see from the two-dimensional (2D) electromagnetic simulation results shown in this figure, the amplitude (Fig. 1c) and phase (Fig. 1d) of this standing wave behave consistently with what is expected in a standard standing wave, demonstrating the zero magnitude for the total electric field and the maximum magnitude for the total magnetic field at the PEC terminal in the right end of this waveguide. So it is obvious that the short circuit occurs only at one location, *i.e.*, at the PEC terminal (and every half wavelength location along the waveguide, where the zero amplitude for the electric field and the maximum magnitude for the magnetic field are repeated periodically.)

Now, we change the geometry by connecting this air-filled 2D PPW to a 2D bound region filled with an EMNZ medium, surrounded by a PEC wall (except at the opening to the PPW) (see Fig. 1e). We have shown in our earlier work that the bound EMNZ may behave as a "point" electromagnetically, even though its physical size may not be small. So, as we mentioned in our earlier work,[28] we can essentially "open" and "stretch" the space using the EMNZ medium, without having any measurable electromagnetic signature for an outside observer. We can examine this feature here for the notion of a short circuit.

Fig. 1f–h show the snapshot, the magnitude, and the phase of the total magnetic fields (oriented normal to the page) inside this structure, respectively. We can see that, in the bound EMNZ region, the entire magnetic field is spatially uniform while dynamically varying with time sinusoidally with the uniform phase across the entire EMNZ region. Fig. 1i–k depict the simulation results for the y-component (*i.e.*, vertical component) of the total electric field, while Fig. 1l–n show the corresponding values for the x-component (*i.e.*, longitudinal component) of the electric field. It is evident that the electric and magnetic fields are essentially decoupled in the EMNZ region, while they vary with time. Moreover, we note that the entire EMNZ region contains the maximum magnitude of the magnetic field with no phase variation throughout the region, while the y-component of the electric field is essentially near zero.

This implies that the entire EMNZ region behaves as an extended "short-circuited volume", and the outside observer in the waveguide sees this region as a "short circuit". In other words, the location of the short circuit has been "stretched" and has "formed" a volume, in the entire region of which the magnetic field attains maximum amplitude, while it is sinusoidally changing with time with the uniform phase in the entire region. This "extended short-circuited volume" may offer an exciting platform for exploring unconventional light–matter interactions.

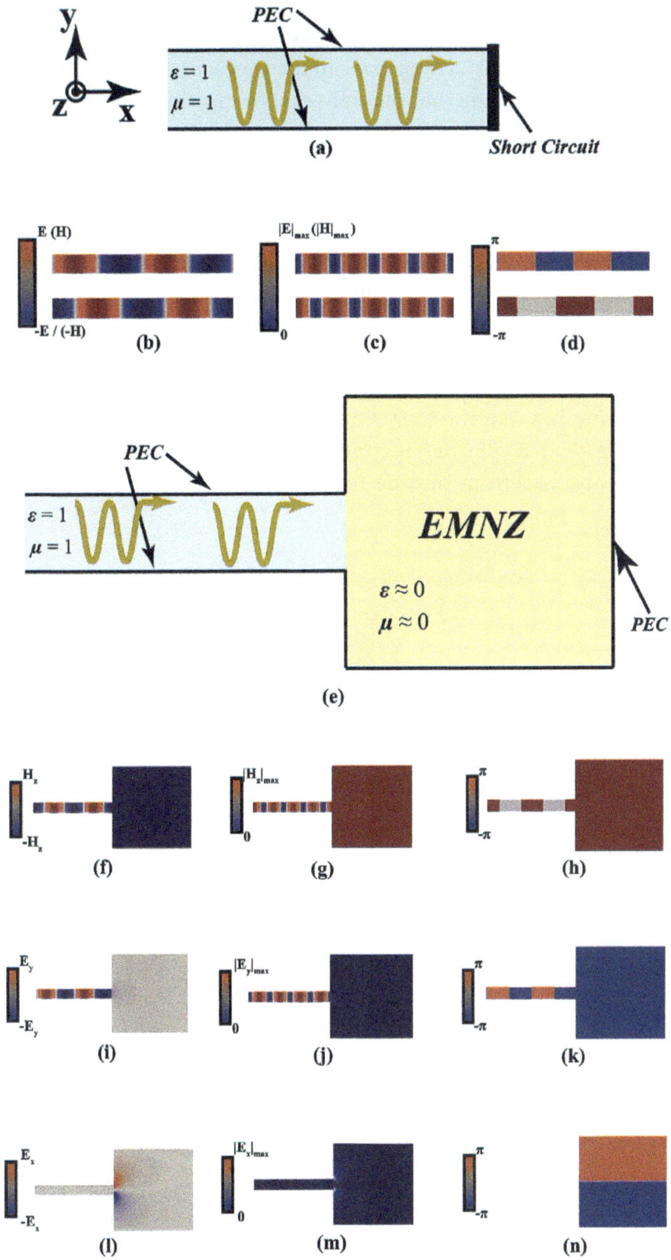

Fig. 1 (a) Geometry of a 2D parallel plate waveguide (PPW) terminated by a PEC wall. (b) Snapshot, (c) magnitude, and (d) phase of the electric/magnetic (top/bottom) fields within the structure shown in (a). (e) Geometry of a 2D parallel plate waveguide (PPW) terminated by a 2D bound region filled with an EMNZ medium, surrounded by a PEC wall. (f) Snapshot, (g) magnitude, and (h) phase of the *z*-component (*i.e.*, normal to the page) magnetic field (H_z) distribution within the structure shown in (e). (i) Snapshot, (j) magnitude, and (k) phase of the *y*-component (*i.e.*, vertical component) of the electric field (E_y) distribution within the structure shown in (e). (l) Snapshot, (m) magnitude, and (n) phase of the *x*-component (*i.e.*, longitudinal) of the electric field (E_x) distribution within the structure shown in (e).

Extended open-circuited volume

By duality, we can also explore an extended region, which electromagnetically behaves as an "open circuit". Going back to the standard 2D PPW with PEC ceiling and floor, let us now assume that the right-end terminal is closed with a perfectly magnetic conducting (PMC) wall. In this case, as seen from Fig. 2a–c, we do have another standing wave pattern with the maximum magnitude of the electric field at the PMC terminal, exhibiting an open circuit. Now, imagine we modify the geometry of the problem by connecting the PPW to an EMNZ region surrounded with the PEC walls on the top and bottom, but the PMC wall at the right side. An analogous 2D full-wave numerical simulation reveals that in the entire EMNZ region, the magnitude of the total magnetic field is essentially zero, while the magnitude of the total electric field is essentially maximum, revealing the fact that the entire region is an open circuit.

This again implies that the EMNZ has essentially "stretched" the open circuit location to an extended volume, creating an extended "open-circuited volume", as viewed by an observer from outside (*e.g.*, an observer sitting in the waveguide port).

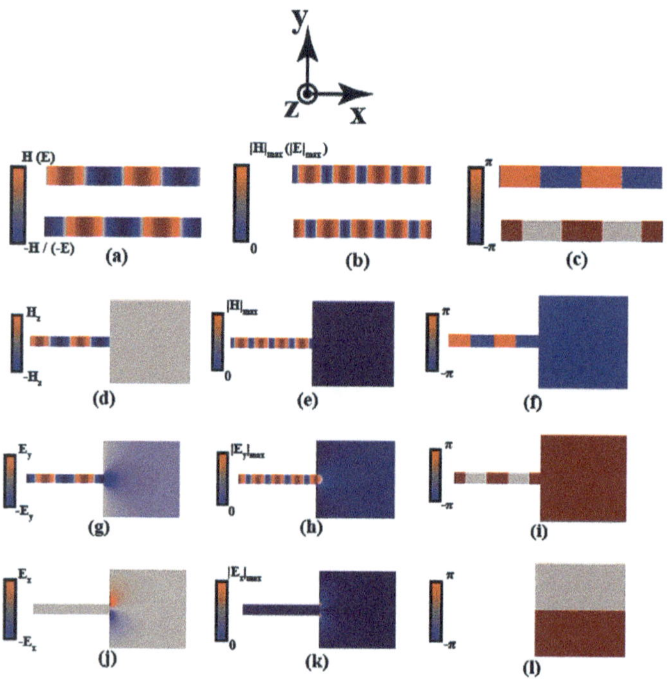

Fig. 2 (a) Snapshot, (b) magnitude, and (c) phase of electric/magnetic (top/bottom) fields within 2D parallel plate waveguide (PPW) terminated with a PMC wall. (d) Snapshot, (e) magnitude, and (f) phase of the z-component magnetic field (H_z) distribution within the 2D PPW terminated by a 2D bound region filled with an EMNZ medium, and terminated by a PMC wall. (g) Snapshot, (h) magnitude, and (i) phase of the y-component of the electric field (E_y) distribution within the 2D PPW terminated by a 2D bound region filled with an EMNZ medium, and terminated by a PMC wall. (l) Snapshot, (m) magnitude, and (n) phase of the x-component of the electric field (E_x) distribution within the 2D PPW terminated by a 2D bound region filled with an EMNZ medium, and terminated by a PMC wall.

End-fire radiating apertures

The fact that the phase inside the extended short-circuited and open-circuited volumetric regions is uniform provides an interesting possibility for coherent radiation from various apertures carved in the walls of these extended regions. Fig. 3(a) shows the sketch of such a 2D geometry. We see that the radiated fields exiting from these apertures are all in phase, regardless of how far apart the apertures are located, as shown in Fig. 3(b). This provides an interesting platform for the design of unconventional aperture distributions on the body of radiating elements that are not necessarily small. It is worth mentioning that while the phase velocity is very large here, owing to the necessary dispersion inherently present in the EMNZ medium, the group and energy velocities are certainly less than the vacuum velocity of light.

Proposed effective EMNZ region for short-circuited volume, open-circuited volume, and end-fire radiating apertures

The findings mentioned above were for a hypothetical EMNZ medium. In this section, we use one of the designs, which effectively behaves as an effective EMNZ for wave interaction from the outside. In one of our previous works,[28] we introduced this design, in which a properly selected dielectric rod with a given permittivity and radius was inserted in an ENZ host medium. Here, we utilize this structure to examine the concepts we have introduced above. Fig. 4 shows the simulation results for the corresponding short-circuited volume, open-circuited volume, and end-fire radiating structure; all using the proposed design that effectively behaves as an EMNZ medium , as viewed from outside. The simulation results clearly reveal the unusual features discussed above, using constructible structures with effective EMNZ properties. We note that, while the field distributions inside this designed structure are different from those inside the

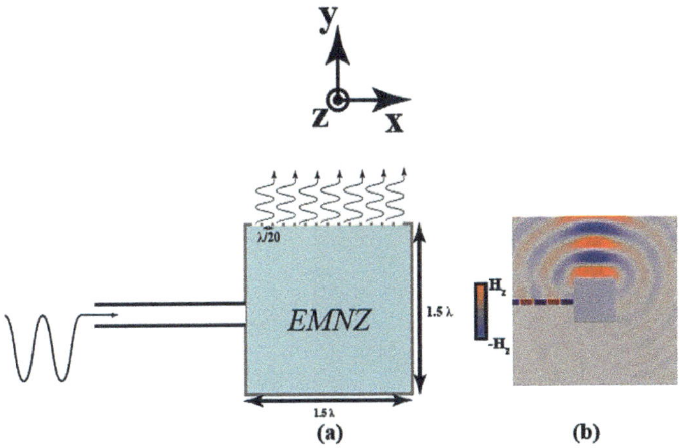

Fig. 3 (a) Geometry of the proposed structure; a PPW feeding a 2D bound region of area $1.5\lambda \times 1.5\lambda$ filled with an EMNZ medium, surrounded by a PEC wall. Fourteen apertures of size $\lambda/20$ each are introduced on the top wall of the EMNZ region. (b) Snapshot of the z-component of the magnetic field radiating from the structure shown in (a) into free space. Coherent radiation from various apertures is observed as expected.

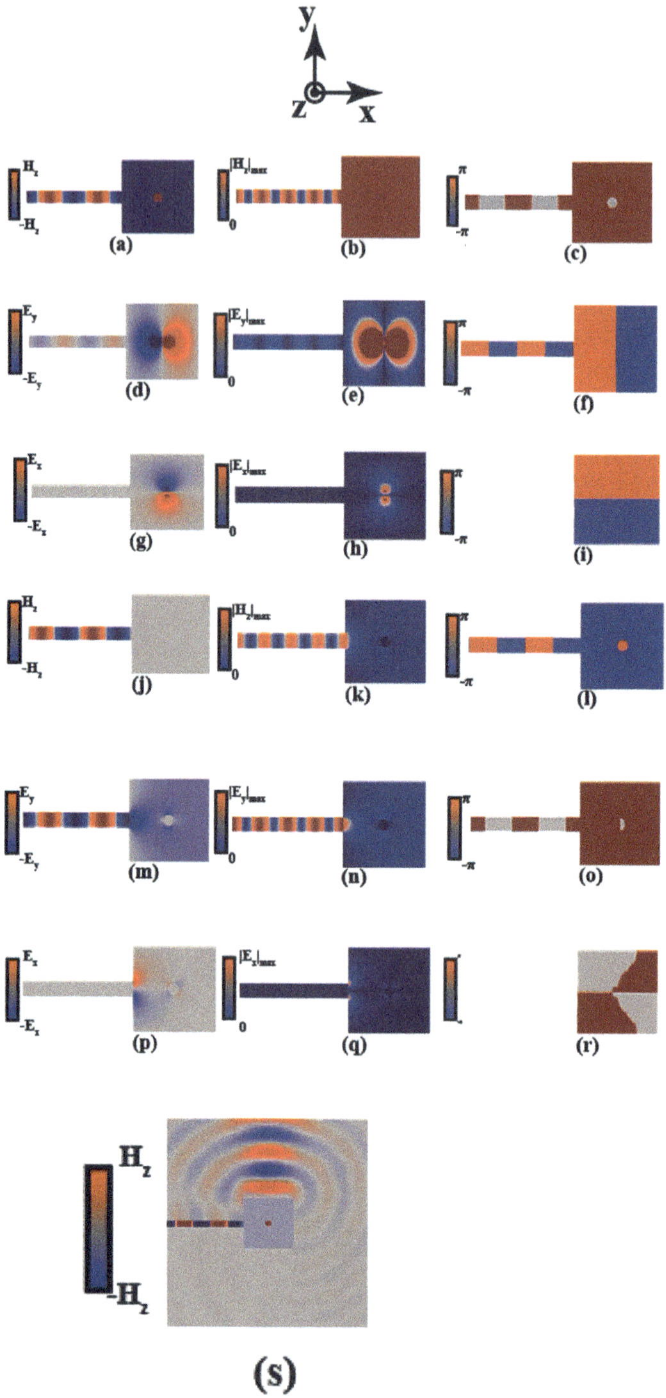

Fig. 4 (a)–(i) Similar to Fig. 1f–n, (j)–(r) similar to Fig. 2d–l, (s) similar to Fig. 3(b) except that here we use the structures proposed in ref. 28 instead of the hypothetical EMNZ medium. We note that while the fields inside this structure may differ from those in the EMNZ media in Fig. 2 and 3, the structure electromagnetically behaves as EMNZ, as viewed from the observer in the outside (*e.g.*, observer sitting in the waveguide port).

hypothetical EMNZ medium shown above, this structure exhibits EMNZ features for the external observers.

Conclusions

Here, we have investigated theoretically and numerically that the notion of short circuit and open circuit can be unconventionally expanded to volumetric regions in which the entire regions behave as short circuit or open circuit. In such regions, the electricity and magnetism are effectively decoupled while they are temporally dynamic, *i.e.*, they vary sinusoidally with time with the uniform phase across the entire regions. The radiation exiting from the apertures on the walls of these regions is also in phase, demonstrating the possibility of coherent radiation from apertures separated by large distances.

Acknowledgements

This work is supported in part by the US Air Force Office of Scientific Research (AFOSR) Multidisciplinary University Research Initiatives (MURI) on Quantum Metaphotonics & Metamaterials, Award no. FA9550-12-1-0488.

Notes and references

1 N. Engheta and R. W. Ziolkowski, *Metamaterials: Physics and Engineering Explorations*, Wiley/IEEE, 2006.
2 G. V. Eleftheriades and K. G. Balmian, *Metamaterials: Fundamental Principles and Applications*, Wiley/IEEE, 2005.
3 W. Cai and V. Shalaev, *Optical Metamaterials: Fundamentals and Applications*, Springer, 2009.
4 M. Silveirinha and N. Engheta, *Phys. Rev. Lett.*, 2006, **97**, 157403.
5 A. Alù, M. Silveirinha, A. Salandrino and N. Engheta, *Phys. Rev. B: Condens. Matter Mater. Phys.*, 2007, **75**, 155410.
6 G. Lovat, P. Burghignoli, F. Capolino, D. R. Jackson and D. R. Wilton, *IEEE Trans. Antennas Propag.*, 2006, **54**, 1017–1030.
7 S. Enoch, G. Tayeb, P. Sabouroux, N. Guérin and P. Vincent, *Phys. Rev. Lett.*, 2002, **89**, 213902.
8 A. Alù and N. Engheta, *Phys. Rev. B: Condens. Matter Mater. Phys.*, 2008, **78**, 045102.
9 E. J. R. Vesseur, T. Coenen, H. Caglayan, N. Engheta and A. Polman, *Phys. Rev. Lett.*, 2013, **110**, 013902.
10 R. Maas, J. Parsons, N. Engheta and A. Polman, *Nat. Photonics*, 2013, **7**, 907–912.
11 I. J. Bahl and K. C. Gupta, *IEEE Trans. Antennas Propag.*, 1974, **22**, 119–122.
12 Y. Pan and S. Xu, *IEEE Trans. Antennas Propag.*, 2009, **3**, 821–825.
13 J. Luo, P. Xu, L. Gao, Y. Lai and H. Chen, *Plasmonics*, 2012, **7**, 353–358.
14 L. Sun, S. Feng and X. Yang, *Appl. Phys. Lett.*, 2012, **101**, 241101.
15 S. Feng, *Phys. Rev. Lett.*, 2012, **108**, 193904.
16 D. A. Powell, A. Alù, B. Edwards, A. Vakil, Y. S. Kivshar and N. Engheta, *Phys. Rev. A*, 2009, **79**, 245135.

17 H. Zhou, S. Qu, Z. Pei, Y. Yang, J. Zhang, J. Wang, H. Ma, C. Gu and X. Wang, *J. Electromagn. Waves*, 2010, **24**, 1387–1396.
18 J. S. Marcos, M. Silveirinha and N. Engheta, submitted for publication.
19 R. W. Ziolkowski, *Phys. Rev. E: Stat., Nonlinear, Soft Matter Phys.*, 2004, **70**, 046608.
20 M. Silveirinha and N. Engheta, *Phys. Rev. B: Condens. Matter Mater. Phys.*, 2007, **75**, 075119.
21 J. Hao, W. Yan and M. Qiu, *Appl. Phys. Lett.*, 2010, **96**, 101109.
22 V. C. Nguyen, L. Chen and K. Halterman, *Phys. Rev. Lett.*, 2010, **105**, 233908.
23 Y. Xu and H. Chen, *Appl. Phys. Lett.*, 2011, **98**, 113501.
24 X. Huang, Y. Lai, Z. H. Hang, H. Zheng and C. T. Chan, *Nat. Mater.*, 2011, **10**, 582–586.
25 L. Zhou, Z. Song, X. Huang and C. T. Chan, *Nanophotonics*, 2012, **1**, 181–198.
26 R. Zhou, H. Zhang and H. Xin, *IEEE Trans. Antennas Propag.*, 2010, **58**, 79–87.
27 Q. He, S. Xiao, X. Li and L. Zhou, *Opt. Express*, 2013, **21**, 28948–28959.
28 A. M. Mahmoud and N. Engheta, *Nat. Commun.*, 2014, **5**, 5638.

Faraday Discussions

PAPER

Dirac dispersion in photonic hypercrystals

Evgenii E. Narimanov*

Received 20th October 2014, Accepted 24th November 2014
DOI: 10.1039/c4fd00207e

Photonic hypercrystals – the recently introduced concept of artificial optical media that combines the properties of hyperbolic metamaterials and photonic crystals [*E. Narimanov, Phys. Rev. X*, 2014, **4**, 041014] – can support Dirac cone dispersion at a finite frequency.

1 Introduction

The cornucopia of new physical phenomena in graphene[1,2] arising from the Dirac cone dispersion of its free carriers, lead to the search for similar behavior in optical systems.[3–6] Dirac cones were found in coupled waveguide arrays[7] as well as in photonic crystals[3–6,8,9] and metamaterials,[8] thus opening the way to the study of such non-trivial phenomena as Klein's tunneling paradox[10] and Shrödinger's "zitterbewegung"[11] that were previously believed to lie entirely within the realm of quantum physics and quantum electrodynamics, and even to possible practical applications of this behavior in actual photonic devices. However, optical systems that are currently known to support Dirac cone dispersion generally rely on spatial variations in at least two or even in all three dimensions – which severely limits their applicability to practical applications.

In the present paper, we demonstrate that planar optical composites can also support propagating waves with Dirac dispersion. This behavior however cannot be found in the "regular" planar optical media, but relies on the new physics introduced by the recently discovered photonic hypercrystals.[12] These novel composite media are essentially hyperbolic metamaterials[13] with a periodic variation of their electromagnetic response properties on a subwavelength scale, and combine the properties of metamaterials and photonic crystals. The resulting effect on the wave propagation and dispersion by band gap formation in what is essentially the metamaterial limit, offers an unprecedented degree of control of light propagation in photonic hypercrystals – and allows for the formation of Dirac dispersion cones in the hypercrystal phase space.

As photonic hypercrystals can be assembled entirely within the restrictions of planar fabrication technology, this offers a unique opportunity to bring the

School of Electrical and Computer Engineering and Birck Nanotechnology Center, Purdue University, West Lafayette, IN 47907, USA

exciting physics offered by Dirac cone dispersion, such as *e.g.* Klein tunneling or Shrödinger's zitterbewegung, to the realm of practical applications in nanophotonics.

2 Photonic hypercrystals: the concept

Metamaterials[14] and photonic crystals[15] currently represent two primary building blocks for advanced nanophotonic devices. With the goal of ultimate control over the light propagation, an artificial optical material must rely on either the effect of a subwavelength pattern that changes the average electromagnetic response of the medium, or on the Bragg scattering of light due to a periodic variation that is comparable to the wavelength. Because of this inherent scale separation, the corresponding metamaterial and photonic crystal concepts are generally considered mutually exclusive within the same environment.

The situation is however dramatically different in the world of hyperbolic metamaterials,[13] where the opposite signs of the dielectric permittivity components in two orthogonal directions ($\mathrm{Re}[\varepsilon_n]\mathrm{Re}[\varepsilon_\tau] < 0$) lead to a hyperbolic dispersion of TM-polarized propagating waves

$$\frac{{k_n}^2}{\varepsilon_\tau} + \frac{{k_\tau}^2}{\varepsilon_n} = \frac{\omega^2}{c^2} \qquad (1)$$

with the wave numbers unlimited by the frequency ω. As a result, a periodic variation in the dielectric permittivity, regardless of how small its period d is (see Fig. 1), will necessarily cause the Bragg scattering of these high-k waves, leading to the formation of photonic band gaps in both the wavenumber and the frequency domains. This behavior is illustrated in Fig. 2, where we show the dispersion diagram for an example of such a photonic hypercrystal. Note that in the entire frequency range shown in this figure, the spatial period of the hypercrystal is below one tenth of the corresponding free-space wavelength.

As long as the hyperbolic dispersion of eqn (1) remains adequate for the description of the hyperbolic metamaterial component of the hypercrystal, the

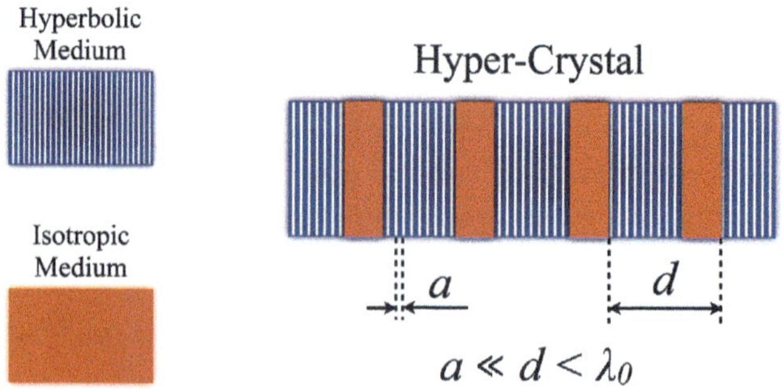

Fig. 1 A planar hypercrystal is formed by introducing a periodic variation in a hyperbolic medium, with the period smaller than the free-space wavelength $d < \lambda_0$, but well above the unit cell size of the hyperbolic (meta)material: $d \gg a$. The desired periodic variation can be achieved by introducing a second medium (which could be either a metal, or a dielectric, or another hyperbolic medium with a different dielectric permittivity tensor) in the design of the composite.

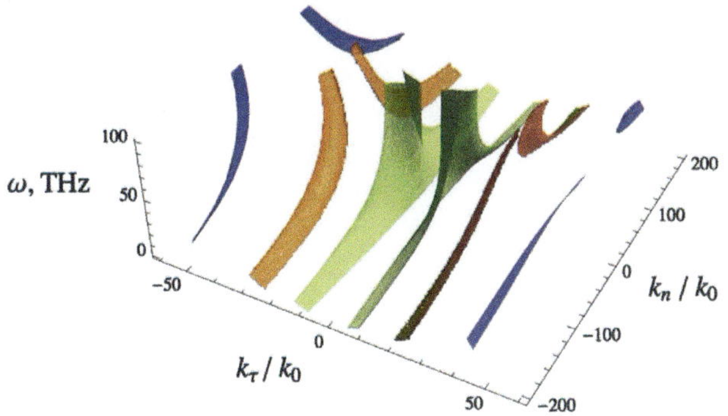

Fig. 2 The dispersion diagram of a photonic hypercrystal. The hypercrystal unit cell is formed by 250 nm of $In_{0.53}Ga_{0.47}As : Al_{0.48}In_{0.52}As$ semiconductor hyperbolic metamaterial[25] with a 5 μm plasma wavelength, followed by 250 nm of an $Al_{0.48}In_{0.52}As$ dielectric layer. k_n and k_τ represent the wavevector components that are respectively normal and tangential to the layers formed in the hypercrystal, $k_0 \equiv \omega/c$ is the free-space wavenumber.

pattern of allowed and forbidden bands seen in Fig. 2 proliferates to ever higher wavenumbers. However, eventually the local wavelength $\lambda \equiv 2\pi/k$ becomes comparable to the unit cell size a of the hyperbolic medium (see Fig. 1), effectively terminating this behavior. The hypercrystal regime is therefore limited to $k \lesssim 1/a$ and can only be observed in the composites with

$$a \ll d < \lambda_0, \tag{2}$$

where λ_0 is the free-space wavelength. The resulting restrictions however can be substantially relaxed with the use of natural hyperbolic media such as *e.g.* sapphire,[16] graphite,[17] TGS[18–20] or bismuth,[21–24] when eqn (2) reduces to simply $d < \lambda_0$.

3 Dirac points in hypercrystals

In its simplest realization, when a planar hypercrystal is formed by interleaving layers of hyperbolic and isotropic media (which can be either dielectric ($\mathrm{Re}[\varepsilon] > 0$) or metallic ($\mathrm{Re}[\varepsilon] < 0$)), the inversion symmetry of the composite allows to classify the TM-propagating modes ($\mathbf{B} = B\hat{\mathbf{n}}$ and $\mathbf{E} \perp \hat{\mathbf{n}}$, where $\hat{\mathbf{n}}$ is the unit vector normal to the layers – see Fig. 1) in terms of the parity "quantum number" $p = \pm 1$, with the standard Bloch states expressed as linear combinations of the even and odd modes $\mathbf{B}_{p=1}(\omega,\mathbf{r}) \pm i\,\mathbf{B}_{p=-1}(\omega,\mathbf{r})$. It is the accidental degeneracy between the states of different parity that is the key to the formation of Dirac cones in photonic hypercrystals.

For the TM-polarized modes propagating parallel to the layers forming the hypercrystal, and thus corresponding to the Bloch momentum $k_n = 0$, we obtain

$$\sin\left(\frac{k_h d_h}{2}\right)\cos\left(\frac{k_i d_i}{2}\right) + \left(\frac{\varepsilon_\tau}{\varepsilon_i}\frac{k_i}{k_h}\right)^p \cos\left(\frac{k_h d_h}{2}\right)\sin\left(\frac{k_i d_i}{2}\right) = 0 \tag{3}$$

where ε_i is the dielectric permittivity of the isotropic medium (see Fig. 1),

$$\varepsilon_h = \begin{pmatrix} \varepsilon_\tau & 0 & 0 \\ 0 & \varepsilon_\tau & 0 \\ 0 & 0 & \varepsilon_n \end{pmatrix} \tag{4}$$

is the dielectric permittivity tensor of the hyperbolic medium, k_h the wavenumber in the hyperbolic layers

$$k_h = \sqrt{\varepsilon_\tau \left(\frac{\omega^2}{c^2} - \frac{{k_\tau}^2}{\varepsilon_n} \right)}, \tag{5}$$

and k_i is the wavenumber in the layers formed by the isotropic medium

$$k_i = \sqrt{\varepsilon_i \frac{\omega^2}{c^2} - {k_\tau}^2}. \tag{6}$$

Here k_τ is the tangential component of the wavevector ($B \sim \exp(ik_\tau x)$), while d_h and d_i are the thicknesses of the hyperbolic and isotropic layers.

As follows from eqn (3), the "even" (parity $p = 1$) and "odd" ($p = -1$) "guided" modes propagating parallel to the layers of the hypercrystal, are degenerate when

$$\frac{\varepsilon_\tau}{\varepsilon_i} \frac{k_i}{k_h} = 1, \tag{7}$$

corresponding to the in-plane momentum

$$k_\tau = k_B \equiv \frac{\omega}{c} \sqrt{\varepsilon_i \frac{1 - \varepsilon_i/\varepsilon_\tau}{1 - {\varepsilon_i}^2/(\varepsilon_n \varepsilon_\tau)}} \tag{8}$$

which for a lossless system corresponds to the Brewster's angle at the interface of the isotropic dielectric ($\varepsilon_i > 0$) and the hyperbolic medium.[25] The frequency corresponding to this accidental degeneracy point is then given by

$$\omega_0 = \frac{2\pi c m}{\varepsilon_\tau d_h + \varepsilon_i d_i} \sqrt{\varepsilon_\tau \frac{1 - {\varepsilon_i}^2/(\varepsilon_n \varepsilon_\tau)}{1 - \varepsilon_i/\varepsilon_n}}, \tag{9}$$

where m is a (positive) integer. Note that the strong dispersion of the dielectric permittivity common in hyperbolic media, makes eqn (9) a nontrivial equation for ω_0. Furthermore, losses inevitable in the hyperbolic media, lead to a nonzero imaginary part of ω_0.

We will illustrate this behavior using the example of a hypercrystal formed by alternating layers of silicon and sapphire – two complimentary materials which form the backbone of the silicon-on-sapphire hetero-epitaxial process.[26] In the mid-IR range, sapphire offers several low-loss hyperbolic bands (see Fig. 3), while silicon behaves as a transparent dielectric with a permittivity of $\varepsilon_{Si} \approx 11.7$. In Fig. 4 we plot the dispersion $\omega(k_\tau)$ of the guided modes for the silicon–sapphire hypercrystal, with the unit cell formed by 676 nm of silicon and 3.55 μm of sapphire.

Expanding eqn (3) near the degeneracy point (k_B,ω_0), we obtain

$$A_0 \frac{\omega - \omega_0}{\omega_0} = (B_0 + pC_0)\frac{k_\tau - k_B}{k_B} + \mathcal{O}\left[\left(\frac{\omega - \omega_0}{\omega_0}\right)^2, \left(\frac{k_\tau - k_B}{k_B}\right)^2\right] \tag{10}$$

where $p = \pm 1$ is the mode parity, and

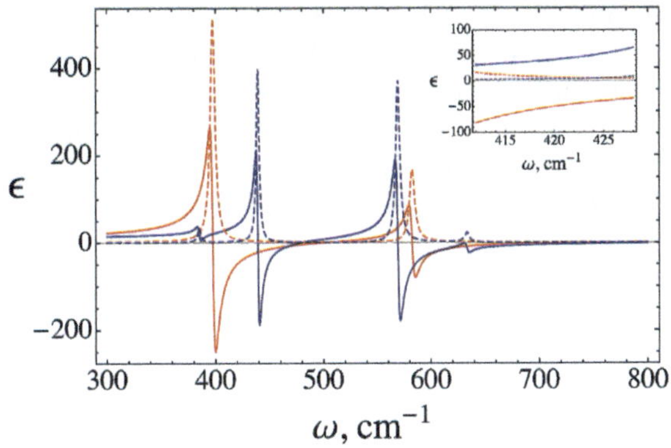

Fig. 3 The dielectric permittivity tensor of sapphire monocrystal in the Restrahlen band.[16] The red and blue lines correspond to the permittivities parallel and perpendicular to the C-axis of the sapphire crystal respectively. The real and imaginary parts of the permittivity are represented by solid and dotted lines. The inset shows the range near the free-space wavelength of 24 μm (≈ 417 cm^{-1}) that supports one of the low-loss hyperbolic bands in sapphire.

$$A_0 = 1 + \frac{\pi mc}{\omega_0}\frac{\varepsilon_\tau d_h}{(\varepsilon_\tau d_h + \varepsilon_i d_i)^2}\sqrt{\varepsilon_\tau \frac{1-\varepsilon_i{}^2/\varepsilon_n\varepsilon_\tau}{1-\varepsilon_i/\varepsilon_n}} \times \left(\frac{\varepsilon_i}{\varepsilon_\tau}\frac{d(\varepsilon_\tau/\varepsilon_i)}{d\omega} + \frac{\varepsilon_i{}^2}{\varepsilon_n\varepsilon_\tau}\frac{\varepsilon_\tau-\varepsilon_i}{\varepsilon_n-\varepsilon_i}\frac{d(\varepsilon_n/\varepsilon_i)}{d\omega}\right), \tag{11}$$

$$B_0 = \frac{2\pi mc}{\omega_0}\frac{\varepsilon_i d_h + \varepsilon_n d_i}{(\varepsilon_\tau d_h + \varepsilon_i d_i)^2}\frac{\varepsilon_\tau{}^{3/2}}{\varepsilon_n\sqrt{\varepsilon_i}}\frac{1-\varepsilon_i{}^2/\varepsilon_n\varepsilon_\tau}{(1-\varepsilon_i/\varepsilon_n)^{3/2}}\sqrt{1-\varepsilon_i/\varepsilon_\tau} \tag{12}$$

$$C_0 = \frac{2c}{\omega_0}\frac{1}{\varepsilon_\tau d_h + \varepsilon_i d_i}\left(\frac{\varepsilon_\tau}{\varepsilon_i}\right)^{3/2}\frac{(1-\varepsilon_i{}^2/\varepsilon_n\varepsilon_\tau)^2}{(1-\varepsilon_i/\varepsilon_n)^{3/2}}\sqrt{1-\varepsilon_i/\varepsilon_\tau} \times \sin\left(\frac{\pi m}{1+\frac{\varepsilon_\tau}{\varepsilon_i}\frac{d_h}{d_i}}\right)\cos\left(\frac{\pi m}{1+\frac{\varepsilon_i}{\varepsilon_\tau}\frac{d_i}{d_h}}\right) \tag{13}$$

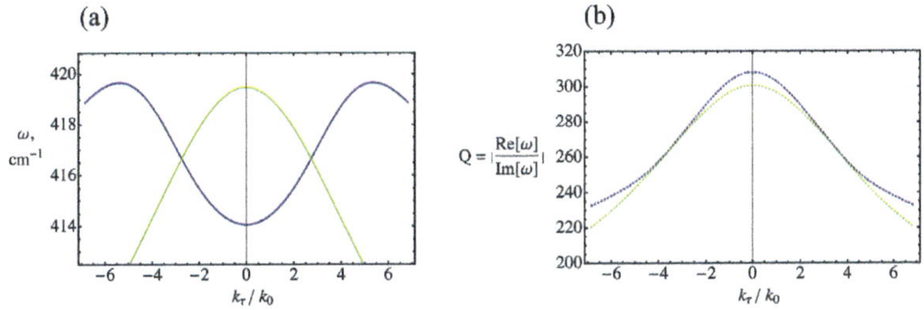

Fig. 4 The frequency ω (panel (a)) and the "Q-factor" (panel (b)) *vs.* in-plane wavenumber k_τ for $m = 1$ even (green line) and odd (blue line) modes propagating parallel to the layers of silicon–sapphire hypercrystal, for $d_h = 3.55$ μm and $d_i = 0.676$ μm.

For a dielectric photonic crystal, we find that $B_0 \pm C_0 > 0$, and the accidental degeneracy point at (k_B,ω_0) corresponds to the standard Brewster angle collapse of the photonic crystal band gap.[15,27] In a hypercrystal however one finds the possibility of $(B_0 + C_0)(B_0 - C_0) < 0$, with the opposite signs of the group velocities of the even and odd guided modes. Furthermore, for $\mathrm{Re}[\varepsilon_n] < 0$ the magnitude of B_0 can be dramatically reduced – all the way to zero in the lossless limit, at

$$\frac{\varepsilon_n(\omega_0)}{\varepsilon_i(\omega_0)} = -\frac{d_h}{d_i}, \tag{14}$$

in which case eqn (10) can be expressed as

$$A_0\frac{\omega - \omega_0}{\omega_0} = \pm C_0\frac{k_\tau - k_B}{k_B} + \mathcal{O}\left[\left(\frac{\omega - \omega_0}{\omega_0}\right)^2, \left(\frac{k_\tau - k_B}{k_B}\right)^2\right] \tag{15}$$

For a lossy system, eqn (9) and (14) cannot be simultaneously satisfied even for a complex ω_0. However, for the modes with $\mathrm{Re}[\omega] \gg \mathrm{Im}[\omega]$ (or equivalently $Q \equiv \mathrm{Re}[\omega]/\mathrm{Im}[\omega] \gg 1$), as in the case of the silicon–sapphire hypercrystal (see Fig. 4b), we can choose

$$\frac{d_i}{d_h} = -\mathrm{Re}\left[\frac{\varepsilon_n(\omega_0)}{\varepsilon_i(\omega_0)}\right], \tag{16}$$

which yields

$$\begin{aligned} A_0\frac{\omega - \omega_0}{\omega_0} &= \pm C_0\frac{k_\tau - k_B}{k_B} \\ &+ \mathcal{O}\left[\left(\frac{\omega - \omega_0}{\omega_0}\right)^2, \left(\frac{k_\tau - k_B}{k_B}\right)^2, \frac{k_\tau - k_B}{k_B}\frac{\mathrm{Im}[\varepsilon]}{\mathrm{Re}[\varepsilon]}\right] \end{aligned} \tag{17}$$

Eqn (9) and (16) define the geometry of the hypercrystal unit cell and the "operational frequency" for the Dirac dispersion point.

For the calculation of the propagating waves in a planar hypercrystal for $k_n \neq 0$, we use the standard T-matrix approach[27] that can be generalized to the case of anisotropic material components.[12] This yields the unit cell transfer matrix

$$T \equiv \begin{pmatrix} T_{11} & T_{12} \\ T_{21} & T_{22} \end{pmatrix} \tag{18}$$

where

$$T_{11} = \left[\cos(k_h d_h) + \frac{i}{2}\left(\frac{k_h}{k_i}\frac{\varepsilon_i}{\varepsilon_\tau} + \frac{k_i}{k_h}\frac{\varepsilon_\tau}{\varepsilon_i}\right)\sin(k_h d_h)\right]\exp(ik_i d_i) \tag{19}$$

$$T_{12} = \frac{i}{2}\left(\frac{k_h}{k_i}\frac{\varepsilon_i}{\varepsilon_\tau} - \frac{k_i}{k_h}\frac{\varepsilon_\tau}{\varepsilon_i}\right)\sin(k_h d_h)\exp(-ik_i d_i) \tag{20}$$

$$T_{21} = -\frac{i}{2}\left(\frac{k_h}{k_i}\frac{\varepsilon_i}{\varepsilon_\tau} - \frac{k_i}{k_h}\frac{\varepsilon_\tau}{\varepsilon_i}\right)\sin(k_h d_h)\exp(ik_i d_i) \tag{21}$$

$$T_{22} = \left[\cos(k_h d_h) - \frac{i}{2}\left(\frac{k_h}{k_i}\frac{\varepsilon_i}{\varepsilon_\tau} + \frac{k_i}{k_h}\frac{\varepsilon_\tau}{\varepsilon_i}\right)\sin(k_h d_h)\right]\exp(-ik_i d_i) \tag{22}$$

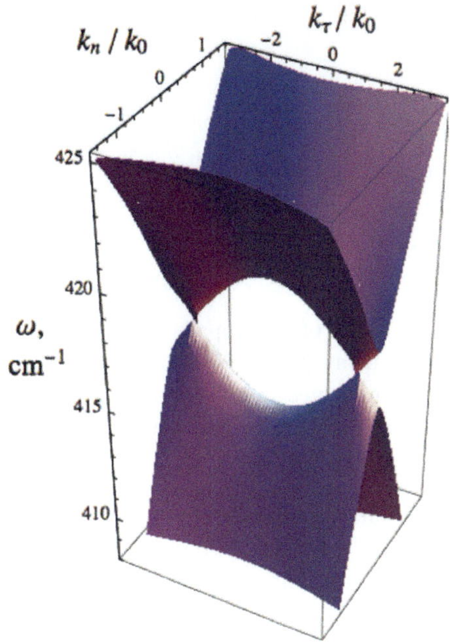

Fig. 5 The dispersion diagram for the silicon–sapphire hypercrystal, introduced in Fig. 4. The thickness of the sapphire layers d_h is 3.55 μm and the width of the silicon layers d_i is 0.676 μm.

In this approach, the magnetic field in the dielectric layer of the unit cell n (so that $(d_i + d_h)(n - 1) < z < (d_i + d_h)(n - 1) + d_i$), is given by

$$\mathbf{B}(\omega,x,z;k_\tau,k_n) = \hat{\mathbf{y}}\, \exp(ik_\tau x + ik_n z) \times \{\alpha \exp[k_n(z - (d_i + d_h)(n - 1))] + \beta \exp[k_n(z - (d_i + d_h)(n - 1))]\} \tag{23}$$

where $\{\alpha,\beta\}$ is the eigenvector of the transfer matrix

$$T\begin{pmatrix}\alpha\\ \beta\end{pmatrix} = \exp[ik_n(d_h + d_i)]\begin{pmatrix}\alpha\\ \beta\end{pmatrix} \tag{24}$$

The dispersion of the propagating waves can then be obtained from the equation

$$\mathrm{Tr}[T] = 2\cos[ik_n(d_h + d_i)] \tag{25}$$

In Fig. 5 we show the dispersion diagram for the example of the silicon–sapphire hypercrystal that we introduced earlier. Note that there are two Dirac cones in the phase space of the system at $(k_\tau,k_n) = (\pm k_B,0)$, as predicted by our earlier analysis.

The new physics introduced by the presence of the Dirac dispersion cones in the phase space of the system can only be explored if the corresponding states represent an example of "good" quasi-particles, with the lifetimes substantially exceeding the period of a single oscillation, $Q \equiv \omega\tau \gg 1$. With the mode lifetime defined by

$$\tau = \frac{1}{|\mathrm{Im}[\omega]|}, \tag{26}$$

for the mode Q-factor we obtain

$$Q = \left|\frac{\mathrm{Re}[\omega]}{\mathrm{Im}[\omega]}\right|. \tag{27}$$

In Fig. 6 we plot the Q-factor of the propagating modes for the frequency range in the proximity of the Dirac point. With Q-factors on the order of several hundred, the corresponding states should be easily accessible experimentally.

In addition to the Dirac points at $k_n = 0$ (see Fig. 5), planar hypercrystals also have similar singularities at the edge of the Brillouin zone ($k_n = \pm\frac{\pi}{d_\mathrm{h} + d_\mathrm{i}}$). These can be analyzed in a similar manner, starting from the "guided modes" that satisfy the boundary condition $B(z + d_\mathrm{h} + d_\mathrm{i}) = \pm iB(z)$, and one finds the behavior similar to the that of Dirac states near $k_n = 0$.

4 The effective Hamiltonian near the Dirac point of a hypercrystal

Focusing on the Dirac point at $(k_\tau, k_n, \omega) = (k_\mathrm{B}, 0, \omega_0)$, we expand eqn (25) in k_n, $k_\tau - k_\mathrm{B}$ and $\omega - \omega_0$. The resulting straightforward calculation leads to the effective Hamiltonian

$$\mathscr{H}_\mathrm{eff} = \begin{pmatrix} 0 & v_\tau(k_\tau - k_\mathrm{B}) - iv_n k_n \\ v_\tau(k_\tau - k_\mathrm{B}) + iv_n k_n & 0 \end{pmatrix} \tag{28}$$

such that

$$\mathscr{H}_\mathrm{eff}\begin{pmatrix}\Psi_1\\ \Psi_2\end{pmatrix} = (\omega - \omega_0)\begin{pmatrix}\Psi_1\\ \Psi_2\end{pmatrix} \tag{29}$$

In terms of the Pauli matrices σ_x and σ_y, the effective Hamiltonian (28) can be expressed as.

$$\mathscr{H}_\mathrm{eff} = \sigma_x v_\tau(k_\tau - k_\mathrm{B}) + \sigma_y v_n k_n \tag{30}$$

Here,

$$v_\tau = c\,\mathrm{Re}\left[\frac{C_0}{A_0}\right], \tag{31}$$

$$v_n = \mathrm{Re}\left[\frac{\omega_0(d_\mathrm{h} + d_\mathrm{i})}{2\pi m A_0}\right], \tag{32}$$

where A_0 and C_0 are respectively defined by eqn (11) and (13).

In Fig. 7 we present the comparison of the exact dispersion diagram of the silicon–sapphire hypercrystal introduced earlier, to the effective Hamiltonian

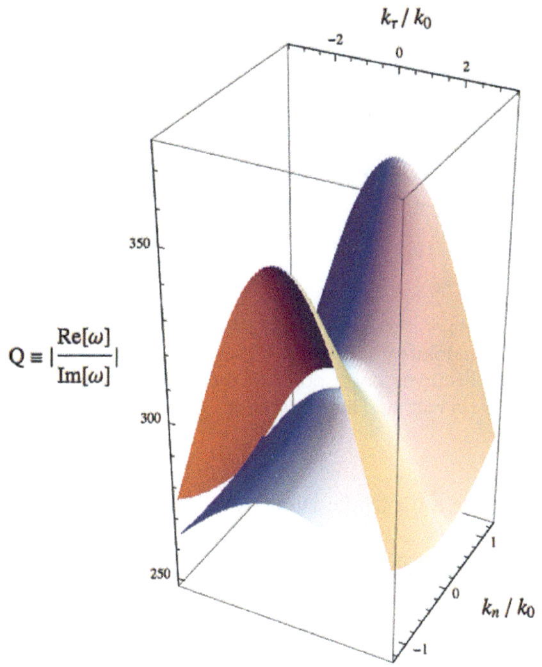

Fig. 6 The mode Q-factors for the silicon–sapphire hypercrystal in Fig. 4 and 5.

approximation of eqn (28)–(32). Note that, aside from the small frequency offset resulting from neglecting second-order cross-terms related to the loss in the hyperbolic media ($\sim\mathcal{O}((k_\tau - k_B)\text{Im}[\varepsilon]/\text{Re}[\varepsilon])$ - see eqn (17)), one finds excellent

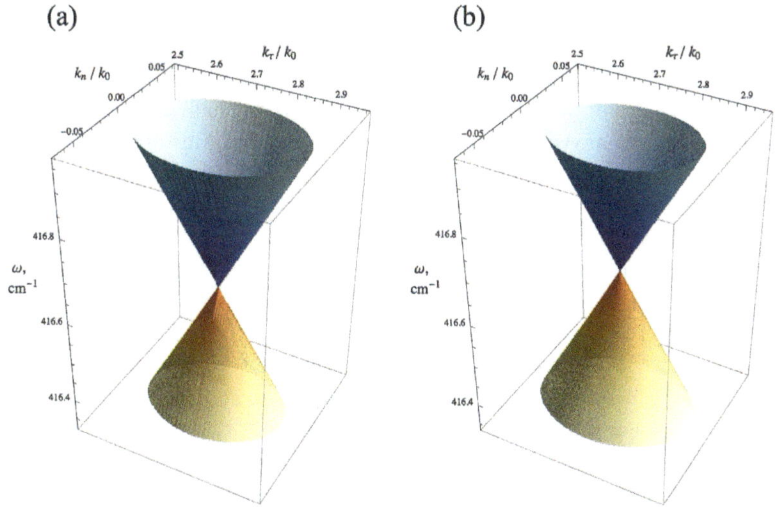

Fig. 7 The comparison of the exact dispersion diagram near the Dirac dispersion point of a silicon–sapphire hypercrystal (a) with the effective Hamiltonian approximation (b). Note the small shift along the frequency axis that results from the losses in the material that are not fully accounted for in the analytical approximation of eqn (28)–(32).

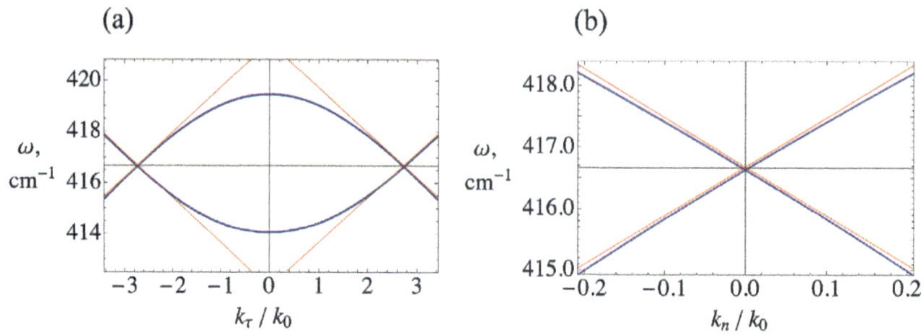

Fig. 8 The cross-sections of the silicon–sapphire hypercrystal dispersion diagram (Fig. 5 and 7), through the Dirac points $k_\tau = \pm k_B$, $k_n = 0$, $\omega = \omega_0$, in the k_τ (panel (a)) and k_n (panel (b)) directions. The blue curves correspond to the exact solution, while the red lines represent the effective Hamiltonian approximation.

agreement between the anisotropic Dirac–Weyl Hamiltonian (28) and the exact solution.

In Fig. 8 we present an alternative representation of this comparison, with the "cross-sections" of the phase space through the Dirac point in k_τ (Fig. 8(a)) and k_n (Fig. 8(b)) directions. Note that Fig. 8 shows a much broader frequency range than Fig. 7.

As the Dirac–Weyl Hamiltonian (28) proves an excellent approximation for the dynamics of the propagating waves in the hypercrystals near the accident degeneracy points, optical experiments in these composites can be used to study the optical equivalents of the many quantum-mechanical phenomena – from the zitterbewegung[11] to Klein tunneling.[10] Furthermore, the corresponding modulation of the transmission and reflection properties of the hypercrystal may possibly find applications in metamaterial devices.

5 Reflectivity near the Dirac point

Near the Dirac point frequency ω_0, a lossless hypercrystal does not support propagating waves anywhere except in the direct proximity of the Brewster's momentum k_B. A study of the electromagnetic reflection from such composites would reveal a nearly 100% reflectivity anywhere aside from a narrow range near the Brewster's angle – see Fig. 9(a). Furthermore, when plotted in the angle–frequency coordinates, the reflection coefficient would clearly reveal the Dirac dispersion cone – as seen in Fig. 9(a).

The presence of material loss however substantially modifies this behavior. While the propagating modes in hypercrystals may have high Q-factors (see *e.g.* Fig. 4 and 6), the Dirac dispersion cones occupy a relatively narrow frequency band $\delta\omega_D \ll \omega_0$. For the silicon–sapphire hypercrystal considered earlier, we find $\Delta\omega_D/\omega_0 \sim 10^{-2}$ (see Fig. 4(a) and 5). As a result, for this material system the imaginary part of the frequency is of the same order as the Dirac bandwidth, Im $[\omega] \sim \Delta\omega_D$ – and the losses lead to a substantial modification of the reflectivity near the Dirac point.

However, while the sharp features of the Dirac dispersion cone in the reflectivity diagram of Fig. 9(a) are indeed washed out by the material loss, the resulting

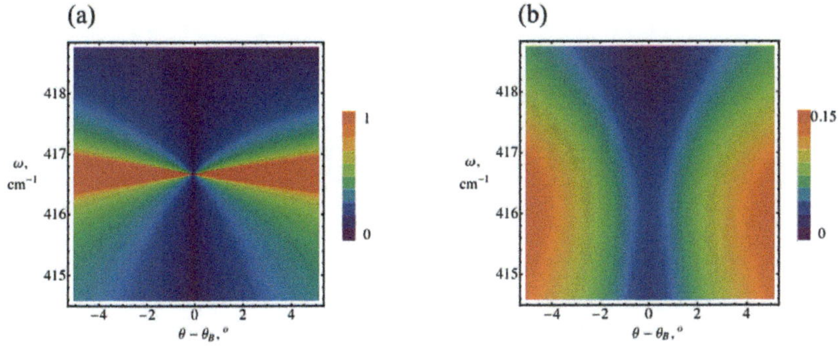

Fig. 9 The reflectivity of the interface of silicon and silicon–sapphire hypercrystal, in a false-color representation. Panel (a) shows the lossless case, while the panel (b) includes the effects of material absorption. The angles θ and θ_B are respectively the incidence angle onto the hypercrystal, and the Brewster's angle for the silicon–sapphire interface.

pattern still shows the signatures of the Dirac point – see Fig. 9(b). As a result, the non-monotonic behavior of the hypercrystal reflectivity observed in an experiment, can be used to locate the Dirac dispersion point.

6 Motti singularity: Dirac points annihilation

While the study of the optical equivalents of both massless (Weyl–Dirac) and massive excitations offers many exciting opportunities[3–9] in fundamental physics research as well as with an eye for possible applications, it may be beneficial to search for optical modes with a higher dispersion complexity. While the Lorentz invariance limits classical elementary particles to two mutually exclusive classes – bradyons[28] (with nonzero rest mass m_0), with the low kinetic energy limit

$$E \simeq \frac{p^2}{2m_0} \tag{33}$$

and lyxons[29] ($m_0 = 0$), with the energy

$$E = cp, \tag{34}$$

a meta-material quasi-particle can combine both kinds of classical motion:

$$E = \sqrt{(cp_z)^2 + \left(\frac{{p_x}^2}{2m_0}\right)^2} \tag{35}$$

Probing such "motion" excitations in optical systems may offer interesting possibilities for *e.g.* novel electromagnetic resonant cavities that confine such modes, polarization control and phase matching in nonlinear-optical processes.

These "motion" waves naturally arise in photonic hypercrystals at the Motti singularity[30] point of "annihilation" of two symmetry-related Dirac cones at $k_\tau = \pm k_B$. In a planar hypercrystal, this happens when the Dirac point frequency ω_0 approaches the frequency ω_M when the in-plane permittivity of the hyperbolic medium matches that of the isotropic component of the hypercrystal.

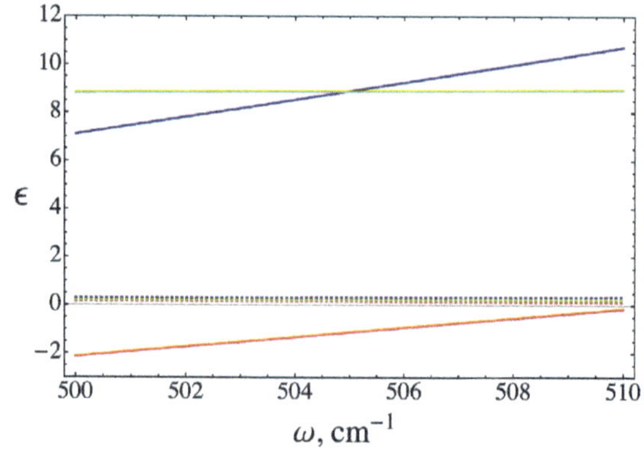

Fig. 10 The dielectric permittivities of doped silicon (green curve) and sapphire (red and blue curves for the components parallel (ε_n) and perpendicular (ε_τ) to the C-axis respectively), for in the low-loss hyperbolic band near the free-space wavelength of 20 μm ($\approx$500 cm^{-1}). Solid curves correspond to the real parts of the permittivities, while dotted lines represent the imaginary parts. Note that at $\omega \approx 505$ cm^{-1} we find $\varepsilon_\tau \approx \varepsilon_{\mathrm{Si}}$.

$$\varepsilon_\tau(\omega_\mathrm{M}) = \varepsilon_\mathrm{i}(\omega_\mathrm{M}) \tag{36}$$

Within the bounds of the silicon–sapphire hypercrystal "platform", we can reach the critical point of (36) by controlled doping of the silicon layers. For n^+-doped silicon, the permittivity can be accurately described by the Drude model[31]

$$\varepsilon_\mathrm{Si} = \varepsilon_\infty \left(1 - \frac{{\omega_p}^2}{\omega(\omega + i\Gamma)}\right) \tag{37}$$

with $\varepsilon_\infty \approx 11.7$, $\omega_\mathrm{p} = 10^{12}\ldots10^{13}$ sec^{-1} (depending on the doping level) and $\Gamma \simeq 10^{11}\ldots10^{12}$ sec^{-1}. For our example in the present section, we will choose $\omega_\mathrm{p} = 250$ cm$^{-1} \equiv 7.5 \times 10^{12}$ sec^{-1}, which leads to the resonance (36) at the frequency $\approx$505 cm^{-1}, and neglect the loss in silicon compared to that in hyperbolic bands of sapphire. This behavior is illustrated in Fig. 10, where we plot the dielectric permittivities of both the doped silicon and sapphire, for the frequency range corresponding to second low-loss hyperbolic band of the latter.

With the proper choice of the dimensions d_h and d_i of the unit cell of the hypercrystal, both Dirac points can be brought to "annihilation" at $(k_\tau,k_n) = (0,0)$, $\omega_0 = \omega_\mathrm{M}$, resulting in the Motti singularity[30] of the photonic density of states of the hypercrystal. The resulting dispersion diagram can be described by the effective Motti–Weyl Hamiltonian

$$\mathscr{H}_\mathrm{MW} = \begin{pmatrix} \dfrac{{k_\tau}^2}{2m_0} & -iv_nk_n \\ iv_nk_n & -\dfrac{{k_\tau}^2}{2m_0} \end{pmatrix} = \sigma_z\frac{{k_\tau}^2}{2m_0} + \sigma_y v_n k_n\ , \tag{38}$$

where σ_y and σ_z are the corresponding Pauli matrices. In Fig. 7 we present two orthogonal cross-sections of the exact dispersion diagram of the doped silicon–sapphire hypercrystal that supports the Motti singular point (36), clearly showing the motion dispersion (35) of the effective Hamiltonian (38), $E(k_\tau) \sim {k_\tau}^2$ and $E(k_n) \sim |k_n|$.

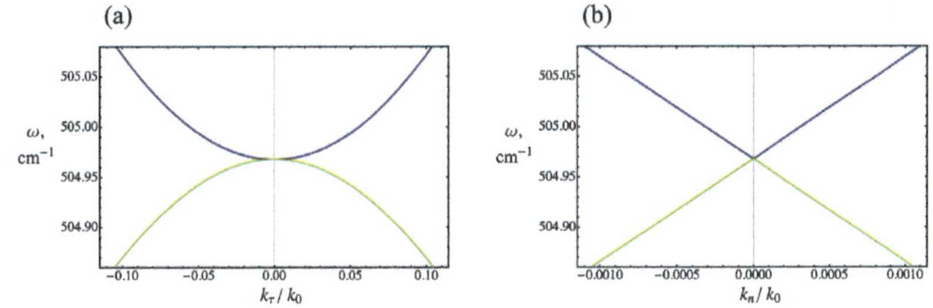

Fig. 11 The cross-sections of the doped silicon–sapphire hypercrystal dispersion that supports the Motti singularity at $\omega_0 = \omega_M \approx 504.97\ \text{cm}^{-1}$, at $k_n = 0$ (panel (a)) and $k_\tau = 0$ (panel (b)), in the lossless limit. The unit cell of the hypercrystal is formed by a $d_i \approx 6.11\ \mu\text{m}$ – thick layer of doped silicon and a $d_h \approx 550$ nm – wide layer of sapphire. Note the "motion" dispersion $E(k_\tau) \sim {k_\tau}^2$, $E(k_n) \sim |k_n|$ of the propagating waves in the hypercrystal near the Motti singularity point.

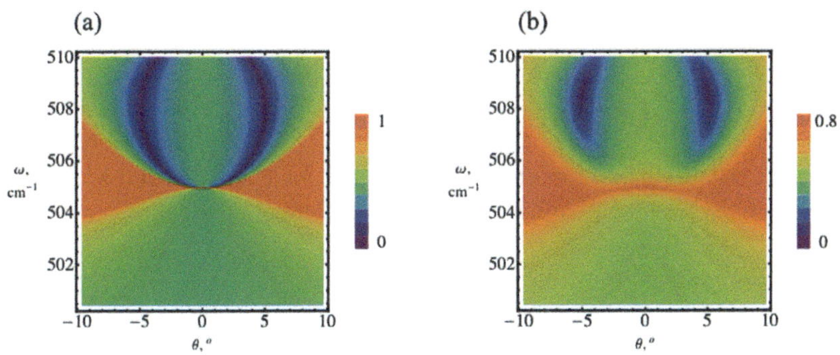

Fig. 12 The reflectivity of the interface of the silicon and silicon–sapphire hypercrystal supporting Motti singularity near the frequency of 505 cm^{-1}, in the lossless approximation (a) and with the sapphire absorption taken into account (b). As in Fig. 9, here θ is the incidence angle onto the silicon–sapphire hypercrystal.

Similar to the Dirac point, the Motti singularity in the dispersion diagram of the hypercrystal also makes an imprint on the reflectivity of the composite. For a lossless system, at the singularity point we expect total reflectivity for any incidence angle $\theta \neq 0$. Furthermore, as a function of the frequency, the reflection coefficient would reveal the quadratic variation of the motion dispersion (36) in the tangential direction,

$$\omega - \omega_M \propto \pm {k_\tau}^2. \tag{39}$$

This behavior is clearly seen in Fig. 11(a) which shows the reflectivity of the doped silicon–sapphire hypercrystal that supports the Motti singularity at the frequency $\omega_M \approx 505\ \text{cm}^{-1}$, in the lossless limit.

While the actual material loss generally reduces the "contrast" of the propagating waves dispersion diagram "imprinted" onto the reflectivity of the hypercrystal, the calculation taking full account of the absorption in sapphire still reveals the motion dispersion (36), (39) – see Fig. 12(b). Note that the width of the peak in reflectivity near the Motti singularity of the actual (lossy) hypercrystal, measured at half-maximum (when it drops from $\approx$80% to $\approx$40%), is only 0.5 cm^{-1}, corresponding to barely 0.1% of the center frequency. Such relatively sharp features in a strongly absorbing system ($Im[\varepsilon]/Re[\varepsilon] \sim 10\%$) give a further indication for the essentially singular nature of the light transmission and reflection in the hypercrystal supporting the motion resonance, and open the route to actual applications of optical hypercrytals in practical nanophotonics.

7 Conclusions

In conclusion, we have demonstrated that photonic hypercrystals allow Dirac–Weyl dispersion, as well as the Motti singularity at the annihilation of symmetry-related Dirac points. Despite relatively high loss in actual hypercrystals imposed by their hyperbolic constituents, the resulting sharp resonance structures in light propagation and reflection allow for both the study of optical analogues of such quantum phenomena as Shrödinger's zitterbewegung and Klein's tunneling, as well as for actual applications of these novel composite media in practical nanophotonics.

Acknowledgements

This work was partially supported by NSF Center for Photonics and Multiscale Metamaterials, ARO MURI and Gordon and Betty Moore Foundation.

References

1 Y. Zhang, Y. W. Tan, H. L. Stormer and P. Kim, Experimental observation of the quantum Hall effect and Berrys phase in graphene, *Nature*, 2005, **438**, 201.
2 A. H. C. Neto, F. Guinea, N. M. R. Peres, K. S. Novoselov and A. K. Geim, The electronic properties of graphene, *Rev. Mod. Phys.*, 2009, **81**, 109.
3 L. G. Wang, Z. G. Wang, J. X. Zhang and S. Y. Zhu, Realization of Dirac point with double cones in optics, *Opt. Lett.*, 2009, **34**, 1510.
4 R. A. Sepkhanov, Y. B. Bazaliy and C. W. J. Beenakker, Extremal transmission at the Dirac point of a photonic band structure, *Phys. Rev. A*, 2007, **75**, 063813.
5 X. Zhang, Observing zitterbewegung for photons near the Dirac point of a two-dimensional photonic crystal, *Phys. Rev. Lett.*, 2008, **100**, 113903.
6 M. Diem, T. Koschny and C. M. Soukoulis, Transmission in the vicinity of the Dirac point in hexagonal photonic crystals, *Phys. B*, 2010, **405**, 2990.
7 F. Dreisow, M. Heinrich, R. Keil, A. Tnnermann, S. Nolte, S. Longhi and A. Szameit, Classical Simulation of Relativistic Zitterbewegung in Photonic Lattices, *Phys. Rev. Lett.*, 2010, **105**, 143902.
8 X. Huang, Y. Lai, Z. H. Hang, H. Zheng and C. T. Chan, Dirac cones induced by accidental degeneracy in photonic crystals and zero-refractive-index materials, *Nat. Mater.*, 2011, **20**, 582.
9 O. Bahat-Treidel, O. Peleg, M. Grobman, N. Shapira, M. Segev and T. Pereg-Barnea, Klein Tunneling in Deformed Honeycomb Lattices, *Phys. Rev. Lett.*, 2010, **104**, 063901.
10 E. Klein, Die Reflexion von Elektronen an einem Potentialsprung nach der relativistischen Dynamik von Dirac, *Zeitschrift für Physik*, 1929, **53**(34), 157.

11 E. Schrödinger, Über die kräftefreie Bewegung in der relativistischen Quantenmechanik (On the free movement in relativistic quantum mechanics), *Berliner Ber.*, 1930, 418; E. Schrödinger, Zur Quantendynamik des Elektrons, *Berliner Ber*, 1931, 63.
12 E. E. Narimanov, Photonic Hypercrystals, *Phys. Rev. X*, 2014, **4**, 041014.
13 Z. Jacob, L. V. Alekseyev and E. E. Narimanov, Optical Hyperlens: Far-field imaging beyond the diffraction limit, *Opt. Express*, 2006, **14**, 8247.
14 W. Cai and V. M. Shalaev, *Optical Metamaterials*, Springer, 2010.
15 J. D. Joannopoulos, S. G. Johnson, J. N. Winn and R. D. Meade, *Photonic Crystals: Molding the Flow of Light*, Princeton University Press, 2008.
16 M. Schubert, T. E. Tiwald and C. M. Herzinger, Infrared dielectric anisotropy and phonon modes of sapphire, *Phys. Rev. B: Condens. Matter Mater. Phys.*, 2000, **61**(12), 8187.
17 J. Sun, J. Zhou, B. Li and F. Kang, Indefinite permittivity and negative refraction in natural material: graphite, *Appl. Phys. Lett.*, 2011, **98**(10), 101901.
18 A. Hadni and X. Gerbaux, Far IR excitation of longitudinal optical phonons in triglycine sulphate, *Ferroelectrics*, 2000, **248**, 15.
19 X. Gerbaux, M. Tazawa and A. Hadni, Far IR transmission measurements on triglycine sulphate (TGS), at 5K, *Ferroelectrics*, 1998, **215**, 47.
20 T. Dumelow, J. A. P. da Costa and V. N. Freire, Slab lenses from simple anisotropic media, *Phys. Rev. B: Condens. Matter Mater. Phys.*, 2005, **72**, 235115.
21 W. S. Boyle, A. D. Brailsford and J. K. Galt, Dielectric anomalies and cyclotron absorption in the infrared: observations on bismuth, *Phys. Rev.*, 1958, **109**(4), 1396.
22 W. S. Boyle and A. D. Brailsford, Far infrared studies of bismuth, *Phys. Rev.*, 1960, **120**(6), 1943.
23 V. D. Kulakovskii and V. D. Egorov, Plasma reflection in bismuth and bismuth-antimony alloys, *Sov. Phys. Solid State*, 1974, **15**(7), 1368.
24 L. V. Alekseyev, V. A. Podolskiy and E. E. Narimanov, Homogeneous hyperbolic systems for terahertz and far-infrared frequencies, *Adv. OptoElectron.*, 2012, **2012**, 267564.
25 A. J. Hoffman, L. V. Alekseyev, S. S. Howard, K. J. Franz, D. Wasserman, V. A. Podolskiy, E. E. Narimanov, D. L. Sivco and C. Gmachl, Negative refraction in semiconductor metamaterials, *Nat. Mater.*, 2007, **6**, 948.
26 H. M. Manasevit and W. J. Simpson, Single-Crystal Silicon on a Sapphire Substrate, *J. Appl. Phys.*, 1964, **35**, 1349.
27 P. Yeh, *Optical Waves in Layered Media*, Wiley, 2005.
28 Which may also referred to as "tardyons" or "ittyons"[29].
29 O.-M. P. Bilaniuk and E. C. G. Sudarshan, Particles beyond the Light Barrier, *Phys. Today*, 1969, **22**(5), 43.
30 P. M. Morse and H. Feshbah, *Methods of Theoretical Physics*, McGraw-Hill, 1953.
31 M. van Exter and D. Grischkowski, Optical and electronic properties of doped silicon from 01. to 2 THz, *Appl. Phys. Lett.*, 1990, **56**(17), 1694.

Faraday Discussions

PAPER

Light emission in nonlocal plasmonic metamaterials

Viktor A. Podolskiy,*[a] Pavel Ginzburg,[bc] Brian Wells[a] and Anatoly V. Zayats[b]

Received 30th September 2014, Accepted 16th October 2014
DOI: 10.1039/c4fd00186a

We present analytical and computational studies of light emission in nonlocal metamaterials formed by arrays of aligned plasmonic nanowires. We demonstrate that the emission lifetime in these composites is a complex function of geometrical and material parameters of the system that cannot be reduced to the "trivial" hyperbolic or elliptical dispersion topology of a homogenised metamaterial. In particular, our studies suggest that the Purcell factor can often be maximized when the composite operates in the elliptic regime, with strong radiation coupling to an additional TM-polarized mode supported by the nonlocal composite, in contrast to the accepted "hyperbolicity related" enhancement.

1. Introduction

Metamaterials, composites designed from dissimilar components of sub-wavelength size, offer new avenues for controlling light generation, propagation, and absorption.[1,2] Recently, plasmonic nanowire arrays have emerged as flexible platforms for confinement and manipulation of optical pulses at deep sub-wavelength scale, along with applications in negative refraction, sensing, and non-linear optics.[3] It has been demonstrated, experimentally and numerically, that the optical response of these composites is strongly affected by the spatial dispersion of the effective permittivity – the dependence of components of the permittivity tensor on a wavevector.[4] Analytical description of optical nonlocality in nanowire arrays has recently been developed.[5] In particular, it was shown that optical nonlocality yields an additional wave, an extra TM-polarized optical mode, that has hyperbolic-like properties at frequency ranges where the main TM wave has elliptic dispersion. In this work, we use the nonlocal effective medium theory to analyse the emission of point dipoles in nanowire media and show that non-locality significantly affects light emission in metamaterials, enhancing the

[a]Department of Physics and Applied Physics, University of Massachusetts Lowell, One University Ave., Lowell, MA, 01854, USA. E-mail: viktor_podolskiy@uml.edu
[b]Department of Physics, Kings College London, Strand, London WC2R 2LS, UK
[c]ITMO University, St. Petersburg 197101, Russia

photonic density of states in the elliptic regime, in contrast with previous theoretical proposals[6] that predict that Purcell enhancement is maximized in the hyperbolic regime. We relate the apparent contradiction to the excitation of plasmonic modes in metamaterial constituents, missed in the local effective medium theory, and argue that the nonlocal effective medium theory provides a more adequate description of the underlying physics.

The manuscript is organized as follows – we first describe the modal structure of the nanowire metamaterial and provide the relationship between the geometrical parameters of the metamaterial, its spatially dispersive permittivity, and dispersion of the optical modes in the composite. The developed nonlocal effective medium theory (EMT) allows solving the boundary condition problem for plane wave incidence. We then employ the plane wave decomposition of a dipolar emitter for the calculation of spontaneous emission rates *via* classical-quantum correspondence and the formalism of the radiation reaction force.

2. Nonlocal optical response of nanowire media

2.1. Optical modes supported by nanowire metamaterials

Plasmonic nanowire arrays represent collections of aligned metallic wires (with permittivity ε_i and average radius r) embedded into a dielectric host (permittivity ε_h) with mean centre-to-centre separation a. Nanowire metamaterials typically operate in the effective medium regime where $r < a \ll \lambda_0$, with $\lambda_0 = 2\pi c/\omega$ being the free-space wavelength, ω being the angular frequency of optical radiation, and c being the speed of light in vacuum. In this regime, the optical properties of the composite may be well described by averaged geometric parameters (concentration $p = \pi r^2/a^2$; [here $p \ll 1$] and, in the case of nonlocal effective medium theory, the shape of the unit cell) and are weakly affected by the details of wire distribution (variations of the wire radius or unit cell size across the composite). In this work, we follow the approach introduced in ref. 5 and consider nanowire metamaterials having square unit cells (see Fig. 1).

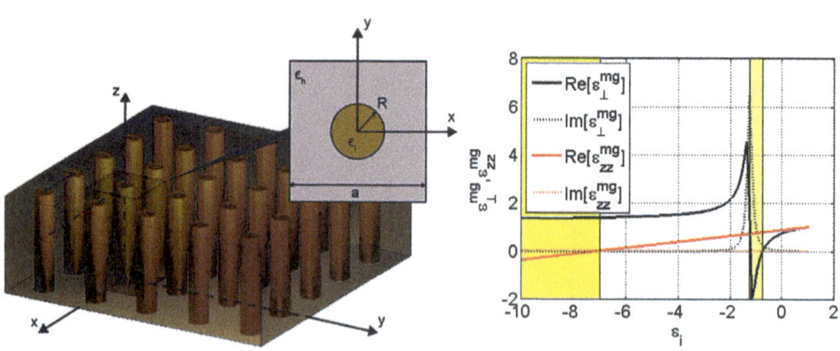

Fig. 1 Geometry of the nanowire composite (left) and the local effective medium parameters (right); Im(ε_i) = 0.1; shaded areas represent the spectral range where the metamaterial operates in the hyperbolic regime; when $\varepsilon_i \gtrsim -1$, the metamaterial is local; an additional wave exists for $\varepsilon_i \lesssim -1$; this wave propagates inside the metamaterial for $-7 \lesssim \varepsilon_i \lesssim -1$ (the main wave has elliptical dispersion); when $\varepsilon_i \lesssim -7$, the longitudinal wave exponentially decays along the wires, while the main wave has hyperbolic dispersion.

As follows from symmetry considerations, nanowire metamaterials have an uniaxial anisotropic response with an optical axis parallel to the wires ($\hat{z}$ axis in this work). In the Cartesian coordinates used in this work, the effective permittivity tensor is diagonal, with components $\varepsilon_{xx} = \varepsilon_{yy} = \varepsilon_{\perp}$, $\varepsilon_{zz} \neq \varepsilon_{\perp}$. In the quasistatic limit, the Maxwell Garnett effective medium theory predicts the following relation between the effective permittivity ε^{mg}, the permittivities of the components of the composite, and the geometrical parameters of the metamaterial:

$$\varepsilon_{\perp}^{\text{mg}} = \varepsilon_{\text{h}} \frac{(1+p)\varepsilon_{\text{i}} + (1-p)\varepsilon_{\text{h}}}{(1+p)\varepsilon_{\text{h}} + (1-p)\varepsilon_{\text{i}}}; \ \varepsilon_{zz}^{\text{mg}} = p\varepsilon_{\text{i}} + (1-p)\varepsilon_{\text{h}} \tag{1}$$

As any uniaxial material, the modes supported by the infinite nanowire metamaterial can be characterized according to their polarization. The modes of the first kind, often referred to as ordinary waves or transverse-electric (TE)-polarized waves, have their electric field in the *xy* plane. Propagation of these waves is not affected by the metamaterial anisotropy and is completely determined by the $\varepsilon_{\perp}$ component of the permittivity tensor.

In contrast to this behaviour, propagation of the waves that have a non-vanishing component of the electric field in the $\hat{z}$ direction is (strongly) affected by the material anisotropy. Depending on the relationship between ε_{i}, ε_{h}, and p, these extraordinary (TM polarized) modes have elliptic (when $\varepsilon_{\perp}^{\text{mg}} \times \varepsilon_{zz}^{\text{mg}} > 0$) or hyperbolic (when $\varepsilon_{\perp}^{\text{mg}} \varepsilon_{zz}^{\text{mg}} < 0$) dispersion (see Fig. 1).

In addition to these transverse electromagnetic waves, a plasmonic nanowire metamaterial may support propagation of longitudinal electromagnetic waves that represent coupled cylindrical surface plasmon (CSP) waves propagating along the wires. As shown in ref. 5, dispersion of these waves can be related to the solution of the eigenvalue problem with eigenvalues and eigenvectors representing propagation constants (k_z^{l}) and sets of amplitudes, respectively, describing contributions of individual cylindrical waves propagating in the nanowires of the composite to the coupled CSP modes. These CSP modes do not exist when wires are dielectric ($\varepsilon_{\text{i}} > 0$), they propagate in the plasmonic elliptic regime ($\varepsilon_{zz}^{\text{mg}} > 0$, $\varepsilon_{\text{i}} < -\varepsilon_{\text{h}}$); and exponentially decay along the wires in the hyperbolic regime ($\varepsilon_{zz}^{\text{mg}} < 0$). In the limit of PEC wires ($\varepsilon_{\text{i}} \rightarrow -\infty$), the propagation constant of the longitudinal wave approaches the limit $k_z^{\text{l}} \rightarrow n_{\infty}^{\text{l}}\omega/c$.

From the effective medium standpoint, the same dispersion relation can be written as $\varepsilon_{zz}(k_z,\omega/c) = 0$ with

$$\varepsilon_{zz}(k_z) = \xi\left(k_z^{\,2} - k_z^{\text{l}^2}\right)\frac{c^2}{\omega^2}; \quad \xi = p\frac{\varepsilon_{\text{i}} + \varepsilon_{\text{h}}}{\varepsilon_{\text{h}} - \left(n_{\infty}^{\text{l}}\right)^2}. \tag{2}$$

Note that the spatial dispersion, the explicit dependence of the permittivity on the wavevector, is a necessary condition for the existence of the longitudinal electromagnetic wave.

Although transverse electromagnetic waves that enter the nanowire material at normal incidence cannot couple to a longitudinal mode, light that obliquely enters the metamaterial can couple to both longitudinal and transverse modes. In a strict effective medium approach, light propagating inside the metamaterial should be represented as a linear combination of three modes. The first of these modes is an ordinary TE wave, described above, whilst the other two represent a mix of the "local" transverse TM and "nonlocal" longitudinal waves. Dispersion of these two waves and the relative contributions of "local" and "nonlocal" components are given by the coupled-oscillator-like expression,

$$\left(k_z^{\,2} - k_z^{l2}\right)\left(k_z^{\,2} - \varepsilon_\perp^{mg}\frac{\omega^2}{c^2}\right) = -\frac{\varepsilon_\perp^{mg}}{\xi}\frac{\omega^2}{c^2}\,k_x^{\,2}. \tag{3}$$

To better illustrate our results, in this work we have fixed the operating wavelength and geometric parameters of the metamaterial, the permittivity of the dielectric host medium, and varied the permittivity of the wire inclusions. For the set of parameters used in this work ($r = 20$ nm, $a = 100$ nm, $\varepsilon_h = 1$, $\lambda_0 = 1.5$ μm), the epsilon near-zero (ENZ) transition occurs at $\varepsilon_i \simeq -7$ with the elliptic regime realized for $\varepsilon_i > -7$, and the hyperbolic regime realized for $\varepsilon_i < -7$ (see Fig. 1 for predictions of local EMT). Dispersions of the three modes propagating inside the nanowire material are shown in Fig. 2. Note that in the regime when the main TM mode is elliptic, the additional wave has hyperbolic-like dispersion.

2.2. Transmission and reflection of light into nonlocal nanowire metamaterials

To understand the transmission and reflection of optical waves by nanowire metamaterials with finite thickness, the solutions to Maxwell's equations in each of the bulk layers are represented as linear combinations of (plane) waves that differ by their polarization and direction (towards the interface or away from the interface; see Fig. 3). The amplitudes of the waves in the neighbouring regions are then related to each other *via* boundary conditions. This relationship is then used to calculate the amplitudes of the waves propagating away from the interface as a function of the amplitude of the waves propagating towards the interface. Such linear relationship between the amplitudes can be conveniently represented in terms of the scattering (S-) matrix.

In conventional local materials, it suffices to enforce the continuity of tangential components of electric and magnetic fields to calculate the S-matrix of the interface. In nonlocal metamaterials, the existence of an additional (TM-

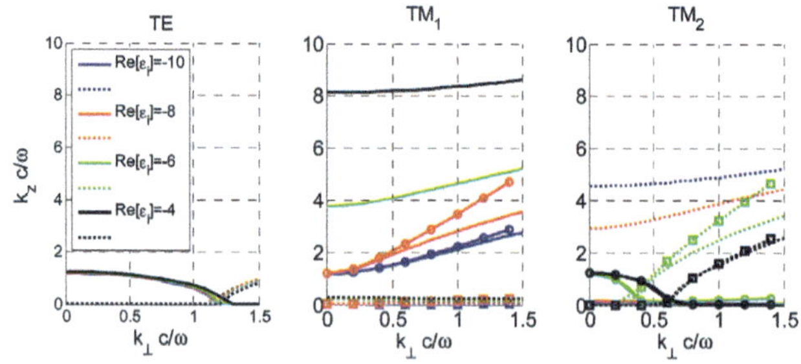

Fig. 2 Dispersion of the three waves supported by the metamaterial for different permittivities of the nanowires: (solid line) Re[k_zc/ω], (dashed line) Im[k_zc/ω], (symbols) dispersion of the "main" TM-polarized mode according to local EMT. All other parameters as in Fig. 1.

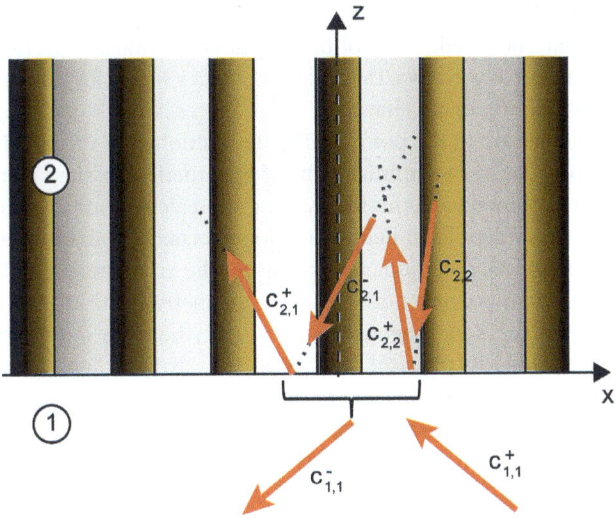

Fig. 3 Schematic of light reflection and transmission at the interface between local and nonlocal media.

polarized) mode requires the introduction of additional boundary conditions (ABCs). Here, we used the methodology of ref. 5 and used the continuity of the tangential electric field, the continuity of the normal electric displacement, and the continuity of the first moment of normal electric displacement (the product of $D(x)\exp 2\pi ix/a$, averaged over the unit cell) to calculate the S-matrix of each interface in the system.

3. Light emission in nanowire metamaterials

In this work, we have followed the well-developed technique that relates the enhancement in the decay rate of the dipole to the regular part of the field generated by this dipole at the origin. Explicitly,

$$\frac{\Gamma}{\Gamma_0} \simeq \frac{3}{2}\frac{\mathrm{Im}(\vec{E}\cdot\vec{P})}{\omega^3|P|^2} \tag{4}$$

where $\vec{E}$ represents the field generated by the point dipole $\vec{P}$. The main objective of this work is to analyse the enhancement of the decay rate in plasmonic nanowire composites and to relate this enhancement to spectral properties of the composites. It is worth noting that placing a dipole directly inside ideal hyperbolic or realistic (lossy) materials leads to unphysical singularities.[7,8] Various real or virtual cavity approaches have been suggested to address the impact of the surrounding media on a host emitter.[9] To address these key issues, in this work we have removed the singularity inside the hyperbolic metamaterial by limiting the wavevector spectrum used in the calculations of the field $\vec{E}$ in eqn (4) and by employing the following three-step strategy that allows us to address the issue of material absorption.

First, we calculated the emission of the point dipole inside hypothetical, lossless, metamaterial. Predictions of both local and nonlocal effective medium theories were analysed, with two TM-polarized modes in nonlocal metamaterials taken as two competing decay channels.

Second, the emission of the point 3D dipole positioned in the small (vacuum-filled) slit surrounded by two infinite slabs of hypothetical lossless metamaterial was analysed. Combined with the first step, these calculations allow us to estimate geometry-dependent local-field-correction effects.[10] Local field correction is the frequently employed technique to account for the arrangement of the emitter and the host matrix. It is worth noting that the geometry of the artificially created cavity could affect the spontaneous emission rate; here we have used the slit geometry, approaching its physical dimensions to zero thickness.

Finally, we calculated the emission of the point dipole positioned in the small vacuum slit surrounded by lossy metamaterial and used the local field correction, estimated above, to calculate the modulation of the dipole lifetime due to the metamaterial.

3.1. Plane wave expansion of Green's function in an homogeneous material

To calculate the modification of a dipole lifetime, we represented Green's function for the anisotropic metamaterial as a set of plane waves, and used eqn (4) to calculate the modification of the dipole lifetime. Since we are considering the emission inside the anisotropic material, the field generated by the dipole needs to be separated into TE and TM-polarized components.

It can be explicitly shown[10] that the field of the $\hat{z}$ polarized dipole is TM-polarized, with the regular part of $E_z^{\text{TM}} = \text{i}P\varepsilon_{\perp}\int_0^{k^{\max}} k^3\text{d}k/(k_z{\varepsilon_{zz}}^2)$, where the integration parameter k represents the in-plane component of the wavevector. In this work, we have taken into account the fact that any effective medium theory fails in the limit when $k \simeq \pi/a$. We therefore constrained the integral above to the range of wavevectors inside the first Brillouin zone of our metamaterial, $k < k^{\max} = \pi/a$.

On the other hand, the $\hat{x}$ polarized dipole emits both TE- and TM-polarized waves, with amplitudes yielding the plane-wave expansion of the field: $E_x^{\text{TM}} = \text{i}P\int_0^{k^{\max}} k_z k\ \text{d}k/(2\varepsilon_{\perp})$, and $E_x^{\text{TE}} = \text{i}P\int_0^{k^{\max}} \omega^2 k\ \text{d}k/(2c^2k_z)$. Due to the uniaxial symmetry of the problem, the y-component of the electric field generated by a y-polarized dipole is identical to the E_x generated by a $\hat{x}$-polarized dipole.

3.2. Emission in lossless metamaterials and local field correction analysis

We began by analysing the emission of a point dipole positioned inside an infinite material described by the local effective medium parameters given by eqn (1), keeping in mind the high-wavevector cut-off due to the finite unit cell size. The results of these calculations are shown in Fig. 4. It is seen that, in agreement with previous studies,[6] the local effective medium theory suggests that the enhancement of the photonic density of states can be linked to (i) plasmon resonance in the wires (corresponding to $\varepsilon_{\text{i}} \simeq -1$) and (ii) the broadband spectral range starting from ENZ ($\varepsilon_{\text{i}} \simeq -7$) and continuing through

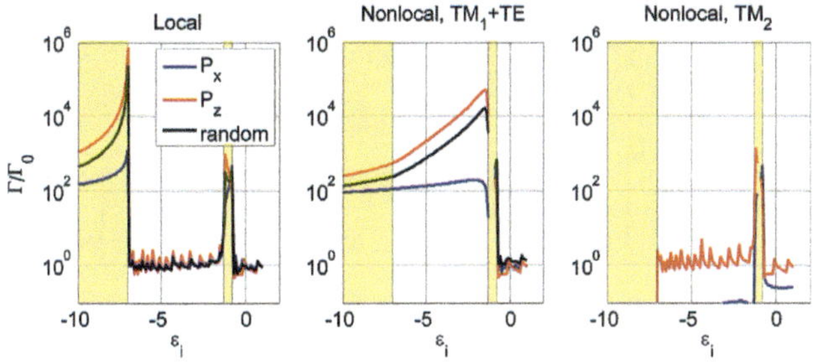

Fig. 4 Enhancement of the decay rate in infinite lossless metamaterials (lines of different colours represent different orientations of the dipole).

the hyperbolic regime of the metamaterial dispersion ($\varepsilon_i < -7$). Interestingly, the ENZ regime seems to provide the maximum reduction of the lifetime.

Calculations of the lifetime modifications were repeated for a situation when a point dipole is positioned inside the homogeneous material described by the nonlocal EMT model [eqn (2) and (3)]. In this case, we assumed that the dipole can independently emit into two TM-polarized channels. As seen in Fig. 4, predictions of the nonlocal EMT drastically differ from predictions of its local counterpart. The main difference, the dramatic enhancement of the decay rate in the elliptic regime can be traced to the existence of an additional electromagnetic mode, originating from the collective excitation of CSPs in the nanowire composite. Note that the hyperbolic-like CSP-rich mode (mode TM_1 in Fig. 2) dominates the emission across both elliptic and hyperbolic regimes. Also note that the nonlocality removes the singularity associated with ENZ regime (at $\varepsilon_i \simeq -7$).

To analyse the emission of the dipole in realistic, lossy, metamaterials, it becomes necessary to place the dipole inside a substantially small cavity carved in the homogeneous material. In this work, we have assumed that the cavity takes the shape of a small planar slit, oriented perpendicular to the wires. The emission of the point dipole was calculated according to the plane-wave expansion, described above, incorporated into the transfer-matrix formalism, as suggested in ref. 11.

For substantially small slit sizes (below ~10 nm), the enhancement of decay rate becomes independent of the slit size due to the high-wavevector k_{max} cut-off described earlier. Furthermore, for a local metamaterial, it becomes possible to eliminate local field correction effects by renormalizing the emission of $\hat{z}$-polarized dipole by $1/(\varepsilon_{zz}^{mg})^2$ (Fig. 5). Similarly, renormalization of the emission of the $\hat{z}$-polarized dipole by $1/\varepsilon_{zz}(k)^2$ positioned inside the slit carved in the nonlocal metamaterial also yields results that qualitatively agree with the total decay rate calculated for a dipole positioned inside the nonlocal medium (the exact nature of the resonance predicted by the nonlocal EMT in the ENZ region will be addressed in our future work). Therefore, we have used the above normalization to take into account local field correction effects.

3.3. Effect of finite material absorption

Finally, we calculated the decay rate of the dipole positioned in the slit between two slabs of lossy nanowire metamaterial with $\mathrm{Im}(\varepsilon_i) = 0.1$ (Fig. 6). As expected,

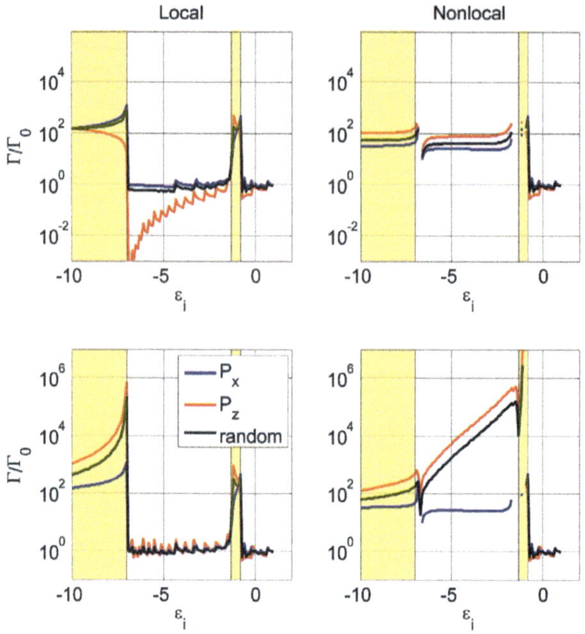

Fig. 5 Emission of the point dipole positioned in the vacuum slit inside the lossless nanowire metamaterial; top and bottom rows correspond, respectively, to results of calculations that are not and are accounted for local field corrections.

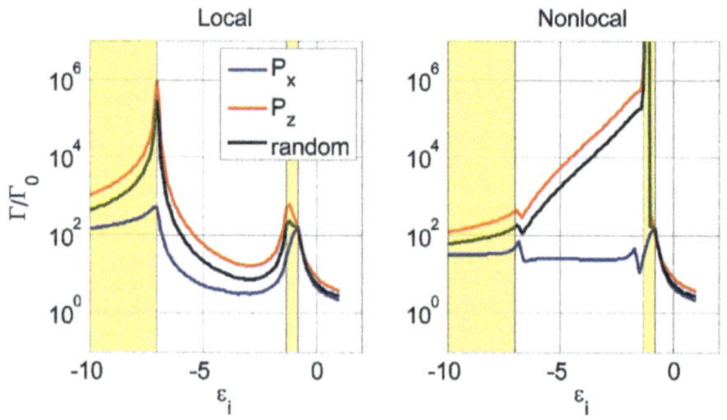

Fig. 6 Enhancement of decay rate in metamaterials with finite losses; the plots take into account local field corrections; $\mathrm{Im}(\varepsilon_i) = 0.1$; see Fig. 1 for effective medium parameters.

the presence of material absorption does not fundamentally alter the dynamics of the emission. In general, an increase in $\mathrm{Im}(\varepsilon_i)$ leads to further enhancement of the decay rate throughout the "plasmonic" spectral range ($\varepsilon_i < -\varepsilon_h$).

4. Discussion

It is clear that predictions by local and nonlocal effective medium theories regarding the enhancement of the decay rate for a dipole emitting inside nanowire metamaterials disagree with each other. It is therefore reasonable to ask which (if any) theory accurately predicts the response of the dipole in real metamaterials. The fundamental reason for the disagreement between the predictions of the two effective medium theories lies in the existence of an additional wave in nonlocal EMT. The existence of this additional wave has been validated in both full-wave numerical solutions of Maxwell's equations in nanowire composites and in experimental studies.[4,5] It is therefore reasonable to assume that nonlocal EMT adequately describes the emission inside the composite. Our main results can be interpreted in the following intuitive manner: the fundamental change in the density of optical states in a nanowire medium is associated not with the onset of hyperbolicity, but rather with the onset of the plasmonic response in the nanowires.

A similar effect is expected in layered materials with low fill-fraction where the layer's transition to plasmonic response is spectrally separated from the composite's transition to hyperbolicity. It should also be mentioned that, in realistic nanowire composites, the enhancement of the decay rate will be affected by standing wave Fabry–Perot type resonances.

5. Conclusions

We have analysed the enhancement of the decay rate of the dipole emitting inside a plasmonic nanowire composite. Predictions of the nonlocal effective medium response, previously demonstrated to adequately describe refraction in nanowire composites,[5] suggest a significant enhancement of the decay rate across the spectral range where nanowires are plasmonic and, in particular, across the spectral range where the composite as a whole exhibits elliptic response.

Acknowledgements

This research was supported by US Army Research Office (Grant #W911NF-12-1-0533), EPSRC (UK), and the ERC iPLASMM project (321268). A. Z. acknowledges support from the Royal Society and the Wolfson Foundation.

References

1 *Tutorials in Metamaterials*, ed. M. Noginov and V. A. Podolskiy, CRC press, 2012.

2 Z. Jacob, L. V. Alekeyev and E. Narimanov, *Opt. Express*, 2006, **14**, 8247; A. Salandrino and N. Engheta, *Phys. Rev. B: Condens. Matter Mater. Phys.*, 2006, **74**, 075103; J. B. Pendry, D. Schurig and D. Smith, *Science*, 2006, **312**, 1780; W. Cai, U. K. Chettiar, A. V. Kildishev and V. M. Shalaev, *Nat. Photonics*, 2007, **1**, 224; R. W. Ziolkowski, *Phys. Rev. E: Stat., Nonlinear, Soft Matter Phys.*, 2004, **70**, 046608; N. Engheta, *Science*, 2013, **340**, 286.

3 A. Kabashin, P. Evans, S. Pastkovsky, W. Hendren, G. A. Wurtz, R. Atkinson, R. Pollard, V. A. Podolskiy and A. V. Zayats, *Nat. Mater.*, 2009, **8**, 867; G. A. Wurtz, R. Pollard, W. Hendren, G. P. Wiederrecht, D. J. Gosztola, V. A. Podolskiy and A. V. Zayats, *Nat. Nanotechnol.*, 2011, **6**, 107.

4 P. A. Belov, R. Marques, S. I. Maslovski, I. S. Nefedov, M. Silverinha, C. R. Simovski and S. A. Tretyakov, *Phys. Rev. B: Condens. Matter Mater. Phys.*, 2003, **67**, 113103; A. L. Pokrovsky and A. L. Efros, *Phys. Rev. B: Condens. Matter Mater. Phys.*, 2002, **65**, 045110; M. Silverinha, *Phys. Rev. E:*

Stat., Nonlinear, Soft Matter Phys., 2006, **73**, 046612; R. J. Pollard, A. Murphy, W. R. Hendren, P. R. Evans, R. Atkinson, G. A. Wurtz, A. V. Zayats and V. A. Podolskiy, *Phys. Rev. Lett.*, 2009, **102**, 127405; K.-T. Tsai, G. Wurtz, J.-Y. Chu, T.-Y. Cheng, H.-H. Wang, A. Krasavin, J.-H. He, B. Wells, V. A. Podolskiy, J.-K. Wang, Y.-L. Wang and A. Zayats, *Nano Lett.*, 2014, **14**, 4971.

5 B. Wells, A. Zayats and V. A. Podolskiy, *Phys. Rev. B: Condens. Matter Mater. Phys.*, 2014, **89**, 035111.

6 Z. Jacob, I. Smolyaninov and E. Narimanov, *Proc. Front. in Opt*, 2009, PDPB6; H. N. S. Krishnamoorthy, Z. Jacob, E. Narimanov, I. Kretzschman and V. M. Menon, *Science*, 2012, **336**, 205; O. Kidwai, S. Zhukovsky and J. E. Sipe, *Opt. Lett.*, 2011, **36**, 2530; C. I. Cortes, W. Newman, S. Molesky and Z. Jacob, *J. Opt.*, 2012, **14**, 063001; A. N. Poddubny, P. A. Belov, P. Ginzburg, A. V. Zayats and Y. S. Kivshar, *Phys. Rev. B: Condens. Matter Mater. Phys.*, 2012, **86**, 035148.

7 W. Vogel and D.-G. Welsch, *Quantum Optics*, Wiley, 3rd, Revised and Extended edn, 2006.

8 S. M. Barnett, B. Huttner and R. Loudon, *Phys. Rev. Lett.*, 1992, **68**, 3698.

9 H. T. Dung, S. Y. Buhmann and D. G. Welsch, *Phys. Rev. A*, 2006, **74**, 023803; S. Scheel, L. Knoll and D. G. Welsch, *Phys. Rev. A: At., Mol., Opt. Phys.*, 1999, **60**, 4094; S. Scheel, L. Knoll and D. G. Welsch, *Phys. Rev. A: At., Mol., Opt. Phys.*, 2000, **61**, 069901.

10 L. Novotny and B. Hecht, *Principles of Nano-optics*, Cambridge U. Press, 2012.

11 S. R. J. Brueck, *IEEE J. Sel. Top. Quantum Electron.*, 2000, **6**, 899.

Faraday Discussions

PAPER

Plasmonics on the slope of enlightenment: the role of transition metal nitrides

Urcan Guler,[a] Alexander V. Kildishev,[ab] Alexandra Boltasseva[ab] and Vladimir M. Shalaev*[ab]

Received 22nd October 2014, Accepted 4th November 2014
DOI: 10.1039/c4fd00208c

The key problem currently faced by plasmonics is related to material limitations. After almost two decades of extreme excitement and research largely based on the use of noble metals, scientists have come to a consensus on the importance of exploring alternative plasmonic materials to address application-specific challenges to enable the development of new functional devices. Such a change in motivation will undoubtedly lead to significant advancements in plasmonics technology transfer and could have a revolutionary impact on nanophotonic technologies in general. Here, we report on one of the approaches that, together with other new material platforms, mark an insightful technology-driven era for plasmonics. Our study focuses on transition metal nitrides as refractory plasmonic materials that exhibit appealing optical properties in the visible and near infrared regions, along with high temperature durability. We take heat-assisted magnetic recording as a case study for plasmonic technology and show that a titanium nitride antenna satisfies the requirements for an optically efficient, durable near field transducer paving the way to the next-generation data recording systems.

1 Introduction

It was the pioneering work by Michael Faraday that reported the "*great power of action*" metal nanoparticles have on light and the effect of parameters such as the nanoparticle dimensions on the plasmonic resonance.[1] In this quintessential lecture he described gold (Au) as "*especially fitted for experiments*" for several reasons mostly based on the ease of observation. Following the same reasoning, noble metals such as Au and silver (Ag) have been widely used over the last few decades for a broad range of activities in the field of plasmonics.[2–11] With the advancement of nanofabrication techniques, the field of plasmonics experienced a burst in research activities, leading to great expectations on outcomes for daily applications mostly associated with optics and photonics. However, between the

[a]*Nano-Meta Technologies, Inc., 1281 Win Hentschel Blvd, West Lafayette, IN 47906, USA. E-mail: uguler@nanometatech.com*
[b]*School of Electrical and Computer Engineering and Birck Nanotechnology Center, Purdue University, West Lafayette, IN 47907, USA. E-mail: shalaev@purdue.edu*

pioneering report of Faraday and the hype in the field, it was the electron beam techniques that revealed plasmonic resonances of metals. In 1959, Powell and Swan reported the observation of bulk and surface plasmon resonances from an aluminium (Al) thin film *via* electron energy loss spectroscopy (EELS).[12] As a low cost material with plasmon resonance covering the UV and visible regions, Al is again attracting the attention of the plasmonics community.[13]

Although other traditional metals, such as copper and aluminium,[14,15] have been utilized for plasmonics applications, their room temperature oxidation and strong dependence of the resonance quality on the oxide remain limitations for their widespread adoption. Similarly, Ag has been claimed as the highest quality plasmonic material that exhibits resonances through the entire visible spectrum, but the use of the material has been hindered due to strong room temperature oxidation and sulfidation; both processes are significantly enhanced when the material is nanostructured.[16,17] The development of chemically stable and biocompatible Ag nanostructures is an exciting challenge with intensive ongoing effort.[18,19] Among the widely used plasmonic metals, Au has attracted a significant amount of attention due to its chemical stability and relatively good plasmonic properties, despite the fact that the resonance quality of Au is significantly lower compared to Ag and Al.[20] However, even this comparison of Au against Ag shows that the parochial approach, which concentrates simply on the resonator strength when evaluating materials, could be misleading in a particular application. In fact, the entire field of plasmonics has reached a stage where the overall functional material properties should be considered instead of plasmonics-only evaluations.

2 Assessment

As a research field that has attracted great attention over the last two decades, giving birth to numerous application proposals, plasmonics is now subject to grand expectations on the deliverables. Relying on the maturity of the field and the fact that technology products are soon expected, here we discuss the brief history of breakthroughs and the current state of the technology from our interpretation of the Gartner Hype Cycle Methodology (HCM). The HCM has been successfully used for a variety of technologies over the last decade and claims to be an efficient way to adopt them to real industrial problems.[21]

Fig. 1 shows the authors' view on the main milestones in plasmonics that have generated a great deal of attention over the past decade. Although the first observations of plasmonic effects are quite old,[22] the systematic investigation of the interaction of light with nanoscale metallic particles clearly starts with the work performed by Faraday in 19th century.[1] After a relatively long period of theoretical studies performed by leading scientists,[23,24] the characterization of electron-beam-excited plasmons are considered the leading experimental efforts in the field.[12,25] Surface enhanced Raman scattering (SERS), first observed in 1974 by Martin Fleischmann *et al.*[26] and attributed to enhanced electric field by David Jeanmaire and Richard Van Duyne in 1977,[27] is perhaps the very first systematic application of plasmonic effects. Soon after the first experimental results were reported, the clear relation between the localized surface plasmon resonance (LSPR) and the enhanced SERS signal was proposed by Martin Moskovits.[28,29] With the advancement of nanofabrication methods and their widespread

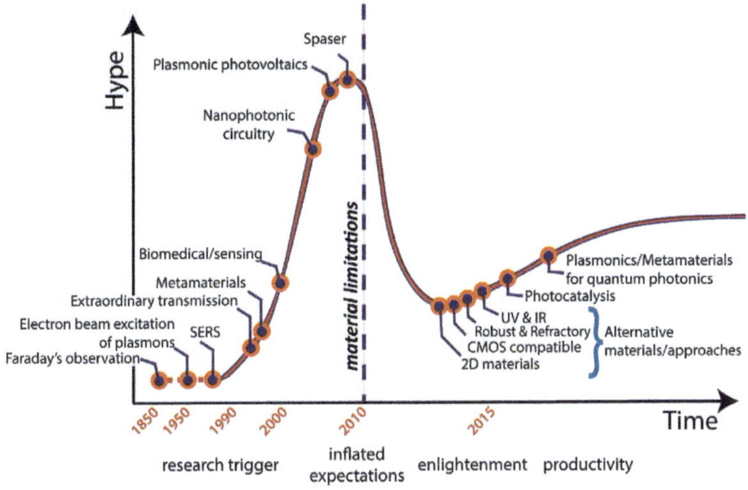

Fig. 1 The hype cycle for the field of plasmonics from the authors' point of view. After a decade of high expectations, material limitations are delaying the realization of the proposed concepts and technology transfer. Merging plasmonics with materials science and chemistry could provide new solutions and bring technological advances across many application spaces.

availability, breakthrough results such as the extraordinary transmission of light through plasmonic nanoholes,[30] and plasmonic metamaterials created a strong impact.[31]

Sensing and biomedical applications have long been the flagship of the field due to the wide range of exciting opportunities offered by bio-compatible, plasmonic nanostructures.[32–34] With a solid research record over the past two decades, a wide variety of sensing and biomedical devices based on plasmonic effects were expected to enter production lines. However, technology transfer has thus far been slow compared to the level of expectations.[35] Beyond biomedical applications, interest in the field was substantially nourished by several recent proposals to develop sub-wavelength, hybrid nanophotonic components that have optical and electronic functionalities,[4,9,36–49] and plasmonic photovoltaics where metal nanoparticles were used as E-field concentrators in thin film devices for enhanced photocurrent generation.[50–54]

In our opinion, the peak of expectations in the field of plasmonics was reached with the breakthrough concept of spaser,[55] the plasmonic counterpart of a laser at nanoscale, that could potentially be the key towards many nanoscale applications such as all-optical circuitry, lithography, probing and microscopy.[56,57] The compensation of losses in plasmonic nanostructures with semiconductor quantum dots,[58] and a demonstration of a spaser based on sub-wavelength resonators made of gold nanoparticles surrounded by a dye-doped silica shell operating in the visible spectrum were reported soon after the proposal.[59] However, despite extensive experimental efforts, losses in metal nanostructures arising from electron scattering have been the major limiting factor for the spaser and the field of plasmonics in general.[60,61] This absorption of electromagnetic energy due to material losses is inherently affecting technologies relying on the field enhancement and lossless propagation of energy, which span the entire list of breakthroughs positioned at the peak of expectations.

The metamaterials research, a field highly correlated with plasmonics, has followed a similar trend over the last decade. Triggered by the negative index materials and superlens,[31,62–64] excitement and expectations peaked with the idea of cloaking and transformation optics.[65,66] In the authors' opinion, after the recognition of material limitations such as losses and scalability, the field is experiencing an enlightenment period with approaches incorporating alternative materials with realistic design parameters.

We believe the breakthroughs that led to the peak of somewhat inflated expectations are now evolving and bringing solutions to earlier problems. For example, the concept of harvesting solar energy using the enhanced interaction of plasmonic nanostructures with light has expanded to the generation of hot electrons for efficient heating, and plasmon induced photocatalysis.[67,68] Mubeen *et al.* recently reported an approach where metal in a plasmonic photovoltaic device is not a field enhancing antenna as in previous concepts, but instead, plays the role of the light absorber that generates carriers.[69] Another approach to beat the metallic losses is to use high refractive index dielectric nanostructures as light trappers.[70] Nanophotonic circuitry is now leaning towards the integration of plasmonics with newly emerging research on 2D materials such as graphene and MoS_2 and quantum photonics.[71–74] Plasmonic metamaterials research is concentrated on alternative materials with application specific characteristics such as low loss, tunability, refractory properties, and CMOS compatibility.[37,75–77] Research into alternative materials for spectral ranges out of the visible are other examples of application-specific efforts in the field of plasmonics.[78,79] In this work, we consider one of the approaches that, together with other new material platforms including semiconductors, 2D materials, phase-changing and ceramic materials, marks an insightful technology-driven era for plasmonics on the slope of enlightenment.

3 Results & discussions

In this paper, we report on a case study for plasmonics technology that could enable the next-generation of data recording systems. In particular, we focus on the potential of refractory plasmonic materials as nanoantennas for field enhancement and local heating. Refractory plasmonic materials provide two unique properties: good plasmonic performance in the visible and near infrared regions, and durability at high temperatures.[76] Although the definition varies with the level of devoutness, we will use the term refractory to describe materials with melting points above 2000 °C. Mechanical hardness and chemical stability are other common features of refractory materials, while metals with refractory properties generally suffer from oxidation at high temperatures.[80] Thus, oxidation is considered as an exception to chemical inertness when classifying refractory metals. Self-passivation of the oxide slows down the process for bulk metals, but nanoscale structures are often too small to accommodate a protective oxide layer.

In contrast to structural deformations such as melting and creep that are usually encountered in the case of soft metals, oxidation of refractory plasmonic materials can be avoided with the integration of optically transparent materials with higher oxidation temperatures. Thus, for applications with extreme operational conditions, refractory plasmonic materials with comparable optical properties can replace noble metals,[81] with the help of optically transparent, oxidation resistant thin layers, and new technologies can be developed.

Refractory metals provide high melting temperatures and hardness along with metallic properties that imply their potential use for plasmonic applications. They have been the material of choice where metallic optical properties and high temperature stability are desired such as spectrally selective photonic crystal emitters for thermophotovoltaic devices.[82,83] However, for applications where enhanced interaction of a sub-wavelength structure with electromagnetic radiation is essential, the requirements go beyond the metallic behaviour and necessitate good plasmonic properties.[77] Although the metrics for a good plasmonic material are functions of a variety of parameters mostly depending on the application,[84] it is usually considered acceptable to reduce the formulation to simplified terms.[77,85] Here, we will limit the discussion to LSPR where nanoparticles are employed as sub-wavelength field enhancers. One such application which has been quite popular over the last few years is the heat assisted magnetic recording (HAMR).[86] In the HAMR approach, magnetic grains made of high coercivity materials, which require an intense magnetic field to be written, are locally heated in order to facilitate the writing process. The requirement of a high coercivity material is simply a result of the motivation to achieve higher density data storage, which causes the use of smaller grains and introduces instabilities.[86] After the writing is accomplished at an elevated temperature, the grain is cooled down to lower temperatures where it can retain the data. It was initially proposed that local heating could be achieved under focused laser illumination.[87] Despite a successful demonstration, laser heating was naturally limited by the diffraction limit which resulted in a spot much larger than the sizes imposed by the dense storage motivation.

In order to achieve sub-wavelength focusing of electromagnetic radiation, plasmonic nanostructures can be used in the system as near field transducers (NFTs).[86] In this scenario, the excitation light is guided onto a plasmonic nanostructure with a resonance around 830 nm, where cost-effective diode lasers are available, and the enhanced near-field is used as a concentrated source of energy for local heating of the magnetic layer. Noble metals are the natural candidates for NFTs as they exhibit strong plasmonic oscillations in the visible region of the electromagnetic spectrum. Although the plasmonic properties of silver (Ag) are superior to Au, chemical instability and consequent performance degradation is a major concern for HAMR and many other applications. Therefore, owing to its chemical stability, Au was used as the plasmonic NFT material in the very first demonstrations of HAMR technology where the idea has been successfully demonstrated.[88,89] However, Au is a soft material with a relatively low bulk melting temperature which is further reduced due to the nanostructured geometries and sharpened edges required for better field confinement.[90] Although operational temperatures around 400 °C are required at the nanoscale near the magnetic grain, NFTs made of pure Au show degradation in their mechanical properties at temperatures around 100 °C.[91] The mechanical instability of Au NFTs under conditions imposed by HAMR's operational principles has been one of the major problems which has slowed the progress towards a reliable product.[92] Among several methods to resolve the problem, alloying and doping of the plasmonic NFT material with a harder dopant, at the expense of lowered optical performance, have attracted attention.[91,93] Although refractory metals can be used as the hard dopants to improve the mechanical properties of plasmonic metals, their use as the plasmonic NFT material has so far not been possible due to poor optical properties.

Transition metal nitrides are refractory materials with high melting points and mechanical hardnesses. They have been widely used as protective coatings in applications where mechanical strength is desired. In addition to their superior mechanical properties, it has been demonstrated that titanium nitride (TiN) and zirconium nitride (ZrN) exhibit plasmonic properties similar to Au, explaining their golden luster.[84,94,95] In fact, these materials have been used in imitation jewellery as a cheaper and more durable replacement for Au.[96] As a bio-compatible material, the use of TiN is common in bio-medical implants.[97,98] Consequently, TiN has been one of the most frequently used transition metal nitrides which has resulted in a good understanding of the deposition techniques, including CMOS compatible processes.[99,100] The optical properties of transition metal nitrides strongly depend on the deposition conditions and can be optimized to nearly match the optical properties of Au.[84,101,102]

Fig. 2 shows the real and imaginary parts of the dielectric permittivities of different metals along with transition metal nitrides. Gold is presented as a reference since it has been widely used as the plasmonic material in many applications, including in HAMR demonstrations. Here we use permittivities experimentally retrieved from Au thin films deposited on a sapphire substrate as described elsewhere.[95] Although the permittivities reported by Johnson and Christy are frequently used in comparison tables, in practice, thin films of metals exhibit optical properties that are inferior to bulk values.[103,104] Transition metal nitride films were also experimentally retrieved from thin films deposited on sapphire and MgO substrates while the permittivities of refractory metals were taken from tabulated data.[77,95,105] Not surprisingly, the magnitude of the real part (ε') is largest for the case of Au. In addition, the imaginary part of the permittivity (ε'') is also one of the smallest among the materials listed. The use of Au as the NFT material would be unquestionable if the plasmonic properties were the only

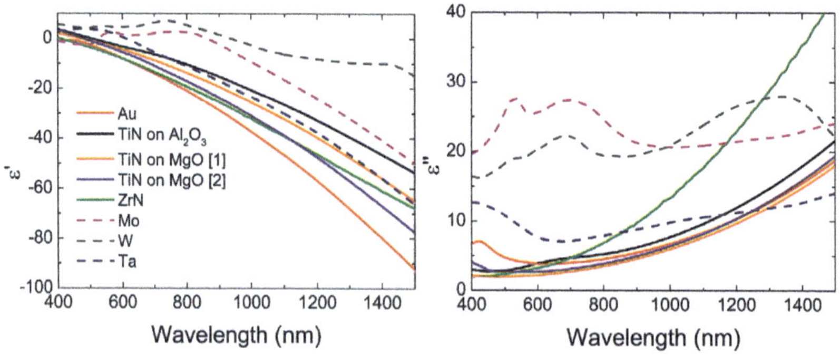

Fig. 2 Dielectric permittivities of refractory transition metal nitrides, refractory metals, and gold. Gold is given as a reference plasmonic material. Transition metal nitrides, which are refractory materials that are durable at high temperatures, exhibit optical properties similar to gold. Refractory metals, on the other hand, are plasmonic in the near infrared region and exhibit larger losses.

concern. However, as stated earlier, the ductility of Au prevents the use of the material for a long period without further processing which degrades its optical properties. On the other hand, refractory metals, which would naturally resolve many issues related to material deformations, do not exhibit plasmonic properties in the spectral region of interest. Tungsten (W) and molybdenum (Mo) are barely plasmonic in the infrared region, however their permittivities are positive in the visible region. Tantalum (Ta), on the other hand, has a permittivity zero cross-over near 610 nm, making the material plasmonic around HAMR operational wavelengths. However, the imaginary part of the permittivity is still significantly higher that Au, although it is much smaller than other refractory metals. For the case of transition metal nitrides, the optical properties are closer to plasmonic metals – representing a significant opportunity to resolve the complications arising due to the harsh operational environment. Both the ε' and ε'' values of ZrN are very close to Au, especially in the visible region, making it a potential candidate to replace noble metals. Fig. 2 presents three different sets of TiN data retrieved from samples grown on sapphire and MgO. Sapphire supports local epitaxial growth of TiN and reduces the losses in the film while MgO provides better lattice matching, leading to an even higher quality epitaxial thin film. Titanium nitride films grown on sapphire substrates have a negative permittivity in the visible region and a loss factor in the same range as the Au film. Thin films of TiN deposited on MgO substrates exhibit plasmonic properties better than the sample grown on sapphire.

The figure of merit (FOM) of a plasmonic nanostructure totally depends on the specific application in mind.[77] The high level of interest in identifying proper plasmonic materials has resulted in a comprehensive collection of reports comparing different material classes for metamaterial, transformation optics, superlens, surface plasmon polariton and LSPR applications.[77,106–108] The general consensus for LSPR applications is to set the ratio of the real and imaginary parts of the dielectric permittivity of a plasmonic material as a FOM.[77,106–108] For a nanoparticle in the quasistatic limit, when the size of the particle is very small compared to the wavelength, the FOM can be conveniently defined as $\varepsilon'/\varepsilon''$ where the real part of permittivity implies a good resonator and the imaginary part represents the loss factor. However, as the particle size becomes comparable to the wavelength of the incident light, the quasistatic approximation loses its validity, and so does the FOM. A straight modification to increase the accuracy of the FOM is to include the size factor which would take into account missing effects such as the existence of the higher order modes, interference effects arising from the multi-modal system, penetration depth of the excitation light through the nanoparticle, *etc.* In addition, even for the case of a single field such as LSPR, there are a variety of operational modes which require different features such as maximized absorption in the particle and subsequent efficient local heating or hot electron generation,[11,109] enhanced field in the near zone for energy confinement,[59,110] or scattering to the far zone for several other applications.[111,112] Previously we have shown that a more comprehensive, yet still convenient, FOM for LSPR applications is the optical cross-section which can be obtained through the Mie formalism.[84] A comparison between the scattering efficiencies of transition metal nitrides and Au shows that the field enhancement at the surface, or at an infinite distance from the particle, for TiN and ZrN are indeed comparable to Au when the nanoparticles are out of the quasistatic regime.[84] Here, we will follow the general consensus and apply the simplified FOM with the quasistatic approximation to compare the LSPR efficiencies of transition metal nitrides, refractory metals, and Au, which is the reference material due to its wide use in the field. In HAMR technology, an NFT is expected to focus electromagnetic energy to a confined volume in the magnetic layer with minimal loss in the plasmonic particle itself. We note that recently reported comparisons between transition metal nitrides and Au employed the FOM in the quasistatic approximation and TiN's optical properties that were not yet optimized for plasmonic applications.[85,113,114]

Table 1 shows the dielectric permittivity values for gold, refractory metals, and refractory transition metal nitrides at a wavelength of 830 nm. Dielectric permittivities retrieved for TiN thin films deposited on c-sapphire and MgO substrates are given to illustrate the variation of the optical properties based on the deposition parameters, (and hence, crystallinity and the amount of defects) and their effect on the FOM. It should be noted that the comparison of the real part to the imaginary part is reasonable due to the fact that (i) the real part illustrates the resonator properties which determines the field enhancement, and (ii) the imaginary part factors in the losses which degrade the resonance and increase the internal heating of the NFT, an undesired effect.[85] Table 1 shows that the experimental data retrieved from the Au thin film deposited on sapphire substrate has the best FOM among all samples. The TiN film deposited on sapphire has a FOM value of 2.14 which is less than half of the Au film, but still better than the previously reported FOM values for TiN.[114] When grown on MgO, TiN films with enhanced optical properties can be achieved due to the better lattice matching between the film and the substrate which promotes epitaxial growth.[102] The TiN samples deposited on MgO have FOMs significantly larger than that of the TiN sample grown on sapphire, values very close to Au. Zirconium nitride has a large normalized real permittivity, however the FOM value is degraded due to large losses. Refractory metals, on the other hand, are either not plasmonic in the spectral region of interest, or exhibit very poor properties, ruling them out for LSPR applications in the visible and near infrared region of the electromagnetic spectrum.

Table 1 Permittivity values for refractory metals, refractory transition metal nitrides and gold at 830 nm. The figure of merit is defined in the quasistatic regime. Titanium nitride samples grown on a MgO substrate outperform other refractory materials and become prominent for high temperature plasmonic applications in the visible and near infrared spectrum

Material	Permittivity (real)	Permittivity (imaginary)	FOM ($\lvert\varepsilon'\rvert/\varepsilon''$)
Au	−23.39	4.67	5.01
TiN (Al_2O_3)	−11.73	5.47	2.14
TiN (MgO) [1]	−15.57	3.68	4.23
TiN (MgO) [2]	−19.08	4.11	4.64
ZrN	−21.15	7.64	2.77
Mo	1.43	23.16	NA
W	4.21	19.41	NA
Ta	−12.89	8.15	1.58

According to the FOM values given in Table 1, TiN thin films grown on MgO substrates exhibit plasmonic properties close to Au, with only 16% and 7% reduction in the optical FOM. However, as a refractory plasmonic material, TiN possesses material properties such as hardness, a high melting point, CMOS-compatibility, and chemical stability that are of significant importance for applications that require extreme operational conditions. Heat assisted magnetic recording, for example, is one such application that employs plasmonic elements successfully, but suffers from ductile material limitations.[92]

Fig. 3 shows FEM simulation results for a "nanobeak antenna" that was previously suggested as an efficient design for near-field generation for HAMR.[115] Here we use this design in order to take a step further in the comparison between TiN and Au. Although the TiN film on a MgO substrate labelled with [2] has a better FOM, we proceed with the data from the sample [1] in our simulations. The calculated absorption efficiencies of the same antenna made of Au and TiN

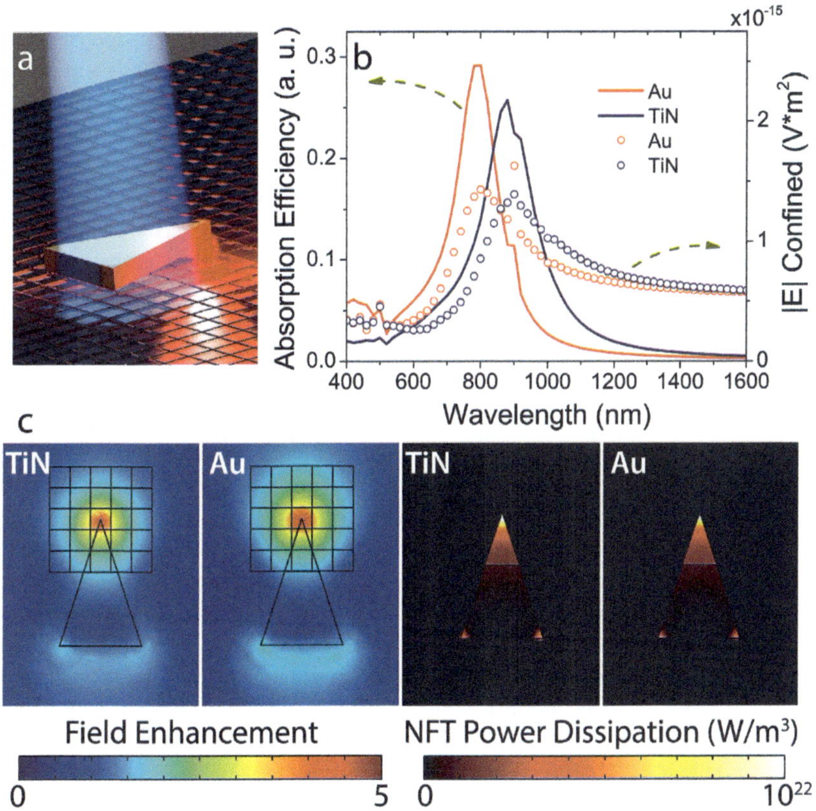

Fig. 3 (a) Illustration of a "nanobeak antenna" design used in simulations where titanium nitride and gold are compared. (b) The absorption efficiencies of gold and titanium nitride nanobeak antennas and the corresponding electric field magnitudes $|E|$ confined in a 25 × 25 × 10 nm^3 volume in the dielectric layer 5 nm below the antenna tip. Dashed arrows indicate the corresponding scale. (c) Electric field magnitudes $|E|$ in the dielectric layer and heat dissipation densities inside the nanoantennas for the peak wavelengths of each material; 880 nm for TiN and 800 nm for gold. Permittivity values are taken from the sample: TiN on MgO [1], as given in Fig. 2.

(sample on MgO [1]), and the corresponding electric field enhancements in a dielectric layer located 5 nm below the NFT are given in Fig. 3(b). Due to the compact design of the NFT and magnetic head, overheating of the plasmonic antenna is an undesired side effect of local heating for HAMR devices.[116–119]

Heat generation inside the plasmonic antenna is a direct result of the absorption of the electromagnetic energy.[120] Thus, for an NFT design, large field enhancement in the magnetic medium is desired while the absorption inside the antenna should be kept minimal. Calculations show that the identical "nanobeak" design with Au and TiN gives comparable field enhancement in the magnetic grain and absorption in the antenna. This result is in agreement with the FOM values given in Table 1 for the quasistatic regime. Fig. 3(c) shows the electric field enhancement in a dielectric layer located 5 nm below the tip of the "nanobeak" antenna. The enhanced field peak is confined within a 25×25 nm^2 area for both TiN and Au antennas. Power dissipation densities inside the antennas are concentrated at the corners, as expected, which illustrates the importance of material durability on the nanometer scale. The calculation results presented in Fig. 3 clearly indicate that a TiN NFT would generate heating in a very similar way that a Au NFT does. In addition to comparable optical properties, TiN offers hardness and high temperature durability. In fact, TiN is already a material of interest as a diffusion barrier between the NFT and the magnetic lip.[121] Therefore, a TiN NFT is a reliable solution to multiple high temperature based problems.

Refractory metals are defined as chemically stable at high temperatures, but the oxidation at elevated temperatures is always an exception.[80] This is a problem for transition metal nitrides as well.[122] A self-passivating thin native oxide of thickness around 1–2 nm is present at the surface of TiN at room temperature. Two stages of oxidation, designated as initial and normal stages, were observed with an order of magnitude difference in the rate constants.[123] Further investigation revealed that the TiO_2 changes from amorphous to crystalline in time, and faster with higher temperatures.[124] For the case of plasmonic applications, where subwavelength nanostructures are the building blocks, the effects of even a thin oxide layer on the performance of the plasmonic component would be significant. The temperatures at which oxidation starts to play a significant role are in the range where plasmonic NFTs in HAMR devices operate (~500 °C).[85] Fortunately, the oxidation is easier to handle compared to other structural problems arising from the softness of Au.[122] Here, we suggest an optically transparent, thin overcoat as an oxidation barrier which provides long term durability at elevated temperatures.[122] Dielectric ceramics are natural candidates as oxidation resistant protective layers over plasmonic transition metal nitrides. Fig. 4 shows reflectance measurements and numerical simulations for thin TiN films grown on a c-sapphire substrate. An overcoat layer with 10 nm thickness is applied as an oxidation resistant barrier. Fig. 4(a) shows the reflectance measurements of a bare TiN film and a sample with a dielectric overcoat. The bare film reflectance matches well with the simulation where a 2 nm native oxide is included. Fig. 4(b) shows the reflectance data for the protected sample after exposure to high temperatures in air for several hours. The sample was cooled down to room temperature after each treatment for optical measurement and heated up to the next level for further testing. The measurements show that the film retains its optical properties at least up to 525 °C which is at the upper limits of proposed

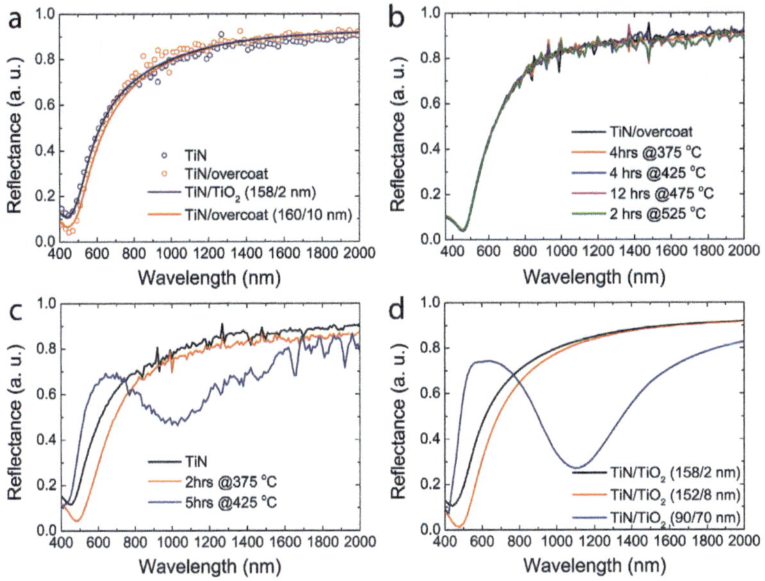

Fig. 4 (a) Reflectance measurements and simulations for TiN thin films with and without an overcoat. (b) The sample with a dielectric overcoat shows significant resistance against oxidation. (c) Bare TiN thin films oxidize at temperatures above 350 °C which causes severe degradation of the optical properties. (d) Changes in the optical properties with increasing temperature can be replicated in simulations by introducing oxide layers with increasing thicknesses.

HAMR operational temperatures. Fig. 4(c) shows the reflectance data of the bare TiN film before and after exposure to 375 and 425 °C in air. The optical properties of the sample degrade significantly due to oxidation. The simulation results given in Fig. 4(d) take the oxide layer into account and match well with the experimental data. For the case of heating at 425 °C, the intensity of the simulated dip shows a deviation from the experimental data likely due to an intermediate oxynitride formed between the oxide and nitride due to oxygen diffusion and nitrogen displacement. Further discussion on the effect was studied earlier by Saha and Tompkins.[124]

4 Conclusions

The field of plasmonics is facing challenges due to materials' limitations. However, we believe that this research field is now entering its technology-driven enlightenment period, exploring alternative materials for application-specific needs and new device concepts. Novel material platforms such as semiconductors, 2D materials, phase-changing and ceramic materials mark a new era for plasmonics and are expected to bring advancements to various photonic technologies such as information and communications, imaging and sensing, healthcare, energy, manufacturing, and national security. Among many plasmonic materials, transition metal nitrides are particularly appealing because they exhibit good plasmonic properties along with excellent refractory characteristics and thus, they could perform well when noble metals fail because of their softness and low melting points. In this work, we take heat assisted magnetic recording as one of the important examples of high temperature plasmonic applications and compare the properties of refractory transition metal nitrides, refractory metals, and gold. Simulation results show that titanium nitride, in terms of its plasmonic properties, can perform very similarly to gold but provides indispensable advantages due to its excellent refractory characteristics. We also show that the oxidation of titanium nitride at high temperatures can be fully avoided through the use of optically transparent, thin ceramic overcoats. Therefore, we believe that refractory plasmonic transition metal nitrides are the key to next generation durable, compact photonic devices. Our findings directly impact the technology of heat-assisted magnetic recording for data storage and could provide harsh-environment capable solutions for defense, intelligence, information technology, aerospace, energy, and oil & gas industries.

Acknowledgements

The authors acknowledge generous support from NSF SBIR grant IIP-1416232, ONR MURI grant N00014-10-1-0942, and NSF MRSEC grant DMR-1120923. We would like to thank Nathaniel Kinsey for his help in the preparation of this manuscript.

References

1 M. Faraday, *Philos. Trans. R. Soc. London*, 1857, **147**, 145–181.
2 W. L. Barnes, A. Dereux and T. W. Ebbesen, *Nature*, 2003, **424**, 824–830.
3 E. Ozbay, *Science*, 2006, **311**, 189–193.
4 H. A. Atwater, *Sci. Am.*, 2007, **296**, 56–63.
5 S. A. Maier, *Plasmonics: Fundamentals and Applications*, Springer, 2007.
6 S. Lal, S. Link and N. J. Halas, *Nat. Photonics*, 2007, **1**, 641–648.
7 J. A. Schuller, E. S. Barnard, W. Cai, Y. C. Jun, J. S. White and M. L. Brongersma, *Nat. Mater.*, 2010, **9**, 193–204.
8 D. K. Gramotnev and S. I. Bozhevolnyi, *Nat. Photonics*, 2010, **4**, 83–91.
9 V. J. Sorger, R. F. Oulton, R.-M. Ma and X. Zhang, *MRS Bull.*, 2012, **37**, 728–738 M723-710.1557/mrs.2012.1170.
10 M. Kauranen and A. V. Zayats, *Nat. Photonics*, 2012, **6**, 737–748.
11 G. Baffou and R. Quidant, *Laser Photonics Rev.*, 2013, **7**, 171–187.
12 C. J. Powell and J. B. Swan, *Phys. Rev.*, 1959, **115**, 869–875.
13 M. W. Knight, N. S. King, L. Liu, H. O. Everitt, P. Nordlander and N. J. Halas, *ACS Nano*, 2013, **8**, 834–840.
14 G. H. Chan, J. Zhao, E. M. Hicks, G. C. Schatz and R. P. Van Duyne, *Nano Lett.*, 2007, **7**, 1947–1952.
15 G. H. Chan, J. Zhao, G. C. Schatz and R. P. V. Duyne, *J. Phys. Chem. C*, 2008, **112**, 13958–13963.
16 U. Guler and R. Turan, *Opt. Express*, 2010, **18**, 17322–17338.
17 M. D. McMahon, R. Lopez, H. M. Meyer, III, L. C. Feldman and R. F. Haglund, Jr, *Appl. Phys. B*, 2005, **80**, 915–921.
18 A. Desireddy, B. E. Conn, J. Guo, B. Yoon, R. N. Barnett, B. M. Monahan, K. Kirschbaum, W. P. Griffith, R. L. Whetten, U. Landman and T. P. Bigioni, *Nature*, 2013, **501**, 399–402.

19 G. A. Sotiriou, G. D. Etterlin, A. Spyrogianni, F. Krumeich, J.-C. Leroux and S. E. Pratsinis, *Chem. Commun.*, 2014, **50**, 13559–13562.
20 M. G. Blaber, M. D. Arnold, N. Harris, M. J. Ford and M. B. Cortie, *Physica B*, 2007, **394**, 184–187.
21 J. Fenn and M. Raskino, *Mastering the Hype Cycle: How to Choose the Right Innovation at the Right Time*, Harvard Business Press, 2008.
22 I. Freestone, N. Meeks, M. Sax and C. Higgitt, *Gold Bull.*, 2007, **40**, 270–277.
23 G. Mie, *Ann. Phys.*, 1908, **330**, 377–445.
24 J. C. M. Garnett, *Philos. Trans. R. Soc., A*, 1904, **203**, 385–420.
25 Y.-Y. Teng and E. A. Stern, *Phys. Rev. Lett.*, 1967, **19**, 511–514.
26 M. Fleischmann, P. J. Hendra and A. J. McQuillan, *Chem. Phys. Lett.*, 1974, **26**, 163–166.
27 D. L. Jeanmaire and R. P. Van Duyne, *J. Electroanal. Chem. Interfacial Electrochem.*, 1977, **84**, 1–20.
28 M. Moskovits, *J. Chem. Phys.*, 1978, **69**, 4159–4161.
29 M. Moskovits, *Notes Record Roy. Soc. Lond.*, 2012, **66**, 195–203.
30 T. W. Ebbesen, H. J. Lezec, H. F. Ghaemi, T. Thio and P. A. Wolff, *Nature*, 1998, **391**, 667–669.
31 J. B. Pendry, *Phys. Rev. Lett.*, 2000, **85**, 3966–3969.
32 L. R. Hirsch, R. J. Stafford, J. A. Bankson, S. R. Sershen, B. Rivera, R. E. Price, J. D. Hazle, N. J. Halas and J. L. West, *Proc. Natl. Acad. Sci. U. S. A.*, 2003, **100**, 13549–13554.
33 C. Loo, A. Lin, L. Hirsch, M. H. Lee, J. Barton, N. Halas, J. West and R. Drezek, *Technol. Cancer Res. Treat.*, 2004, **3**, 33–40.
34 A. M. Gobin, M. H. Lee, N. J. Halas, W. D. James, R. A. Drezek and J. L. West, *Nano Lett.*, 2007, **7**, 1929–1934.
35 P. D. Howes, R. Chandrawati and M. M. Stevens, *Science*, 2014, 346.
36 R. F. Oulton, V. J. Sorger, D. A. Genov, D. F. P. Pile and X. Zhang, *Nat. Photonics*, 2008, **2**, 496–500.
37 A. Boltasseva and H. A. Atwater, *Science*, 2011, **331**, 290–291.
38 R. Zia, J. A. Schuller, A. Chandran and M. L. Brongersma, *Mater. Today*, 2006, **9**, 20–27.
39 M. L. Brongersma and V. M. Shalaev, *Science*, 2010, **328**, 440–441.
40 S. I. Bozhevolnyi, V. S. Volkov, E. Devaux, J.-Y. Laluet and T. W. Ebbesen, *Nature*, 2006, **440**, 508–511.
41 T. W. Ebbesen, C. Genet and S. I. Bozhevolnyi, *Phys. Today*, 2008, **61**, 44.
42 R. Charbonneau, C. Scales, I. Breukelaar, S. Fafard, N. Lahoud, G. Mattiussi and P. Berini, *J. Lightwave Technol.*, 2006, **24**, 477.
43 A. Silva, F. Monticone, G. Castaldi, V. Galdi, A. Alù and N. Engheta, *Science*, 2014, **343**, 160–163.
44 P. Berini and I. De Leon, *Nat. Photonics*, 2012, **6**, 16–24.
45 J. R. Krenn, M. Salerno, N. Felidj, B. Lamprecht, G. Schider, A. Leitner, F. R. Aussenegg, J. C. Weeber, A. Dereux and J. P. Goudonnet, *J. Microsc.*, 2001, **202**, 122–128.
46 K. F. MacDonald, Z. L. Samson, M. I. Stockman and N. I. Zheludev, *Nat. Photonics*, 2009, **3**, 55–58.
47 A. G. Curto, G. Volpe, T. H. Taminiau, M. P. Kreuzer, R. Quidant and N. F. van Hulst, *Science*, 2010, **329**, 930–933.

48 M. Abb, P. Albella, J. Aizpurua and O. L. Muskens, *Nano Lett.*, 2011, **11**, 2457–2463.
49 M. Staffaroni, J. Conway, S. Vedantam, J. Tang and E. Yablonovitch, *Photonics Nanostruct. Fundam. Appl.*, 2012, **10**, 166–176.
50 H. R. Stuart and D. G. Hall, *Appl. Phys. Lett.*, 1998, **73**, 3815–3817.
51 D. M. Schaadt, B. Feng and E. T. Yu, *Appl. Phys. Lett.*, 2005, **86**, 063106.
52 D. Derkacs, S. H. Lim, P. Matheu, W. Mar and E. T. Yu, *Appl. Phys. Lett.*, 2006, **89**, 093103.
53 H. A. Atwater and A. Polman, *Nat. Mater.*, 2010, **9**, 205–213.
54 A. Polman and H. A. Atwater, *Nat. Mater.*, 2012, **11**, 174–177.
55 D. J. Bergman and M. I. Stockman, *Phys. Rev. Lett.*, 2003, **90**, 027402.
56 M. I. Stockman, *Nat. Photonics*, 2008, **2**, 327–329.
57 N. I. Zheludev, S. L. Prosvirnin, N. Papasimakis and V. A. Fedotov, *Nat. Photonics*, 2008, **2**, 351–354.
58 E. Plum, V. A. Fedotov, P. Kuo, D. P. Tsai and N. I. Zheludev, *Opt. Express*, 2009, **17**, 8548–8551.
59 M. A. Noginov, G. Zhu, A. M. Belgrave, R. Bakker, V. M. Shalaev, E. E. Narimanov, S. Stout, E. Herz, T. Suteewong and U. Wiesner, *Nature*, 2009, **460**, 1110–1112.
60 J. B. Khurgin and G. Sun, *Appl. Phys. Lett.*, 2010, **96**, 181102–181103.
61 D. Pile and J. Krenn, *Nat. Photonics*, 2012, **6**, 714–715.
62 V. G. Veselago, *Sov. Phys. Usp.*, 1968, **10**, 509.
63 R. A. Shelby, D. R. Smith and S. Schultz, *Science*, 2001, **292**, 77–79.
64 V. M. Shalaev, W. Cai, U. K. Chettiar, H.-K. Yuan, A. K. Sarychev, V. P. Drachev and A. V. Kildishev, *Opt. Lett.*, 2005, **30**, 3356–3358.
65 J. B. Pendry, D. Schurig and D. R. Smith, *Science*, 2006, **312**, 1780–1782.
66 U. Leonhardt, *Science*, 2006, **312**, 1777–1780.
67 O. Neumann, A. S. Urban, J. Day, S. Lal, P. Nordlander and N. J. Halas, *ACS Nano*, 2012, **7**, 42–49.
68 S. Linic, P. Christopher and D. B. Ingram, *Nat. Mater.*, 2011, **10**, 911–921.
69 S. Mubeen, J. Lee, W.-r. Lee, N. Singh, G. D. Stucky and M. Moskovits, *ACS Nano*, 2014, **8**, 6066–6073.
70 M. L. Brongersma, Y. Cui and S. Fan, *Nat. Mater.*, 2014, **13**, 451–460.
71 K. M. Goodfellow, R. Beams, C. Chakraborty, L. Novotny and A. N. Vamivakas, *Optica*, 2014, **1**, 149–152.
72 N. P. de Leon, M. D. Lukin and H. Park, *IEEE J. Sel. Top. Quantum Electron.*, 2012, **18**, 1781–1791.
73 F. J. García de Abajo, *Science*, 2013, **339**, 917–918.
74 P. Alonso-González, A. Y. Nikitin, F. Golmar, A. Centeno, A. Pesquera, S. Vélez, J. Chen, G. Navickaite, F. Koppens, A. Zurutuza, F. Casanova, L. E. Hueso and R. Hillenbrand, *Science*, 2014, **344**, 1369–1373.
75 N. Strohfeldt, A. Tittl, M. Schäferling, F. Neubrech, U. Kreibig, R. Griessen and H. Giessen, *Nano Lett.*, 2014, **14**, 1140–1147.
76 U. Guler, A. Boltasseva and V. M. Shalaev, *Science*, 2014, **344**, 263–264.
77 G. V. Naik, V. M. Shalaev and A. Boltasseva, *Adv. Mater.*, 2013, **25**, 3264–3294.
78 C. Langhammer, M. Schwind, B. Kasemo and I. Zorić, *Nano Lett.*, 2008, **8**, 1461–1471.
79 A. Frölich and M. Wegener, *Opt. Mater. Express*, 2011, **1**, 883–889.

80 H. O. Pierson, *Handbook of Refractory Carbides & Nitrides: Properties, Characteristics, Processing and Apps*, William Andrew, incorporated, 1996.
81 W. Li, U. Guler, N. Kinsey, G. V. Naik, A. Boltasseva, J. Guan, V. M. Shalaev and A. V. Kildishev, *Adv. Mater.*, 2014, **26**, 7959–7965.
82 V. Rinnerbauer, Y. X. Yeng, W. R. Chan, J. J. Senkevich, J. D. Joannopoulos, M. Soljacic and I. Celanovic, *Opt. Express*, 2013, **21**, 11482–11491.
83 K. A. Arpin, M. D. Losego, A. N. Cloud, H. Ning, J. Mallek, N. P. Sergeant, L. Zhu, Z. Yu, B. Kalanyan, G. N. Parsons, G. S. Girolami, J. R. Abelson, S. Fan and P. V. Braun, *Nat. Commun.*, 2013, **4**, 2630.
84 U. Guler, G. V. Naik, A. Boltasseva, V. M. Shalaev and A. V. Kildishev, *Appl. Phys. B*, 2012, **107**, 285–291.
85 N. Zhou, X. Xu, T. Hammack Aaron, C. Stipe Barry, K. Gao, W. Scholz and C. Gage Edward, *J. Nanophotonics*, 2014, **3**, 141–155.
86 M. H. Kryder, E. C. Gage, T. W. McDaniel, W. A. Challener, R. E. Rottmayer, G. Ju, Y.-T. Hsia and M. F. Erden, *Proc. IEEE*, 2008, **96**, 1810–1835.
87 W. Challener, C. Mihalcea, C. Peng and K. Pelhos, *Opt. Express*, 2005, **13**, 7189–7197.
88 W. Challener, C. Peng, A. Itagi, D. Karns, W. Peng, Y. Peng, X. Yang, X. Zhu, N. Gokemeijer and Y.-T. Hsia, *Nat. Photonics*, 2009, **3**, 220–224.
89 B. C. Stipe, T. C. Strand, C. C. Poon, H. Balamane, T. D. Boone, J. A. Katine, J.-L. Li, V. Rawat, H. Nemoto, A. Hirotsune, O. Hellwig, R. Ruiz, E. Dobisz, D. S. Kercher, N. Robertson, T. R. Albrecht and B. D. Terris, *Nat. Photonics*, 2010, **4**, 484–488.
90 Q. Jiang, S. Zhang and M. Zhao, *Mater. Chem. Phys.*, 2003, **82**, 225–227.
91 T. Zhao, M. C. Kautzky, W. A. Challener and M. A. Seigler, U.S. Pat., US8427925 B2, 2013.
92 N. C. Lindquist, J. Jose, S. Cherukulappurath, X. Chen, T. W. Johnson and S.-H. Oh, *Laser Photonics Rev.*, 2013, **7**, 453–477.
93 V. P. S. Rawat, B. C. Stipe, U.S. Pat., US20120127839 A1, 2012 .
94 G. V. Naik, J. L. Schroeder, X. Ni, A. V. Kildishev, T. D. Sands and A. Boltasseva, *Opt. Mater. Express*, 2012, **2**, 478–489.
95 U. Guler, J. C. Ndukaife, G. V. Naik, A. G. A. Nnanna, A. V. Kildishev, V. M. Shalaev and A. Boltasseva, *Nano Lett.*, 2013, **13**, 6078–6083.
96 Y. Nagato, U.S. Pat., 4589917, 1986.
97 A. Wisbey, P. J. Gregson and M. Tuke, *Biomaterials*, 1987, **8**, 477–480.
98 S. Dong, X. Chen, L. Gu, L. Zhang, X. Zhou, Z. Liu, P. Han, H. Xu, J. Yao, X. Zhang, L. Li, C. Shang and G. Cui, *Biosens. Bioelectron.*, 2011, **26**, 4088–4094.
99 T. E. Tang, W. Che-Chia, R. A. Haken, T. C. Holloway, L. R. Hite and T. G. W. Blake, *IEEE Trans. Electron Devices*, 1987, **34**, 682–688.
100 H. Wakabayashi, Y. Saito, K. Takeuchi, T. Mogami and T. Kunio, *IEEE Trans. Electron Devices*, 2001, **48**, 2363–2369.
101 G. V. Naik, J. Kim and A. Boltasseva, *Opt. Mater. Express*, 2011, **1**, 1090–1099.
102 G. V. Naik, B. Saha, J. Liu, S. M. Saber, E. A. Stach, J. M. K. Irudayaraj, T. D. Sands, V. M. Shalaev and A. Boltasseva, *Proc. Natl. Acad. Sci. U. S. A.*, 2014, **111**, 7546–7551.
103 W. Chen, M. D. Thoreson, S. Ishii, A. V. Kildishev and V. M. Shalaev, *Opt. Express*, 2010, **18**, 5124–5134.

104 M. Bosman, L. Zhang, H. Duan, S. F. Tan, C. A. Nijhuis, C. W. Qiu and J. K. W. Yang, *Sci. Rep.*, 2014, **4**, 5537.
105 X. Ni, Z. Liu and A. V. Kildishev, 2008, http://nanohub.org/resources/3692.
106 P. R. West, S. Ishii, G. V. Naik, N. K. Emani, V. M. Shalaev and A. Boltasseva, *Laser Photonics Rev.*, 2010, **4**, 795–808.
107 M. G. Blaber, M. D. Arnold and M. J. Ford, *J. Phys.: Condens. Matter*, 2010, **22**, 143201.
108 M. G. Blaber, M. D. Arnold and M. J. Ford, *J. Phys. Chem. C*, 2009, **113**, 3041–3045.
109 C. Clavero, *Nat. Photonics*, 2014, **8**, 95–103.
110 V. M. Shalaev, *AIP Conf. Proc.*, 2001, **560**, 237–258.
111 E. Stratakis and E. Kymakis, *Mater. Today*, 2013, **16**, 133–146.
112 R. Bardhan, S. Lal, A. Joshi and N. J. Halas, *Acc. Chem. Res.*, 2011, **44**, 936–946.
113 B. X. Xu, Z. H. Cen, J. F. Hu and J. W. H. Tsai, *J. Appl. Phys.*, 2015, **117**, 17C112.
114 T. Zhao, S. Sahoo, M. C. Kautzky, A. V. Itagi, U.S. Pat., 20130279315, 2013.
115 T. Matsumoto, K. Nakamura, T. Nishida, H. Hieda, A. Kikitsu, K. Naito and T. Koda, *Appl. Phys. Lett.*, 2008, **93**, 031108.
116 S. Xiong, J. Kim, Y. Wang, X. Zhang and D. Bogy, *J. Appl. Phys.*, 2014, **115**, 17B702.
117 S. Bhargava and E. Yablonovitch, *Proc. SPIE*, 2014, 9201.
118 X. Baoxi, T. Yeow Teck, C. Cheow Wee, L. Jianming, Z. Jing, Y. Kaidong and A. Chengwu, *IEEE Trans. Magn.*, 2012, **48**, 1789–1793.
119 B. X. Xu, Z. J. Liu, R. Ji, Y. T. Toh, J. F. Hu, J. M. Li, J. Zhang, K. D. Ye and C. W. Chia, *J. Appl. Phys.*, 2012, **111**, 07B701.
120 A. O. Govorov and H. H. Richardson, *Nano Today*, 2007, **2**, 30–38.
121 J. A. Katine, V. P. S. Rawat, N. L. Robertson and B. C. Stipe, Google Patents, 2013, US8351151 B2.
122 U. Guler, A. V. Kildishev, V. M. Shalaev, A. Boltasseva, D. Stocks, G. Naik, WO Pat., PCT/US14/38561, 2014
123 H. G. Tompkins, *J. Appl. Phys.*, 1992, **71**, 980–983.
124 N. C. Saha and H. G. Tompkins, *J. Appl. Phys.*, 1992, **72**, 3072–3079.

Faraday Discussions

PAPER

Plasmonics in atomically thin materials

F. Javier García de Abajo*[ab] and Alejandro Manjavacas[c]

Received 10th November 2014, Accepted 18th November 2014
DOI: 10.1039/c4fd00216d

The observation and electrical manipulation of infrared surface plasmons in graphene have triggered a search for similar photonic capabilities in other atomically thin materials that enable electrical modulation of light at visible and near-infrared frequencies, as well as strong interaction with optical quantum emitters. Here, we present a simple analytical description of the optical response of such kinds of structures, which we exploit to investigate their application to light modulation and quantum optics. Specifically, we show that plasmons in one-atom-thick noble-metal layers can be used both to produce complete tunable optical absorption and to reach the strong-coupling regime in the interaction with neighboring quantum emitters. Our methods are applicable to any plasmon-supporting thin materials, and in particular, we provide parameters that allow us to readily calculate the response of silver, gold, and graphene islands. Besides their interest for nanoscale electro-optics, the present study emphasizes the great potential of these structures for the design of quantum nanophotonics devices.

I. Introduction

Plasmons – the collective electron oscillations in nanostructured conductors – allow us to control light at the nanometer scale, particularly using the large concentration and enhancement of electromagnetic intensity that they generate.[1] Additionally, and unlike other optical excitations in small systems (*e.g.*, atomic and molecular quantum emitters), plasmons display a powerful combination of two appealing properties: they are robust (*i.e.*, they are not destroyed by the presence of a dielectric environment) and they interact strongly with light (*e.g.*, they display excitation cross-sections typically exceeding the projected area of the nanostructures that sustain the plasmons). These features have facilitated the use of plasmons in applications as varied as nonlinear optics,[2–5] ultrasensitive detection down to the single-molecule level *via* surface-enhanced Raman scattering (SERS),[6–9] cancer diagnosis and therapy,[10–15] quantum information processing,[16–19] improved photovoltaics,[20,21] and subwavelength lithography.[22] Optical metamaterials also largely

[a]*ICFO-Institut de Ciencies Fotoniques, Mediterranean Technology Park, 08860 Castelldefels, Barcelona, Spain. E-mail: javier.garciadeabajo@icfo.es*
[b]*ICREA-Institució Catalana de Recerca i Estudis Avançats, Passeig Lluís Companys, 23, 08010 Barcelona, Spain*
[c]*Department of Physics and Astronomy and Laboratory for Nanophotonics, Rice University, Houston, Texas 77005, USA*

rely on subwavelength plasmons to display properties that are not available in naturally occurring materials.[23–25] These efforts are due in part to the impressive progress made in nanofabrication[26] and colloid chemistry[27,28] techniques, as well as in the theoretical understanding of the response of nanometallic structures.[29,30]

The field of plasmonics has been quite focused on noble metals, which are generally regarded as prototypical plasmonic materials, although they suffer from relatively large inelastic losses that limit the lifetime of plasmons down to a few optical cycles in deep-subwavelength structures. In this context, a search for better plasmonic materials has been initiated with a view to reducing absorption.[31–34] Recently, highly doped graphene has emerged as a promising alternative,[35–44] combining huge field confinement and enhancement with comparatively lower losses,[43,45] as well as large electrical tunability of its optical response.[46–49] These properties hold great potential for electro-optics applications, such as fast light modulation *via* electrostatic gating,[38,39,41–44] which has been demonstrated with the achievement of frequency variations spanning a whole octave.[43]

Unfortunately, plasmons in graphene, as well as in other so-called two-dimensional crystals[50] and in topological insulators,[51] have so far been observed at mid-infrared (mid-IR) and lower frequencies, as they are limited by the low carrier densities in these materials. In contrast, atomically thin metals already possess a substantial conduction electron density in their undoped state, thus sustaining plasmons in the visible and near-infrared (vis-NIR), which are spectral ranges with better prospects for technological applications. Additionally, atomically thin noble metal nanoislands can undergo strong interaction with light and exhibit significant electrical tunability,[52] as the doping levels that are currently attainable using gating technology can produce substantial fractional changes in the conduction electron density.

Plasmons in metal clusters of atomic dimensions have been examined and optically characterized for a long time,[53] and they have even been used as a toolbox to test the ability of different first-principles computational methods to simulate optical and electron-based spectroscopic measurements.[54] In a separate effort, atomic self-assembly has been used to produce monoatomic gold wires,[55] which were later shown to sustain extremely confined plasmons.[56] Similar low-dimensional plasmons have been experimentally characterized using electron spectroscopy in ultrathin indium[57] and silicide[58] wires, as well as in few-atomic-layer silver films[59] and monolayer $DySi_2$.[60] Unfortunately, no further exploration has been pursued towards the coupling of propagating light to these systems and their application to nanophotonics.

Motivated by the availability of these atomically thin materials and their potential for nanophotonics applications, we present here a simple analytical study of the optical properties of disks and ribbons, accompanied by a discussion of their ability to achieve tunable complete optical absorption and quantum strong coupling between plasmons and optical emitters.

II. Optical response and tunability of 2D metallic nanoislands

We describe thin metals in terms of a frequency-dependent 2D conductivity $\sigma(\omega)$, which is related to the bulk dielectric function of the material through $\varepsilon(\omega) = 1 +$

$4\pi i\sigma(\omega)/\omega t$, where t is the film thickness. This local approximation works well for atomically thin islands of noble metals with a lateral extension above ~10 nm, as shown by comparison with quantum-mechanical simulations based upon the random-phase approximation.[52] In the low-frequency limit, the dielectric function is well approximated by the Drude model $\varepsilon(\omega) = 1 - \omega_{bulk}^2/\omega(\omega + i\gamma)$, where ω_{bulk} is the bulk classical plasmon frequency and γ is a phenomenological relaxation rate. Combining these two expressions for ε, we find that the 2D conductivity reduces to

$$\sigma(\omega) = \frac{\omega_{bulk}^2 t}{4\pi} \frac{i}{\omega + i\gamma}. \tag{1}$$

This formula can even be applied to include the full ω dependence of the measured dielectric function by simply allowing ω_{bulk} and γ to depend on ω. In noble metals, these parameters are relatively independent of frequency over the NIR spectral range (see Fig. 5 in Appendix D). The present formalism can also describe graphene, where ω_{bulk} depends on the Fermi energy E_F relative to the so-called Dirac point as $\omega_{bulk} = (2e/\hbar)\sqrt{E_F/t}$ (for example, for a nominal graphene thickness $t = 0.34$ nm, as extracted from the interlayer distance in graphite, and considering a realistic value of the Fermi energy $E_F = 1$ eV,[49] we have $\hbar\omega_{bulk} = 4.1$ eV).

The far-field response of islands that are small compared with the light wavelength can be expressed in terms of their polarizability $\alpha(\omega)$, which admits simple approximate expressions under the reasonable assumption that the lowest-order dipole mode dominates the spectral strength. More precisely, using the expressions derived in Appendix A, we find

$$\alpha(\omega) \approx \frac{tA}{4\pi} \frac{\omega_{bulk}^2}{\omega_p^2 - \omega(\omega + i\gamma)}, \tag{2}$$

where A is the area of the island and ω_p is its lowest-order plasmon frequency (see Table 1 for disks and ribbons). In particular, for a disk of diameter D and thickness t placed at the planar interface between two media of permittivities ε_1 and ε_2, we have (see Table 1 and derivation in Appendix A)

$$\omega_p \approx \frac{\omega_{bulk}}{n_{eff}} \sqrt{\frac{3\pi t}{8D}}, \tag{3}$$

where $n_{eff} = \sqrt{(\varepsilon_1 + \varepsilon_2)/2}$.

Fig. 1(a) shows the values of ω_p predicted by eqn (3) for gold, silver, and graphene disks embedded in silica. We consider noble metal disks consisting of 1, 2, or 5 atomic monolayers, which can clearly reach the NIR. In contrast, the shaded area shows that the plasmon energies for graphene nanodisks lie in the mid-IR, even for relatively high doping levels ($E_F \leq 1$ eV).

We note that the quality factor Q of the plasmon resonances (*i.e.*, 2π times the number of optical cycles after which the intensity has decayed by $1/e$) is given by ω_p/γ. Here, γ is the Drude damping of eqn (1), which depends on frequency and the material as shown in Fig. 5 (Appendix D), leading to the dependence of Q on ω_p illustrated by Fig. 1(b). High-quality (mobility $\mu \leq 10\,000$ cm^2 V^{-1} s^{-1}) highly doped ($E_F \leq 1$ eV) graphene exceeds the performance of gold but is below that of silver for plasmon energies above ~1.6 eV, which are only reachable with graphene structures that are smaller than those considered in Fig. 1,[61] although edge effects can then introduce important corrections.[62,63]

The range of electro-optical tunability of graphene disk plasmons is illustrated by the shaded area in Fig. 1(a). For a given disk diameter, the plasmon energy can be moved up to the upper value of that area when the doping is increased up to $E_F = 1$ eV. For silver and gold, the range of tunability is lower than in graphene, although it has the advantage that the plasmons are in the NIR. Nonetheless, using currently available gating technology, under the same doping conditions that allow achieving a graphene Fermi energy $E_F = 1$ eV, corresponding to a charge carrier density $n = 7 \times 10^{13}$ cm^{-2}, we obtain a fractional variation of the plasmon energy $\approx \pm n/2n_0 = 2.5\%$ in gold and silver, where $n_0 = 1.39 \times 10^{15}$ cm^{-2} is the areal density of conduction electrons in a (111) atomic layer. This produces an overall fractional variation of the plasmon that is resolvable with a quality factor $Q \sim n_0/n$, represented by the dashed vertical line of Fig. 1(b), which is clearly within reach with silver.

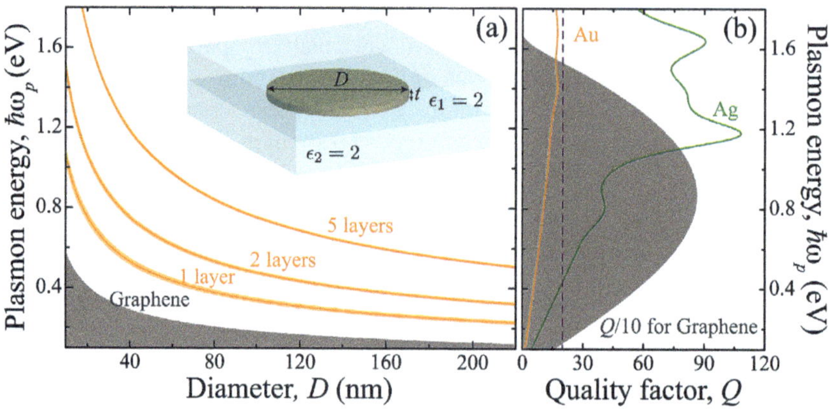

Fig. 1 Energy, quality factor, and electrical tunability of thin-disk plasmons. (a) Dipole plasmon energy as a function of diameter for disks formed by 1–5 atomic layers of gold or silver embedded in an $\varepsilon = 2$ dielectric, as predicted from the analytical model of Table 1. The thickness of each atomic layer is set to 0.236 nm (*i.e.*, the separation between the (111) atomic planes in these materials, which also have similar values of $\hbar\omega_{bulk} \approx 9$ eV). Graphene-disk plasmon energies are shown as well for Fermi energies $E_F = 0$–1 eV, illustrating the large degree of electrical tunability of this material. The shaded regions for noble metal disks give the variation of the plasmon energy when electrically doping the disks up to additional carrier densities in the $\pm 7 \times 10^{13}$ cm^{-2} range (*i.e.*, the same as when doping graphene to $E_F = 1$ eV). (b) Quality factor $Q = \omega_p/\gamma$ of gold and silver plasmons in the electrostatic limit. The effective plasmon damping rate γ depends on frequency as shown in Appendix D (see Fig. 5). The quality factor of graphene is obtained from the random-phase approximation conductivity in the local limit (local-RPA,[61] see eqn (D1) in Appendix D), which includes temperature ($T = 300$ K) and interband transition effects. We assume Fermi energies $E_F \leq 1$ eV and an intrinsic impurity-limited lifetime estimated for DC mobilities $\mu \leq 10\,000$ cm^2 V^{-1} s^{-1} as $\tau = \mu E_F/e v_F^2$, where $v_F = 10^6$ m s^{-1} is the Fermi velocity of graphene. The dashed vertical line indicates the value of Q that equals the full-width fractional variation of the plasmon energy when single-atom gold or silver disks are doped with carrier densities $\pm 7 \times 10^{13}$ cm^{-2}.

III. Coupling to quantum emitters

The large concentration of electromagnetic energy associated with the plasmons of atomically thin structures can lead to strong interaction with nearby quantum emitters. This idea has been recently explored in graphene[36,64,65] and we elaborate on it here to produce a semi-analytical model that is directly applicable to any thin conducting material. For this purpose, we introduce the 2D charge density $\rho_p(\mathbf{R})$ associated with the plasmon as a function of position $\mathbf{R} = (x, y)$ along the metal island. This quantity can be conveniently normalized for one plasmon, as discussed in Appendix B, where analytical expressions are given for the lowest-order dipole modes of disks and ribbons (see Table 1). Intuitively, ρ_p plays a similar role as the charge density $-e\phi_f^*(\mathbf{r})\phi_i(\mathbf{r})$ associated with the transition of one electron between the bound states ϕ_i and ϕ_f of a confined system. We now consider a two-level quantum emitter (*e.g.*, an atom or molecule) of transition dipole $\mathbf{d}_0$. Taking the metal island to lie in the $z = 0$ plane and the emitter at position $\mathbf{r}_0$, the electrostatic emitter-plasmon interaction is simply given by

$$\hbar g = \frac{1}{{n_{\text{eff}}}^2}\int d^2\mathbf{R}\rho_p(\mathbf{R})\frac{\mathbf{d}_0\cdot(\mathbf{R}-\mathbf{r}_0)}{|\mathbf{R}-\mathbf{r}_0|^3}, \quad (4)$$

where the integral is extended over the area of the island, n_{eff} is defined right after eqn (3), and $\mathbf{d}_0$ is the effective emitter transition dipole, which is related to its radiative lifetime τ_0 in the absence of the island through ${\tau_0}^{-1} = 4n_{\text{eff}}{\omega_0}^3{d_0}^2/3\hbar c^3$. Incidentally, $\mathbf{d}_0$ is the transition dipole in a vacuum multiplied by a local-field correction $3{n_{\text{eff}}}^2/(2{n_{\text{eff}}}^2 + 1)$.[66]

The quantum evolution of the emitter-plasmon system can be described by the Hamiltonian[36,64]

$$H = \hbar[\omega_p a^+ a + \omega_0\sigma^+\sigma + g(a^+\sigma + a\sigma^+)] + \mathbf{d}_p\cdot\mathbf{E}^{\text{ext}}(t)(a^+ + a), \quad (5)$$

where a and σ (a^+ and σ^+) are the annihilation (creation) operators of the plasmon and the emitter excitation of energies $\hbar\omega_p$ and $\hbar\omega_0$, respectively. Let us stress that we are using the same rate g of emitter-plasmon coupling as defined by the electrostatic energy of eqn (4). In the Hamiltonian (5) we are neglecting the direct interaction of the time-dependent external field $\mathbf{E}^{\text{ext}}$ with the emitter, as its dipole $\mathbf{d}_0$ is assumed to be small compared with the plasmon dipole

$$\mathbf{d}_p = \int d^2\mathbf{R}\mathbf{R}\rho_p(\mathbf{R}).$$

Incidentally, the normalization of ρ_p for a single plasmon is actually based on this dipole, as explained in Appendix B.

The lifetime of the emitter τ_0 and the plasmon decay rate γ can be introduced in the quantum description of the combined system through the density matrix ρ, which follows the equation of motion[67,68]

$$\frac{d\rho}{dt} = \frac{i}{\hbar}[\rho, H] + \frac{1}{2\tau_0}[2\sigma\rho\sigma^+ - \sigma^+\sigma\rho - \rho\sigma^+\sigma] + \frac{\gamma}{2}[2a\rho a^+ - a^+ a\rho - \rho a^+ a]. \quad (6)$$

In this formalism, we can calculate the polarizability $\alpha(\omega)$ by first obtaining the expected value of the induced dipole from $\text{tr}\{\mathbf{d}_p(a^+ + a)\rho\}$ upon illumination with a weak external field $\mathbf{E}^{\text{ext}}(t) = \mathbf{E}_0 e^{-i\omega t}$ + c.c. We find the induced dipole to admit the form $\alpha(\omega)\mathbf{E}_0 e^{-i\omega t}$ + c.c., thus defining $\alpha(\omega)$. In the absence of the emitter (*i.e.*, taking $g = 0$), we recover a polarizability $\alpha(\omega)$ as given by the $j = p$ term of eqn (B1), thus demonstrating the self-consistency of our plasmon-normalization scheme. Additionally, when the combined system is considered, the linear polarizability becomes $\alpha(\omega) = \alpha_0(\omega) + \alpha_0^*(-\omega)$ with

$$\alpha_0(\omega) = \frac{d_p^{\,2}}{\omega_p - \omega - i\gamma/2 - g^2(\omega_0 - \omega - i/2\tau_0)^{-1}}.$$

For $\omega_p = \omega_0$, this expression exhibits two poles at frequencies $\omega = \omega_p \pm g$, yielding a vacuum Rabi splitting given by $2g$.

For the vacuum Rabi splitting to be observable, it must be larger than the width of the plasmon peak, that is, $g/\gamma > 1$. This condition signals the so-called strong-coupling regime, which has been argued to be achievable in graphene.[36] In this regime, the bosonic plasmon state mixes with the fermionic two-level emitter to produce a Jaynes–Cummings ladder of hybridized states,[69] which has been predicted to produce non-classical statistics of the plasmon population upon external illumination, as well as a nonlinear optical response[64] (*i.e.*, the nonlinearity of the quantum emitter is inherited by the combined plasmon-emitter system).

It should be noted that the decay rate of the excited emitter is enhanced by the coupling to the plasmon and becomes for $\gamma\tau_0 \gg 1$

$$\Gamma = \frac{1}{\tau_0} + \gamma\frac{g^2}{(\omega_p - \omega_0)^2 + \gamma^2/4} \tag{7}$$

under the condition that the fraction in this expression is small (weak coupling). This well-known result is rederived in Appendix C from eqn (5) and (6), and we also show that the dielectric formalism of Appendix A reproduces eqn (7) with g as defined by eqn (4), provided the plasmon charge density $\rho_p(\mathbf{R})$ is normalized as prescribed in Appendix B, thus demonstrating the self-consistency of the theoretical methods elaborated in this work.

Equipped with the analytical model for the plasmons of thin conductor islands discussed in the Appendix, we examine in Fig. 2 the ratio g/γ, where g is calculated from eqn (4) using the analytical expression of ρ_p for a disk plasmon given in Table 1. We find g/γ to depend on the lifetime of the emitter τ_0, the Drude parameters of the conducting disk ω_{bulk} and γ (see eqn (1)), the disk diameter and thickness D and t, and the index of refraction of the surrounding medium n_{eff} only through a multiplicative coefficient $(n_{eff}c)^{3/2}/(\gamma D\omega_{bulk}\sqrt{\tau_0 t})$. We plot g/γ in Fig. 2 expressed in units of that coefficient (left scale) as a function of the distance z_0 between the emitter and the center of the disk. The emitter dipole is assumed to be parallel to the disk. The right scale shows the ratio calculated for $\hbar\omega_{bulk} = 9$ eV and $\hbar\gamma = 0.07$ eV, typical of gold in the NIR, with $\tau_0 = 1$ ns, $D = 10$ nm, $t = 0.236$ nm (*i.e.*, one (111) atomic layer), and $n_{eff}^{\,2} = 2$. This result indicates that the strong-coupling regime is reachable over a wide range of distances using gold islands. Silver structures should produce larger coupling because γ is smaller in that material (see Fig. 5). Further confinement of the plasmons in structures that display hotspots, such as bowtie antennas,[70] could lead to even larger values of g/γ.

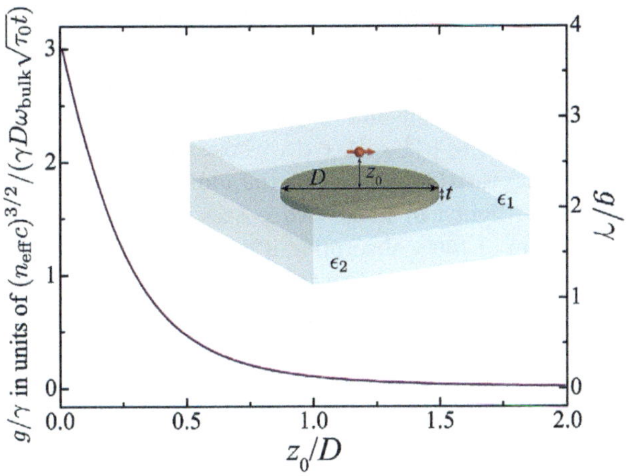

Fig. 2 Achieving quantum strong coupling between the lowest-energy plasmon of a thin conducting disk and an optical emitter. We represent the ratio g/γ between the coupling and plasmon decay rates for an emitter oriented parallel to the graphene (see inset), as a function of its separation z_0 from the center of a disk of diameter D and thickness t. The disk material is characterized by a Drude plasma frequency $\omega_{\rm bulk}$ and it is placed at the interface between two dielectric media that define the effective permittivity $n_{\rm eff} = \sqrt{(\varepsilon_1 + \varepsilon_2)/2}$. The lifetime of the emitter in the absence of the disk is τ_0. The left vertical axis is given in units of the dimensionless quantity $(n_{\rm eff}c)^{3/2}/[\gamma D\omega_{\rm bulk}\sqrt{\tau_0 t}]$, whereas the right axis corresponds to the choice $\hbar\omega_{\rm bulk} = 9$ eV and $\hbar\gamma = 0.07$ eV (gold in the NIR), with $\tau_0 = 1$ ns, $D = 10$ nm, $t = 0.236$ nm (*i.e.*, a single (111) atomic layer), and $\varepsilon_1 = \varepsilon_2 = 2$ (glass).

Decoherence produced by inelastic transitions can severely damage the efficiency of quantum emitters when they are placed in a solid-state environment, although close to 100% efficiencies can be achieved with organic molecules under cryogenic conditions.[71] Fortunately, the enhancement of the coupling rate from the emitter to the plasmon at the frequency of the latter can decrease the relative importance of inelastic decay channels in the emitter (*e.g.*, coupling to phonons of the surrounding material, Auger processes, *etc.*), so that in practice the coherent part of the decay in emitters such as nitrogen-vacancies in diamond, in which the zero-phonon elastic channel accounts for only a small fraction of the emission, can be enhanced by coupling to the nanoisland, and we are thus under similar conditions as those considered in this study (*i.e.*, the emitter decay through coupling to a plasmon dominates over other inelastic channels).

IV. Complete optical absorption

Complete optical absorption has been studied and observed over many frequency ranges in disordered metal films,[72,73] through lattice resonances in gratings and planar metamaterials,[74–80] assisted by localized plasmonic resonances,[81,82] using multilayer structures,[83] and in overdense plasma.[84] However, the possibility of achieving complete optical absorption in atomically thin films offers additional advantages, as we discuss below.

It is well known that the maximum absorbance produced by an optically thin film in a homogeneous environment is 50%: the incident light induces charges and currents that have no memory of where light is coming from, and therefore, they radiate symmetrically towards both sides of the film with a scattered wave amplitude r; consequently, the reflected and transmitted amplitudes are r and $1 + r$, where the first term in the transmission is the incident field of unit amplitude; the absorbance is thus $1 - |r|^2 - |1 + r|^2$, whose maximum value is 1/2 as a function of the complex variable r. Now, with the addition of a reflecting screen on one side of the film, light can make two passes through the thin material, producing a maximum of 100% absorption if both incident and reflected waves are in phase at that plane. This is the so-called Salisbury screen configuration, in which the film/metal-screen separation should be roughly $\lambda/4n$[85,86] (*i.e.*, a phase of π is produced by the metallic reflection and another π contribution comes from phase associated with the round trip propagation between the film and the screen, assumed to be embedded in an environment of refractive index n).

These ideas have been recently explored for graphene, leading to the prediction of complete optical absorption by a suitably patterned carbon layer,[87] under the condition that the extinction cross-section per unit cell element is of the order of the unit cell area. The observation of electrically tunable large absorbance in patterned graphene has been recently accomplished.[88,89] We argue here that similar levels of tunable absorption are achievable using noble metals.

It is instructive to first examine the maximum extinction of a thin island in vacuum. The corresponding cross section is[90] $\sigma^{\rm ext} = 4\pi(\omega/c){\rm Im}\{\alpha(\omega)\}$, which upon insertion of eqn (2) is found to exhibit a maximum at $\omega = \omega_{\rm p}$, given approximately by $\sigma^{\rm ext} = (\omega_{\rm bulk}{}^2 t/\gamma c) \times A$. Remarkably, this maximum extinction is independent of shape for a given area A of the island, under the assumption that an individual plasmon mode dominates the extinction. In particular, for single atomic layers of gold (silver) films ($t = 0.236$ nm), considering the plasmon energy $\hbar\omega_{\rm p}$ to be in the NIR, we have $\hbar\omega_{\rm bulk} \sim 9$ eV and $\hbar\gamma \sim 70$ meV ($\hbar\gamma \sim 20$ meV), so that the maximum cross section is 1.4 (4.8) times the area of the island. For highly doped graphene, this number is even larger due to the comparatively lower losses of this material (see Fig. 5).

Complete optical absorption is achievable in periodic arrays placed above a Salisbury screen. The normal-incidence reflection coefficient of a doubly-periodic array of small period a compared with the light wavelength, surrounded by a homogeneous environment of refractive index n, reduces to[87]

$$r = \frac{{\rm i}S}{\alpha^{-1} - G},$$

where $S = (2\pi\omega/nA_{\rm c}c)$, $G = \mathfrak{g}/n^2a^3 + {\rm i}S$, $A_{\rm c}$ is the unit cell area, and $\mathfrak{g}$ is a number that depends on symmetry (*e.g.*, $\mathfrak{g} \approx 5.52$ and $\mathfrak{g} \approx 4.52$ for hexagonal and square arrays, respectively, assuming that all islands interact through their induced dipoles; corrections due to nearest-neighbor interactions beyond dipolar terms are possible for closely spaced islands, in which case the coefficient $\mathfrak{g}$ can depend on their shape). With a Salisbury screen of reflectivity $r_0 = |r_0|{\rm e}^{{\rm i}\varphi_0}$ separated a distance d from the array, the incident and reflected waves are exactly on phase when $\varphi_0 + 2\omega dn/c$ is a multiple of 2π. Complete absorption is then produced under the condition $|r_0| = -r/(1 + 2r)$, which is satisfied at a frequency ω given by $\omega^2 = \omega_{\rm p}{}^2 - (\mathfrak{g}/4\pi n^2)(tA/a^3)\omega_{\rm bulk}{}^2$, where $\omega_{\rm p} \approx (\omega_{\rm bulk}/n)\sqrt{3\pi t/8D}$, provided we have

$$\frac{2n\gamma c}{\omega_{\rm bulk}{}^2 t}\frac{A_{\rm c}}{A} = 1 + \frac{1}{|r_0|}. \tag{8}$$

Interestingly, this condition for complete optical absorption is also independent of the shape of the island. Using the approximate values of $\omega_{\rm bulk}$ and γ noted above for gold and silver in the NIR, and considering for simplicity a non-absorbing Salisbury screen ($|r_0| = 1$) and a glass environment ($n^2 = 2$), the condition (8) for perfect absorption in gold (silver) arrays is fulfilled with a fraction $A/A_{\rm c} = 1.02$ (0.29) between the areas of the island and the unit cell. Consequently, this condition can be easily met using silver atomic monolayers, and also with multilayers of either gold or silver.

Incidentally, similar results are obtained for ribbon arrays of period a.[61] Then, the condition (8) remains unchanged, with $A/A_{\rm c} = D/a$, to where D is now the ribbon width. Because of the ribbon translational symmetry, we work with 2D rather than 3D scattering, so we need to redefine $\alpha(\omega) \approx (tD/4\pi)\omega_{\rm bulk}{}^2/[\omega_{\rm p}{}^2 - \omega(\omega + {\rm i}\gamma)]$, $G = 2\pi^2/3n^2a^2 + {\rm i}S$ and $S = 2\pi\omega/nac$. The frequency at which complete absorption occurs is then $\omega = \omega_{\rm p}\sqrt{1-(\pi^2/24)(D/a)^2}$, where $\omega_{\rm p} \approx (\omega_{\rm bulk}/n)\sqrt{4t/\pi D}$ (see Table 1).

It is important to note that the value of $\omega_{\rm bulk}$ can be modulated by $\approx\pm2.5\%$ in noble metals using currently available gating technology (see above), and this in particular produces peak shifts larger than the peak width in silver (see Fig. 1(b)), as shown in Fig. 3 for an illustrative example. This type of structure is convenient because the ribbons can be contacted at a large distance away from the region in which the optical modulation is pursued. Besides, the strong absorption only occurs for polarization across the ribbons, thus suggesting a possible application as tunable polarizers.

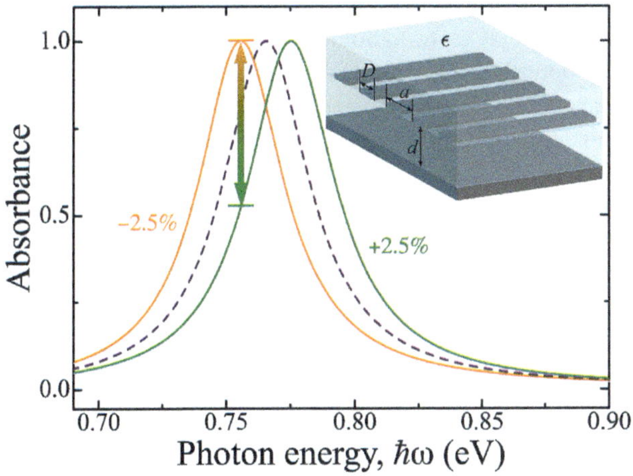

Fig. 3 Tunable complete optical absorption. We show the absorbance of a single-atomic-layer silver ribbon array embedded in silica ($\varepsilon = n^2 = 2$) and placed at a distance $d = 286$ nm above a perfectly reflecting mirror. The ribbon width and period are $D = 20$ nm and $a = 68$ nm, respectively. The response of silver is described with $\hbar\omega_{\rm bulk} = 9$ eV and $\hbar\gamma = 0.02$ eV. The dashed curve shows the result for undoped silver ribbons, whereas the solid curves correspond to a fractional variation of $\pm2.5\%$ in the areal conduction electron density (*i.e.*, $\pm7 \times 10^{13}$ cm^{-2}), enabling a large change in absorption, as indicated by the double vertical arrow.

V. Outlook and perspectives

Graphene plasmons are receiving much attention due in part to the demonstrated ability to modulate their frequencies by electrically doping the carbon layer through suitably engineered gates.[38,39,41–44] This modulation can be potentially realized at high speeds, as the number of charge carriers that are needed to produce changes in the Fermi energy of the order of the electronvolt is relatively small, and consequently, so are the inductance and capacitance associated with the graphene itself. Unfortunately, graphene plasmons have only been observed at mid-infrared and lower frequencies. Moving to the vis-NIR is challenging and requires patterning structures with sizes <10 nm under realistically attainable doping conditions. In this respect, molecular self-assembly provides a viable way of synthesizing nanographenes in this size range.[91–94] Doped carbon nanotubes have also been predicted to display plasmons that are rather insensitive to their degree of chirality,[61] and therefore, they provide a viable route towards the fabrication of large-scale tunable plasmonic structures operating in the vis-NIR regime. Polycyclic aromatic hydrocarbons also sustain excitations at vis-NIR frequencies that behave as graphene plasmons[95] and constitute promising candidates to advance towards atomic-scale tunable plasmonics.

Although plasmons in atomically thin metals have been observed in several systems,[56–60] these studies have focused on extended surfaces whose plasmon dispersion relations are far from the light cone, thus averting the possibility of direct coupling to propagating light. Further patterning of these types of surfaces into disks and ribbons such as those considered here could facilitate the coupling to optical probes. An alternative option consists of decorating atomically thin films with dielectric colloids to provide periodic optical contrast. Substrate pre-patterning of disks, ribbons, or other morphologies, followed by atomic layer deposition constitutes yet another possibility.

We conclude that atomically thin materials hold great potential for the manipulation of light at truly nanometer scales and for the development of applications to optical signal processing, quantum optics, and sensing. We should emphasize that small nanoparticles, not necessarily atomically thin, can produce similar levels of strong-coupling and optical absorption as discussed above for thin films, although they are not tunable using electrical gates because the injected charge carriers have to compete with a much larger number of bulk conduction electrons. The great opportunities offered by these materials are however accompanied by formidable challenges to produce the islands at designated positions and with controlled morphology, possibly requiring a combination of top-down patterning and bottom-up self-assembly methods similar to those mentioned above.

Appendix A: optical response of thin metal islands in the electrostatic limit

The scale-invariant character of the electrostatic problem (*i.e.*, the absence of a frequency-dependent length scale imposed by the wavelength) has been used on several occasions to express the solutions in terms of modal expansions.[96–99] Here, we formulate a suitable decomposition for optically thin structures. We consider islands of small characteristic size D (*e.g.*, the diameter for disks or the width for ribbons) compared with the light wavelength, such that the optical electric field $\mathbf{E} = -\nabla\phi$ can be expressed in terms of a scalar potential ϕ. The islands are however taken to be large enough to be described as infinitesimally thin domains characterized by a local, frequency-dependent 2D conductivity $\sigma(\omega)$. Following previous analyses for graphene,[43,61,100] we write the self-consistent potential at positions $\mathbf{R} = (x, y)$ in the plane of the island as

$$\phi(\mathbf{R}) = \phi^{\text{ext}}(\mathbf{R}) + \frac{1}{{n_{\text{eff}}}^2}\frac{\mathrm{i}}{\omega}\int\frac{d^2\mathbf{R}'}{|\mathbf{R}-\mathbf{R}'|}\nabla_{\mathbf{R}'}\cdot\sigma(\mathbf{R}',\omega)\nabla_{\mathbf{R}'}\phi(\mathbf{R}'), \tag{A1}$$

which is the sum of the external perturbation ϕ^{ext} and the contribution produced by the induced charges (integral term). The island is chosen to lie at the planar interface between two media of permittivities ε_1 and ε_2, which contribute to the above expression through a $1/{n_{\text{eff}}}^2$ factor multiplying the in-plane Coulomb potential, where

$$n_{\text{eff}} = \sqrt{(\varepsilon_1+\varepsilon_2)/2}. \tag{A2}$$

Now, using the definitions $\vec{\theta} = \mathbf{R}/D$ and

$$\eta = \frac{1}{{n_{\text{eff}}}^2}\frac{\mathrm{i}\sigma(\omega)}{\omega D},$$

taking the gradient in both sides of eqn (A1), and multiplying by $-\sqrt{f}$, we find[100]

$$\vec{\varepsilon}(\vec{\theta},\omega) = \vec{\varepsilon}^{\,\text{ext}}(\vec{\theta},\omega) + \eta(\omega)\int d^2\vec{\theta}'\mathbf{M}(\vec{\theta},\vec{\theta}')\cdot\vec{\varepsilon}(\vec{\theta}',\omega), \tag{A3}$$

where

$$\vec{\varepsilon}(\vec{\theta},\omega) = -\sqrt{f(\vec{\theta})}\nabla_{\vec{\theta}}\phi(\vec{\theta},\omega),$$

whereas $f(\vec{\theta})$ is a filling function that is 1 if $\vec{\theta}$ lies on the metal and zero otherwise, so that the frequency and spatial dependences of the conductivity are separated as $\sigma(\mathbf{R}, \omega) = f(\mathbf{R})\sigma(\omega)$. Notice that this formalism is also valid for inhomogeneous layers by allowing f to take values different from 0 or 1.[101] Here, we have defined $\mathbf{M}(\vec{\theta},\vec{\theta}') = \sqrt{f(\vec{\theta})f(\vec{\theta}')}\nabla_{\vec{\theta}}\otimes\nabla_{\vec{\theta}}(1/|\vec{\theta}-\vec{\theta}'|)$, which is a real, symmetric operator that admits a complete orthonormal set of real eigenvectors $\vec{\varepsilon}_j$ and eigenvalues $1/\eta_j$ satisfying the relations

$$\text{eigensystem}\rightarrow\eta_j\int d^2\vec{\theta}'\mathbf{M}(\vec{\theta},\vec{\theta}')\cdot\vec{\varepsilon}_j(\vec{\theta}') = \vec{\varepsilon}_j(\vec{\theta}),$$

$$\text{orthogonality}\rightarrow\int d^2\vec{\theta}\vec{\varepsilon}_j(\vec{\theta})\cdot\vec{\varepsilon}_{j'}(\vec{\theta}) = \delta_{jj'},$$

$$\text{closure}\rightarrow\sum_j\vec{\varepsilon}_j(\vec{\theta})\otimes\vec{\varepsilon}_j(\vec{\theta}') = \delta(\vec{\theta}-\vec{\theta}')\mathbf{I}_2.$$

Then, the solution to eqn (A3) reduces to

$$\vec{\varepsilon} = \sum_j[c_j/(1-\eta/\eta_j)]\vec{\varepsilon}_j, \tag{A4}$$

where $c_j = \int d^2\vec{\theta}\vec{\varepsilon}^{\,\text{ext}}(\vec{\theta})\cdot\vec{\varepsilon}_j(\vec{\theta})$.

Applying these results to a uniform external field $\mathbf{E}_0$ aligned with a symmetry direction of the island $\hat{\mathbf{x}}$ (*i.e.*, for $\vec{\varepsilon}^{\text{ext}} = \sqrt{f}DE_0\hat{\mathbf{x}}$), we obtain the polarizability along that direction $\alpha(\omega) = E_0^{-1}\int d^2\mathbf{R}x\rho^{\text{ind}}(\mathbf{R})$ from the induced density

$$\rho^{\text{ind}}(\mathbf{R}) = \frac{\mathrm{i}\sigma}{\omega}\nabla_{\mathbf{R}}\cdot f(\mathbf{R})\nabla_{\mathbf{R}}\phi(\mathbf{R}) = \frac{-\mathrm{i}\sigma}{\omega D^2}\nabla_{\vec{\theta}}\cdot\sqrt{f\left(\vec{\theta}\right)}\vec{\varepsilon}\left(\vec{\theta}\right). \tag{A5}$$

Inserting eqn (A4) into this expression, we obtain

$$\alpha(\omega) = D^3\sum_j \frac{A_j}{\dfrac{-1}{{n_{\text{eff}}}^2\eta_j} - \dfrac{\mathrm{i}\omega D}{\sigma(\omega)}}, \tag{A6}$$

where j runs over eigenmodes of the system and

$$A_j = \left|\int d^2\vec{\theta}\sqrt{f\left(\vec{\theta}\right)}\varepsilon_{jx}\left(\vec{\theta}\right)\right|^2 \tag{A7}$$

are dimensionless coupling coefficients. Using the conductivity of eqn (1), we can recast eqn (A6) as

$$\alpha(\omega) = \frac{tD^2}{4\pi}\sum_j \frac{A_j{\omega_{\text{bulk}}}^2}{{\omega_j}^2 - \omega(\omega + \mathrm{i}\gamma)}, \tag{A8}$$

where the plasmons of the nanoisland can be identified with modes j of negative eigenvalues η_j and frequencies

$$\omega_j = \frac{\omega_{\text{bulk}}}{n_{\text{eff}}}\frac{1}{\sqrt{-4\pi\eta_j}}\sqrt{\frac{t}{D}}, \tag{A9}$$

corresponding to the $\omega \approx \omega_j - \mathrm{i}\gamma/2$ poles of eqn (A8).

Eqn (A8) involves coefficients that are subjected to two useful sum rules:[61]

(i) For any arbitrarily shaped island of area A, we have $\sum_j A_j = A/D^2$. This result is readily obtained from the definition of the A_j coefficients in eqn (A7) upon using the closure relation for $\vec{\varepsilon}_j$ (see above). Applying this sum to a Drude metal (*i.e.*, for frequency-independent ω_{bulk}), we conclude that the integral of the extinction cross-section ($\propto \omega\alpha(\omega)$) is actually proportional to $\omega_{\text{bulk}}At$, which is in turn proportional to the number of electrons (*i.e.*, it fulfills the *f*-sum rule[102]).

(ii) Another sum rule follows from eqn (A6) in the $\omega \to 0$ limit (*i.e.*, when the island behaves as a perfect conductor, so that $\omega/\sigma \to 0$). Without loss of generality, we can consider a freestanding island ($n_{\text{eff}} = 1$), so we have $-\sum_j \eta_j A_j = \alpha(0)/D^3$. Now, for in-plane polarization of a disk of diameter D, we have $\alpha(0) = D^3/6\pi$ (this result can be derived from the polarizability of an ellipsoid of vanishing height[103]), which leads to $-\sum_j \eta_j A_j = 1/6\pi$. Likewise, from the transversal polarizability of a thin metal ribbon of width D[90] ($\alpha(0) = D^2L/16$, where $L \to \infty$ is the length), we find $-\sum_j \eta_j A_j = L/16D$.

Interestingly, for these types of structures and polarizations, we find one single mode $j = p$ to be dominant and to absorb most of the weight in the above sums.[61] More precisely, this is the lowest-order dipole plasmon. Neglecting all other modes, these sum rules lead to the values of ω_p, η_p, and A_p listed in Table 1 and extensively used throughout this work to produce analytical estimates of plasmonic behavior.

Appendix B: charge induced by a single plasmon

We introduce a purely electrostatic scheme to normalize the induced charge density $\rho_j(\mathbf{R})$ associated with a single plasmon j. From linear-response theory,[102] the polarizability reads

$$\alpha(\omega) = \frac{1}{\hbar}\sum_j d_j^2\left(\frac{1}{\omega_j - \omega - i\gamma/2} + \frac{1}{\omega_j + \omega + i\gamma/2}\right), \tag{B1}$$

where

$$d_j = \int d^2\mathbf{R}\, x\rho_j(\mathbf{R}) \tag{B2}$$

is the dipole moment associated with mode j for polarization along a symmetry direction $\hat{\mathbf{x}}$. Now, in order to compare eqn (B1) with eqn (A8), we neglect γ^2 in front of ω_j^2 and approximate eqn (B1) as

$$\alpha(\omega) \approx \frac{2}{\hbar}\sum_j \frac{\omega_j d_j^2}{\omega_j^2 - \omega(\omega + i\gamma)}. \tag{B3}$$

Now, inserting the Drude conductivity of eqn (1) into eqn (A6), comparing the result with eqn (B3), and taking eqn (A9) into account, we find the normalization condition

$$d_j^2 = A_j\hbar\omega_{\text{bulk}}n_{\text{eff}}\sqrt{\left(\frac{-\eta_j}{16\pi}\right)tD^5}. \tag{B4}$$

Within the single-mode approximation noted at the end of the previous paragraph, writing the charge density associated with the lowest-order disk dipole plasmon as $\rho_p(\mathbf{R}) = \rho_p(R)\cos\varphi$, where $\rho_p(R)$ gives the radial dependence, we find the normalization condition

$$\left|\int_0^{D/2} R^2\, dR\rho_p(R)\right|^2 = \frac{\hbar\omega_{\text{bulk}}n_{\text{eff}}}{16\pi^2}\sqrt{\frac{2tD^5}{3\pi}}.$$

Similarly, the plasmon charge density $\rho_p(x)$ along the transversal direction $\hat{\mathbf{x}}$ of a ribbon contained in the $|x| < D/2$ region satisfies

$$\left|\int_0^{D/2} x\, dx\rho_p(x)\right|^2 = \frac{\hbar\omega_{\text{bulk}}n_{\text{eff}}}{64A}\sqrt{\frac{tD^5}{\pi}},$$

where $A = LD$ is the ribbon area (with $L \to \infty$) and we have utilized the symmetry $\rho_p(-x) = -\rho_p(x)$. Using these normalizations, we find that the analytical expressions for ρ_p that are given in Table 1, where the density profile is taken to fit previous calculations[101,104] based upon the boundary-element method. Actually, these formulas reproduce the calculated density profiles rather well, as shown in Fig. 4.

Table 1 Analytical approximations for the parameters ω_p, η_p, and A_p corresponding to the lowest-order dipole resonance of a disk of diameter D and a ribbon of width D and length $L \to \infty$. The dependence on the dielectric environment is entirely contained in a factor $n_{eff} = \sqrt{(\varepsilon_1+\varepsilon_2)/2}$ dividing the frequency when the film is placed at the interface between two media of permittivities ε_1 and ε_2. The rightmost column shows approximations to the radial and transversal dependences of the charge density induced by a single plasmon in disks and ribbons, respectively, which involve the coefficients $C_d^2 = 4.58 n_{eff}(\hbar\omega_{bulk}/D^3)\sqrt{t/D}$ and $C_r^2 = 3.10 n_{eff}(\hbar\omega_{bulk}/LD^2)\sqrt{t/D}$. The ribbon edges are taken at $x = \pm D/2$

	ω_p	η_p	A_p	$\rho_p(\mathbf{R})$
Disk	$(\omega_{bulk}/n_{eff})\sqrt{3\pi t/8D}$	$-2/3\pi^2$	$\pi/4$	$C_d \frac{x}{D}\left(1+\frac{e^{-5(1-2R/D)}}{4\sqrt{1-2R/D}}\right)$
Ribbon	$(\omega_{bulk}/n_{eff})\sqrt{4t/\pi D}$	$-1/16$	L/D	$C_r \frac{x}{D}\left(1+\frac{e^{-5[1-(2x/D)^2]}}{4\sqrt{1-(2x/D)^2}}\right)$

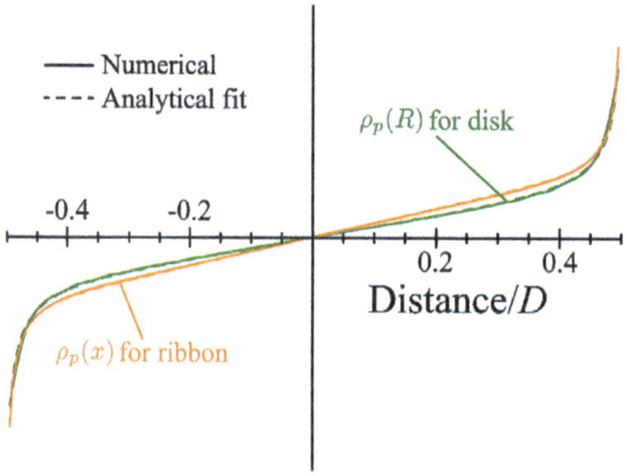

Fig. 4 Induced charge of a single plasmon in thin disks and ribbons. We show the charge densities associated with the lowest-order dipolar plasmon in a disk of diameter D ($m = 1$ azimuthal symmetry) and in a ribbon of width D (transversal polarization with wave vector $k_\parallel = 0$ along the direction of translational symmetry). The solid curves correspond to numerical results taken from the literature for disks[104] and ribbons.[101] The dashed curves represent the fitting functions of Table 1.

Appendix C: plasmon-enhanced emitter decay rate

1. Density-matrix approach

It is convenient to expand the density matrix of eqn (6) as

$$\rho = \sum_{ln,l'n'} \rho_{ln,l'n'}(t)\, e^{-i(l-l')\omega_0 t}\, e^{-i(n-n')\omega_p t}\, e^{-(1/2\tau_0)(l+l')t}\, e^{-(\gamma/2)(n+n')t} |ln\rangle\langle l'n'|, \quad (C1)$$

where $\rho_{ln,l'n'}(t)$ are time-dependent coefficients, while $|ln\rangle$ denotes a state with n plasmons accompanied by the excited (de-excited) emitter for $l = 1$ ($l = 0$). Inserting this expression into eqn (6), we find

$$\begin{aligned}\dot{\rho}_{ln,l'n'} = &-\mathrm{i}g\left[\rho_{0n+1,l'n'}\sqrt{n+1}\mathrm{e}^{-\mathrm{i}\Delta t}\delta_{l,1} + \rho_{1n-1,l'n'}\sqrt{n}\mathrm{e}^{\mathrm{i}\Delta t}\delta_{l,0}\right.\\ &\left.- \rho_{ln,0n'+1}\sqrt{n'+1}\mathrm{e}^{\mathrm{i}\Delta^* t}\delta_{l',1} - \rho_{ln,1n'-1}\sqrt{n'}\mathrm{e}^{-\mathrm{i}\Delta^* t}\delta_{l',0}\right]\\ &+ \tau_0^{-1}\rho_{1n,1n'}\mathrm{e}^{-t/\tau_0}\delta_{l,0}\delta_{l',0} + \gamma\rho_{ln+1,l'n'+1}\sqrt{(n+1)(n'+1)}\mathrm{e}^{-\gamma t},\end{aligned} \quad \text{(C2)}$$

where $\Delta = \omega_p - \omega_0 - \mathrm{i}(\gamma - \tau_0^{-1})/2$. We now argue that, for an initial density matrix in which all terms of eqn (C1) with $l + n > N$ or $l' + n' > N$ are zero (*i.e.*, a density matrix involving a maximum number of N excitations in the combined plasmon-emitter system), the last two terms of eqn (C2) vanishes because it involves states that are never populated. We are then left with a self-contained subset of equations involving coefficients $\rho_{ln,l'n'}$ with $l + n = l' + n' = N$. For this manifold of N excitations, we trivially find solutions $\rho_{ln,l'n'} = a_{ln}a^*_{l'n'}$, where the coefficients a_{ln} satisfy the equations

$$\dot{a}_{0N} = -\mathrm{i}ga_{1N-1}\sqrt{N}\mathrm{e}^{\mathrm{i}\Delta t}$$

$$\dot{a}_{1N-1} = -\mathrm{i}ga_{0N}\sqrt{N}\mathrm{e}^{-\mathrm{i}\Delta t}$$

and admit the familiar Jaynes–Cummings solutions[69]

$$\begin{pmatrix} a_{0N} \\ a_{1N-1} \end{pmatrix} \propto \begin{pmatrix} -\omega_\pm \mathrm{e}^{-\mathrm{i}\omega_\mp t} \\ g\sqrt{N}\mathrm{e}^{\mathrm{i}\omega_\pm t} \end{pmatrix}$$

with $\omega_\pm = -\Delta/2 \pm \sqrt{\Delta^2/4 + g^2N}$. In the $|\Delta| \gg g$ limit, we have $\omega_+ \approx g^2N/\Delta$ and $\omega_- \approx -\Delta$, so the solution with the upper (lower) signs has $|a_{0N}| \ll |a_{1N-1}|$ ($|a_{0N}| \gg |a_{1N-1}|$) at $t = 0$, and therefore it corresponds to the initially excited (de-excited) emitter. The decay rate of the emitter when it is initially excited and the plasmon is not populated (*i.e.*, starting from $l = 1, n = 0$) is then given by the decay of the $|10\rangle\langle 10|$ term of eqn (C1) in the upper-sign solution. We find $\Gamma = \tau_0^{-1} - \dot{\rho}_{10,10}(0)/\rho_{10,10}(0) = \tau_0^{-1} + 2\mathrm{Im}\{\omega_+\} \approx \tau_0^{-1} + 2g^2\mathrm{Im}\{1/\Delta\}$, which reduces to eqn (7) under the condition $\gamma\tau_0 \gg 1$.

2. Dielectric approach

The decay rate of an emitter placed at a position $\mathbf{r}_0$ in the vicinity of a plasmonic structure can be related to its transition dipole $\mathbf{d}_0$ as[105]

$$\Gamma = \frac{1}{\tau_0} + \frac{2}{\hbar}\,\mathrm{Im}\{\mathbf{d}_0^*\cdot\mathbf{E}^{\mathrm{ind}}\}, \quad \text{(C3)}$$

where $\mathbf{E}^{\mathrm{ind}}$ is the self-induced electric field produced by a dipole $\mathbf{d}_0$ located at $\mathbf{r}_0$. We can calculate this field from the dielectric formalism of Appendix A using the induced density of eqn (A5), but now the coefficients c_j of eqn (A4) have to be obtained from the external dipole field $\mathbf{E}^{\mathrm{ext}} = (1/n_{\mathrm{eff}}^2)(\mathbf{d}_0\cdot\nabla_0)\nabla_0(1/|\mathbf{r}_0 - \mathbf{R}|)$, where n_{eff} is defined in eqn (A2) (notice that the potential produced by the dipole at the planar interface between media of permittivities ε_1 and ε_2 is the same as in a vacuum multiplied by $1/n_{\mathrm{eff}}^2$). After some algebra, we find

$$\mathbf{d}_0^*\cdot\mathbf{E}^{\mathrm{ind}} = \frac{t}{4\pi n_{\mathrm{eff}}^4 D^4}\sum_j \frac{\omega_{\mathrm{bulk}}^2}{\omega_j^2 - \omega(\omega + \mathrm{i}\gamma)}\left|\int \mathrm{d}^2\mathbf{R}\left[\nabla_{\vec{\theta}}\cdot\sqrt{f\left(\vec{\theta}\right)}\vec{\varepsilon}_j\left(\vec{\theta}\right)\right] \times \left(\mathbf{d}_0\cdot\nabla_0\frac{1}{|\mathbf{r}_0 - \mathbf{R}|}\right)\right|^2. \quad \text{(C4)}$$

Inserting eqn (C4) into eqn (C3), and retaining only the $j = p$ term, we recover the emitter decay rate given by eqn (7) with the exact same value of the coupling rate g as given by eqn (4), provided we define

$$\rho_j(\vec{\theta}) = \sqrt{\frac{\hbar\omega_{\text{bulk}}{}^2 t}{8\pi\omega_j D^4}}\left[\nabla_{\vec{\theta}}\cdot\sqrt{f(\vec{\theta})}\vec{\varepsilon}_j(\vec{\theta})\right].$$

Finally, inserting this expression into eqn (B2), integrating by parts, and keeping in mind the definition of A_j in eqn (A7), we recover eqn (B4) for the plasmon transition strength. Therefore, we conclude that the normalization of the plasmon charge density discussed in Appendix B, based upon the polarizability of the plasmonic structure, produces the same decay rate of a neighboring emitter when calculated either following the semi-classical dielectric formalism described in this paragraph or using the density-matrix formalism with the intuitive coupling rate defined by eqn (4).

Appendix D: Drude parameters for noble metals and graphene

We show in Fig. 5 the Drude parameters ω_{bulk} and γ for silver, gold, and graphene. For noble metals, we obtain these parameters by fitting the measured dielectric function of the material[106] to the expression $\varepsilon(\omega) = 1 - \omega_{\text{bulk}}{}^2/\omega(\omega + i\gamma)$. For graphene, we use the local-RPA conductivity,[36,107] which we correct in the following expression to simultaneously account for inelastic attenuation and finite temperature T in both intraband and interband transitions:[61,108,109]

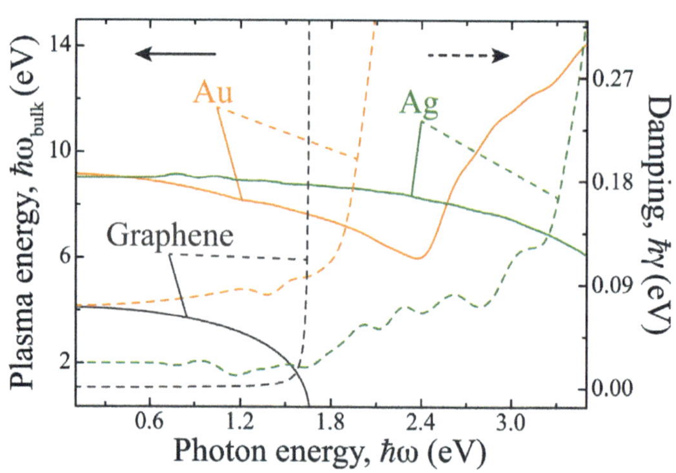

Fig. 5 Drude parameters for the response of noble metals. We represent the ω-dependent parameters ω_{bulk} (solid curves) and γ for Au and Ag as obtained from $\varepsilon(\omega) = 1 - \omega_{\text{bulk}}{}^2/\omega(\omega + i\gamma)$, where $\varepsilon(\omega)$ is the measured dielectric function of these materials.[106] The graphene parameters are obtained by fitting eqn (1) to match the local-RPA conductivity (eqn (D1)) with the Fermi energy, mobility, and temperature $E_F = 1$ eV, $\mu = 10\,000$ cm^2 V^{-1} s^{-1}, and $T = 300$ K, respectively, assuming a film thickness $t = 0.34$ nm equal to the interatomic plane distance in graphite. Notice that ω_{bulk} is no longer real for $\hbar\omega > 1.66$ eV in graphene (*i.e.*, above the range of plasmonic response for the chosen doping level).

$$\sigma(\omega) = \frac{-e^2}{\pi\hbar^2}\frac{\mathrm{i}}{\omega + \mathrm{i}\tau^{-1}}\int_{-\infty}^{\infty}\mathrm{d}E\left[|E|\frac{\partial f_\mathrm{E}}{\partial E} + \frac{(E/|E|)}{1 - 4E^2/\left[\hbar^2(\omega + \mathrm{i}\tau^{-1})^2\right]}f_\mathrm{E}\right], \quad \text{(D1)}$$

where $f_\mathrm{E} = 1/[1 + \mathrm{e}^{(E-E_\mathrm{F})/k_\mathrm{B}T}]$ is the Fermi-Dirac distribution as a function of electron energy E and Fermi energy E_F. The first term inside the integral of eqn (D1), which corresponds to intraband electron–hole pair transitions within the partially occupied Dirac cones of the doped carbon layer, can be integrated analytically to yield a contribution $\mathrm{i}e^2E_\mathrm{F}^\mathrm{eff}/\pi\hbar^2(\omega + \mathrm{i}/\tau)$ with $E_\mathrm{F}^\mathrm{eff} = E_\mathrm{F} + 2k_\mathrm{B}T\log(1 + \mathrm{e}^{-E_\mathrm{F}/k_\mathrm{B}T})$. The second term, which originates in interband transitions between lower and upper Dirac cones, needs to be integrated numerically. In Fig. 5, we represent ω_bulk and γ for graphene by fitting eqn (1) to the values of $\sigma(\omega)$ calculated from eqn (D1).

Acknowledgements

This work has been supported in part by the European Commission (Graphene Flagship CNECT-ICT-604391 and FP7-ICT-2013-613024-GRASP). A. M. acknowledges financial support from the Welch foundation through the J. Evans Attwell-Welch Postdoctoral Fellowship Program of the Smalley Institute of Rice University (Grant no. L-C-004).

References

1 K. R. Li, M. I. Stockman and D. J. Bergman, *Phys. Rev. Lett.*, 2003, **91**, 227402.
2 M. Danckwerts and L. Novotny, *Phys. Rev. Lett.*, 2007, **98**, 026104.
3 A. R. Davoyan, I. V. Shadrivov and Y. S. Kivshar, *Opt. Express*, 2008, **16**, 21209.
4 S. Palomba and L. Novotny, *Phys. Rev. Lett.*, 2008, **101**, 056802.
5 B. Metzger, M. Hentschel, T. Schumacher, M. Lippitz, X. Ye, C. B. Murray, B. Knabe, K. Buse and H. Giessen, *Nano Lett.*, 2014, **14**, 2867.
6 K. Kneipp, Y. Wang, H. Kneipp, L. T. Perelman, I. Itzkan, R. R. Dasari and M. S. Feld, *Phys. Rev. Lett.*, 1997, **78**, 1667.
7 S. Nie and S. R. Emory, *Science*, 1997, **275**, 1102.
8 H. Xu, E. J. Bjerneld, M. Käll and L. Börjesson, *Phys. Rev. Lett.*, 1999, **83**, 4357.
9 L. Rodríguez-Lorenzo, R. A. Álvarez-Puebla, I. Pastoriza-Santos, S. Mazzucco, O. Stéphan, M. Kociak, L. M. Liz-Marzán and F. J. García de Abajo, *J. Am. Chem. Soc.*, 2009, **131**, 4616.
10 D. P. O'Neal, L. R. Hirsch, N. J. Halas, J. D. Payne and J. L. West, *Cancer Lett.*, 2004, **209**, 171.
11 C. Loo, A. Lowery, N. J. Halas, J. L. West and R. Drezek, *Nano Lett.*, 2005, **5**, 709.
12 A. M. Gobin, M. H. Lee, N. J. Halas, W. D. James, R. A. Drezek and J. L. West, *Nano Lett.*, 2007, **7**, 1929.
13 P. K. Jain, I. H. El-Sayed and M. A. El-Sayed, *Nanotoday*, 2007, **2**, 18.
14 P. K. Jain, X. H. Huang, I. H. El-Sayed and M. A. El-Sayed, *Acc. Chem. Res.*, 2008, **41**, 578.
15 X. Qian, X.-H. Peng, D. O. Ansari, Q. Yin-Goen, G. Z. Chen, D. M. Shin, L. Yang, A. N. Young, M. D. Wang and S. Nie, *Nat. Biotechnol.*, 2008, **26**, 83.
16 D. E. Chang, A. S. Sörensen, P. R. Hemmer and M. D. Lukin, *Phys. Rev. Lett.*, 2006, **97**, 053002.

17 D. Dzsotjan, A. S. Sörensen and M. Fleischhauer, *Phys. Rev. B: Condens. Matter Mater. Phys.*, 2010, **82**, 075427.
18 S. Savasta, R. Saija, A. Ridolfo, O. Di Stefano, P. Denti and F. Borghese, *ACS Nano*, 2010, **4**, 6369.
19 A. Manjavacas, F. J. García de Abajo and P. N. Nordlander, *Nano Lett.*, 2011, **11**, 2318.
20 H. A. Atwater and A. Polman, *Nat. Mater.*, 2010, **9**, 205.
21 T. P. White and K. R. Catchpole, *Appl. Phys. Lett.*, 2012, **101**, 073905.
22 J. Dong, J. Liu, G. Kang, J. Xie and Y. Wang, *Sci. Rep.*, 2014, **4**, 5618.
23 N. Engheta, *Science*, 2007, **317**, 1698.
24 N. I. Zheludev, *Science*, 2010, **328**, 582.
25 A. D. Boardman, V. V. Grimalsky, Y. S. Kivshar, S. V. Koshevaya, M. Lapine, N. M. Litchinitser, V. N. Malnev, M. Noginov, Y. G. Rapoport and V. M. Shalaev, *Laser Photonics Rev.*, 2010, **5**, 287.
26 P. Nagpal, N. C. Lindquist, S. H. Oh and D. J. Norris, *Science*, 2009, **325**, 594.
27 M. Grzelczak, J. Pérez-Juste, P. Mulvaney and L. M. Liz-Marzán, *Chem. Soc. Rev.*, 2008, **37**, 1783.
28 J. A. Fan, C. H. Wu, K. Bao, J. M. Bao, R. Bardhan, N. J. Halas, V. N. Manoharan, P. Nordlander, G. Shvets and F. Capasso, *Science*, 2010, **328**, 1135.
29 N. J. Halas, S. Lal, W. Chang, S. Link and P. Nordlander, *Chem. Rev.*, 2011, **111**, 3913.
30 D. M. Solís, J. M. Taboada, F. Obelleiro, L. M. Liz-Marzán and F. J. García de Abajo, *ACS Nano*, 2014, **8**, 7559.
31 P. R. West, S. Ishii, G. V. Naik, N. K. Emani, V. M. Shalaev and A. Boltasseva, *Laser Photonics Rev.*, 2010, **4**, 795.
32 E. Feigenbaum, K. Diest and H. A. Atwater, *Nano Lett.*, 2010, **10**, 2111.
33 A. Boltasseva and H. A. Atwater, *Science*, 2011, **331**, 290.
34 G. V. Naik, V. M. Shalaev and A. Boltasseva1, *Adv. Mater.*, 2013, **25**, 3264.
35 A. Vakil and N. Engheta, *Science*, 2011, **332**, 1291.
36 F. H. L. Koppens, D. E. Chang and F. J. García de Abajo, *Nano Lett.*, 2011, **11**, 3370.
37 A. Y. Nikitin, F. Guinea, F. J. García-Vidal and L. Martín-Moreno, *Phys. Rev. B*, 2011, **84**, 161407.
38 L. Ju, B. Geng, J. Horng, C. Girit, M. Martin, Z. Hao, H. A. Bechtel, X. Liang, A. Zettl, Y. R. Shen, *et al.*, *Nat. Nanotechnol.*, 2011, **6**, 630.
39 Z. Fei, G. O. Andreev, W. Bao, L. M. Zhang, A. S. McLeod, C. Wang, M. K. Stewart, Z. Zhao, G. Dominguez, M. Thiemens, *et al.*, *Nano Lett.*, 2011, **11**, 4701.
40 Y. V. Bludov, N. M. R. Peres, and M. I. Vasilevskiy, *Phys. Rev. B*, 2012, **85**, 245409.
41 J. Chen, M. Badioli, P. Alonso-González, S. Thongrattanasiri, F. Huth, J. Osmond, M. Spasenović, A. Centeno, A. Pesquera, P. Godignon, *et al.*, *Nature*, 2012, **487**, 77.
42 Z. Fei, A. S. Rodin, G. O. Andreev, W. Bao, A. S. McLeod, M. Wagner, L. M. Zhang, Z. Zhao, M. Thiemens, G. Dominguez, *et al.*, *Nature*, 2012, **487**, 82.

43 Z. Fang, S. Thongrattanasiri, A. Schlather, Z. Liu, L. Ma, Y. Wang, P. M. Ajayan, P. Nordlander, N. J. Halas and F. J. García de Abajo, *ACS Nano*, 2013, **7**, 2388.
44 V. W. Brar, M. S. Jang, M. Sherrott, J. J. Lopez and H. A. Atwater, *Nano Lett.*, 2013, **13**, 2541.
45 A. Woessner, M. B. Lundeberg, Y. Gao, A. Principi, P. Alonso-González, M. Carrega, K. Watanabe, T. Taniguchi, G. Vignale, M. Polini, J. Hone, R. Hillenbrand and F. H. L. Koppens, *Nat. Mater.*, 2015, DOI: 10.1038/NMAT4169.
46 F. Wang, Y. Zhang, C. Tian, C. Girit, A. Zettl, M. Crommie and Y. R. Shen, *Science*, 2008, **320**, 206.
47 K. F. Mak, M. Y. Sfeir, Y. Wu, C. H. Lui, J. A. Misewich and T. F. Heinz, *Phys. Rev. Lett.*, 2008, **101**, 196405.
48 Z. Q. Li, E. A. Henriksen, Z. Jian, Z. Hao, M. C. Martin, P. Kim, H. L. Stormer and D. N. Basov, *Nat. Phys.*, 2008, **4**, 532.
49 C. F. Chen, C. H. Park, B. W. Boudouris, J. Horng, B. Geng, C. Girit, A. Zettl, M. F. Crommie, R. A. Segalman, S. G. Louie, *et al.*, *Nature*, 2011, **471**, 617.
50 A. Scholz, T. Stauber and J. Schliemann, *Phys. Rev. B: Condens. Matter Mater. Phys.*, 2013, **88**, 035135.
51 P. Di Pietro, M. Ortolani, O. Limaj, A. Di Gaspare, V. Giliberti, F. Giorgianni, M. Brahlek, N. Bansal, N. Koirala, S. Oh, *et al.*, *Nat. Nanotechnol.*, 2013, **8**, 556.
52 A. Manjavacas and F. J. García de Abajo, *Nat. Commun.*, 2014, **5**, 3548.
53 W. Ekardt and J. M. Pacheco, *Clustering Phenomena in Atoms and Nuclei*, Springer, Berlin, 1992.
54 G. Onida, L. Reining and A. Rubio, *Rev. Mod. Phys.*, 2002, **74**, 601.
55 R. Losio, K. N. Altmann, A. Kirakosian, J.-L. Lin, D. Y. Petrovykh and F. J. Himpsel, *Phys. Rev. Lett.*, 2001, **86**, 4632.
56 T. Nagao, S. Yaginuma, T. Inaoka and T. Sakurai, *Phys. Rev. Lett.*, 2006, **97**, 116802.
57 H. V. Chung, C. J. Kubber, G. Han, S. Rigamonti, D. Sanchez-Portal, D. Enders, A. Pucci and T. Nagao, *Appl. Phys. Lett.*, 2010, **96**, 243101.
58 E. P. Rugeramigabo, C. Tegenkamp, H. Pfnür, T. Inaoka and T. Nagao, *Phys. Rev. B: Condens. Matter Mater. Phys.*, 2010, **81**, 165407.
59 F. Moresco, M. Rocca, T. Hildebrandt and M. Henzler, *Phys. Rev. Lett.*, 1999, **83**, 2238.
60 E. P. Rugeramigabo, T. Nagao and H. Pfnür, *Phys. Rev. B: Condens. Matter Mater. Phys.*, 2008, **78**, 155402.
61 F. J. García de Abajo, *ACS Photonics*, 2014, **1**, 135.
62 S. Thongrattanasiri, A. Manjavacas and F. J. García de Abajo, *ACS Nano*, 2012, **6**, 1766.
63 A. Manjavacas, S. Thongrattanasiri and F. J. García de Abajo, *Nanophotonics*, 2013, **2**, 139.
64 A. Manjavacas, P. Nordlander and F. J. García de Abajo, *ACS Nano*, 2012, **6**, 1724.
65 P. A. Huidobro, A. Y. Nikitin, C. González-Ballestero, L. Martín-Moreno and F. J. García-Vidal, *Phys. Rev. B*, 2012, **85**, 155438.
66 E. Yablonovitch, T. J. Gmitter and R. Bhat, *Phys. Rev. Lett.*, 1988, **61**, 2546.
67 Z. Ficek and R. Tanas, *Phys. Rep.*, 2002, **372**, 369.

68 P. Meystre and M. Sargent, *Elements of Quantum Optics*, Springer-Verlag, Berlin, 2007.
69 E. Jaynes and F. Cummings, *Proc. IEEE*, 1963, **51**, 89.
70 S. Thongrattanasiri and F. J. García de Abajo, *Phys. Rev. Lett.*, 2013, **110**, 187401.
71 Y. L. A. Rezus, S. G. Walt, R. Lettow, A. Renn, G. Zumofen, S. Götzinger and V. Sandoghdar, *Phys. Rev. Lett.*, 2012, **108**, 093601.
72 O. Hunderi and H. P. Myers, *J. Phys. F*, 1973, **3**, 683.
73 V. G. Kravets, S. Neubeck, A. N. Grigorenko and A. F. Kravets, *Phys. Rev. B*, 2010, **81**, 165401.
74 M. C. Hutley and D. Maystre, *Opt. Commun.*, 1976, **19**, 431.
75 D. Maystre and R. Petit, *Opt. Commun.*, 1976, **17**, 196.
76 E. Popov and L. Tsonev, *Surf. Sci.*, 1990, **230**, 290.
77 W. C. Tan, J. R. Sambles and T. W. Preist, *Phys. Rev. B*, 2000, **61**, 13177.
78 J. J. Greffet, R. Carminati, K. Joulain, J. P. Mulet, S. Mainguy and Y. Chen, *Nature*, 2002, **416**, 61.
79 N. I. Landy, S. Sajuyigbe, J. J. Mock, D. R. Smith and W. J. Padilla, *Phys. Rev. Lett.*, 2008, **100**, 207402.
80 F. J. García de Abajo, *Rev. Mod. Phys.*, 2007, **79**, 1267.
81 S. Kachan, O. Stenzel and A. Ponyavina, *Appl. Phys. B*, 2006, **84**, 281.
82 T. V. Teperik, F. J. García de Abajo, A. G. Borisov, M. Abdelsalam, P. N. Bartlett, Y. Sugawara and J. J. Baumberg, *Nat. Photon.*, 2008, **2**, 299.
83 S. Collin, F. Pardo, R. Teissier and J. L. Pelouard, *Appl. Phys. Lett.*, 2004, **85**, 194.
84 Y. P. Bliokh, J. Felsteiner and Y. Z. Slutsker, *Phys. Rev. Lett.*, 2005, **95**, 165003.
85 R. L. Fante and M. T. McCormack, *IEEE Trans. Antennas Propag.*, 1988, **36**, 1443.
86 N. Engheta, in *Digest of the 2002 IEEE AP-S International Symposium*, San Antonio, TX, (IEEE, 2002), June 16–21, 2002, vol. 2, pp. 392–395.
87 S. Thongrattanasiri, F. H. L. Koppens and F. J. García de Abajo, *Phys. Rev. Lett.*, 2012, **108**, 047401.
88 Z. Fang, Y. Wang, A. Schlather, Z. Liu, P. M. Ajayan, F. J. García de Abajo, P. Nordlander, X. Zhu and N. J. Halas, *Nano Lett.*, 2014, **14**, 299.
89 M. S. Jang, V. W. Brar, M. C. Sherrott, J. J. Lopez, L. Kim, S. Kim, M. Choi and H. A. Atwater, *Phys. Rev. B: Condens. Matter Mater. Phys.*, 2014, **90**, 165409.
90 H. C. van de Hulst, *Light Scattering by Small Particles*, Dover, New York, 1981.
91 X. Li, X. Wang, L. Zhang, S. Lee and H. Dai, *Science*, 2008, **319**, 1229.
92 B. Li, K. Tahara, J. Adisoejoso, W. Vanderlinden, K. S. Mali, S. De Gendt, Y. Tobe and S. D. Feyter, *ACS Nano*, 2013, **7**, 10764.
93 J. Cai, P. Ruffieux, R. Jaafar, M. Bieri, T. Braun, S. Blankenburg, M. Muoth, A. P. Seitsonen, M. Saleh, X. Feng, *et al.*, *Nature*, 2010, **466**, 470.
94 K. Müllen, *ACS Nano*, 2014, **8**, 6531.
95 A. Manjavacas, F. Marchesin, S. Thongrattanasiri, P. Koval, P. Nordlander, D. Sánchez-Portal and F. J. García de Abajo, *ACS Nano*, 2013, **7**, 3635.
96 D. J. Bergman, *Phys. Rev. B*, 1979, **19**, 2359.
97 K. Ghosh and R. Fuchs, *Phys. Rev. B*, 1988, **38**, 5222.
98 F. Ouyang and M. Isaacson, *Philos. Mag. B*, 1989, **60**, 481.
99 D. J. Bergman and K. J. Dunn, *Phys. Rev. B*, 1992, **45**, 13262.
100 F. J. García de Abajo, *ACS Nano*, 2013, **7**, 11409.

101 S. Thongrattanasiri, I. Silveiro and F. J. García de Abajo, *Appl. Phys. Lett.*, 2012, **100**, 201105.
102 D. Pines and P. Nozières, *The Theory of Quantum Liquids*, W. A. Benjamin, Inc., New York, 1966.
103 F. J. García de Abajo, *Rev. Mod. Phys.*, 2007, **79**, 1267.
104 I. Silveiro and F. J. García de Abajo, *Appl. Phys. Lett.*, 2014, **104**, 131103.
105 L. Novotny and B. Hecht, *Principles of Nano-Optics*, Cambridge University Press, New York, 2006.
106 P. B. Johnson and R. W. Christy, *Phys. Rev. B: Solid State*, 1972, **6**, 4370.
107 L. A. Falkovsky and A. A. Varlamov, *Eur. Phys. J. B*, 2007, **56**, 281.
108 V. P. Gusynin, S. G. Sharapov and J. P. Carbotte, *Phys. Rev. Lett.*, 2006, **96**, 256802.
109 V. P. Gusynin, S. G. Sharapov and J. P. Carbotte, *New J. Phys.*, 2009, **11**, 095013.

ROYAL SOCIETY OF CHEMISTRY

PAPER

Ultimate limit of field confinement by surface plasmon polaritons

Jacob B. Khurgin*

Received 6th October 2014, Accepted 16th October 2014
DOI: 10.1039/c4fd00193a

We show that electric field confinement in surface plasmon polaritons propagating at metal/dielectric interfaces enhances the loss due to Landau damping, which effectively limits the degree of confinement itself. We prove that Landau damping, and associated with it surface collision damping, follow directly from the Lindhard formula for the dielectric constant of a free electron gas. Furthermore, we demonstrate that even if all of the conventional loss mechanisms, caused by phonons, electron–electron interactions, and interface roughness scattering, were eliminated, the maximum attainable degree of confinement and the loss accompanying it would not change significantly compared to the best existing plasmonic materials, such as silver.

Introduction

Steady advances in nanofabrication made in the last decade have inspired research in nano-plasmonics, a field that carries exciting promise in various areas of technology such as sub-wavelength imaging, sensing, nano-scale optical interconnects and active devices.[1,2] In one way or another, these promises are all hinged upon the ability to concentrate an optical field into the sub-wavelength dimensions that is a salient feature of surface plasmon polaritons (SPPs), whose nature is a combination of an electro-magnetic field with charge waves of free carriers in a metal (or semiconductor). When the dimensions are reduced way below the wavelength the magnetic field is greatly diminished (static limit), and the energy which is normally stored in the form of magnetic energy is instead stored in the form of kinetic energy of the carriers (kinetic inductance) which makes sub-wavelength oscillation mode sustainable. Unfortunately and inevitably, the free carrier oscillations dissipate energy at a very high rate, of the order of $\gamma_m \sim 10^{14}\ s^{-1}$ in noble metals and $10^{13}\ s^{-1}$ in highly doped semiconductors. As a result, the losses in the SPP which are significantly sub-wavelength (in the case of propagating SPPs it means the SPP propagation constant β is much larger than the wave-vector in dielectric k_d) are always on the scale of γ_m, and are independent on the shape and exact size as long as it is significantly sub-wavelength.[3] Due to these high losses the promise of nanoplasmonics, which includes miniature

Johns Hopkins University, Baltimore, MD 21218, USA. E-mail: jakek@jhu.edu

efficient sources and detectors of radiation, nano-scale optical interconnects, super-resolution imaging, and others has not been fully realized yet.[4]

The reason the optical losses in metals are high is unfortunately the same reason that makes support charge oscillations at optical frequencies sustainable in the first place. The high free carrier density (Fermi level high in the conduction band), which is requisite to maintain a high plasma frequency also assures that the density of states into which these electrons can scatter is also very high, with strong scattering following from this unfortunate fact. In optical range the metal loss is caused by more than one mechanism[5] – scattering by phonons, carrier–carrier Umklapp processes, residual intra-band absorption and scattering on surface imperfections can all contribute on more or less equal scales, hence their reduction is not simple. Nevertheless, some strides in that direction are being made. Most obviously, the reduction of metal surface roughness achieved with epitaxial growth have yielded reduced loss in Ag,[6] while using Al in place of Ag[7] reduced parasitic interband absorption in the blue part of the spectrum. The phonon-assisted absorption can be reduced, but not eliminated by going to cryogenic temperatures because the spontaneous phonon emission is possible even at absolute zero, while the temperature-independent Umklapp scattering may be somewhat mitigated in the materials with different (less spherical) Brillouin zone shapes, although, once again, the numerous efforts with materials as diverse as ITO and TiN[8] so far have not shown substantial improvement over ubiquitous noble metals.

Nevertheless, the hope is alive that sooner or later, a novel material with a negative ε and substantially lower losses in the optical range will emerge, and in anticipation of these developments, it is worthwhile to estimate their practical impact, *i.e.* what would be the maximum attainable degree of field concentration if all scattering-related losses in the metal in the optical range were essentially eliminated. Obviously, this question has been raised before. For instance, it has been long known that since SPP is a combination of field and electronic oscillation, the maximum SPP wave vector $K_{\max}$ cannot possibly exceed the Fermi wave vector $k_F \sim 1.2 \times 10^8$ cm^{-1} for noble metals, *i.e.* restricting the degree of field concentration to about 5 A. This is also the scale at which electron tunneling in the nanogap in the plasmonic dimers commences[9] that further assures that the electric field cannot be confined to sub-nanometer dimensions. Yet this limit, usually referred to as the "quantum limit" has not been achieved experimentally, and more recently a different explanation which has put the limit of field concentration in the range of a few nanometers has been put forward. The explanation was based on the non-locality phenomenon,[10] or, in simpler terms on special dispersion of the dielectric constant $\varepsilon(\boldsymbol{k})$. Using the hydrodynamic model of nonlocality Mortensen *et al.*[11] have shown that when the characteristic dimensions of the system become comparable with the characteristic length $l_c = v_F/\omega$, where v_F is a Fermi velocity and ω is a frequency, the nonlocal effects prevent the light from tight confinement and broaden the resonances, particularly for the dimer structures. Larkin and Stockman[12] pointed out that spatial dispersion effects limit the resolution of the plasmonic superlens to about 5 nm in the visible range, *i.e.* comparable to a few l_c. They also pointed out that spatial dispersion is intimately related to Landau damping – *i.e.* direct absorption of electromagnetic waves by electrons below the Fermi level. Landau damping in nanoscale metallic objects can also be interpreted as a quantum confinement effect or as absorption assisted by surface collisions as explained by Kreibig and Vollmer,[13] who introduced the phenomenological expression for this process as a "surface collision scattering rate" $\gamma_s \sim v_F/a$ where v_F is the Fermi velocity and a is the characteristic dimension of the metallic object. According to this phenomenological treatment the additional damping is caused by the limited physical dimension of the system and is the result of the restriction of the mean free path of electrons.

However, to the best of our knowledge,[14,15] there is no unified treatment of nonlocality (*i.e.* the spatial dispersion of the real part of ε) and Landau damping (*i.e.* the spatial dependence of the imaginary part of ε) which would allow one to provide an unambiguous answer on which of the two phenomena exert a stronger influence on the SPP properties. In this work we develop this unified treatment and show that it is Landau damping, *i.e.* the loss induced by the field confinement itself that is responsible for the most dramatic limitations in plasmonics. We show that in the case of a propagating SPP on a single metal–dielectric interface, the limitations arise due to the field (and not the electron) confinement and thus are not influenced by the metal dimensions. Landau damping is shown to become important when characteristic dimensions are on the scale of 10 nm which is at least an order of magnitude larger than l_c, and, finally, and to some degree surprisingly, it is shown that the total elimination of all other loss mechanisms in metals will not yield noticeable improvement in field confinement and power dissipation compared to the best plasmonic materials of today.

The damping rate of SPP due to surface collisions

We start with the case of the SPP propagating at the interface of metal with $\varepsilon_m < 0$ and dielectric $\varepsilon_d > 0$ as shown in Fig. 1a with the propagating constant β and two components of electric field, the normal component

$$E_x = \begin{cases} E_0\,\mathrm{e}^{-qx}\cos(\beta z - \omega t) & x > 0 \\ \dfrac{\varepsilon_m}{\varepsilon_d} E_0\,\mathrm{e}^{q_d x}\cos(\beta z - \omega t) & x < 0 \end{cases} \tag{1}$$

that is subjected to damping by surface collisions, and also the tangential component

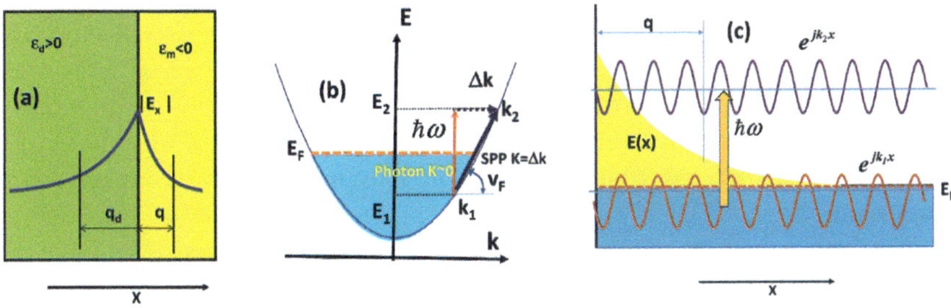

Fig. 1 (a) Sketch of the fields in the propagating SPP. (b) Phonon or impurity assisted absorption of a photon with a wavevector $K \sim 0$ and a direct absorption of SPP with a large wavevector K (Landau damping). (c) Wavefunctions involved in the absorption of SPP.

$$E_z = \begin{cases} \dfrac{q}{\beta} E_0 e^{-qx} \sin(\beta z - \omega t) & x > 0 \\ \dfrac{q_m}{\beta} E_0 e^{q_d x} \sin(\beta z - \omega t) & x < 0 \end{cases} \quad (2)$$

which is not damped by surface collisions. For the absorption in the metal to take place, the electrons with energy E_1 and wavevector $\boldsymbol{k}_1$ below the Fermi level make a transition to a state above it with energy $E_2 = E_1 + \hbar\omega$ and wavevector $\boldsymbol{k}_2$ resulting in a wavevector mismatch Δk that makes the transition forbidden (Fig. 1b). The value of the mismatch is on the scale of $\Delta k_c \sim \omega/v_F \sim l_c^{-1}$ where the Fermi velocity v_F is about 1.4×10^8 cm s^{-1} for noble metals, *i.e.* for a 1 eV photon energy the wave-vector mismatch is about 1 nm^{-1} and is thus much larger than wavevector of the electromagnetic wave. As a result, absorption usually involves the additional act of scattering, due to phonons, electron–electron scattering, or imperfections that occur at the rate γ_0 which occurs on the scale of a few tens of femtoseconds in most plasmonic metals. At first glance, in the SPP both β and q are much less than 1 nm^{-1} and one should not expect much absorption. But if one takes a look at the normal direction, one can see that electric field contains all kinds of wavevectors – indeed by taking the one dimensional Fourier transform of (1) we obtain the power spectrum of it.

$$|E(K)|^2 = E_0{}^2 \frac{q/\pi}{K^2 + q^2} \quad (3)$$

Thus the fraction of power of the wave with wavevectors exceeding Δk_c is of the order of $F(K > \Delta k_c) \approx (2/\pi)(q/\Delta k_c)$. The penetration depth $L_p \sim (2q)^{-1}$ of the SPP is typically on the scale of a few tens of nanometers, hence a few percent of the SPP energy does have wavevector sufficient for Landau damping to take place. Landau damping is shown schematically in Fig. 1b as a direct "diagonal" transition between the states $\boldsymbol{k}_1$ and $\boldsymbol{k}_2$ caused by the electromagnetic wave with wave-vector $\boldsymbol{K} = \Delta \boldsymbol{k}_c$. As a result of Landau damping the imaginary part of the dielectric constant of the metal, describing the energy loss by the electromagnetic wave (and hence by SPP) to the individual electronic excitations will increase. The phenomenological "surface collision rate" γ_s introduced by Kreibig modifies the expression for the effective dielectric constant of the metal as

$$\varepsilon''_{\text{eff}}(\omega) = \frac{\omega_p{}^2(\gamma_0 + \gamma_s)}{\omega^3} \quad (4)$$

where γ_0 is the momentum relaxation rate due to phonons, electron–electron scattering and defects, usually referred to as the bulk scattering rate. The surface collision rate may be introduced phenomenologically,[10,13] or quantum-mechanically,[16] but it is simply added to the expression for the dielectric constant using the Matthiessen rule, and is not directly connected to nonlocality.

We now evaluate the energy damping rate of SPP polaritons that is due to the localization of the field in the vicinity of the metal interface with the dielectric. As one can see from the Fig. 1c the two free electron wavefunctions $\psi_{1(2)} = L^{-1/2} e^{i\mathbf{k}_\parallel \cdot \mathbf{r}_\parallel} e^{ik_{x1(2)}}$ (L being the normalization length) are orthogonal and optical transition between them is not allowed. But since the electric field is localized according to (1), the square of the interaction Hamiltonian between the two states can be found as

$$|H_{12}|^2 = \frac{e^2|\langle 1|\boldsymbol{p}\cdot\mathbf{A}|2\rangle|^2}{m^2} = \frac{e^2\hbar^2 k_{x1}k_{x2}E_0{}^2}{4m^2\omega^2L^2(\Delta k^2+q^2)} \tag{5}$$

Note that the fact that electrons get reflected by the surface is not used in (5) – the existence of the transition is strictly due to the confinement of the field. Since both states are close to the Fermi energy, *i.e.* $\hbar\omega \ll E_F$, we can make two important approximations, that $\hbar\omega \approx \hbar v_{Fx}\Delta k$ where v_{Fx} is the transverse component of velocity on the Fermi surface, and, furthermore $\hbar^2 k_{1x}k_{2x}/2m \approx mv_{Fx}^2/2m$, which upon substitution into (5) yields

$$|H_{12}|^2 = \frac{e^2 v_{Fx}{}^4 E_0{}^2}{4\omega^4L^2(1+q^2v_{Fx}{}^2/\omega^2)} \tag{6}$$

We now invoke Fermi's golden rule to evaluate the total rate of the field induced upward transitions from state 1, $R_1 = 2\pi|H_{12}|^2 L\rho_x(E_F)$, where $\rho_x(E_F) = (2\pi\hbar|v_{Fx}|)^{-1}$ is the one-dimensional density of the final states, evaluated under consideration that neither spin nor direction of propagation change as the transition takes place. We thus obtain

$$R_1 = \frac{e^2|v_{Fx}{}^3|E_0{}^2}{4\hbar^2\omega^4L(1+q^2v_{Fx}{}^2/\omega^2)} \tag{7}$$

Next we perform a summation over all of the states 1 inside the Fermi sphere from which transitions into unoccupied states can take place. That involves integration over the Fermi surface and then multiplying by $\hbar\omega$ as well as the normalization length L to obtain the surface rate of excitation of hot electrons per unit of area. Integration over the Fermi surface is simply multiplication by the three-dimensional density of sates $\rho_{3D} = m^2v_F/\pi^2\hbar^3$ and averaging over the angles which yields $|v_{Fx}{}^3| = v_x{}^3/4$, so we obtain

$$\frac{dN_{2D}}{dt} = \frac{e^2m^2v_F{}^4E_0{}^2}{16\pi^2\hbar^4\omega^3(1+q^2/\Delta k_c{}^2)} \tag{8}$$

Since $q \ll \Delta k_c$ we can write the expression for the energy loss at the surface as

$$-\frac{dU_{2D}}{dt} = \frac{e^2m^2v_F{}^4E_0{}^2}{16\pi^2\hbar^3\omega^2} \tag{9}$$

where U_{2D} is the time-averaged two-dimensional density of kinetic energy of electrons in the SPP which can be found as the integral of the three-dimensional density of energy

$$U_{2D}(x) = \int_0^\infty \frac{1}{4}\left(\frac{\partial(\omega\varepsilon_m)}{\partial\omega} - \varepsilon_b\right)E_x{}^2(x)dx = \frac{N_e}{8q}\frac{e^2E_0{}^2}{m\omega^2} \tag{10}$$

where ε_b is the part of the dielectric constant due to bound electrons, and the electron density in a parabolic band is $N_e = k_F{}^3/3\pi^2 = m^3v_F{}^3/3\pi^2\hbar^3$, hence

$$U_{2D} = \frac{1}{24\pi^2 q}\frac{e^2 m^2 v_F{}^3 E_0{}^2}{\hbar^3 \omega^2}. \tag{11}$$

Comparison of (9) and (11) immediately results in the energy relaxation rate

$$\frac{1}{U_{2D}}\frac{dU_{2D}}{dt} = -\frac{3}{2}qv_F = -2\gamma_s \tag{12}$$

where γ_s is the momentum scattering rate due to surface collisions

$$\gamma_s = \frac{3}{4}qv_F \tag{13}$$

that enters into expression (4) for the dielectric constant. This expression is not much different from the one in previous works, where the scattering rate for a nanoparticle with radius R is $\gamma_s = Av_F/R$ but in this work we have obtained this expression using a full quantum mechanical derivation. The fact that our coefficient A is less than values obtained for nanoparticles is easy to explain by the fact that our problem is one dimensional, hence not every electron contributes to the surface absorption.

It is very important to stress that the "surface collision damping rate" obtained by us does not require the electrons to be confined, or even reflected – simple confinement of light on the scale of the penetration length $L_{pen} = 1/2q$ is sufficient to overcome momentum conservation rules and cause substantial absorption even if the electrons are considered to be free. In other words the confinement of light is what causes the damping, hence the proper name for γ_s should be the "Landau damping rate", or, better, "time-of-flight" broadening. Nevertheless, we shall use the term "surface collisions damping" to conform to the existing literature.

Surface collision damping as a non-local effect

Next we demonstrate that result (13) can also be obtained by using a simple, phenomenological picture of simply "shifting" the resonance frequency in the Drude formula to account for the possibility of a "diagonal in k-space" transition as shown Fig. 1b. In the classical Drude formula, obtained in the long wavelength limit of the Lindhard approximation, the energy difference between two states involved in the optical transition is zero, but for the electromagnetic wave with the finite wavevector K the resonance is shifted by roughly $\Delta\omega = \alpha v_F K$ where α is on a scale of unity. Using the Klimontovich–Silin–Lindhard approximate formula[17] for the dielectric constant of the metal one can obtain $\alpha = \sqrt{3/5}$ and

$$\varepsilon(\omega, K) = \varepsilon_b + \frac{3}{K^2 v_F{}^2}\left[1 - \frac{\omega}{2Kv_F}\ln\frac{\omega + Kv_F}{\omega - Kv_F}\right] \approx \varepsilon_b - \frac{\omega_p{}^2}{\omega^2 - \frac{3}{5}K^2 v_F{}^2 + i\omega\gamma_0} \tag{14}$$

where γ_0 is the "intrinsic" scattering due to phonons, defects and electron–electron scattering, and ε_b is the interband contribution to the dielectric constant ($\varepsilon_b \approx 4.1$ for Ag) eqn (14) is easy to interpret as a modification of the Drude formula. For negligibly small wave vectors the resonant energy of the transition between the two free electron states is zero, but as the wave-vector increases the resonance shifts upward, by roughly $\Delta\omega \approx \sqrt{3/5}Kv_F \ll \omega$ where Landau damping takes place (the factor of 3/5 can be traced to the averaging over the Fermi surface). As shown in Fig. 2a the whole dispersion curve shifts towards higher frequencies resulting in a small change of the dielectric constant. Separating the dielectric constant into the real and imaginary parts one can write

$$\begin{aligned}\varepsilon_r(\omega, K) &= \varepsilon_b - \frac{\omega_p^2}{\omega^2}\frac{K_0^2(K_0^2 - K^2)}{(K_0^2 - K^2)^2 + K_0^4\gamma_0^2/\omega^2} \\ \varepsilon_i(\omega, K) &= \frac{\omega_p^2}{\omega^2}\frac{K_0^4\gamma_0/\omega}{[K_0^2 - K^2]^2 + K_0^4\gamma_0^2/\omega^2}\end{aligned} \tag{15}$$

where $K_0 = \sqrt{5/3}\Delta k_c$ is roughly the wave-vector at which the "diagonal" absorption of light (Landau damping) commences.

The plot of the wavevector–dielectric constant is shown in Fig. 2b for $\omega/\gamma_0 = 20$ and for the imaginary part of the dielectric constant it consists of essentially a flat response for $K \ll K_0$ followed by the sharp delta-function like Lorentzian peaks near. $K = \pm K_0$

$$\varepsilon_i(\omega, K) \approx \begin{cases} \dfrac{\omega_p^2\gamma_0}{\omega^3} & |K| \ll K_0 \\ \dfrac{\omega_p^2}{\omega^2}\dfrac{K_0^2\gamma_0/4\omega}{[K \mp K_0]^2 + K_0^2\gamma_0^2/4\omega^2} \approx \dfrac{\omega_p^2}{\omega^2}\dfrac{\pi K_0}{2}\delta(K \mp K_0) & K \approx \pm K_0 \end{cases} \tag{16}$$

The effective dielectric constant can be obtained by integrating (16) over the normalized power density spectrum of the electric field inside the metal $|E(K)|^2 = q/\pi(K^2 + q^2)$, which is also of Lorentzian shape as plotted in Fig. 2b for $q = K_0/10$, which immediately gives us

$$\varepsilon_{i,\mathrm{eff}}(\omega, q) = \frac{\omega_p^2\gamma_0}{\omega^3} + \frac{\omega_p^2}{\omega^2}\frac{qK_0}{K_0^2 + q^2} \approx \frac{\omega_p^2\gamma_0}{\omega^3} + \frac{\omega_p^2\sqrt{3/5}qv_F}{\omega^3} = \frac{\omega_p^2(\gamma_0 + \gamma_s)}{\omega^3} \tag{17}$$

where the surface collision broadening is $\gamma_s = \sqrt{3/5}qv_F \approx 0.77qv_F$ *i.e.* a result that is very close to the quantum mechanical derivation (13). Besides providing a

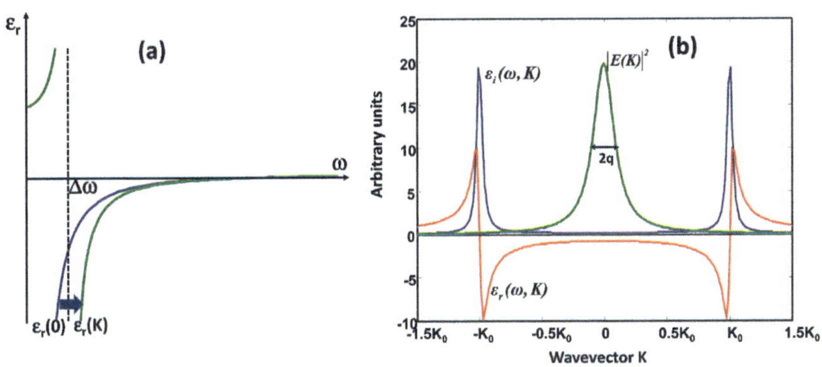

Fig. 2 (a) Sketch of the resonance shift due to non-locality (eqn (14)). (b) The effective dielectric constant of SPP can be obtained by overlapping the spatial power spectrum of SPP $|E(k)|^2$ with the wave-vector dependent dielectric constant (real and imaginary parts).

simple physical interpretation, eqn (21) also confirms Matthiessen's rule of the strength of absorption induced by surface reflection (γ_s) being added to the strength of absorption induced by all other means (γ_0).

If we now turn our attention to the real part of the dielectric constant, then one can see that over the region where the $K \ll K_0$, the power density spectrum $|E(K)|^2$ looks like a delta function, while integrating over the real part of the Lorentzian leads to cancellation, and, as a result $\varepsilon_{r,\mathrm{eff}}(\omega) \approx \varepsilon_r(\omega,0)$. In other words, the nonlocality effects are dominated by the change in the imaginary part of the dielectric constant, *i.e.* the surface collision damping (a.k.a. time-of-flight broadening).

Ultimate limit of field confinement imposed by surface collision damping

We now want to see how the increased damping due to field confinement manifests itself for the propagating SPPs of Fig. 1a. The propagation constant of SPP can be found as

$$\beta(\omega) = k_\mathrm{D}(\omega)\sqrt{\frac{\varepsilon_\mathrm{m}(\omega)}{\varepsilon_\mathrm{d} + \varepsilon_\mathrm{m}(\omega)}} \tag{18}$$

where $k_\mathrm{D}(\omega) = 2\pi\varepsilon_\mathrm{d}^{1/2}\omega/c$ is the wavevector of the free propagating electromagnetic wave in the dielectric with frequency-independent dielectric constant ε_d and the metal dielectric constant, according to (14) and (17) given by

$$\varepsilon_\mathrm{eff}(\omega, q) \approx \varepsilon_\mathrm{b} - \frac{{\omega_\mathrm{p}}^2}{\omega^2 + \mathrm{i}\omega[\gamma_0 + \gamma_s(q)]} \tag{19}$$

and the collision-induced damping rate being

$$\gamma_\mathrm{s} = \frac{3}{4}q(\omega)v_\mathrm{F}f_x(\omega) \tag{20}$$

where

$$q(\omega) = \sqrt{\beta^2(\omega) - \varepsilon_\mathrm{m}{k_\mathrm{D}}^2} \tag{21}$$

and

$$f_x = \frac{|E_x|^2}{|E_z|^2 + |E_x|^2} = \frac{\beta^2}{\beta^2 + q^2} \tag{22}$$

is a fraction of the energy contained in the normal component of electric field. If we now introduce the frequency of SP resonance, $\omega_\mathrm{SP} = \omega_\mathrm{p}/\sqrt{(\varepsilon_\mathrm{b} + \varepsilon_\mathrm{d})}$ and normalize the frequencies to it, $\tilde{\omega} = \omega/\omega_\mathrm{SP}$, and introduce the effective index of SPP as $\tilde{\beta} = \beta/k_\mathrm{D}$ and the normalized decay constant $\tilde{q} = q/k_\mathrm{D}$ we obtain from (18) and (19)

$$\tilde{\beta}(\tilde{\omega}) = \sqrt{\frac{\varepsilon_\mathrm{b}(\tilde{\omega}^2 - 1) - \varepsilon_\mathrm{d} + \mathrm{i}\varepsilon_\mathrm{b}\tilde{\omega}Q_0^{-1} + \mathrm{i}\varepsilon_\mathrm{b}\tilde{\omega}^2{Q_\mathrm{s}}^{-1}}{(\varepsilon_\mathrm{b} + \varepsilon_\mathrm{d})(\tilde{\omega}^2 - 1 + \mathrm{i}\tilde{\omega}{Q_0}^{-1} + \mathrm{i}\tilde{\omega}^2{Q_\mathrm{s}}^{-1})}} \tag{23}$$

where $Q_0 = \omega_\mathrm{SP}/\gamma_0$ and $Q_\mathrm{s}(\tilde{\omega}) = 4c/3{\varepsilon_\mathrm{d}}^{1/2}v_\mathrm{F}f_x\tilde{q}_\mathrm{r}$ and $\tilde{q}_\mathrm{r}(\tilde{\omega})$ is the real part of the normalized decay constant. Since $Q_\mathrm{s}(\tilde{\omega})$ itself is a function of $\tilde{\beta}(\tilde{\omega})$ eqn (23) can be solved self-consistently, by iterations, but prior to that a few important observations can be made.

First, the impact of surface damping would become noticeable when $\gamma_s \sim \gamma_0$, which will happen not far from the SP resonance, hence when $Q_s \sim Q_0$, and since in the vicinity of SPP resonance $\tilde{q} \approx \tilde{\beta}$ and $f_x \approx 1/2$ we obtain the value of the effective index at which the surface dumping mast be taken into account,

$$\tilde{\beta}_s \approx \frac{8c}{3{\varepsilon_d}^{1/2} v_F Q_0} \tag{24}$$

If we consider the combination of Ag and GaN (${\varepsilon_d}^{1/2} = 2.3$, $\gamma_0 = 3.2 \times 10^{13}\ s^{-1}$, $\omega_{SP} = 4.5 \times 10^{15}\ s^{-1}$; $Q_0 \approx 140$)[18] we obtain $\tilde{\beta}_s \approx 1.6$ while for the GaN–Au combination ($\gamma_0 = 1.2 \times 10^{14}\ s^{-1}$; $Q_0 \approx 43$)[18] we obtain $\tilde{\beta}_s \approx 5$. As expected, it is for good metal like silver that the surface collision role becomes important early on, while for the less perfect metals like gold, the influence of surface collision does not become important until much later.

Secondly, one can use (23) to find the ultimate value of the effective index, and hence confinement, of SPP which could have been obtained in the hypothetical "ideal" metal with $\gamma_0 = 0$, *i.e.* free of defects, phonon scattering, electron–electron interaction, and residual interband absorption. Obviously, such a metal does not exist, however it is useful to see what kind of improvement can be achieved by reducing the loss in the metal. By inserting $\tilde{\omega} = 1$ and ${Q_0}^{-1} = 0$; into (23) we obtain

$$\tilde{\beta}_{max} = \sqrt{\frac{\varepsilon_d i Q_s}{(\varepsilon_b + \varepsilon_d)}} \approx \sqrt{i \frac{{\varepsilon_d}^{1/2}}{(\varepsilon_b + \varepsilon_d)} \frac{8c}{3 v_F \tilde{\beta}_{max,r}}} \tag{25}$$

Therefore we obtain a rather simple expression for the maximum effective index (real part) attainable with the "ideal" metal

$$\tilde{\beta}_{max,r} \approx \left(\frac{{\varepsilon_d}^{1/2}}{\varepsilon_b + \varepsilon_d} \frac{4c}{3 v_F} \right)^{1/3} \tag{26}$$

Then for a wide variety of dielectrics transparent in the visible and near IR with refractive indices between 1.5 and 3, $4 < \tilde{\beta}_{max,r} < 4.5$, so one arrives at a rather surprising result – one can reduce the wavelength of the SPP propagating on the metal–dielectric interface by no more than a factor of 4.5 relative to the plane wave propagating in dielectric, no matter how low the loss in the bulk metal is.

Numerical results

We now calculate the dispersion curves (23) for the SPP propagating on the boundary between the metal with silver-like dispersion ($\varepsilon_b = 4.1$, $\hbar\omega_{SP} = 9.3$ eV;) and GaN ${\varepsilon_d}^{1/2} = 2.3$ resulting in a SPP resonance near 415 nm. When it comes to the scattering constant in the metal we shall consider 5 different cases:

A. The best bulk silver with bulk damping constant $\gamma_0 = 3.2 \times 10^{13}\ s^{-1}$ and no surface collision damping taken into account ($\gamma_s = 0$).

B. The best bulk silver with bulk damping constant $\gamma_0 = 3.2 \times 10^{13}\ s^{-1}$ with surface collision damping taken into account.

C. "Dirty silver" with bulk damping constant $\gamma_0 = 1.2 \times 10^{14}\ s^{-1}$ similar to that of gold and no surface collision damping taken into account ($\gamma_s = 0$). The reason for using "dirty silver" instead of gold is that one cannot observe interface SP resonance in gold in combination with any dielectric due to high interband absorption, but it is important to see how the surface collision damping affects metals with reasonably high bulk loss.

D. "Dirty silver" with bulk damping constant $\gamma_0 = 1.2 \times 10^{14}\ \mathrm{s}^{-1}$ similar to that of gold with surface collision damping taken into account.

E. An "ideal metal" with no bulk damping ($\gamma_0 = 0$), with surface collision damping taken into account.

The results are shown in Fig. 3a, as well as the light line representing the propagation of plane electromagnetic waves in GaN. As expected, not taking into account surface damping for the best silver (curve A) leads to very large propagation constant, with an effective index exceeding 7, but once surface collision damping has been included (curve B) the propagation constant is reduced almost two-fold. If we now consider the more realistic silver, full of defects due to the surface deposition process, whose damping rate is comparable to gold, the curves without (C) and with (D) surface collision damping, the latter's impact is less significant, although still prominent. But it is curve (E) which is most telling – if one starts with the best available silver and then hypothetically gets rid of all damping processes, be it defects, phonons or electron–electron interactions, then, even if the surface is atomically smooth the increase of the attainable propagation constant $\beta_{\mathrm{max,r}}$ (and hence the degree of confinement q) will be only about 12%, and the maximum effective index will not exceed roughly 4.4, just as predicted by (26). That means the minimum confinement depth in the normal direction $d_{\mathrm{min},x} = (q_{\mathrm{d}}{}^{-1} + q^{-1})/2$ will not be less then roughly $\lambda_{\mathrm{D}}/25n_{\mathrm{d}}$ where, $n_{\mathrm{d}} = \varepsilon_{\mathrm{d}}{}^{1/2}$. Furthermore, if we now consider the imaging using SPPs, the superlens,[19,20] and apply the analysis of ref. 21, the minimum spot size (confinement in lateral direction) achievable in this configuration would be $d_{\mathrm{min},x} \approx 2\pi/\beta_{\mathrm{max,r}} \approx \lambda/4.4n$, *i.e.*, close to the result obtained in ref. 12.

To show how surface collision damping affects losses we also plot in Fig. 3b the figure of merit, defined as the ratio $\mathrm{FOM} = \mathrm{Re}(\beta)/2\mathrm{Im}(\beta)$, or as one can say, the

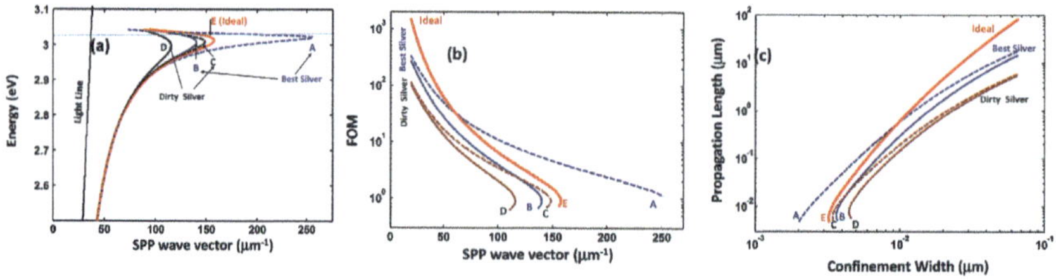

Fig. 3 (a) Dispersion curves. (b) Figure of merit *vs.* wavevector. (c) Propagation length *vs.* confinement width of SPP on the interface between Ga–N and Ag-like metals for the following cases: (A) the best bulk silver with bulk damping constant $\gamma_0 = 3.2 \times 10^{13}\ \mathrm{s}^{-1}$ and no surface collision damping taken into account ($\gamma_s = 0$). (B) The same but with surface collision damping taken into account. (C) "Dirty silver" with bulk damping constant $\gamma_0 = 1.2 \times 10^{14}\ \mathrm{s}^{-1}$ similar to that of gold and no surface collision damping taken into account ($\gamma_s = 0$). (D) The same but with bulk damping constant $\gamma_0 = 1.2 \times 10^{14}\ \mathrm{s}^{-1}$ similar to that of gold with surface collision damping taken into account. (E) An "ideal metal" with no bulk damping ($\gamma_0 = 0$), with surface collision damping taken into account.

phase shift accumulated over one absorption length. Once again we can see that for the high quality silver (curves A and B) the impact of surface collisions becomes important at large wavevectors with almost an order of magnitude difference, achieved at $\beta_r = 150\ \mu m^{-1}$ while for the higher loss metal (curves C and D) the difference is somewhat less. But the most important is the fact that for the "ideal" metal the FOM improves by only a factor of two relative to the "best silver".

Yet another way to show the effect of surface collision damping is to plot the propagation length $L_{prop} = 1/2\text{Im}(\beta)$ *vs.* the field penetration width in dielectric $L_{con} = 1/2\text{Re}(q_d)$ as demonstrated in Fig. 3c. Reducing the bulk losses γ_0 helps to increase the propagation length by about an order of magnitude for confinements wider than 50 nm, but for tighter confinements, as surface collision damping becomes dominant, and curves B, D, and E get close to each other, getting rid of all bulk losses results in only a marginal increase of the propagation length.

To further emphasize this point we consider the case of a gap plasmon,[22] Fig. 4a, with a GaAs core and metal cladding made either of Ag or Au. Once again we consider the same 5 cases as above, except cases C and D now refer to gold, as well as to "dirty silver". The dispersion curves for all 5 cases, shown in Fig. 4b, are essentially identical, but the amount of loss differs dramatically, as shown in Fig. 4c where the propagation length is plotted *versus* the gap width. As one can see the inclusion of surface collision damping greatly reduces the propagation length for silver (curves A and B) and somewhat less than that for gold. The most significant observation to be made from this figure is that hypothetically avoiding all bulk loss in the "ideal" metal (curve E) does increase the propagation length for the weakly confined gap plasmons with gap size over 100 nm, but for the tightly confined ones, with gap size less than 50 nm the effect is marginal. It appears that surface collision damping alone makes the propagation length shorter than 1 μm, which makes the gap plasmon impractical for applications as, say, an interconnect.

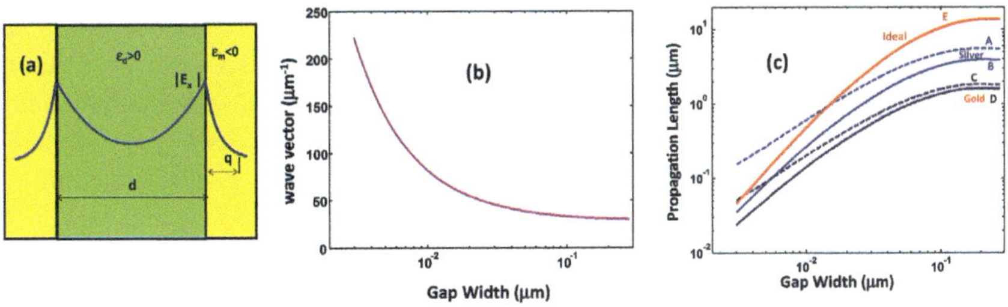

Fig. 4 (a) Sketch of gap SPP. (b) Propagation length *vs.* confinement width of gap SPP with GaN core and metal cladding. (c) Propagation length *vs.* confinement width of gap SPP. (A) The best bulk silver with bulk damping constant $\gamma_0 = 3.2 \times 10^{13}\ s^{-1}$ and no surface collision damping taken into account ($\gamma_s = 0$). (B) The same with surface collision damping taken into account. (C) Gold with bulk damping constant $\gamma_0 = 1.2 \times 10^{14}\ s^{-1}$ similar to that of gold and no surface collision damping taken into account ($\gamma_s = 0$). (D) The same with bulk damping constant $\gamma_0 = 1.2 \times 10^{14}\ s^{-1}$ similar to that of gold with surface collision damping taken into account. (E) An "ideal metal" with no bulk damping ($\gamma_0 = 0$), with surface collision damping taken into account.

Conclusions

In this work we have considered the impact of non-local effects on the properties of propagating SPPs. We have shown that the increase in loss is due to the final extent of the optical field, rather than due to collisions with the surface. In other words, surface collision damping/broadening is better described as time-of-flight broadening. We have obtained a full quantum-mechanical expression for the damping rate and corresponding changes in the imaginary part of the dielectric constant. We have confirmed that the increased damping is a non-local effect that follows naturally from the Lindhard theory of the wavevector-dependent dielectric constant and does not have to be introduced phenomenologically. We have shown that nonlocality – engendered change in the imaginary part of dielectric constant exerts a much stronger influence on properties of plasmonic structures than the dispersion of the real part.

We then applied the theory to the case of propagating SPPs and have shown that not only does surface collision damping increase loss, but it actually prevents the field from being concentrated into the tight regions. As a result, even if the bulk scattering of the metal had been completely eliminated, one would have not been able to concentrate the field into the regions that are substantially tighter than roughly $\lambda/4.5n_d$ *i.e.* exceeding the diffraction limit only by a factor of few. One can show that similarly, in the case of dimers, not only will the line width broaden as two nanoparticles get closer,[14] but the mode itself will expand. This can be simply understood using a coupled mode analysis[23] in which the gap mode in the dimer consists of a superposition of multipole modes. Since higher order modes are strongly confined near a surface, they are damped relative to lower order modes, and as a result they do not get excited as easily as lower order modes.

These results raise a very important question about the impact of the efforts to reduce bulk loss in metals. It appears that eliminating the defects and imperfections of fabrication process, and, hypothetically, reducing phonon and electron–electron scattering will not reduce loss significantly relative to today's best silver for the substantially (factor of few) sub-wavelength structures. There are of course other compelling reasons for looking at different materials, such as cost and compatibility with CMOS processes for integration of plasmonics with electronics. Furthermore, when it comes to the structures that are not tightly confined such as long range SPPs [Bern] the already relatively low loss can be further reduced using materials with a smaller bulk loss. But once the confinement gets really tight, surface collision damping makes losses high and is nearly independent of bulk losses. It seems that as long as there are two electronic states separated by photon energy, one occupied and one empty, there will always be a transition between two of them, and hence absorption. The only way to avoid all losses, including Landau damping is to develop materials with narrow conduction bands, such that there is no empty state within $\hbar\omega$ from the bottom of the band,[24] so that a transition cannot take place by any means. Perhaps that is where the effort to develop low loss materials should be directed.

Acknowledgements

The author acknowledges fruitful discussions with A. V. Uskov.

References

1 M. Stockman, Nanoplasmonics: past, present, and glimpse into future, *Opt. Express*, 2011, **19**, 22029–22106.

2 W. L. Barnes, A. Dereux and T. W. Ebbesen, Surface plasmon subwavelength optics, *Nature*, 2003, **424**, 824.

3 J. B. Khurgin and G. Sun, Scaling of losses with size and wavelength in nanoplasmonics, *Appl. Phys. Lett.*, 2011, **99**, 211106.

4 J. B. Khurgin and A. Boltasseva, Reflecting upon the losses in plasmonics and metamaterials, *MRS Bull.*, 2012, **37**, 768–779.

5 *Optical Properties of Solids*, ed. Abeles, North-Holland, Amsterdam, 1972.

6 Y. Wu, C. Zhang, N. M. Estakhri, Y. Zhao, J. Kim, M. Zhang, X.-X. Liu, G. K. Pribil, A. Alù, C.-K. Shih and X. Li, Intrinsic optical properties and enhanced plasmonic response of epitaxial silver, *Adv. Mater.*, 2014, **26**, 6106–6110.

7 P. M. Schwab, *et al.* Linear and nonlinear optical characterization of aluminum nanoantennas, *Nano Lett.*, 2013, **13**, 1535–1540.

8 A. Boltasseva and H. A. Atwater, Low-loss plasmonic metamaterials, *Science*, 2011, **331**, 290.

9 J. Zuloaga, E. Prodan and P. Nordlander, Quantum description of the plasmon resonances of a nanoparticle dimer, *Nano Lett.*, 2009, **9**, 887.

10 F. J. Garcıa de Abajo, Nonlocal effects in the plasmons of strongly interacting nanoparticles, dimers, and waveguides, *J. Phys. Chem. C*, 2008, **112**, 17983–17987.

11 N. A. Mortensen, S. Raza, M. Wubs, T. Søndergaard and S. I. Bozhevolnyi, A generalized nonlocal optical response theory for plasmonic nanostructures, *Nat. Commun.*, 2014, **5**, 3809.

12 A. Larkin and M. I. Stockman, Imperfect perfect lens, *Nano Lett.*, 2005, **5**, 339–343.

13 U. Kreibig and M. Vollmer, *Optical Properties of Metal Clusters*, Springer-Verlag, Berlin, 1995.

14 G. Toscano, S. Raza, W. Yan, C. Jeppesen, S. Xiao, M. Wubs, A.-P. Jauho, S. I. Bozhevolnyi and N. A. Mortensen, Nonlocal response in plasmonic waveguiding with extreme light confinement, *J. Nanophotonics*, 2013, **2**, 161–166.

15 N. A. Mortensen, Nonlocal formalism for nanoplasmonics: phenomenological and semi-classical considerations, *Phot. Nanostr.*, 2013, **11**, 302–308.

16 A. V. Uskov, I. E. Protsenko, N. A. Mortensen and E. P. O'Reilly, Broadening of plasmonic resonance due to electron collisions with nanoparticle boundary: a quantum mechanical consideration, *Plasmonics*, 2014, **9**, 185–192; I. E. Protsenko and A. V. Uskov, Photoemission from metal nanoparticles, *Phys.-Usp.*, 2012, **55**, 508–518.

17 G. D. Mahan, *Many-particle physics*, Kluwer Academics, New York, 2000.

18 P. Johnson and R. Christy, Optical constants of the noble metals, *Phys. Rev. B: Condens. Matter Mater. Phys.*, 1972, **6**, 4370–4379.

19 J. B. Pendry, Negative refraction makes a perfect lens, *Phys. Rev. Lett.*, 2000, **85**, 3966.

20 N. Fang, H. Lee, C. Sun and X. Zhang, Sub-diffraction-limited optical imaging with a silver superlens, *Science*, 2005, **308**, 534.

21 B. Zhang and J. B. Khurgin, Eigen mode approach to the sub-wavelength imaging with surface plasmon polaritons, *Appl. Phys. Lett.*, 2011, **98**, 263102.
22 H. T. Miyazaki and Y. Kurokawa, Squeezing visible light waves into a 3-nm-thick and 55-nm-long plasmon cavity, *Phys. Rev. Lett.*, 2006, **96**(9), 097401.
23 G. Sun, J. B. Khurgin and A. Bratkovsky, Coupled-mode theory of field enhancement in complex metal nanostructures, *Phys. Rev.B: Condens. Matter Mater. Phys.*, 2011, **84**, 045415.
24 J. B. Khurgin and G. Sun, In search of the elusive lossless metal, *Appl. Phys. Lett.*, 2010, **96**, 181102.

Faraday Discussions

DISCUSSIONS

Plasmonic and new plasmonic materials: general discussion

F. Javier García de Abajo, Riccardo Sapienza, Mikhail Noginov, Felix Benz, Jeremy Baumberg, Stefan Maier, Duncan Graham, Javier Aizpurua, Thomas Ebbesen, Anatoliy Pinchuk, Jacob Khurgin, Katarzyna Matczyszyn, James T. Hugall, Niek van Hulst, Paul Dawson, Christopher Roberts, Michael Nielsen, Luca Bursi, Michael Flatté, Jun Yi, Ortwin Hess, Nader Engheta, Mark Brongersma, Viktor Podolskiy, Vladimir Shalaev, Evgenii Narimanov and Anatoly Zayats

DOI: 10.1039/c5fd90022k

Jeremy Baumberg opened a general discussion of the Introductory Lecture by Mark Brongersma: Can you see the dynamics of the forming process of these Ag nanowires in real time using EELS?

Mark Brongersma answered: This is a great question. We have not yet been able to see the forming process in real time, but this would potentially provide exciting new insights into the optical properties of small metallic junctions.

Jeremy Baumberg commented: Can we expect to see quantum effects at the final stages of the metal nanowire connection in these memristor devices at room temperature?

Mark Brongersma replied: Quantum mechanical effects are expected at the final stages of the filament growth based on recent theoretical and experimental work. We have not been able to see this yet.

Niek van Hulst remarked: Following the question by Jeremy Baumberg, your plasmonic EELS maps show that the phase can shift up to even $\pi/4$ upon gradual growth of the junction to the second rod. How can the phase effect be so big? What might be happening in the gap: is one junction at a specific hotspot dominating or rather are many smaller junctions building up the effect?

Mark Brongersma answered: The phase pickup is indeed very large. This is due to the fact that an electrical connection between the two wires can dramatically change the flow of electron (plasmon) currents in the structure. As a result, large changes in the reflection phase can be obtained.

F. Javier García de Abajo commented: The history of plasmonics has received an important contribution from U. Fano,[1] where he explained and classified Wood anomalies according to the source originating them. He discussed lattice resonances and surface modes (*i.e.*, plasmons) in the description of those anomalies.

1 U. Fano, *J. Opt. Soc. Am.*, 1941, **31**, 213.

Mark Brongersma answered: Thank you for providing this valuable reference, which I had not seen before.

Stefan Maier asked: Can you comment on your view of the application potential of plasmonics in photovoltaics? I note that you did not mention it prominently in your roadmap.

Mark Brongersma replied: The focus in the PV industry is currently on very high efficiency devices. In my opinion, plasmonics can not boost the efficiency of such devices. There are however great opportunities for enhancing the performance of photoelectrochemical cells.

Jacob Khurgin opened a general discussion of the paper by Nader Engheta: Nader, regarding longitudinal resonances, I think one should distinguish between "natural" epsilon-near-zero (ENZ) materials, like doped semiconductors where longitudinal resonances are always excited, and metamaterials where one has to be very careful and probably perform analysis with exact structures rather than an effective medium.

Nader Engheta answered: If you are referring to the emission from a dipole in these media, then you are partially right. For the emission from an emitter in the medium, when one deals with "natural" ENZ materials, the situation is different from the case of dipole in the "engineered structures" and "metamaterials" that "effectively" behave as ENZ according to homogenization methods. In the case of "engineered structures", the emitter is influenced by the full local density of states, which depends on many factors including the local medium and local constituent parameters (which is not zero), the scattering and diffraction from all the inclusions in the vicinity of, and far from, the emitters, *etc.* This full density of states is not based on the effective medium, and has to be calculated based on the information from all the local inclusions (near and far), *etc.* That is what we have done. However, when you deal with the interaction of waves coming from outside of the structure formed by subwavelength inclusions and subwavelength periodicities, and then being scattered by the block of structures with "effective" ENZ behavior, then the reflection and transmission can be evaluated based on the "homogenized" values of the structure. However, please note that for natural ENZ materials, the longitudinal resonances do not have to be always excited when the wave is illuminating the material from outside.

Michael Flatté asked: How sensitive are your calculations to the transition region between the region where ε and μ are 1, and the region where they are 0?

Given that a phase velocity for $\varepsilon = \mu = 0$ diverges, have you taken retardation into account in your calculations?

Nader Engheta responded: The calculation is not sensitive to such transition region, and indeed the retardation has been taken into account. We have taken into account the dispersion characteristics of the epsilon-and-mu-near-zero (EMNZ) material at the frequency for which we have the EMNZ properties, and thus indeed the retardation is taken into account. While the phase velocity in such medium is high at this frequency of interest, the group velocity is indeed lower than the vacuum velocity of light (as we have examined in our transient simulations not discussed here), and thus the results indeed satisfy the causality and the Kramers–Kronig relations.

Ortwin Hess said: Placing an emitter inside an epsilon-near-zero and/or mu-near-zero environment is a very interesting concept where many traditional assumptions such as Fermi's Golden Rule many need to be re-visited. Will the emitter, as considered in your present study, be assumed to "see" the environment at the moment of the emission?

Nader Engheta responded: This is a very good question. Indeed, this is a very interesting problem. In our ongoing work, we are studying various aspects of the problem of emitters in the presence of these "extreme" metamaterials. In general, the emission depends on the full density of states at the location of emitters, for which one needs to take into account many factors such as local environment parameters, the inclusions in the vicinity of the emitter, the scattering and diffraction of waves from the near- and far-zone inclusions (*i.e.*, the so called "back-action" on the dipole), *etc.* So the near- and far-field interaction among inclusions and emitters is important in evaluating the emission from such emitters. An interesting point to keep in mind is that, as mentioned above by Jacob Khurgin, one should distinguish between the case of "natural ENZ materials" and the case of "engineered structures" (*i.e.*, "metamaterials") that behave "effectively" as ENZ when one does homogenization. For the emitter, one needs to evaluate the full local density of states, which as I mentioned earlier, should take many factors into account. However, for the "outside" observer and for the reflection and transmission of waves from this structure, then the entire structure looks like behaving "effectively" as ENZ.

Christopher Roberts commented: These simulations presented in this work appear to be two-dimentional simulations. Were any full-wave three-dimensional simulations performed? How would this affect the results presented?

Nader Engheta answered: Yes, the simulation results we have in the *Faraday Discussions* manuscript (*Faraday Discuss.*, 2015, DOI: 10.1039/C4FD00205A) are all two-dimensional. However, as part of our ongoing work on this subject, we have also done full-wave three-dimensional simulations in various cases. For the most parts, the results are analogous.

Jeremy Baumberg opened a general discussion of the paper by Evgenii Narimanov: What is the effect of vertical and/or lateral disorder in these multilayers?

Does it collapse the bandgap or induce additional coupling between bands? How difficult will it be to make such multilayers in practise? What are the key parameters to hit?

Evgenii Narimanov answered: Let's start with the "vertical" disorder - *i.e.* thickness variations of layers forming the (planar) hypercrystal. Yes, such a disorder would indeed collapse the bandgap. However, the hybrid Tamm-plasmon surface states, in my view, that the most interesting features will survive. If anything, the disorder that drives the system to localization will only "help" to localize them even stronger - thus in fact helping to form the surface states rather than destroying them.

The "lateral"/"in-plane" disorder is a very different story. Looking at the resulting behavior from the perturbative point of view, we expect it to result in out-of-plane scattering, thus reducing the surface mode lifetimes. However, with "metallic" elements being one of the necessary building blocks for a hypercrystal (we need a negative component in the permittivity for hyperbolic dispersion!), strong material absorption is inevitable in such media. As a result, for a modest disorder, the resulting mode lifetime is dominated by the material loss, with the scattering contribution being a (relatively) small correction.

Mikhail Noginov remarked: Please, describe the Dirac diffraction in hypercrystals.

Evgenii Narimanov replied: Planar hypercrystals may support an even number of Dirac dispersion points. A pair of such Dirac cones is created at the point of photonic bandgap collapse in what I refer to as the Motti singularity.

Mark Brongersma opened a general discussion of the paper by Viktor Podolskiy: How the nonlocal effects in your presented metal/dielectric nanostructures will show up in real experiments? How can they be measured and how can they be controlled to create optical devices with an improved performance?

Viktor Podolskiy replied: The nonlocality introduces additional TM-polarized waves to metamaterials. As a result, transmission and reflection spectra for TM-polarized light carry signatures of interference of multiple TM-polarized waves. In transmission, this effect was demonstrated in metamaterials experimentally in ref. 1 and 2; in reflection, the effect is detailed theoretically in ref. 3. In homogeneous materials, the effect of nonlocality on optical properties dates back to 1960s; see *e.g.* the work of Hopfield and Thomas;[4] a good summary of nonlocality in bulk materials can be found in the book by Agranovich and Ginzburg, Crystal Optics with Spatial Dispersion and Excitons.[5] The drastic difference between nonlocality in homogeneous media studied previously and nonlocality in metamaterials lies in the origin of additional waves. In homogeneous materials, the nonlocal response is attributed to the spatial dispersion of the material itself. In metamaterials, however, nonlocality appears at the "effective medium" level; the response of every component of the metamaterial may remain local, while the granularity of the composite leads to the spatial dispersion of the effective permittivity. Whether nonocal effects are advantageous or detrimental for applications depend on a particular application. They are advantageous for, *e.g.*, nonlinearity designs, as shown experimentally, and perhaps for multiplexing in optical signal processing, as theoretised, where an additional wave can be used as an additional communication channel.

1 R. J. Pollard, A. Murphy, W. R. Hendren, P. R. Evans, R. Atkinson, G. A. Wurtz, A. V. Zayats and V. A. Podolskiy, *Phys. Rev. Lett.*, 2009, **102**, 127405.
2 G. A. Wurtz, R. Pollard, W. Hendren, G. P. Wiederrecht, D. J. Gosztola, V. A. Podolskiy and A. V. Zayats, *Nat. Nanotechnol.*, 2011, **6**, 107.
3 S. Inampudi, D. C. Adams, T. Ribaudo, D. Slocum, S. Vangala, W. D. Goodhue, D. Wasserman and V. A. Podolskiy, *Phys. Rev. B*, 2014, **89**, 125119.
4 J. J. Hopfield and D. G. Thomas, *Phys. Rev.*, 1963, **132**, 563.
5 V. M. Agranovich and V. L. Ginzburg, Crystal Optics with Spatial Dispersion and Excitons, Springer Series in Solid-State Sciences, Springer-Verlag, Berlin, 1984, Vol. 42.

Mark Brongersma asked: How does the emission of a quantum emitter change as it is moved from outside of the presented material to the inside? What happens when emitters are placed at the interfaces?

Viktor Podolskiy responded: When the dipole is positioned in the far-field of the metamaterial, the effect of the metamaterial is essentially limited to the reflection of the dipole radiation. When the dipole is inside the metamaterial, it is able to couple to the strongly enhanced density of states that exists in the metamaterial. As the dipole moves through the interface, there is a transition from one regime to the other. Qualitatively, the effect is somewhat similar to what is happening when the dipole is moved from air to the proximity of the plasmonic layer. Quantitatively, the situation is different, of course. The exact dependence of the emission rate on the height is a function of multiple parameters (geometry of the unit cell, material parameters, wavelength) that should be analyzed for a particular system.

Duncan Graham remarked: What was the material used to fabricate the nanowires?

Viktor Podolskiy answered: This work is the theoretical study of dipole emission in wire media. Here we consider a model situation where the operating frequency and geometrical parameters of the wires are fixed, and the permittivity of the wire is changed. Real nanowire composites with both gold and silver wires have been experimentally demonstrated. See, *e.g.* ref. 1 and references therein.

1 A. V. Kabashin, P. Evans, S. Pastkovsky, W. Hendren, G. A. Wurtz, R. Atkinson, R. Pollard, V. A. Podolskiy and A. V. Zayats, *Nat. Mater.*, 2009, **8**, 867.

Ortwin Hess asked: There is a question related to the geometry of the setup: does the emission occur towards the top and the bottom? What is the propagation direction?

Viktor Podolskiy answered: Yes, emission occurs all around the dipole; on the implementation level, we use plane-wave expansion of Green's function to calculate the emission spectrum of the dipole. When the emission into lossy materials is analyzed, the dipole is positioned in the planar cavity cut-out inside the metamaterial. In this situation, the transfer matrix formalism is used to take into account the reflection and transmission of individual plane waves from the cavity into the metamaterial stack.

Ortwin Hess asked: Which assumptions have been made concerning the boundary conditions? Is there anisotropy in any direction?

Viktor Podolskiy answered: There is anisotropy and nonlocality in the metamaterial; as a result, the metamaterial supports two transverse-magnetic (TM)-polarized modes and one transverse-electric (TE)-polarized mode for each wavenumber; in general, all three waves have different dispersions. When the dipole is positioned inside the lossless metamaterial, the two TM-polarized modes are assumed to represent two independent emission channels. When the dipole is positioned inside the cavity, the continuity of the tangential component of the electric field and the continuity of the normal component of the electric displacement are used as the foundation for the boundary conditions. Additional boundary conditions (required to calculate reflection/transmission of light through the interface between local and nonlocal regions of space) arise from enforcing the continuity of $\langle D(x)\exp(2\pi ix/a)\rangle$.

Jeremy Baumberg enquired: Are these effects influenced by the length of the wires? Are they assumed to be infinite? What about the position of the dipoles relative to the wires, both laterally and, more importantly, vertically? If they are near the end, would the effects be very different?

Viktor Podolskiy replied: These are important questions. In this work, wires are assumed to be infinite. When wires have finite length, the overall shape of rate enhancement will be modulated by the Fabry–Pérot type resonances of the composite. The formalism presented in this work can be used to analyze the dependence of lifetime dynamics on the vertical position of the dipole. A detailed study of these phenomena will be presented elsewhere. As far as the lateral position is concerned, the position of the dipole inside the unit cell should modulate the coupling of the emitted radiation to the plasmonic modes of the nanorods. Hand-wavingly, dipoles positioned closer to the wires should exhibit stronger coupling to cylindrical surface plasmons and a stronger Purcell factor. However, the current implementation of the nonlocal EMT presented here is not sensitive to the lateral position of the dipoles. Therefore, the emission rate enhancement calculated in this work represents some weighted average of the emission rate reductions of dipoles positioned at different lateral locations within the unit cell. Further studies are needed to determine the exact lifetime dynamics as a function of the lateral position of the dipole.

Jeremy Baumberg remarked: Is there anything that you can see from inside that will be different from this structure in the hyperbolic regime? What property of the dipole might actually change when placed inside, especially in this regime?

Viktor Podolskiy replied: Probably the most interesting result of our work is the fact that there appears to be virtually no difference between dipole emission in hyperbolic *vs.* elliptic regime. We understand this fact as follows: enhancement of the photonic density of states originates not from hyperbolicity but rather from coupling of light to the plasmonic modes supported by the composite. From the effective medium standpoint, plasmonic modes are represented by a combination of longitudinal and transverse modes. When the metamaterial operates in the elliptic regime, the "additional mode" is "dark"; it weakly couples to plane waves with relatively low transverse wavenumbers; it becomes "bright" in the hyperbolic regime. Dipole emission, however, presents a significant contribution to high-transverse-wavenumber spectral components. Therefore, dipole radiation can strongly couple to plasmonic modes in both the elliptic and hyperbolic regimes, which is reflected in the predictions of the nonlocal effective medium theory shown in this work.

Jacob Khurgin said: I doubt the validity of the effective medium theory when it comes to light emission properties of metamaterials, in particular the hyperbolic ones. It is not that difficult to use one of the methods of solid state physics to determine, not just the exponential part of the wave function, but also the periodic "atomic" function – which would look like a periodic array of coupled slab surface plasmon polariton (SPPs). Then, one can get all the needed properties. In solid state physics, the band structure (dispersion curves) may be useful to determine transport properties but for everything else you must know the "atomic" or "tight binding" orbitals.

Evgenii Narimanov commented: It really depends on the nature of the emitter. If it is a quantum dot that is large compared to the metamaterial unit cell size, the effective medium would do quite fine.

Anatoly Zayats added: In the conventional, local, effective medium description, the electromagnetic field inside the metamaterial is homogenous. The nonlocal effective medium theory (EMT) used in the presented paper takes into account the mode-specific distribution of the electromagnetic fields across the unit cell that essentially play the role of the atomic wavefunctions mentioned in this question. In the case of nanorod-based metamaterials, these are cylindrical surface plasmons of individual nanorods. In the case of a multilayer slab, it will be a coupled SPPs.

Riccardo Sapienza addressed Viktor Podolskiy and Nader Engheta: Effective medium theories are very accurate when dealing with waves transmitted and reflected from a metamaterial as, in particular, for hyperbolic media. Instead, a dipolar emission probes the local properties inside the medium and the full local density of states, including the dipole field back-action on itself. How far can we extend the metamaterial averaging without losing in the approximation the most important features of light emission?

Viktor Podolskiy responded: Local effective medium theories (EMTs) are relatively accurate when analyzing refraction/transmission of waves that enter a material at close to normal incidence. As the angle of incidence (transverse wavenumber) increases, the accuracy of EMTs is diminished. Dipoles (and other small objects) tend to generate waves with a broad spectrum of transverse wavenumbers. Therefore, conventional local EMTs tend to be inaccurate when describing the emission of dipoles inside metamaterials. Quantitative deviation between predictions of EMTs and properties of real composites depends on multiple parameters that include the geometry of the unit cell, the operating frequency, and the material parameters (in particular, losses). In nanowire composites, small losses increase the deviations from local EMTs (see, *e.g.*, ref 1). Interestingly, in the low-loss limit, reflection/transmission, even for moderate angles (20–40 degrees from the normal), is not described by local EMTs, especially in the elliptic regime. However, nonlocal EMT does a good job in this situation. From a microscopic perspective, a periodic nanorod composite should be considered as a photonic crystal with subwavelength period; the field inside this composite should be represented as a linear combination of the modes of this crystal, and boundary conditions should be used to calculate the amplitudes of the modes. Nonlocal effective medium theory takes into account two of these modes. Our results show that these two modes dominate the reflection and transmission of light by the composites in the limit when the unit cell is much smaller than the (vacuum) wavelength. As the unit cell size becomes comparable to the wavelength, higher-order modes, corresponding to higher bands of the photonic crystal, need to be considered.

1 B. M. Wells, A. V. Zayats and V. A. Podolskiy, *Phys. Rev. B*, 2014, **89**, 035111.

Nader Engheta added: This is a very good point. When a dipole is inserted in a "structure" or in a metamaterial, which is formed by collections of inclusions, the emission of the dipole is influenced by the full local density of states that take into account everything there, *i.e.*, local properties, near-field and far-field scattering and diffraction from all the inclusions (*i.e.*, as you say the "back-action" of the wave on the dipole), *etc.* So, to evaluate these features of the emission, one needs to know all the features (near and far) and the full local density of states. From the "outside", however, when we evaluate the reflection and transmission of waves from the structure formed by sub wavelength inclusions and sub wavelength periodicities, then the effective-medium theory can be useful, as you mentioned.

Ortwin Hess asked: Placing an emitter in the system does trigger the question of whether the emitter does "see" the environment at the time of the emission. How is the emission coupled back to the emission state of the emitter? Is the emitter strongly coupled to the environment?

Viktor Podolskiy replied: The emitter obviously "sees" the environment. From the quantum electrodynamics (QED) standpoint, the lifetime of the excited state is inversely proportional to the density of optical states in the surrounding medium. High-index materials, as well as plasmonic systems, are known to provide an increase of the density of states and, thus, an increased emission rate. In the quasiclassic (weak coupling) limit, the enhancement of the emission rate can be calculated using Green's function approach. The effect of the environment is, thus, explicitly included. In particular, in the model used in this work, the effect of the environment is related to the dispersion of the modes supported by the composite. The broader the wavenumber spectrum of the propagating modes, the stronger the emission rate enhancement. We have not considered the strong coupling regime.

F. Javier García de Abajo addressed Viktor Podolskiy and Nader Engheta: When studying the emission from an emitter placed inside a homogeneous material, it is important to consider the emission dipole, which takes an emitter-dependent value in free space, whereas the effective dipole inside the material is corrected by the local field factor $3/(2+1/\varepsilon)$, as pointed out by Yablonovitch *et al.*[1] This factor is derived, for example, when analysing the emitter inside an empty spherical hole surrounded by the material, in the limit in which the radius vanishes. There is also an index-of-refraction factor correcting the emission rate. How much are these factors affecting the strength of the emission? Is it suppressed in the $\varepsilon \rightarrow 0$ limit? Is this perhaps indicating that any emission is produced by deviations from homogeneity, and therefore, the emission characteristics can be used to sense the local environment?

1 E. Yablonovitch *et al.*, *Phys. Rev. Lett.*, 1988, **61**, 2546.

Viktor Podolskiy answered: In our work, we aim to calculate the emission rate increase as compared to the dipole emission in vacuum. For a dipole positioned inside lossless materials, such rate enhancement can be calculated by considering the imaginary part of Green's function of the dipole. Nonlocal field corrections are required in this case. When dipole is positioned inside a lossy material, the Green's function technique is not applicable since any loss yields a singularity of the imaginary part of Green's function. To avoid this singularity, the dipole is then positioned inside a cavity cut-out within the (lossy) material. The local field correction is then required to take into account the effect of such cavity. For spherical cavities, the correction is the expression in the question. For slit-shaped cavities, as used in our work, the correction is $1/(\varepsilon_{zz})^2$ for z-polarized dipoles and 1 for x-polarized dipoles. The validity of such correction can be seen by comparing Fig. 4 and the bottom row of Fig. 5 of our *Faraday Discussions* manuscript (*Faraday Discuss.*, 2015, DOI:10.1039/C4FD00186A).

As far as the emission in the ENZ regime is concerned, in the absence of a longitudinal wave, the (local) effective medium theory predicts a singularity of rate enhancement when $\varepsilon_{zz} \rightarrow 0$. However, when a longitudinal wave is taken into account, this singularity is removed (see Fig. 4–6, *Faraday Discuss.*, 2015, DOI:10.1039/C4FD00186A). In the effective medium approach presented here, the emitter positions are averaged over the unit cell of the nanorod lattice. In the microscopic approach, the rate depends on the position of the emitter within the cell. Surely, sensing the local environment by analyzing the emission of dipoles embedded into the (meta)materials is interesting and an important study. We are planning to report on this in the near future.

Nader Engheta answered: We are currently studying the various issues related to the interaction of an emitter with its environment, particularly the case of "extreme metamaterials". Clearly the full local photon density of states, which depends on many factors including the local relative permittivity, influences the emission of the emitter. An interesting point to consider is that, in our study, the epsilon-near-zero (ENZ) regions are limited in space (*i.e.*, they are not infinitely extended spatially), and this issue also affects the local density of states and the emission of the emitter. When you are referring to the suppression of the emission in the limit of $\varepsilon \rightarrow 0$, one needs to take into account that the structures are finite in size. So the emission of the emitter may also be affected (*e.g.*, not fully suppressed) due the finiteness of the structure.

F. Javier García de Abajo addressed Nader Engheta: In the study of the interaction of electron beams with plasmons, Ferrel predicted a particular kind of coupling to bulk plasmons in a thin film,[1] which should be observed in both the energy-loss/momentum distribution of the inelastically scattered electron and also in the emission of radiation, as these losses are confined within the light cone. These losses, associated with the excitation of bulk plasmons by the electron, were later observed in experiments (*e.g.*, see the work of Vincent and Silcox[2]). As these plasmons correspond to $\varepsilon = 0$, and have essentially longitudinal nature, these studies bear some relevance on the presented work and the ensuing discussion: they essentially reveal the existence of longitudinal modes within the light cone, and therefore, their ability to couple to radiation.

1 R. A. Ferrel, *Phys. Rev.*, 1958, **111**, 1214.
2 R. Vincent and J. Silcox, *Phys. Rev. Lett.*, 1973, **31**, 1487.

Nader Engheta answered: This is an interesting point, and we thank you for bringing to our attention these references, which can be useful in our ongoing study of light emission and interaction with ENZ and MNZ media.

Mark Brongersma asked Viktor Podolskiy: How should fluorence decay rates be calculated in a nanostructured metamaterial? Can one use conventional techniques such as finite-difference time-domain simulations (FTDT) or are such simulations fundamentally flawed for this type of calculation? If they are flawed, what should be the approach to calculate decay rates?

Viktor Podolskiy answered: FDTD and finite element method (FEM) calculations, in principle, can be used to estimate emission rates. However, extreme care should be taken when using these techniques due to complex scale separation. In particular, one should ensure that interaction of a single dipole with the metamaterial is analyzed. Therefore, multiple unit cells of the metamaterial (and a single dipole) should be included in the geometry, and the convergence of results with respect to the metamaterial "size" should be analyzed. The mesh should be small enough to resolve both the dipole and metamaterial components. As a rule, such computations are prohibitively resource-intensive even on today's computers. The nonlocal effective medium theory presented in our work provides an estimation of emission rates for composite media, avoiding the need for resolving the structure of the composite in multiscale computations.

Javier Aizpurua enquired: One needs to be careful when near-field properties are addressed with the use of nonlocal effective medium responses. What would happen, for example, if one had a surface separating vacuum and one of these nonlocal metamaterials, and an emitter were located in front of it? What kind of modes would that emitter excite? Would it excite for example the longitudinal modes from outside? That would not make too much sense, I believe, but maybe those modes could arise artificially from the construction of the nonlocal response. Barrera and Fuchs performed such a study some years ago, and this kind of strange situations were generated. How do you see this evaluation of modes in the near-field based on metamaterial nonlocal responses?

Viktor Podolskiy replied: From the microscopic perspective, the transverse and longitudinal modes of the nanorod-based metamaterial are related to coupled cylindrical surface plasmons (CSPs) of the nanorods. If the dipole is positioned in the near-field proximity of such metamaterial, its emission would couple to the plasmonic modes of the nanorods. The developed nonlocal effective medium description explicitly takes into account the plasmonic modes when constructing the field in the metamaterial and is in agreement with the microscopic picture. According to the effective medium picture, radiation from a dipole is coupled to both "main" and "additional" waves supported by the metamaterial nanowire. We find such coupling not only physical, but also necessary. This behavior is consistent with the work by Barrera and Fuchs[1] that considers nanosphere-based metamaterials. Similar to the above paper, in the $k_x = 0$ limit, our nonlocal EMT for nanowires recovers Maxwell Garnett results. However, the crucial difference between nanosphere and nanowire metamaterials is seen for higher wavenumbers. While sphere-based metamaterials exhibit discrete [in k] multipolar resonances, coupling to a longitudinal CSP-based mode in nanowire metamaterials exists for any obliquely propagating beam and depends on the polarization of the incident light. Additional boundary conditions are required to calculate the portion of radiation coupled to main and additional waves. Our approach presents a first-principle recipe to implement additional boundary conditions in metamaterials by requiring the continuity of the averaged components of the tangential component of the electric field, the normal component of the electric displacement, and the product of the normal component of the electric displacement multiplied by the block wave profile. Such implementation of additional boundary conditions explicitly takes into account the distribution of electromagnetic fields of both modes across the unit cell of the metamaterial.

1 R. G. Barrera and R. Fuchs, *Phys. Rev. B*, 1995, **52**, 3256.

F. Javier García de Abajo addressed Nader Engheta: The homogeneous electron gas provides a good description of some good metals, such as aluminium. This model provides an example of a system that is homogenous (so, it does not include spatial corrugations) and is however nonlocal. The decay of an excited atom or molecule placed inside one of those metals should therefore provide insight into the influence of nonlocal effects on the transition rate and optical emission characteristics, for example when the emission is in the region close to the bulk plasmon predicted in the mentioned model, and particularly depending on the distance to the surface and the shape of the metal (*e.g.*, a thin film).

Nader Engheta responded: This is a very good point, and indeed such a study will be very useful in exploring the light-matter interaction for the "extreme" metamaterials we are studying.

Nader Engheta added: A question was asked about the group velocity of waves in ENZ media. Considering the Drude dispersion for an ENZ medium, at first one may suspect that the group velocity may be zero at frequencies near the plasma frequency. However, this is not so for the following reasons: (1) we do not operate "exactly" at the plasma frequency, and instead we are slightly above this frequency, and thus the group velocity is not exactly zero, and (2) and more importantly, when one adds a small amount of inevitable material loss, one would notice that at the frequency near the plasma frequency (but slightly above the plasma frequency), the group velocity is not zero. This can be easily derived analytically using the Drude model with a small loss included. Furthermore, when the ENZ medium has a finite extent (*i.e.*, not too large compared to the free space wavelength), the structural dispersion should also be taken into account, in addition to the material dispersion.

Katarzyna Matczyszyn opened a general discussion of the paper by Vladimir M. Shalaev: Can you make nanoparticles out of nitrides? Do they need a special coating apart from the oxide layer? The biocompatibility of titanium was mentioned but as a bulk material, what about the cytotoxicity of nitride nanoparticles? Can you make different shapes and sizes of those?

Vladimir Shalaev responded: Yes, colloidal titanium nitride samples can be produced and the shape of the particles are spherical or cubic. The size of the particles can vary from five nanometer to a few hundred nanometers, depending on the fabrication method. We recently reported a native oxide layer of thickness around 1-2 nm and they do not need further coating.[1] Nanoparticles of titanium nitride have not been studied extensively in terms of biocompatibility. However, there are some studies with positive results on the chemical resistance of titanium nitride powders in the body.[2]

1 U. Guler, S. Suslov, A. V. Kildishev, A. Boltasseva and V. M. Shalaev, arXiv:1410.3920, 2014.
2 V. A. Lavrenko, V. A. Shvets, N. V. Boshitskaya and G. N. Makarenko, *Powder Metall. Met. Ceram.*, 2001, **40**, 630–636.

Thomas Ebbesen commented: Beyond solar thermal-voltaics, I think thermal management in general (*e.g.* recovery of energy from very hot engines) could benefit from titanium nitride since it is so stable. So how easy is it to grow the material on various substrates?

Vladimir Shalaev answered: Yes, both selective absorbers and emitters can be fabricated with refractory plasmonic materials, thus they can be used for thermophotovoltaics (TPV) as well. Waste heat recovery is a very interesting application of TPV systems and dictates the coating of a selective emitter over irregular surfaces. This brings a two-fold concern: i) the shape and ii) the crystalline structure of the surface. Fortunately, due to their hardness, transition metal nitrides have been widely used in industry as protective surface coatings and a variety of techniques are available for the deposition of thin films over irregular surfaces. On the other hand, high optical quality films require epitaxial growth which is possible for substrates with matching crystalline structures and lattice constants. Although the latter is a serious limitation for some applications, the case for TPV is different. In short, titanium nitride and zirconium nitride can be easily deposited on irregular surfaces for waste heat recovery with TPV. Further details can be found in a recent review paper.[1]

1 U. Guler, V. M. Shalaev and A. Boltasseva, *Mater. Today*, 2015, **18**, 227–237.

Paul Dawson asked: In addition to titanium nitride (TiN), have you also considered aluminium titanium nitride? This material is also well known in other contexts (*e.g.* coating on tool bits) and looks quite silvery, just as TiN looks very golden. My second comment concerns the good optical properties of thin film TiN, which appear to depend critically on it being highly epitaxial with the MgO or sapphire substrate. Is that correct? You also mentioned TiN nanoparticles/colloids – are these highly crystalline in nature, so that they can have ascribed to them the same optical properties as the thin films (aside from any of the usual (small) size-dependent effects on the optical data)?

Vladimir Shalaev answered: Regarding your first question, it is possible that other metal nitrides are also good plasmonic alternatives. So far, we have studied TiN, ZrN, and HfN as plasmonic transition metal nitrides.

The answer to your second question is yes, thin films with single crystalline properties have better optical properties mostly due to the reduced number of traps that would damp the plasmonic oscillations if they were present. In our studies on nanoparticles of TiN, transmission electron microscopy results confirm the single crystalline properties of colloidal samples. Optical measurements show that the plasmon resonance occurs at the spectral region similar to that of epitaxial thin films.[1]

1 U. Guler, S. Suslov, A. V. Kildishev, A. Boltasseva and V. M. Shalaev, arXiv:1410.3920, 2014.

Felix Benz said: One of the big advantages of gold is that the surface chemistry is extremely versatile. For example, thiol based compounds can be attached to the surface and, starting from these, a large number of molecules can be tethered by simple click chemistry. These molecules can be designed to bind specifically to cancer cells to direct nanoparticles to them. Can you please comment on what is known about the surface chemistry of TiN and how it compares to gold?

Vladimir Shalaev responded: Titanium nitride has a biologically inert surface which makes it the material of choice as a biomedical surface coating. However, the surface chemistry of titanium nitride at the nanometer scale has not been studied widely. We see this as an exciting research opportunity. On the other hand, in our examination of colloidal titanium nitride nanoparticles, we observed a native oxide shell with a thickness of 1–2 nm around the core particle.[1] As a frequently used nanomaterial, TiO_2 surface chemistry has been extensively studied and we believe the presence of the thin layer is likely to provide the versatility required for the surface modification of plasmonic titanium nitride nanoparticles. Considering the advantages of titanium nitride over gold, we believe studies on the surface chemistry of these nanoparticles will follow soon.

1 U. Guler, S. Suslov, A. V. Kildishev, A. Boltasseva and V. M. Shalaev, arXiv:1410.3920, 2014.

James T. Hugall asked: Despite gold being an excellent plasmonic material in the visible EM region, it is clear that unprotected gold nanostructures degrade under only modest increases in temperature, making them unsuitable for a wide range of applications. However, if we were to embed gold plasmonic nanoparticles or structures within a dielectric of a high melting point, we could protect the structures from degradation under temperature rises. Indeed, you could imagine these embedded plasmonic structures maintaining their shape, whilst being able to undergo a phase transition to a liquid or semi-molten state. As the temperature increases, the imaginary part of the dielectric constant of the metal will increase, making the material less favourable for plasmons, but as the metal enters a semi-molten/molten state, can plasmons still exist in such states and what properties might they have?

Vladimir Shalaev answered: In fact, it has been shown that the thermal stability of gold nanoparticles can be enhanced to some extend by encapsulating the particle with a dielectric layer.[1,2] However, the support that can be provided by a coating layer is not sufficient to make gold useful for high temperature applications. Plasmonics with metals in the liquid state is an interesting subject mostly due to the potential for reconfigurable and tunable applications. Studies with metals that experience phase transitions at lower temperatures have shown that plasmonic properties still exist in the liquid phase with some modifications mostly due to electronic band structure broadening.[3,4]

1 C. Radloff and N. J. Halas, *Appl. Phys. Lett.*, 2001, **79**, 674.
2 M. Bosman, L. Zhang, H. Duan, S. F. Tan, C. A. Nijhuis, C. W. Qiu and J. K. W. Yang, *Sci. Rep.*, 2014, **4**, 5537.
3 S. R. C. Vivekchand, C. J. Engel, S. M. Lubin, M. G. Blaber, W. Zhou, J. Yong Suh, G. C. Schatz and T. W. Odom, *Nano. Lett.*, 2012, **12**, 4324–4328.
4 J. Wang, S. Liu, Z. V. Vardeny and A. Nahata, *Opt. Express*, 2012, **20**, 2346–2353.

Michael Nielsen commented: In your paper, you show that TiN lattice-match grown on MgO offers a figure-of-merit (FOM) of absolute value of the real part of epsilon over the imaginary part of ~90% that of Au at a wavelength of 800nm. My two part question is as follows:

1. How does the FOM of TiN compare to Au in other frequency regimes (telecommunications, near infrared, terahertz, etc)?
2. You also show that if TiN is grown on Al_2O_3 instead of MgO, the FOM drops to ~43% of that of Au at 800nm. How much does the growth substrate affect the quality of the TiN films for plasmonic applications? And, since you mention that one of the advantages of TiN over noble metals, beside its refractory nature, is its complementary metal-oxide-semiconductor (CMOS) compatibility (since it is already used as a diffusion barrier for copper interconnections), what is the FOM/quality of the TiN films grown on standard CMOS dielectrics SiO_2 and Si_3N_4 where it does not have the advantage of lattice matching?

Vladimir Shalaev responded: In the near infrared region, including the telecommunication wavelengths, FOM comparison between TiN and Au is mostly the same. We have not studied the properties of TiN in the terahertz regime yet.

Regarding your second question, the quality difference between samples grown on MgO and sapphire is not necessarily so big. In the article, we provide published data for films grown on sapphire. When the growth parameters are optimized, the sapphire sample quality gets much closer to samples grown on

MgO, as our current results indicate. For other substrates where epitaxial growth is not possible, film quality degrades. However, as it was shown earlier, it is possible to obtain lattice matched films on substrates such as Si.[1]

1 J. Narayan, P. Tiwari, X. Chen, J. Singh and R. Chowdhury, *Appl. Phys. Lett.*, 1992, **61**, 1290.

Jeremy Baumberg opened a general discussion of the paper by F. Javier García de Abajo: For monolayers of Au, do we not expect that the real band structure should play a distinct role in the plasmons? For instance spin–orbit coupling *etc.* Quantum confined states have been seen in monolayers of Fe within spintronic devices, and perhaps then we cannot use the simple formalism directly without some subtle modifications?

F. Javier García de Abajo replied: Actually, we do not expect that the real band structure should play a distinct role in the plasmons for the systems that we are considering. In a recent publication,[1] we have theoretically analyzed the plasmons of atomically thin gold islands through the random-phase approximation (RPA) response formalism (see, for example, ref. 2 for a description of the RPA), using as input the electron wave functions of a box potential. These calculations agree well with a classical, Drude description of the material when the size of the islands exceeds ~15 nm. Likewise, in atomic gold wires decorating the steps of vicinal silicon surfaces, the experimentally observed plasmon dispersion agrees reasonably well with both types of descriptions (quantum-mechanical RPA and classical Drude model), as shown by Nagao *et al.*[3] These studies reveal that the details of the band structure are not important at plasmon energies that are large compared to those details, and in particular, the measured spin–orbit splitting in atomically thin gold structures (see, for example, the work by Losio *et al.*[4]) is much smaller than the plasmon energies in the visible and near-infrared. These experimental and theoretical results indicate that the subtle modifications that you mention play a minor role in noble-metal plasmons, although perhaps other materials could show stronger effects.

1 A. Manjavacas and F. J. García de Abajo, *Nat. Commun.*, 2014, **5**, 3548.
2 D. Pines and P. Nozieres, The Theory of Quantum Liquids, W. A. Benjamin, Inc., New York, 1966.
3 T. Nagao *et al.*, *Phys. Rev. Lett.*, 2006, **97**, 116802.
4 R. Losio *et al.*, *Phys. Rev. Lett.*, 2001, **86**, 4632.

Ortwin Hess commented: Fundamentally there is a question that one can and should ask in this circumstance. How many electrons do we actually need for a plasmon? This is perhaps also related to the previous question, not least as in the community one normally would regard and treat optical properties of plasmons or surface plasmon polaritons in terms of a dielectric with loss and a dispersion, rather than a system of many electrons. Thereby, one clearly does avoid to explicitly take on board the influence of dynamic many body aspects, band structure, and scattering processes. It does now seem that even with methods such as DFT one would need to go beyond random phase approximation to give us insight into quantisation elements and optical properties.

F. Javier García de Abajo answered: I refer to my answer to the previous question. Additionally, it is time-dependent density functional theory (TDDFT) what guarantees a full description of the dynamical response of the system, as DFT alone only yields ground state properties. Now, the exact functional of TDDFT is unknown, but there are several public-domain TDDFT codes that accurately explain plasmons, down to systems composed of only a few atoms. Also, I believe that the RPA is generally sufficient to describe the response of many materials of interest. For example, aluminium is well described by the Lindhard formula for a non-interacting electron gas; noble metals then need to be supplemented with a smooth background to describe the d-band polarisation; and the RPA also gives a satisfactory level of description of graphene nanostructures down to very small sizes.

Jeremy Baumberg asked: I am very worried about the rebadging of molecular excited states as plasmons. In the determinant description, when I add an electron and charge the system, do not all the modes change? Is it ever possible to just change the collective mode and not any of the others? Why would we use a "plasmon" nomenclature then for the shifts in all the energy states. Particularly for delocalised electronic states, such as in conjugated polymers, the excited state absorption is well known and not described by any plasmon description. Should we aim to reverse this description?

F. Javier García de Abajo answered: Thank you for this question, because it brings out a very important issue that we did not address in the present paper,

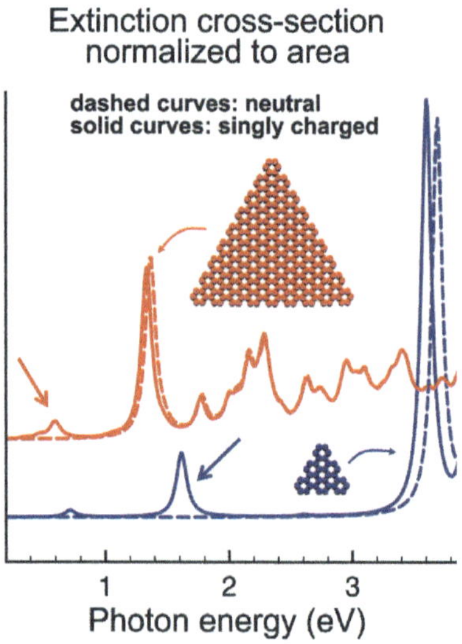

Fig. 1 Extinction cross-section of graphene nanotriangles in neutral (dashed curves) and singly-charged (solid curves) states. The triangles have armchair edges passivated with hydrogens. The number of atoms in the triangles is 36 and 270, respectively. The high-energy parts of the spectra undergo comparatively small variations when adding one electron to the system. In contrast, new plasmon modes emerge at low energy (see thick arrows). These spectra are calculated in the RPA, using tight-binding electronic wave functions, as explained in ref. 5.

although I mentioned it during my talk to illustrate the limit of very small systems. As atomic aspects of graphene and the low-size limit where we encounter polycyclic aromatic hydrocarbons (PAHs) are not covered at all in this paper, I believe it is pertinent to provide the readers with some supporting information.

Just to add some perspective, let me recall that in graphene we find that new plasmon modes emerge if it is electrically doped. Besides, the plasmon energy blueshifts as we depart further from the neutrality point. We also find that these plasmons suffer strong quenching when their energy is below the change in chemical potential due to the coupling to electronic zig-zag edge states (see ref. 1 for a tutorial description of these effects). This behaviour persists in structured graphene down to islands consisting of a small number of carbon atoms (with hydrogen-passivated edges), eventually reaching sizes for which they are known as polycyclic aromatic hydrocarbons. As a matter of fact, we cannot establish a clear criterium based on the plasmon characteristics that I have just outlined allowing us to separate molecular sizes from macroscopic graphene islands. It is in this context that we refer to some of the excitations in molecular radicals as plasmons.[2] But which of the states are plasmons?

To answer this question, it is convenient to bear in mind that the Hartree–Fock (HF) approximation permits defining different many-body electron configurations as Slater determinants, which describe reasonably well many of the molecular excited states, but not some of the states that we refer to as molecular plasmons. The latter require going beyond HF by including the Coulomb interaction between different configurations (*e.g.*, using time-dependent HF or TDDFT), they involve sizeable contributions from more than one of those configurations,[3] and they disappear if the Coulomb interaction is switched off.[2,4] These characteristics are also shared by the plasmons of large graphene islands and metallic nanoparticles. I believe that this new insight, and most importantly, the identification of these modes as the low-size limit of a more general kind of optical excitations that are sustained by electrically doped carbon allotropes,[1] justifies why we refer to them as plasmons. The term "molecular plasmons" provides a link to a wealth of previously unnoticed physics in plasmonics. We explain these arguments in more detail in ref. 2, where we discuss previous literature on optical excitations in neutral and charged molecules.

Regarding the last part of your question, I include a figure showing that only minor changes are produced in the excitation spectra of neutral nanographene structures upon addition of one electron (Fig.1). In particular, new plasmon modes emerge in the charged clusters. This behaviour is qualitatively similar both for a molecule consisting of 36 atoms (blue curves) and for a larger island formed by 270 atoms (red curves). The emerging features have a plasmonic character according to the above criteria, whereas the less electrically tunable features include both high-energy plasmons and excitonic modes of the neutral structure.

As you can see, the identification of this general class of excitations as plasmons enables us to think of new structures capable of sustaining them (from PAHs to nanoribbons and other nanographene structures), thus providing a powerful, innovative platform for electro-optics over a broad spectral range, from the infrared to the visible, with potential applications for light modulation and quantum optics, as we argue in the present paper. I sincerely hope that your worries about the rebadging of molecular excited states as plasmons are dissipated by these arguments, and I refer to the publications cited above for more details.

1 F. J. García de Abajo, *ACS Photonics*, 2014, **1**, 135.
2 A. Manjavacas *et al.*, *ACS Nano*, 2013, 7, 3635.
3 C. M. Krauter *et al.*, *J. Chem. Phys.*, 2014, **141**, 104101.
4 S. Bernadotte *et al.*, *J. Phys. Chem. C*, 2013, **117**, 1863.
5 A. Manjavacas *et al.*, *ACS Nano*, 2012, **6**, 1766.

Javier Aizpurua asked: In connection with the collective electronic state in molecules that is always present as a solution, as you showed in your analogy with the collective modes in nuclear physics, I am wondering whether this collective mode is captured in any standard quantum mechanical calculation based on typical quantum chemistry methods (Gaussian for example giving typical HOMO–LUMO levels). If so, then how can we identify the distinctive nature of this collective mode technically with respect to the other single particle modes?

F. Javier García de Abajo responded: I believe that the term "standard quantum mechanical calculation" is too ambiguous, as there are so many levels of approximation (and methods) for solving the many body problem in condensed matter. Now, plasmons are not captured in the well-known Hartree–Fock approximation, but they emerge in time-dependent Hartree–Fock (TDHF), which leads to the random-phase approximation response function. Essentially, plasmon modes arise when switching on the Coulomb interaction between excited configurations, excited in spring-like motion of the induced charge. In TDHF and TDDFT, plasmons are identifiable as states composed of multiple configurations (Slater determinants); this provides a practical procedure for separating them from excitonic modes.[1] Additionally, plasmons emerge when the Coulomb interaction ν between excited states is switched on (*e.g.*, by moving from the non-interacting susceptibility χ_0 to the RPA susceptibility $\chi = (1 - \chi_0\nu)^{-1}\chi_0$). This is clearly illustrated by the Lindhard response function for the homogeneous electron gas,[2] and it also provides a practical criterium to distinguish plasmons in molecules.[3,4] Incidentally, you mention Gaussian, which is a general platform rather than a method, allowing us to perform calculations with many different methods, including HF and TDDFT, where the above criteria can be readily evaluated. In brief, plasmons are captured in some of the quantum chemistry methods, but not in others, and we can identify them with the above criteria.

1 C. M. Krauter *et al.*, *J. Chem. Phys.*, 2014, **141**, 104101.
2 J. Lindhard, *K. Dan. Vidensk. Selsk. Mat. Fys. Medd.*, 1954, **28**, 1.
3 S. Bernadotte *et al.*, *J. Phys. Chem. C*, 2013, **117**, 1863.
4 A. Manjavacas *et al.*, *ACS Nano*, 2013, 7, 3635.

Javier Aizpurua commented: You asked the audience how many electrons are needed to have a plasmon, and I would like to add a second complementary question to that one: how many more electrons can stand a localized plasmon? We know that small particles can host very few additional electrons. Are the atomic-thick metallic layers that you propose different to that regard to 3D systems because of the different morphology? In other words, how many electrons would a finite monoatomic metallic layer be able to host before the next electron is expelled out from the structure?

F. Javier García de Abajo answered: Experiments have been reported for gold nanoparticles in which the plasmons are shown to shift when injecting a number of additional electrons of the order of 10% of the existing valence electrons.[1] This experimentally demonstrated fraction of injected electrons is larger than what we are considering in our work. Additionally, it is important to note that the added charges tend to pile up in the outermost atomic layer of a metal,[2] at least in a linear doping regime, so these experiments for 3D particles provide a conservative estimate of the doping densities supported by our atomically thin metal structures.

Obviously, the electrical doping density that is reachable in a metallic nanostructure depends on the chemical and physical environment (for instance, passivated structures should be more robust), but based both on the stability confirmed by the above experiments, as well as on demonstrated devices for gate-based doping,[3] we are considering attainable conditions in the present paper, as we assume similar doping densities as those produced with gating devices in nanographene studies.[4]

I believe the real problem is not that the electrons are expelled out from the structure, but instead, that electrical breakdown might occur when trying to reach higher doping densities.

1 C. Novo *et al.*, *J. Am. Chem. Soc.*, 2009, **131**, 14664.
2 N. D. Lang and W. Kohn, *Phys. Rev. B*, 1973, 7, 3541.
3 C. F. Chen *et al.*, *Nature*, 2011, **471**, 617.
4 Z. Fang *et al.*, *ACS Nano*, 2013, 7, 2388.

Michael Flatté remarked: you asked of the community, how many electrons are required to make a plasmon. Ordinarily, one would consider a single electron to be sufficient to make a plasmon, and the oscillation would be of that electron relative to a background positive compensating charge. In your description of molecular plasmons, it would be helpful to be more precise about your definition of a plasmon – for example, in benzene, one has molecular orbitals involving multiple electrons, and it seems unlikely for it to be useful to describe the coherent, collective motion of such electrons as plasmons.

F. Javier García de Abajo replied: According to Wikipedia, "a plasmon is a quantum of plasma oscillation", and it "can be described in the classical picture as an oscillation of free electron density with respect to the fixed positive ions in a metal".[1] I believe that this is a commonly accepted view in our community. Then, a single electron does not constitute a plasma, and therefore, it cannot sustain plasmons. The hydrogen atom is an instance of the type of system that you are considering, and I do not believe that its atomic states qualify as plasmons. But, what is a plasmon in small molecular systems?

Plasmons have been identified in molecules for a long time (for example in C_{60}[2]), while more insight has been recently obtained by examining the role of Coulomb interactions[3] and plasmon electronic wave functions consisting of several Slater determinants.[4] Additionally, plasmons in charged polycyclic aromatic hydrocarbons share similar characteristics to those of highly doped graphene. This is useful because it allows us to foresee *a priori* which molecules should exhibit plasmons and assist us in the design of applications to electro-optical modulation[5] and sensing based on the large field enhancement associated with these excitations.[6]

1 http://en.wikipedia.org/wiki/Plasmon.
2 J. W. Keller and M. A. Coplan, *Chem. Phys. Lett.*, 1992, **193**, 89.
3 S. Bernadotte *et al.*, *J. Phys. Chem. C*, 2013, **117**, 1863.
4 C. M. Krauter *et al.*, *J. Chem. Phys.*, 2014, **141**, 104101.
5 A. Manjavacas *et al.*, *ACS Nano*, 2013, 7, 3635.
6 L. Bursi *et al.*, *ACS Photonics*, 2014, **1**, 1049.

Thomas Ebbesen asked: The fact that the spectrum of a molecule changes upon addition of an electron is not surprising since you modify the electronic structure. Such electrochromic behavior (change in colour upon reduction or oxidation of molecules) has been observed since at least the first half of the 20th century. Such colour changes can be induced electrochemically but also by electron transfer from another compound or by ultrafast light induced charge transfer. Here you are considering the addition of an electron to the delocalized π-orbitals of aromatic compounds, such as naphtalene, and you call it a molecular plasmon. I do not see what you are bringing new here except perhaps renaming something well established. Perhaps you can clarify.

F. Javier García de Abajo answered: Your comment is related to a more detailed question formulated by Jeremy Baumberg (see above). I refer to my answer to that question, in which I explain why we use the term "molecular plasmons", what is the new insight that we provide and why it enables us to explore a whole range of electro-optical applications that are fundamentally different from previous electrochromic behaviour.

Paul Dawson commented: Could I just clarify that the graphs of plasmon energy *versus* diameter of metal discs (Fig. 1 in your article, *Faraday Discuss.*, 2015, DOI:10.1039/C4FD00216D), although plotted as continuous lines, are really lines joining points depicting fundamental dipole modes of the discs? The modes are presumably very closely spaced in energy for the larger disc diameters. What is the spacing in energy at the lowest end of the diameter range shown (~10–20 nm)?

F. Javier García de Abajo replied: That is correct: there is no continuous variation with diameter in actual materials, where the number of atoms increases in discrete steps. However, for the sizes that we are considering, the variation in the dipole frequency and strength between contiguous discrete sizes is very small. In the present paper, we use a continuous dielectric formalism, similar to that employed in many studies of metal nanoplasmonics, where the diameter enters as a continuous parameter. This level of description compares well with the RPA, using particle-in-a-box electronic wave functions[1] for sizes above ~15 nm, whereas the discreteness of the electronic structures plays a noticeable role only for smaller sizes. I expect that, although finite-size effects become increasingly important at small sizes, the change in dipole characteristics when adding a few atoms to an atomically-thin noble-metal disk is still small down to a diameter of 10 nm (*i.e.*, consisting of ~1100 atoms). The composition of the plasmon (*i.e.*, how many transitions contribute to it) adds another source of discreteness, also below ~10 nm.[2]

1 A. Manjavacas and F. J. García de Abajo, *Nat. Commun.*, 2014, **5**, 3548.
2 F. J. García de Abajo, *Nature*, 2012, **483**, 417.

Paul Dawson said: My question concerns edge effects. In the case of very thin metal discs, these appear to be modelled as having atomically abrupt edges. How important is this to the results that you observe? What difference would it make if the edges were more graded/had a more rounded profile in going from the substrate to the film surface? Would you still see plasmonic activity *i.e.* would distinct plasmon modes still be supported?

F. Javier García de Abajo answered: Within the RPA description that I mentioned earlier,[1] there is only a mild dependence on the edge profile across the disk (*i.e.*, the shape of the electron wave functions extending above and below the disk), which does not affect the main characteristics of the plasmons. We have not explored the dependence on the profile in the other direction (*i.e.*, the lateral 1D edge profile), although I expect this effect to be small for sizes above 10 nm (*i.e.*, similar to what happens in round metal nanoparticles). Additionally, the type of description that we are using works well even in more extreme structures consisting of atomic rows of gold atoms, as shown by the comparison between theory and experiment by Nagao *et al.*[2]

1 A. Manjavacas and F. J. García de Abajo, *Nat. Commun.*, 2014, **5**, 3548.
2 T. Nagao *et al.*, *Phys. Rev. Lett.*, 2006, **97**, 116802.

Paul Dawson asked: In the case of graphene, especially some of the very small entities, for example, the triangle with the 7 nm sides, does it make a difference whether the edges are terminated in armchair or zig-zag configuration? If so, can you give brief explanation of why?

F. Javier García de Abajo replied: Thank you for this question, which deals with an important issue in small systems. Indeed, there is a big difference between zig-zag and armchair edges: in an armchair-edged triangle of 7 nm side length, a new near-infrared plasmon mode is clearly resolvable in the absorption spectrum upon addition of one electron or hole starting from a neutral structure; in contrast, no new plasmon is observed in that energy range if the edges are in zig-zag. This is due to the existence of electronic zig-zag edge states of zero energy, which provide additional channels of plasmon decay for the size under consideration. You can find a more detailed description of this effect in the work by Manjavacas *et al.*[1]

1 A. Manjavacas *et al.*, *Nanophotonics*, 2013, **2**, 139.

Mikhail Noginov remarked: You have some surface waves in graphene. What qualifies them to be called surface plasmons?

F. Javier García de Abajo responded: From the electronic band structure of this material, one readily observes that a gap region exists in the energy-momentum dispersion diagram of doped graphene, flanked by intraband (within the partially populated Dirac cones) or interband (between cones below and above the Dirac points) electron-hole pair excitations. A plasmon band emerges in this gap as a result of Coulomb interactions and is composed of multiple virtual excitations of electron-hole pairs. This situation is analogous to what happens in traditional plasmonic metals (*e.g.*, aluminium, gold or silver): the plasmon emerges in the region outside the electron-hole pair continuum, driven by a self-consistent Coulomb interaction between several electrons. In fact, in both graphene and traditional good metals such as aluminium, the Drude response model (*i.e.*, the solution of the local equation of classical electron motion driven by both the external field and the self-consistent Coulomb interaction) provides a fair description of these excitations, with differences arising in the linear (graphene) or parabolic (aluminium) electron dispersion relations, but otherwise embodying collective electron motion, which in my view qualifies them as plasmons.

Michael Flatté asked: Why is it surprising that you cannot write a plasmon as a single Slater determinant? Slater determinants are Fock states, and any coherent, collective motion requires a coherent superposition of Fock states. For example, the classical electric field of a laser is constructed through a coherent superposition of different photon numbers (different Fock states), and one should expect a plasmon to be constructed similarly.

F. Javier García de Abajo replied: It is indeed not surprising at all that a plasmon cannot be written as a single Slater determinant. This has been neatly discussed by Krauter *et al.*[1] Your analogy with classical collective motion as a superposition of Fock states (*i.e.*, to form a coherent state) is very instructive.

1 C. M. Krauter *et al.*, *J. Chem. Phys.* 2014, **141**, 104101.

Luca Bursi said: Thank you for your very nice paper. My question is about the relatively new concept of "molecular plasmons". You showed that adding or removing a single electron switches on/off the molecular plasmons in polycyclic aromatic hydrocarbons. Those plasmons can be seen, if I understood well, as the molecular limit of the localized surface plasmon resonances (LSPRs) taking place in doped graphene nanoflakes, upon interaction with light. In these systems, because of the doping, the occurrence of LSPRs seems accepted, but molecular plasmons are also observed in nanoflakes as well as in polyaromatic hydrocarbons without doping and, as you said, in all these systems the plasmonic picture is the same. Could you please comment on this point?

F. Javier García de Abajo replied: Thank you for the compliment. I share your view on molecular plasmons in polycyclic aromatic hydrocarbons (PAHs) as being the molecular limit of plasmons in graphene nanoislands. Indeed, there are plasmons already in the neutral structures, which are less tunable upon addition of charge carriers, as shown in Figure 1. In all cases, the plasmons are identified because they consist of more than one configuration, and also because they are strongly dependent on the Coulomb interaction between excited states. However, it should be noted that small PAHs are very nonlinear compared to larger nanographenes, and in particular, a state with two plasmons can have an energy substantially different from twice the energy of one plasmon. In other words, the plasmon only obeys Boson communication relations in the limit of a large number of electrons. I believe that these systems are thus great for advancing towards nonlinear plasmon-mediated optics.[1]

1 J. D. Cox and F. J. García de Abajo, *Nat. Commun.*, 2014, **5**, 5725.

Jun Yi opened a general discussion of the paper by Jacob Khurgin: The confinement of a field on the nanoscale is sufficient to overcome momentum conservation rules and cause direct absorption by electrons. I wonder whether the assisted momentum from SPPs would help the inelastic light scattering of electrons in the plasmonic structures. For example, could we expect strong photoluminescence signals from a highly confined system, such as a small nanoparticle or particles coupled with a small gap?

Jacob Khurgin responded: One has to distinguish between photoluminescence (PL) due to band-to-band transitions, which are independent of the confinement, and PL due to transitions between conduction electrons in the s-band which requires confinement. As far as PL goes, it can never be really strong because the radiative decay rate, even with Purcell enhancement, is on the 10 ps scale while a non-radiative decay occurs on the femtosecond scale.

Javier Aizpurua asked: I strongly disagree with the use of the word "ultimate" to define the limit of confinement imposed by the dynamical screening of electrons to plasmons. It is of course true that the nonlocality of the response imposes a limit to localization, as many authors have correctly pointed out during the last years, but I think that one should be very careful when a direct transfer of the thresholds of momentum exchange, valid for the bulk metal, are transferred to surfaces and nanoantennas, where other effects are known to be even more important to establish the ultimate confinement of plasmons. Can you comment on the sense and meaning of "ultimate" in this context?

Jacob Khurgin responded: I agree with your comment and should have probably used a different word. Note that the term "ultimate" is correct when applied to propagating SPPs on the metal/dielectric interface – as shown in my work, no matter what you do, the dispersion curve "bends over" at the wave vector that is about 5 times the wave vector in the dielectric. It is not entirely true for other shapes. For instance for the gap SPP, as shown, the confinement *per se* is not influenced by the Landau damping, only the loss is. But what it does is limiting the extent to which the electromagnetic field gets enhanced in gap SPPs. For the different shapes, like dimers, in my view the Landau damping both broadens the resonance and also "squeezes" the mode out of the gap. Once again, it might not limit the size of the mode but will definitely limit the enhancement to the field. Essentially, I have tried to show that the "diffusion" term in the nonlocal theory comes from Landau damping. and can be found in a rather straightforward way. I agree that the "ultimate" limit is the Fermi wavelength which results in Friedel oscillations.

Javier Aizpurua commented: I tend to think that nonlocality does not determine the most restrictive limitation to the localization of plasmons. Our atomistic *ab initio* calculations of near-fields around metallic clusters, performed by the group of Daniel Sánchez Portal in San Sebastián, show extreme localization at the atomic scale. Don't you think that the quantum nature of the atoms and the charge density profile at the interfaces is more determinant to establish the limits to plasmon localization? Is thus the ≈ 5 nm localization limit that you mention in the paper a realistic limit?

Jacob Khurgin replied: As I mention in the reply to the previous comment, Landau damping eventually morphs into the absorption by individual atomic states on an atomic scale, but note that the oscillator strength is conserved, hence one can get some very rough idea of the strength of absorption by individual metal atoms. Clearly exact calculations will give you a much better idea. But either way, on the atomic scale, a metal atom absorbs light strongly and that broadens the reasonance and reduces Q and the enhancement. The field will be localized, I agree, but getting the radiation in and out of this mode will be difficult. Perhaps one should talk about the "ultimate" limit to enhancement rather than localization. Or about an "ultimate loss", even if you get rid of phonons, Coulomb interactions, roughness *etc.*, you still have this "last" source of loss.

Anatoliy Pinchuk said: When you compared Landau damping to other damping mechanisms, such as Kreibig scattering, what numbers did you use for the A-parameter and what sizes of the particles did you consider?

Jacob Khurgin answered: Kreibig scattering and Landau damping are essentially two faces of the same process. You have an optical transition between two free electron states with different wave vectors. This transition is not allowed in infinite space with a uniform field. When the space is limited due to either the size of the nanoparticle (Kreibig) or due to the confinement of the field (*i.e.* presence of high *k*-vector components), the transition becomes allowed. Either way, you end up with the scattering on the scale of v_F/l where l is the characteristic size. The parameter A in my case was 3/4 and I did not consider any specific shape or size for the Kreibig nanoparticle, I just mentioned that I get the correct order of one.

Mark Brongersma addressed Jacob Khurgin and Javier Aizpurua: How can one experimentally probe the ultimate limits of confinement in real structures? There are beautiful experiments with high concentrations of fields and tunnelling effects. However, how do I verify experimentally that the field distribution calculated by *e.g.* the Aizpurua group is correct? Can this only be assessed indirectly by for example studying the tunneling of charge between closely spaced nanoparticles? The beautiful experiments from Baumberg and co-workers suggest that one atom in a nanoscale gap between two metallic structures can impact the optical properties. Their experimental results seem to provide consistent results with quantum theory. Since this system is so sensitive, how does it not exclude other possible experimental configurations (*e.g.* impurities, local reconstructions, *etc.* in the gap) that could explain the same result? It would be valuable to have more surface scientists enter our community.

Jacob Khurgin answered: I think that cathodo-luminescence would be a valuable technique. As far as the fact that single molecule changes the luminescence, it is not necessarily the sign of confinement on that scale.

Javier Aizpurua responded: The ultimate limits of confinement are really challenging to address experimentally, and different theoretical approaches, including ours, might not be sufficient to interpret the experimental features. Up to now, and within the limits of quantum approaches, strong dynamical screening, spill out of electrons, and quantum tunneling have been identified, as key elements modifying the optical response, unambiguously, as far as we believe in Quantum Mechanics. The problem, as you correctly point out, is that our predictions are usually addressing quite restrictive (and sometimes academic) situations, such as smooth surfaces, or "too clean" interfaces, or tiny unpractical systems, whereas in reality, experiments involve more complex situations with a relevant influence of the actual atomistic structuring of facets, of the presence of molecules, or of the modification of potential barriers by external perturbations and charging effects. Along these lines, we are progressively including all these elements in more complete theoretical frameworks that identify all these features, however, the experimental connection and identification of quantum features at optical frequencies is still very hard to make experimentally. I totally agree that the role of surface physicists should be key in that sense. For example, the community of Scanning Tunneling Microscopy (STM) provides a unique tool to explore the quantum regime in plasmonics, being the STM junction a natural plasmonic cavity itself. The difficulty of isolating such an extreme quantum regime and connecting it with the optical response, relies on the identification of the source of conductance. Sometimes, as you point out, a conductance channel can be established just by simple and direct physical contact of a few atoms. In other situations, it might be assisted by a complex distribution of the electronic molecular structure, therefore, I agree that it is hard to identify the ultimate limits of quantum confinement and quantum effects by just looking at the conductivity, where all those extra effects might be present; however, there are new experiments that approach and address this regime with increasing accuracy. We have witnessed in the last months how some experimental groups have approached the quantum regime by directly probing the near field signal through SERS and TERS techniques. One of the examples is the experiment by Prof. Dong in Hefei where the plasmonic response is controlled at Angstrom separation distances, providing subnanometric molecular resolution.[1] Other examples of the access to the quantum regimen, as a limit to the near-field confinement, are experiments addressing the SERS signal[2] or the nonlinear response in cavities,[3] where it is identified that the near-field in the cavity is screened and quenched due to quantum effects that are triggered out at these subnanometric separation distances. Accessing signals that depend on the near fields can thus be an important approach to address quantum effects in cavities. Considering the experimental need for accurate models of these extreme situations, we are making an effort to develop more complete frameworks and schemes of calculation within quantum mechanics that include the effects of the environment, atomistic features and molecular spacers that could be extremely relevant in determining the actual response and limits of optical confinement. I think that it would be great that surface physicists gets fully involved in the future of this extreme regime in Nanophotonics.

1 P. Zhang *et al.*, *Nature*, 2013, **498**, 82.
2 Z. Wenqi *et al.*, *Nature Commun.*, 2014, **5**, 5228.
3 G. Hajisalem *et al.*, *Nano Lett.*, 2014, **14**, 6651.

Jeremy Baumberg commented: I agree with some of the other comments here about "ultimate limits". It is clear from experiments that we are able to confine plasmons enough to get single molecule SERS reliably. We also see plasmonic resonances that are not broadened away, but sharp. It seems that the limit you discuss which seems nicely laid out, would give the maximum possible smearing when the wavevector change is set by the size of a single atom. So I do not understand how this limit can hold. Perhaps it is so strongly geometry dependent that it is not so general?

Jacob Khurgin responded: Yes, "ultimate confinement" is wrong (it is correct for interface SPPs but then becomes geometry dependent). Let us settle on the term "ultimate loss", even if we wash away phonons, carrier–carrier scattering, surface roughness, defects, *etc.* We will always have this last (ultimate) loss mechanism which will affect the degree of field enhancement.

Javier Aizpurua remarked: An important aspect in setting the localization of the plasmons in finite nanoantennas is the geometry. Nonlocality establishes a limitation for localization, as you point out; however, in small particles, tips, and cavities, we can find surface facets and vertices at the atomic scale that can probe molecules. Results in TERS, for instance, show subnanometric optical resolution when a molecule is located at a metallic tip or edge. Wouldn't your theory require to include this aspect of the geometry, at least for finite-sized nanoantennas?

Jacob Khurgin responded: I absolutely agree with this comment. My article does not include these phenomena. On the atomic scale, the whole concept of metal becomes moot. One should probably look at actual electronic states in the metal atoms. Then, what I call "nonlocality caused by Landau damping" becomes simply absorption between individual atomic states which will, of course, limit the enhancement, but should be treated in a more precise way than it has been done in this article.

F. Javier García de Abajo said: At low frequencies, the Landau mechanism that you investigate in this work essentially introduces strong quenching associated with the coupling to electron–hole pairs at distances below v_F/ω. I think it is interesting to note that the Fermi velocity v_F has a similar order of magnitude in most noble metals, also close to that of graphene. However, the strength of the field enhancement depends on the dimensionality, and for example it reaches values well above those of noble metals in graphene, as predicted using the random-phase approximation as well.[1] What are then the characteristics that make a plasmonic material better to achieve large field enhancement and localization? Is dimensionality an important factor?

1 S. Thongrattanasiri and F. J. García de Abajo, *Phys. Rev. Lett.*, 2013, **110**, 187401.

Jacob Khurgin responded: You are correct, in 2D materials the Landau damping is much weaker and the effect I have described will occur when transverse confinement occurs on the scale of a single lattice space. In graphene nanodisks, the effect will reappear due to lateral confinement.

Ortwin Hess asked: It is clearly interesting to see that, when metals are brought in close proximity in the order of the lattice constant of the material, then quantum effects are clearly expected to be relevant. On the other hand, the strong absorption mode that you seem to be seeing in your theory is one that you would expect, you would probably not be able to see from any optical experiments on the outside. One would, however, be able to see an effect if one had an emitter inside the gap. The emitted excitation would very strongly coupled with that mode and one would, with appropriate structuring of the plasmonic nanoparticle acting as an antenna, be able to actually extract a signal. Is this a scenario that would be included in your assumptions?

Jacob Khurgin answered: Yes, the experimental proof of the material I talked about would require an emitter in the gap. I also think my theory does break down on the scale of a single atom, by how much it breaks down is the subject I am working on now.

Faraday Discussions

PAPER

A classical treatment of optical tunneling in plasmonic gaps: extending the quantum corrected model to practical situations

Rubén Esteban,[ab] Asier Zugarramurdi,[cd] Pu Zhang,[ef] Peter Nordlander,[g] Francisco J. García-Vidal,[bf] Andrei G. Borisov[bc] and Javier Aizpurua*[ab]

Received 6th October 2014, Accepted 16th October 2014
DOI: 10.1039/c4fd00196f

The optical response of plasmonic nanogaps is challenging to address when the separation between the two nanoparticles forming the gap is reduced to a few nanometers or even subnanometer distances. We have compared results of the plasmon response within different levels of approximation, and identified a classical local regime, a nonlocal regime and a quantum regime of interaction. For separations of a few Ångstroms, in the quantum regime, optical tunneling can occur, strongly modifying the optics of the nanogap. We have considered a classical effective model, so called Quantum Corrected Model (QCM), that has been introduced to correctly describe the main features of optical transport in plasmonic nanogaps. The basics of this model are explained in detail, and its implementation is extended to include nonlocal effects and address practical situations involving different materials and temperatures of operation.

1 Introduction

Surface Plasmon Polaritons, the collective oscillations of free electrons that are excited resonantly in metals by an optical electromagnetic field, can be localized in finite structures acting as optical nanoantennas. The resulting Localized Surface Plasmon Polaritons (LSPPs) allow the manipulation of light at the

[a]*Materials Physics Center CSIC-UPV/EHU, Paseo Manuel de Lardizabal 5, 20018, Donostia-San Sebastián, Spain. E-mail: aizpurua@ehu.es; Fax: +34 94301 5800; Tel: +34 94301 8830*
[b]*Donostia International Physics Center DIPC, Paseo Manuel de Lardizabal 4, 20018, Donostia-San Sebastián, Spain*
[c]*Institut des Sciences Moléculaires d'Orsay, CNRS-Université Paris-Sud, UMR 8214, France*
[d]*COMP, Department of Applied Physics, Aalto University, P. O. Box 11100, FIN-00076 Aalto, Espoo, Finland*
[e]*Department of Photonics Engineering, Technical University of Denmark, DK-2800 Kgs., Lyngby, Denmark*
[f]*Departamento de Física Teórica de la Materia Condensada and Condensed Matter Physics Center (IFIMAC), Universidad Autónoma de Madrid, E-28049 Madrid, Spain*
[g]*Department of Electrical and Computer Engineering, Laboratory of Nanophotonics, Rice University, MS378, Houston Texas 77005, USA*

nanoscale and the achievement of strong and very confined local field enhancements.[1–6] This opens a route to a variety of applications, in *e.g.* Surface Enhanced Raman Spectroscopy (SERS),[7–11] control of radiation from single quantum emitters,[12–17] or generation of high harmonic pulses at extreme ultraviolet frequencies *via* non-linear processes.[18] The optical response of many plasmonic systems[3,4,7–9,19–22] has been successfully addressed using Maxwell's equations. Within this classical approach different materials are separated by sharp boundaries and the material's properties are given by the position and frequency dependent dielectric function $\varepsilon(\mathbf{r}, \omega)$. Within the local, linear dielectric response, we can relate the dielectric displacement $\mathbf{D}$ and the electric field $\mathbf{E}$ at a position $\mathbf{r}$ through $\mathbf{D}(\mathbf{r}, \omega) = \mathbf{E}(\mathbf{r}, \omega) + 4\pi\mathbf{P}(\mathbf{r}, \omega) = \varepsilon(\mathbf{r}, \omega)\mathbf{E}(\mathbf{r}, \omega)$, where the medium polarisation $\mathbf{P}$ verifies $4\pi\mathbf{P}(\mathbf{r}, \omega) = (\varepsilon(\mathbf{r}, \omega) - 1)\mathbf{E}(\mathbf{r}, \omega)$ (in atomic units). Equivalently, $\mathbf{P}(\mathbf{r}, \omega) = \chi(\mathbf{r}, \omega)\mathbf{E}(\mathbf{r}, \omega)$, where $\chi(\mathbf{r}, \omega) = (\varepsilon(\mathbf{r}, \omega) - 1)/4\pi$ is the medium polarisability. For an isotropic and homogeneous medium, ε and χ can be described as scalars while, in the general case, these are tensors of rank 2. The described linear relationship between $\mathbf{E}$ and $\mathbf{P}$ excludes possible non-linearities that occur for strong fields,[23–25] and it only depends on the particular point of evaluation $\mathbf{r}$.

Among the different plasmonic configurations, gap-nanoantennas composed by two metal particles separated by a narrow dielectric gap, as those schematically depicted in Fig. 1(a) and (b), are particularly interesting as they serve to probe many fundamental optoelectronic processes at the nanoscale. The local classical approach of metallic nanogaps predicts that arbitrarily large charge densities can be induced at the opposite sides of a vanishingly narrow plasmonic gap, leading to extremely intense fields at the gap[3–5,7–9,19–22,26,27] and to strongly red-shifted hybridized bonding LSSP resonances that arise from Coulomb coupling between the modes. The lowest energy gap mode is known as Bonding Dimer Plasmon (BDP) and its redshift with narrowing gap is usually a good fingerprint of the plasmonic cavity.[21,22,28–32] Within the local classical treatment, the transition from the separated to the overlapping nanoparticles is characterised by the discontinuity of the energy of the resonances, with a sharp change of the optical response at the touching point. After contact, a set of Charge Transfer Plasmons (CTPs) that blueshift with increasing overlap appears.[21,29,33,34] This behaviour is outlined by the solid blue lines in Fig. 1(c).

Local classical treatments, however, disregard nonlocal and quantum effects[35,36] that lead, for example, to the size dependence of the plasmon energy of individual nanoparticles.[37–45] In addition to quantum-size effects, the complex quantum interactions between electrons can also change dramatically the optical response of very narrow, nanometer and subnanometer gaps; a regime progressively available for experimental testing, where the limitations of classical descriptions are manifested.[46–55] A local classical description of the response of metallic nanoparticles does not properly account for the correlated motion of the conduction electrons and their spill out the nanoparticle surfaces. Thus, the classical local model fails to capture: (i) the spatial profile of the screening charge[56] with smooth transition of the electron density at the interfaces, instead of sharp discontinuities, (ii) strong nonlocal interactions so that, in general, the centroid of the screening charge density does not coincide with the geometrical interface, (iii) the tunneling between the metal nanoparticles, where the conductive contact can be established prior to the direct geometrical contact.

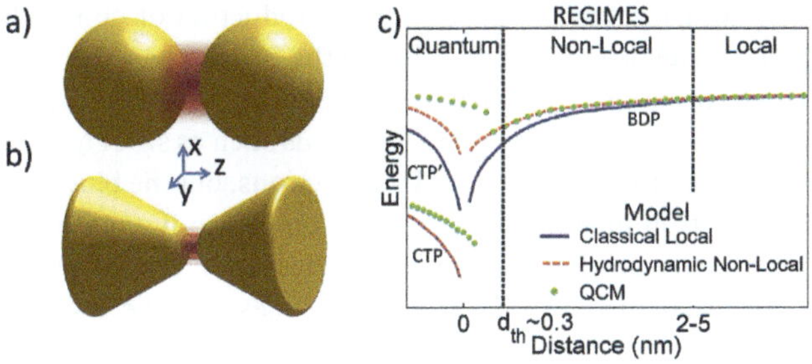

Fig. 1 General consideration of quantum effects and the Quantum Corrected Model (QCM). (a and b) Example of the implementation of QCM in metallic gaps. In (a), a spatially inhomogeneous effective medium whose properties depend continuously on the separation distance is introduced in the gap between two metallic spheres. (b) A bow-tie plasmonic nanoantenna where the effective medium in the gap is described by a set of discrete homogeneous shells of effective material. Each of these shells is characterized by a constant permittivity that corresponds to a representative local separation distance. The permittivity always remains constant in the longitudinal *z* direction along the axis of the dimers, and only changes in the horizontal *xy* plane. (c) Sketch of the evolution of the resonant wavelength of the Bonding Dimer Plasmon (BDP) and two Charge Transfer Plasmons (CTP and CTP′) of a metallic dimer in vacuum. Three different interaction regimes can be distinguished: a classical local regime ($d \gtrsim 2-5$ nm), a classical nonlocal regime ($0.3-0.4 \text{ nm} \lesssim d \lesssim 2-5$ nm), and a quantum regime ($d \lesssim 0.3-0.4$ nm). Behaviour of a local treatment (solid blue line), a nonlocal hydrodynamical model (dashed red line) and the Quantum Corrected Model (QCM, green dots) are displayed.

To be able to capture these effects, one possibility is to perform rigorous quantum mechanical calculations of the optical response using the Time Dependent Density Functional Theory (TDDFT).[34,57–61] In contrast to classical calculations, quantum results for very narrow gaps show pronounced effects of electron tunneling such as a strong reduction of the field enhancement, a continuous transition of modes between the non-touching and contact regimes and the appearance of a Charge Transfer Plasmon (CTP) mode before the nanoparticles touch.[33,62] These effects are outlined by dotted green lines in Fig. 1(c).

Full-quantum approaches are very useful in understanding the physics involved, but show limitations when trying to model realistic materials and structures. Because of the heavy computational load, only relatively small structures have been addressed so far. For example, using a free-electron jellium description of the metal allows the simulation of systems with a few thousand electrons, corresponding to dimensions of only a few nanometers. Attempting the full atomistic approach further reduces the feasible size of the system to several hundreds of electrons.[63] In contrast, typical plasmonic systems are much larger and can easily contain many millions of electrons. It is thus worthwhile to invest the theoretical effort to elaborate approximate methods that effectively account for nonlocal and quantum effects and allow for efficient computational schemes.

During the last years, different models have been developed where nonlocal effects have been accounted for with model dielectric functions, or within the theory of surface screening.[2,64–73] The nonlocal hydrodynamical (NLHD) description has attracted considerable interest because of its numerical efficiency for arbitrarily-shaped objects[47,74–84] and the possibility to obtain semi-analytical solutions using transformation optics.[85] The introduction of nonlocality *via* the standard NLHD model produces a blueshift of the plasmonic modes with respect to the local classical solution, as schematically depicted with the dotted red line in Fig. 1(c). The NLHD results are adequate for metals such as silver and gold, whose optical response is significantly affected by d-electrons, but the blueshift obtained for simple free-electron metals is in contradiction with full quantum results.[86,87] Some approaches have proposed the recovery of nonlocal results by a convenient rescaling of the local distances[86,87] and thicknesses[88] at the metal interfaces, providing good descriptions of the plasmon energies and dispersions in nanometric gaps. Recently, the inclusion of the realistic density profile above the surface into the NLHD description has allowed the retrieval of full quantum and experimental results both for d-electron and simple metals,[89] at an increased computational cost. While the above treatments address nonlocal screening, it has only been recently that the charge transfer between the particles due to quantum tunneling could be accounted for within the classical treatment appropriate for large systems. The Quantum Corrected Model[62] (QCM) was introduced to account for the tunneling current across the gap *via* the insertion of an effective conductive medium in the gap [red areas sketched in the dimers of Fig. 1(a) and (b)]. The conductivity of this effective medium is set from the electron transmission properties of the interparticle barrier. The method was first successfully tested against TDDFT calculations of the optical response of nanogaps of small particles,[62] and latter used to interpret experimental results for subnanometric realistic gaps.[46,90]

As sketched in Fig. 1(c), we can distinguish between three different separation distance regimes in plasmonic gaps. The classical regime corresponds to wide gaps, down to a separation that depends on the size of the particles but can be estimated as ~2–5 nm, where a classical local approach gives basically the same results as a nonlocal treatment. We identify this range of large separations as a local classical regime. As the gap becomes smaller, both local and non-local classical descriptions predict a similar qualitative behaviour: a continuous redshift of the plasmonic modes and a rise of the near fields. However, quantitative differences emerge, as observed in Fig. 1(c). We thus define the range of separation distances below ~2–5 nm down to ~0.3–0.5 nm as the one corresponding to strong nonlocal effects. For separation distances below ~0.3–0.5 nm, the electron tunneling between the nanoparticles quenches the strong charge concentration across the gap and thus critically affects the plasmonic response.[60] This pure quantum effect can become significant even for larger distances under intense lasers.[34] We thus identify a third range of plasmonic gap sizes, below ~0.3–0.5 nm, where pure quantum effects take over. As depicted in Fig. 1(c), in this quantum regime, electron tunneling leads to a gradual transition from modes that redshift towards modes that blueshift with narrowing gaps (left hand side of the figure). The three ranges of distances identified in this introductory figure correspond to the three levels of theoretical modelling required to correctly address the optical response of the strongly interacting plasmonic nanogap system.

The objective of this paper is to discuss comprehensively the QCM as a method that can incorporate quantum effects and in particular electron tunneling in many practical situations of interest. Notably, QCM typically uses a local framework, but can incorporate nonlocality as demonstrated for 2D-particles.[91] Here, we have developed a nonlocal QCM and applied it to dimers of realistic size. We have analyzed the plasmonic response with and without inclusion of nonlocality and/or tunneling to reveal the relative importance of each contribution for gaps of different sizes. This is highly relevant for the interpretation of experimental data obtained in plasmonic nanogap systems.

In the following, we first describe the technical details of QCM,[46,62] and discuss the relevance of the long-range image charge interactions for the correct description of the electron tunneling barrier. We detail the step-by-step procedure to define the effective medium required in the QCM for the case of arbitrary metals surrounded by a dielectric. We also show how nonlocality can be incorporated into the model. We then discuss local classical, nonlocal classical, local QCM and nonlocal QCM results of the optical response of Au dimers to identify separately the effects of tunneling and nonlocality. Last, we discuss the implementation of the QCM for different materials, and consider the influence of temperature. Throughout the paper atomic units (a.u.) are used unless otherwise stated.

2 Implementation of the local QCM

In this section, we first describe the quantum mechanical calculation of the static tunneling between two closely located metallic surfaces (Subsection 2.1). The metal-to-metal tunneling barrier is obtained within the simple jellium model (JM) description of the leads. A typical Scanning Tunneling Microscopy (STM) approach is then used to obtain the conductance of the junction (Subsection 2.2). Within the JM, the metal is described as a gas of interacting electrons moving in a uniform positive background charge representing ionic cores. The interface between the background charge and the vacuum defines the so-called jellium edge that delimits the free electron system. For flat metal surfaces, the jellium edge is located in front of the surface atomic layer at a distance typically given by half the separation between successive atomic planes in the direction perpendicular to the surface. Under the assumption of a linear potential drop across the gap, the calculated conductance yields the static conductivity. Despite its simplicity, the JM captures the collective plasmonic modes of the conduction electrons at the surface in individual nanoparticles and nanoparticle dimers.[56,59,60,92] The JM has also been successfully used to model effects associated with conduction electrons in a variety of metallic systems such as in electronic and optical properties of metal clusters and surfaces,[93–96] charge transfer reactions between atoms and surfaces,[97] conductances of molecular junctions,[98] and strong-field effects.[99]

The second part of this section will discuss the basis of the QCM, *i.e.*, how to incorporate the optical conductivity derived from a quantum mechanical calculation into the classical calculations of the plasmonic response (Subsection 2.3). Upon an adiabatic assumption, typical for the description of the strong field ionisation in atomic and molecular physics,[100–102] the static conductivity (obtained quantum mechanically) is extended towards optical frequencies *via* a simple Drude model. We initially consider free electron metals in vacuum, but the more general scenario incorporates the contribution of d-electrons to the response of the metal and the presence of a dielectric as the surrounding medium (Subsection 2.4).

The key aspect of the QCM with respect to previous classical models is the introduction of an effective medium of adequate permittivity (that relates to the optical conductivity) at the gap between metallic interfaces, to account for the possibility of charge transfer across the gap. Significantly, the definition of the effective medium incorporates the resistive nature of the gap impedance. This effective material can then be inserted into a conventional solver of Maxwell's equations to obtain the optical response for an arbitrary geometry (Subsection 2.5).

We note that the QCM is compatible with different quantum approaches, as the only input needed is the conductivity of the junction, which can be obtained with a jellium model of the metal nanoparticles, or with a full atomistic description. Atomistic studies[63] of plasmonic dimers have pointed out some quantitative differences with respect to the jellium model for small clusters, but the main trends of the results are qualitatively similar.

2.1 Electron tunneling

For typical plasmonic systems where the size of the gap between the nanoparticles is much smaller than the radius of curvature, the local geometry of the junction can be approximated as two flat semi-infinite parallel metallic surfaces separated by a dielectric gap of size ℓ. To calculate the energy Ω- and separation ℓ-dependent electron transmission $T(\Omega,\ell)$ through the potential barrier between the flat free-electron metal leads, we have used a wave packet propagation method detailed previously.[103] Briefly, a "probe" electron wave packet is launched from the inside of one of the metal leads onto the junction. The transmitted and reflected electron fluxes are collected in two virtual detectors placed in the asymptotic regions so that the transmission T and reflection R coefficients can be extracted. Because of the symmetry of the model system, we consider only the electron motion along z, the direction perpendicular to the surfaces. The one-dimensional time-dependent Schrödinger equation, describing the evolution of the "probe" electron wave packet, is then directly solved on a grid of equidistant points.

The potential acting on the conduction electrons in the system is derived by transforming the well established model potentials proposed in ref. 104. Specifically, a constant potential inside the metal is imposed, which is consistent with the JM description. As a result of this simplification, the effects of a contact between particular crystallographic faces[63] cannot be addressed. However, for realistic plasmonic systems, the contact area is large enough so that the contact can be assumed to occur between poly-crystal surfaces rather than mono-crystalline ones. Provided that the pertinent parameters, such as the surface work function, are set from empirical data, the JM is well suited for the description of the electric field-induced inter-particle tunneling of conduction electrons across the ultra small plasmonic junction. Note that the use of the JM is not a limitation of the QCM. The properties of the effective medium that fills the gap could be derived on the basis of full atomistic quantum calculations, provided their feasibility.

The explicit form of the model potential for the electron interaction with a semi-infinite metal surface is given by:

$$V_s(z - z_{im}) = \frac{\exp(-\xi(z - z_{im})) - 1}{4(z - z_{im})}, \quad z > z_{im}, \tag{1}$$

$$-\frac{U_0}{A\exp(B(z-z_{\rm im}))+1}, \quad z \le z_{\rm im}, \tag{2}$$

where the U_0, $z_{\rm im}$, and ξ parameters are set as in ref. 104 for (111) metal surfaces. For Na, we can use the tabulated data for the (110) surface (see data summarized in Table 1) or the potential calculated with density functional theory for a spherical dimer.[34] The numerical values for A and B are obtained by requiring continuity of the potential and its derivative at $z = z_{\rm im}$. Within the JM of metals, the tunneling barrier is mainly determined by the vacuum part of the interface potential, thus the calculated transmission of the junction is robust with respect to the choice of the surface orientation.

For an electron on the vacuum side above the surface ($z > z_{\rm im}$), the potential $V_{\rm s}$ given by eqn (1) converges to the classical attractive image potential, which is of paramount importance for the correct description of the tunneling barrier.[60,105,106] When two metal surfaces are brought in front of each other, the image potential has to be corrected by the inclusion of the multiple image term $V_{\rm mi}(z)$ accounting for the cross terms in the screening interactions:[105,106]

$$V_{\rm mi}(z) = \frac{1}{4}\sum_{n=1}^{\infty}\left\{\frac{-1}{\left(z-z_{\rm im}^{(l)}\right)+nZ}+\frac{-1}{\left(z-z_{\rm im}^{(r)}\right)+nZ}\right\}, \quad z_{\rm im}^{(l)} \le z \le z_{\rm im}^{(r)}, \tag{3}$$

where the indexes (l) and (r) stand for the left and right metal surface, respectively. For z inside the metals, $z < z_{\rm im}^{(l)}$ or $z > z_{\rm im}^{(r)}$, the multiple image term vanishes $V_{\rm mi}(z) = 0$.

Table 1 connects the position of the image plane $z_{\rm im}$ and jellium edge $z_{\rm g}$ to the position of the surface atomic layer $z_{\rm al}$. The distance ℓ between the two surfaces is given by the separation between their jellium edges $\ell = z_{\rm g}^{(r)} - z_{\rm g}^{(l)}$, thus defining the contact at $\ell = 0$. In this section, we will give a detailed description of the results obtained for a Au–Au junction, but Table 1 also gives the parameters of the model for Ag, Cu, and Na.

Finally, the total potential accounting for the long-range image-potential tail of the electron–surface interaction is given by

$$U(z) = V_{\rm s}(z - z_{\rm im}^{(l)}) + V_{\rm s}(z_{\rm im}^{(r)} - z) + V_{\rm mi}(z). \tag{4}$$

In order to illustrate the role of long-range image–charge interactions, we occasionally use the effective one electron potential $U^{1e}(z)$ as derived from the density functional theory (DFT) studies within the local density approximation

Table 1 Jellium model potential parameters: U_0, ξ, work function ϕ, jellium edge $z_{\rm g}$ and image plane position $z_{\rm im}$ with respect to the surface atomic layer $z_{\rm al}$ for different metals. Values are given in units of Bohr radius, 1 a_0 = 0.529 Å

	U_0 (eV)	ξ (a_0^{-1})	ϕ (eV)	$(z_{\rm g} - z_{\rm al})(a_0)$	$(z_{\rm im} - z_{\rm al})(a_0)$
Ag	9.64	1.17	4.56	2.215	2.22
Au	11.03	1.33	5.55	2.225	2.14
Cu	11.89	1.27	4.94	1.97	2.11
Na	6.1	1.0	2.9	1.52	2.12

(LDA). In this case, we performed static DFT calculations of a plasmonic dimer formed by spherical nanoparticles, extracted the potential along the dimer axis, and then used it in the quantum transport calculations.[62] The known defect of LDA is that it misses long-range correlations and fails to reproduce the image potential tail of electron surface interactions.[107] In Fig. 2, we show the potential obtained at the vacuum junction between Au surfaces for different separation distances. We have compared the potentials obtained including long-range image–charge interactions and using the DFT LDA calculations. The latter choice leads to overestimating significantly the barrier between the two surfaces.[60] The QCM calculations presented in this paper use the potential model given by eqn (1)–(4) that account for the image–charge interactions. The results in Fig. 2, which were obtained for a relatively simple static case of flat interfaces, represent all the information that will ultimately be required to implement the QCM, as we discuss in the next subsections. The probability of transmission through the junction for an electron moving perpendicular to the interface $T(\Omega,\ell)$ is determined by the potentials described here, and calculated as the ratio between the transmitted and incident electron fluxes across the gap.

2.2 Conductivity of the gap

It is possible to convert the calculated electron transfer probability $T(\Omega,\ell)$ into the static conductivity σ_0 of the plasmonic gap. Within the standard approach used in the description of *e.g.* STM junctions,[106] the tunneling current density $J(V,\ell)$ through the junction between two planar jellium surfaces for a (small) applied bias V is given by:[108]

$$J(V,\ell,\theta)=\frac{k_{\mathrm{B}}\theta}{2\pi^2}\int_0^\infty \mathrm{d}\Omega\,T(\Omega,\ell)\ln\left(\frac{1+\mathrm{e}^{(E_{\mathrm{F}}+V/2-\Omega)/k_{\mathrm{B}}\theta}}{1+\mathrm{e}^{(E_{\mathrm{F}}-V/2-\Omega)/k_{\mathrm{B}}\theta}}\right). \tag{5}$$

$T(\Omega,\ell)$ depends on the electron energy Ω and separation distance ℓ. The zero energy is placed at the bottom of the conduction band. The logarithmic factor accounts for temperature θ effects arising from the Fermi–Dirac energy statistics in the leads, where k_{B} is the Boltzmann constant.

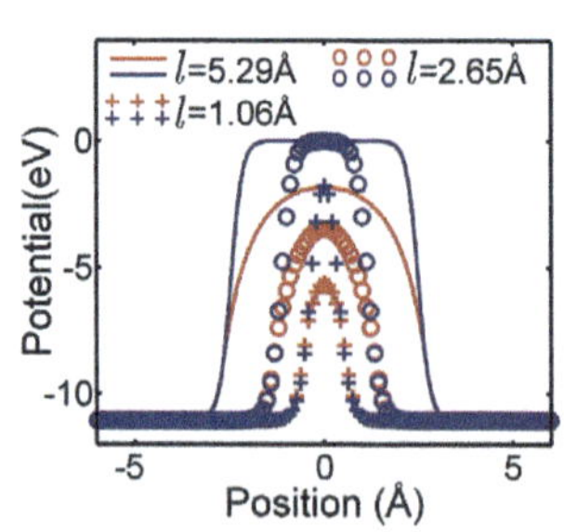

Fig. 2 Quantum mechanical input to the QCM. Potential across the gap for several values of separation ℓ as a function of the position along z, where the zero position corresponds to the gap center. The results were obtained for gold surfaces separated by vacuum when considering long range interactions in the potentials given by image charges [eqn (3) and (4)] (red lines and symbols) and without considering these long range interactions within the DFT LDA (blue lines and symbols).

The conductance per unit area is given by:

$$G(\ell,\theta)=\left(\frac{\partial J}{\partial V}\right)\Big|_{V=0},\tag{6}$$

where, from eqn (5), one obtains the final expression for the conductance $G(\ell,\theta)$:

$$G(\ell,\theta)=\frac{1}{2\pi^2}\int_0^\infty \mathrm{d}\Omega T(\Omega,\ell)\frac{1}{1+\mathrm{e}^{(\Omega-E_\mathrm{F})/k_\mathrm{B}\theta}}.\tag{7}$$

We discuss the effect of temperature θ in Subsection 5.1. We have otherwise assumed $\theta=0$ K in the following, and eqn (7) is simplified to

$$G(\ell)=\frac{1}{2\pi^2}\int_0^{E_\mathrm{F}} T(\Omega,\ell)\mathrm{d}\Omega.\tag{8}$$

Finally, assuming that the bias is given by $V=E\ell$, where E is the electric field in the vacuum gap between the metal surfaces, the conductivity of the junction $\sigma_0(\theta,\ell)=\frac{\partial J}{\partial E}$ is given by

$$\sigma_0(\ell)=\ell G(\ell),\tag{9}$$

which holds only for not too small ℓ values. For very narrow gaps, the described approach is not valid primarily because the fields in the junction can no longer be considered as homogeneous, so that the $V=E\ell$ approximation breaks.

Fig. 3 shows the electron transmission probability $T(E_\mathrm{F},\ell)$ between left and right leads, as calculated at the electron Fermi energy E_F for the Au junction. These results are obtained both with the model potential that accounts for long-range interactions [given by eqn (3) and eqn (4)] as well as with the DFT LDA potential that misses the long-range contribution (potentials displayed in Fig. 2). Except for very narrow gaps $\ell \lesssim 2$ Å, $T(\ell)$ shows in both cases a nearly exponential dependence with the junction size ℓ, typical of tunneling processes. For small ℓ, the potential barrier becomes low and the electrons are transferred from metal to metal *via* direct transitions over the barrier. The transmission probabilities tend

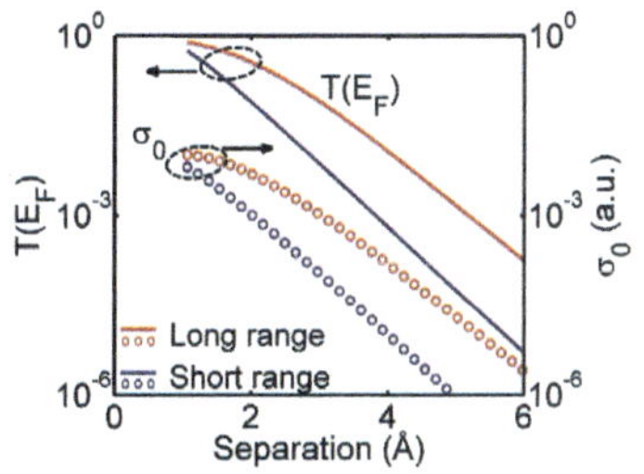

Fig. 3 Quantum mechanical input to the QCM. Transmission at the Fermi Energy $T(E_\mathrm{F},\ell)$ (continuous line) and static conductivity σ_0 (open circles), both calculated quantum-mechanically, as a function of the separation distance ℓ. The results are obtained for gold surfaces separated by vacuum when considering long range image–charge interactions in the potentials [eqn (3) and (4)] (red lines and symbols) and without considering these long range interactions within the DFT LDA (blue lines and symbols).

then to 1 in a non-exponential manner as the gap closes, reflecting the formation of a continuous solid. Fig. 3 also demonstrates that, consistent with the overestimation of the potential barrier, the DFT LDA description results in too small transmission probabilities. The junction width where the tunneling process becomes important is underestimated by about 1 Å within DFT LDA.

Fig. 3 also shows that, similarly to the transmission probability $T(E_F,\ell)$, the static conductivity σ_0 (open circles) approximately follows an exponential dependence with ℓ except at the shortest distances. Indeed, because of the rapid decrease of the transmission probability and lowering of the electron energy, only a small energy region around the Fermi level contributes to the tunneling current (conductivity σ_0) which is determined by $T(E_F,\ell)$ [see eqn (8) and (9)]. The decay lengths of these two exponential functions are of the order of one atomic unit ($\approx$0.53 Å). The long range image–charge interactions also influence σ_0 considerably, as manifested by the faster decay of the conductivity with ℓ obtained using the LDA approximation (higher potential barrier in Fig. 2, blue lines and symbols) as compared to the more accurate model that includes the image potential (red lines and symbols).

2.3 Effective medium within the QCM: Drude metals in vacuum

We describe now how to account for the tunneling between plasmonic nanoparticles within the framework of classical Maxwell's equations which is the essence of the QCM. The interest in such approach is obvious since standard electromagnetic solvers can be applied so that the impact of quantum effects on the optical properties of practical plasmonic systems can be efficiently calculated. To some extent, the underlying idea here is similar to the text book examples relating *e.g.* the macroscopic dielectric constant of the medium to the polarizability of individual molecules.[109] In this subsection, we concentrate on particles made of a free-electron metal and surrounded by vacuum.

The typical curvature radius R of plasmonic nanoparticles is of some tens of nanometers, much larger than the gap widths at which tunneling effects become important. Under this condition, the quantum effects for a region of the nanogap characterized by a local separation distance ℓ can be modeled by assuming that quantum effects in that position show similar behaviour to a pair of planar metal surfaces separated by the same separation ℓ. According to the discussion introduced in the previous sections, in the static case the current density between these two surfaces is given by

$$J(\ell) = \sigma_0(\ell)E. \tag{10}$$

Eqn (10) describes the tunneling regime typical for STM with small σ_0 and static homogeneous electric field E inside the junction. For time varying fields, we have used the adiabatic assumption that has proven its validity, for example, in the description of strong field ionisation in atomic and molecular physics.[100–102] Under this assumption, the frequency-dependent currents and fields at the gap can be related through the static conductivity as

$$J(\ell,\omega) = \sigma_0(\ell)E(\omega), \tag{11}$$

where, similar to eqn (10), homogeneity of the optical fields in the junction is assumed. Note that σ_0 is a *real* number so that the junction has resistive character. Similar equations have been recently derived[110,111] within a framework of the Tien–Gordon theory developed for microwave fields,[112] however, this approach does not account for field penetration into the nano-object neither it allows for a simple interpolation of the tunneling and contact regimes.

Since the typical sizes of the tunneling gaps are at the nanometer scale, *i.e.*, 100 times smaller than the wavelength of optical frequencies, eqn (11) can be generally considered as a local one. One can now consider the relationship between the permittivity of an effective medium that fills in the gap of two plasmonic nanoparticles and the conductivity across the gap within *classical* Maxwell's equations:

$$\varepsilon_g(\ell) = 1 + i4\pi\sigma_0(\ell)/\omega, \tag{12}$$

and realize that this relationship connects classically the current density across the gap and the optical field inside the gap, exactly in the same way as the *quantum* relationship between the tunneling current and the field in the gap. We note that at this stage any description of the dielectric properties of the metal can be adopted (Drude, empirical,…), and $\sigma_0(\ell)$ can be calculated with quantum methods, as discussed earlier or with any other level of sophistication in the description of the junction.

While the simple model resulting in eqn (12) describes the STM or tunneling regime with well separated nanoparticles, it does not allow to directly address all gap sizes down to the touching regime, where the junction progressively behaves as a continuous metal bridge. We seek an effective dielectric medium that has the permittivity described by eqn (12) for large separations and that becomes identical to the surrounding metal at small ℓ. In this respect, assuming particles made of a free-electron Drude metal is instructive as it allows to obtain simple analytical expressions. Within the Drude model, the frequency-dependent permittivity of the particles is given by:

$$\varepsilon_m(\omega) = 1 - \frac{{\omega_p}^2}{\omega(\omega + i\gamma_p)} \tag{13}$$

where ω_p is the plasmon frequency and γ_p describes the damping. In this paper we have used $\omega_p = 9.065$ eV and $\gamma_p = 0.0708$ eV for gold, as extracted from a fit to empirical data at low energies.[113]

To describe the effective material at the gap, and taking into account that, for small separations, electrons are easily transferred and the gap becomes metallic, we have also used a Drude-type expression, which allows an easy interpolation to the continuous metal case at $\ell = 0$. We write the permittivity ε_g of the effective medium at the gap as:

$$\varepsilon_g(\omega, \ell) = 1 - \frac{{\omega_p}^2}{\omega(\omega + i\gamma_g(\ell))}, \tag{14}$$

which is related to the conductivity by

$$\sigma(\omega,\ell) = -i\omega(\varepsilon_g(\omega,\ell) - 1)/4\pi. \tag{15}$$

The separation-dependent loss parameter γ_g allows to set the permittivity (and thus the conductivity) to a desired value for each ℓ. If we now require that the tunneling regime, given by eqn (12), must be reproduced by eqn (14) for large separation distances, $\ell \to \infty$, we obtain

$$\gamma_g(\ell) = \frac{{\omega_p}^2}{4\pi\sigma_0(\ell)}, \tag{16}$$

where we have approximated $\gamma_g(\ell) - i\omega \approx \gamma_g(\ell)$, since $\sigma_0(\ell)$ becomes exponentially small when ℓ increases (Fig. 3). Notice that, within this approximation, due to the exponential increase of γ_g for large ℓ, eqn (12) is recovered independently of the gap plasmon frequency assumed in eqn (14). We take it as equal to the value describing the particles ω_p to facilitate the recovery of the metal permittivity ε_m when contact is established. In principle, one could directly apply eqn (16) for each sufficiently large value of ℓ. Nonetheless, looking at the behaviour of σ_0 in Fig. 3 using a simple exponential function is suggested.

For vanishing gaps, the permittivity of the effective medium ε_g should tend towards the permittivity of the metals ε_m for all wavelengths, as the potential barrier between the particles disappears. Thus, for $\ell \to 0$, we should recover the dielectric constant of the Drude metal, or equivalently $\gamma_g \to \gamma_p$. However, eqn (16) does not necessarily extrapolate directly to γ_p at $\ell = 0$ due to the limitations on the quantum calculation of the conductivity σ_0 for small separations. We then proceed as follows. Taking into account the exponential dependence of σ_0 with separation found for large enough ℓ, we impose

$$\gamma_g(\ell) = \gamma_p e^{\ell/\ell_c}. \tag{17}$$

The description of the effective material given by this simple exponential expression is a key aspect of the implementation of the QCM. ℓ_c is chosen to verify eqn (16) for sufficiently large ℓ values. In our implementation, we select ℓ_c such that eqn (16) is exactly verified for an specific $\ell = \ell_0$ at which the transmission at the Fermi energy $T(E_F,\ell_0)$ is one percent. We explicitly checked that the final results are robust with respect to the choice of the matching point ℓ_0, as far as it is located in the deep tunneling regime.[62] Fig. 4 shows the results for γ_g obtained for Au (ω_p = 9.065 eV, γ_p = 0.0708 eV), where we consider the calculation of σ_0 including long range interactions [eqns (3) and (4), red colours] or only short range interactions within the DFT LDA approach [blue colours], as explained in previous sections. In the tunneling regime for $\ell \gtrsim 3$ Å, the exponential dependence given by eqn (17) provides a reasonable approximation to $\gamma_g(\ell)$, obtained directly from the calculated tunneling characteristics using eqn (16), particularly when the image charge interactions (long range) are accounted for. We obtain in this case $\ell_c = 0.4$ Å. The discrepancy between eqn (16) and the exponential fit becomes rather large at short distances, where the STM-type model resulting in eqn (16) does not apply.

Following our procedure, γ_g becomes very large for large separation distances, obtaining resistive contacts, *i.e.*, the effective material has mostly real conductivity. Indeed, at optical frequencies for sufficiently large separation distances, $\gamma_g(\ell) \gg \omega$ and the purely static conductivity is recovered, $\sigma(\omega,\ell) = \sigma_0(\ell)$. As the gap closes, γ_g decreases, $\sigma(\omega,\ell)$ becomes complex and the effective material acquires inductive character becoming purely metallic. At very small ℓ, the value of the

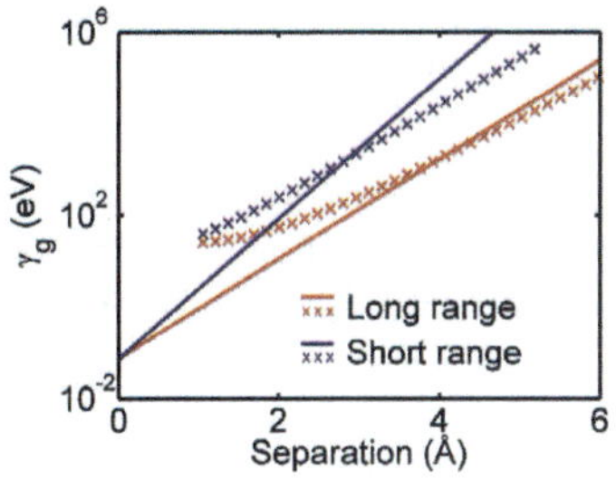

Fig. 4 Description of the effective medium used in the QCM. The loss parameter γ_g required for the effective medium in the gap is plotted as a function of the separation distance ℓ. Solid lines correspond to the exponential expression used for the QCM implementation, while crosses are obtained from eqn (16) using the static conductivity σ_0 calculated quantum-mechanically. The results are obtained for gold surfaces separated by vacuum with [red] and without [blue] taking into account long-range image–charge interactions.

metal ε_m at both sides of the gap is recovered. The resistive to inductive transition is not described within the tunneling STM theory, thus a direct comparison between the predictions of the QCM and full quantum calculations is then needed in order to validate our approach.

We performed such an analysis for a Na nanosphere dimer. Na is a prototype of a free-electron metal with $\omega_p = 5.16$ eV and $\gamma_p = 0.218$ eV. The JM approximation can then be used, allowing to perform full quantum Time-Dependent Density Functional Theory (TDDFT) calculations of the plasmonic response of a spherical nanoparticle dimer comprising 2148 electrons.[34,62] Each Na sphere has a closed shell structure with 1074 electrons so that the nanoparticle response exhibits a well developed plasmon resonance. The sphere radius is $R_{cl} = 2.17$ nm. This is small compared to typical plasmonic nanostructures, but treating the correlations between the 2148 electrons contained in the dimer is already extremely heavy computationally. Fig. 5(a) and (b) display the amplitude and phase of the gap conductivity σ as obtained within the QCM (effective medium from eqn (15), blue lines) and from TDDFT calculations (red stars), for a varying separation distance ℓ under the excitation of radiation with energy $\hbar\omega \approx 2.2$ eV,

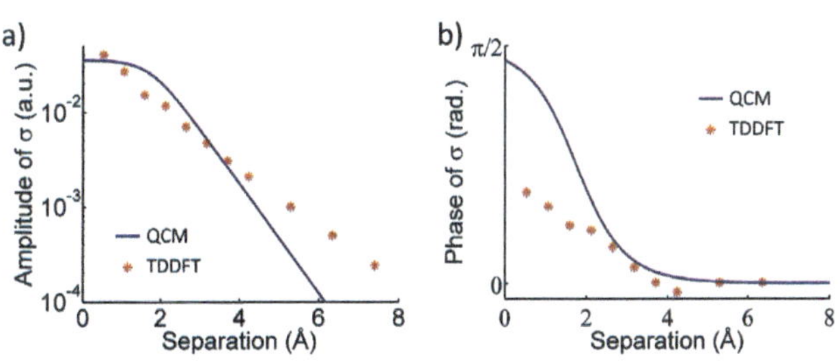

Fig. 5 Conductivity of the junction between two spherical Na nanoparticles in vacuum. Results are shown as function of the width of the junction ℓ. (a) Amplitude and (b) phase of the conductivity, as given by TDDFT calculations (red stars) and as calculated with QCM (blue solid lines). The frequency corresponds to an energy of ~2.2 eV.

close to the bonding dimer plasmon. In TDDFT, the conductivity is defined as the ratio between the current and the electric field at the central point between the nanoparticles.

For sufficiently large ℓ, both approaches give nearly real conductivity σ, manifesting the resistive nature of the contact and the validity of the adiabatic approach, according to which the tunneling current should have the same phase than the optical field due to the instantaneous response. As the distance decreases, the junction acquires inductive character, with a dephasing between the currents and the fields. The difference between the TDDFT and QCM data for the phase of σ is only significant at very narrow gaps, where the TDDFT gives larger efficient losses, probably linked with the finite cross-section of the junction, resulting in a smaller phase. The agreement between QCM and TDDFT is also good for the absolute value of the conductivity, with a somewhat larger disagreement for $\ell \gtrsim 6$ Å. In this regime, where $\ell \gtrsim R_{\rm cl}$ for the small dimer considered in Fig. 5, the description of the junction as a composition of contacts between planar surfaces obviously fails. However, we expect that the agreement will improve for common systems with larger $R_{\rm cl}$ values. We also note that tunneling is sufficiently small for $\ell \gtrsim 6$ Å so that it does not affect the optical response. In general, the agreement between the two approaches is surprisingly good, in particular considering the small radius of curvature of the plasmonic dimer considered for this comparison.

2.4 Generalization of the QCM to noble metals and dielectric gaps

In the previous section, we described the plasmonic material as a free-electron gas using the classical Drude permittivity given by eqn (13). This description is valid for materials such as Na but for noble metals such as gold or silver it does not account for the significant contribution of d-electrons to the optical response. Further, the surrounding medium was considered to be vacuum. The QCM can be straightforwardly extended to the case of arbitrary metals surrounded by any dielectric medium.[46] The permittivity of metals $\varepsilon_{\rm m}$, as obtained from the literature,[113] can be described as:

$$\varepsilon_{\rm m}(\omega) = \varepsilon_{\rm m}^{\rm d}(\omega) - \frac{{\omega_{\rm p}}^2}{\omega(\omega + i\gamma_{\rm p})} \tag{18}$$

which explicitly distinguishes between the contribution from interband transitions involving d-electrons $\varepsilon_{\rm m}^{\rm d}$, and that from the free electron gas (right-hand side term). In contrast to eqn (13), $\varepsilon_{\rm m}^{\rm d}$ does not need to be equal to 1 and can be frequency-dependent. For the calculations of the gold dimers presented here, $\omega_{\rm p} = 9.065$ eV and $\gamma_{\rm p} = 0.0708$. $\varepsilon_{\rm m}^{\rm d}$ is then directly obtained from eqn (18) using empirical data.[113]

To describe the permittivity of the QCM effective medium at the gap $\varepsilon_{\rm g}$, we treat the free electrons as in the previous section and introduce the contribution of the d-electrons, which decays exponentially for increasing gaps, with a decay length $\ell_{\rm d}$.

$$\varepsilon_{\rm g}(\omega,\ell) = \varepsilon_0(\omega) + \left(\varepsilon_{\rm m}^{\rm d}(\omega) - \varepsilon_0(\omega)\right){\rm e}^{-\ell/\ell_{\rm d}} - \frac{{\omega_{\rm p}}^2}{\omega\left(\omega + i\gamma_{\rm g}(\ell)\right)}. \tag{19}$$

ε_0 is the relative permittivity of the surrounding dielectric medium, which does not need to be vacuum and can depend on the frequency. $\gamma_g(\ell)$ in the Drude term is obtained as in the previous section [eqn (16) and (17)], where we continue using the jellium approach to obtain the quantum value of the static conductivity, σ_0, since the quantum charge transfer is dominated by the free electrons. A more complex *ab initio* calculation including d-electrons and the actual dielectric material in the plasmonic gap could also be used if feasible. ε_g thus tends to the permittivity of the particles for vanishing gaps, and to the value of the surrounding medium for large separations $(\gamma_g(\ell \rightarrow \infty) \rightarrow \infty)$.

The decay length ℓ_d of the d-electron contribution $(\varepsilon_m^d - \varepsilon_0)e^{-\ell/\ell_d}$ is of the order of the Bohr radius since the electron density probability associated with the d-electrons is confined to the proximity of the metal. The exact value of ℓ_d can be set in several ways. For example, from the radial decay of the 5d orbital,[114] we can set $\ell_d = 1.5$ a.u ≈ 0.8 Å for gold and other typical materials. This is the value used in the calculations in this paper. We could also consider that the d-electron contribution should decay faster than the free-electron component, due to the higher localisation of the d-orbitals, and use a ℓ_d somewhat smaller than ℓ_c. Nonetheless, it is the Drude term that gives the critical conductive contribution to the permittivity setting the regime of conductive coupling between the nanoparticles across the gap. Therefore, as we have explicitly observed for local QCM calculations of Au dimers, the calculated optical response of the nanoparticle assembly is robust with respect to the specific choice of ℓ_d.

2.5 Implementation of the QCM in classical electrodynamics

We describe in this section how to implement the QCM using an electrodynamic solver of Maxwell's equations. The metallic particles are characterized as a medium of permittivity ε_m (possibly obtained from empirical data) that is delimited by sharp interfaces. The position of the interface is chosen at the jellium edge in the JM description of the metal, as indicated in Table 1. This choice is not unique, and one can also consider the classical metal boundary defined by the position of the centroid of the screening charge at the surface.[86,87] For the implementation of the QCM, however, it is important that the classical definition of ℓ corresponds to the one used in the quantum calculations of the tunneling current. This definition will be also relevant when comparing experiments and calculations with the aim of establishing a criterion of absolute separation distances.

The distance-dependent effective medium of permittivity ε_g described in the previous section needs to be inserted in the gap. In the general case of the nanoparticle dimer, the junction is not between two planar surfaces so that there is not a single global distance. For each pair of opposite points separated by the local distance ℓ, the local ε_g can be set to the same value as the permittivity calculated for the corresponding ℓ-wide flat gaps. This simple approach assumes that the nanoparticle surfaces are of sufficiently large radius of curvature to be considered as locally flat, an assumption that should be valid considering typical experimental geometries. ε_g remains constant along the line joining opposite points, but can change as we move in the orthogonal direction of the gap, leading in general to an spatially inhomogenous ε_g.

A general prescription to determine how to define opposite points and the connecting line is not necessarily straightforward, but typical structures often present a 'natural' option. For example, for configurations rotationally symmetric with respect to a given axis [z in Fig. 1(a) and (b)], we define the local ℓ as the distance between opposite points linked by a line parallel to this axis. Conveniently, the expected polarization of the fields, which is critical for tunneling, is also often expected to be along the same axis. For a more complex configuration, it may be beneficial to define opposite points, as those connected by a line parallel to the local electric field vector, and the distance as the length of this line.

Fig. 1(a) sketches the insertion in the gap of the effective medium with spatially-inhomogeneous permittivity ε_g, for a dimer composed of two metallic spheres. The plasmonic response can then be found using a classical solver of Maxwell's equations. For the results obtained here, two different solvers have been used: a finite element solver (COMSOL Multiphysics) and a boundary element method (BEM).[115] BEM requires the gap to be divided into different areas characterized by a mean distance to avoid the inhomogenous distribution of the material. A possible implementation for the case of a bowtie antenna is sketched in Fig. 1(b). For rotationally symmetric configurations, we have often found sufficient to consider ~3–8 shells placed in the region of the gap where the local distance is smaller than around 6 Å, but different scenarios may require different number of shells. For example, if field discontinuities need to be minimized at the contact regions between two shells and one of the particles, it can be convenient to use more shells.

The described implementation of the QCM comprises a series of steps to define the permittivity ε_g of the effective medium to be inserted in the gap. It is, however, possible to conceive modifications of the given procedure that would change to some extent the value of ε_g. In a previous work considering free-electron metals,[62] we observed very little change of the plasmonic response after modification of the exact definition. We have also verified that ignoring the contribution of the d-electrons to ε_g for two Au spheres does not essentially modify the extinction and near field enhancement spectrum within the local QCM model.

Thus, the QCM seems generally robust towards modifications on the implementation details. Nonetheless, two conditions should likely be respected: the effective medium should have a strong resistive component, except possibly for very narrow gaps, and ε_g should model correctly the exponential dependence of the absolute value of the gap impedance with the distance determined from the quantum calculations of the transmission probability.

3 Extension of the QCM to include nonlocal effects

A critical parameter for classical calculations is the relationship between the polarization induced in a material and the electric fields. Up to here, we have considered a local description of a metal. It assumes that the polarization at a point is proportional to the local fields at this same position, which introduces a certain error by ignoring the influence of the surrounding charge distribution, in particular for strong interactions between interfaces. This is precisely the case of small metallic particles and/or very narrow gaps.

Possibly, the most common approach to introduce nonlocality when studying plasmonic systems within a classical framework is the hydrodynamical model,[116,117] which considers an extra term in the dynamics of the free electron gas related to its pressure and thus, the deformation to an external perturbation. In this model, nonlocality is introduced in k-space by using a permittivity that is not only a function of frequency, but also of the wavevector k. In real space, this description corresponds to the introduction of an additional contribution to the polarization from the immediate vicinity of the non-uniform charge density, so that the polarization at each point is obtained *via* an integral comprising the electric fields over the surrounding region.

We have recently discussed how, for a pure free electron gas such as Na, the hydrodynamical model does not correctly reproduce certain aspects of the charge densities associated to the localized surface plasmons.[87] The model predicts an electron density that decays considerably before the jellium edge defining the limit of the particles, *i.e.*, the centroid of charge is displaced towards the interior within this approach. In contrast, full quantum simulations reveal that, for these free electron metals, the electron density is pushed a short distance towards the outside of the particles, *i.e.* there is a spill-out of electrons. Nonetheless, for the typical noble metals used in plasmonic experiments, the centroid of the charge is indeed displaced towards the interior of the particles due to the influence of the bound d-electrons, and the hydrodynamic approach can then be applied for practical purposes.[47]

In the following, we describe briefly how we introduce the hydrodynamical approach into both the classical local and QCM theoretical frameworks. First, we discuss how the nonlocality is implemented into the optical response. Again, we consider separately the contribution from bound and free electrons, and write the electric displacement vector $\vec{D}$ as:

$$\vec{D} = \vec{E} + 4\pi\vec{P}_{\mathrm{d}} + 4\pi\vec{P}_{\mathrm{c}} \tag{20}$$

with $\vec{P}_{\mathrm{d}}$ and $\vec{P}_{\mathrm{c}}$ the polarization vectors for bound (d-electrons) and conduction electrons, respectively, and the electric field vector $\vec{E}$. As in standard studies of the optical response of metal nanostructures, $\vec{P}_{\mathrm{d}}$ is described *via* a local approximation:

$$4\pi\vec{P}_{\mathrm{d}}(\omega) = \varepsilon_{\mathrm{m}}^{\mathrm{d}}(\omega)\vec{E}(\omega) \tag{21}$$

whereas nonlocal effects are incorporated by means of the polarization field $\vec{P}_{\mathrm{c}}$. This free-electron polarization field is directly related to the induced current density $\vec{J}_{\mathrm{c}} = \partial\vec{P}_{\mathrm{c}}/\partial t$. For the implementation of nonlocal effects into the COMSOL Multiphysics package, it is much more convenient to work with $\vec{J}_{\mathrm{c}}$ instead of $\vec{P}_{\mathrm{c}}$. In the hydrodynamical model, the time evolution of the induced current density due to conduction electrons is described by the linearized Navier–Stokes equation:

$$\frac{\partial\vec{J}_{\mathrm{c}}}{\partial t} + \gamma_{\mathrm{p}}\vec{J}_{\mathrm{c}} = \frac{{\omega_{\mathrm{p}}}^2}{4\pi}\vec{E} - \beta^2\vec{\nabla}n \tag{22}$$

where $n = -\vec{\nabla}\vec{P}_c$ is the induced charge density. One immediately recognizes the nonlocal contribution from the free electrons, *i.e.*, the dependence on the gradient of the induced charge density. Typically, $\beta = \sqrt{3/5}v_F$, where v_F is the Fermi velocity. In particular, one can show that in the frequency domain:

$$\beta^2\vec{\nabla}(\vec{\nabla}\vec{J}_c) + \omega(\omega + i\gamma_p)\vec{J}_c = i\omega\frac{\omega_p{}^2}{4\pi}\vec{E}. \tag{23}$$

This equation for the induced current density and Maxwell's equations are related through their mutual dependence on the electric field $\vec{E}$ and $\vec{J}_c$. They can be solved simultaneously in COMSOL by choosing appropriate boundary conditions for $\vec{J}_c$, such as continuity of its normal component. This implementation of nonlocal effects into our numerical framework reduces it to a local response when β is simply set to zero, making the integration of a local and nonlocal description of different parts of a plasmonic system easy to realize.

Inclusion of the effective gap material in the hydrodynamical model leads to what we previously referred to as nonlocal QCM. An equation similar to eqn (23) can be written within the gap region:

$$\beta^2\vec{\nabla}(\vec{\nabla}\vec{J}_c) + \omega(\omega + i\gamma_g)\vec{J}_c = i\omega\frac{\omega_p{}^2}{4\pi}\vec{E} \tag{24}$$

where γ_g (see Section 2.3 for its definition) replaces γ_p. At the current stage, for the sake of simplicity, we assume that the optical response within the gap region is local, *i.e.*, $\beta = 0$ in the region. Regarding the polarization contribution coming from bound electrons, it is treated in the same way as in the local QCM and, therefore, it does not appear explicitly in eqn (24).

Complementary to the hydrodynamical model, we have also considered a simple approach to the previously introduced nonlocality,[86,87] based on a simple rescaling of the separation distances to take into account the spatial distribution of the electrons. To calculate the nonlocal results for two particles at physical separation distances d, as defined by the jellium edges, it is simply required to obtain the classical local response at $d + d_n$ separation, where d_n is the distance given by the position of the two centroids of charge, $d_n = -1.8$ Å for Na and d_n ~1.7–3.0 Å for Au. This approach proved to be successful when predicting the observed peak positions for two Na wires and distances up to a few Ångströms.[87] Unfortunately, it is not evident how to implement this method when a non-negligible quantum charge transfer between the particles is considered, and thus it is only used here for classical nonlocal calculations before contact. Other approaches to introduce nonlocality have been recently discussed, such as adding an extra dielectric layer of well defined properties to mimic the nonlocal displacement of the centroid of charge at the surface,[88] including extra nonlocal dissipation terms,[84] or considering more accurately the electron density profile at the interfaces.[89]

4 Nonlocality and quantum effects in the plasmonic response

In the following, we analyze the response of a plasmonic dimer as a function of the separation distance to explore the different regimes sketched in Fig. 1(c): local, nonlocal, and quantum. We have calculated the optical response of a dimer composed by two Au spheres of $R = 25$ nm, using a local or nonlocal hydrodynamic description and with or without inclusion of the interparticle electron tunneling within the QCM. In all the calculations, we solved the classical electromagnetic Maxwell's equations to obtain the optical response of the system. However, to distinguish between different approaches, we have denoted as *classical local* the calculations where the optical response of the metal is described by a local dielectric constant without charge transfer across the gap, *classical nonlocal* refers to calculations where the metal response is described within the hydrodynamical approach, also without considering charge transfer between particles, *QCM local* calculations use a local dielectric constant for the metal and include the electron tunneling within the QCM, and last, *QCM nonlocal* calculations consider the hydrodynamical approach for the metal response and the QCM to include charge transfer due to tunneling.

The system holds cylindrical symmetry with respect to the z axis that connects the particle centers, located at $z_{\pm} = R \pm d/2$. Thus, d indicates the distance between the two closest points between the particles (different from the local separation ℓ) and $z_{\pm}$ is measured from the center of the junction. $d < 0$ corresponds to overlapping particles. Experimental values of the gold permittivity are used[113] and a plane-wave polarized along the dimer axis illuminates the system. The near-field enhancement $|E_g/E_0|$ is defined as the ratio between the electric field amplitude at the center of the gap E_g and the amplitude of the incident field E_0. After contact, the fields inside the metallic junction are strongly screened, therefore we have only shown the field enhancement for positive distances d. Nonlocal results were obtained with the Finite Element Method (COMSOL Multiphysics), and the local calculations were performed using the Boundary Element Method (BEM). We have confirmed that both methods provide the same results for the case of QCM local calculations.

The classical local response of the dimers has been studied by many different groups and is generally well understood.[21,118,119] We show the extinction spectra and near-field enhancement at the gap of the Au dimer for several representative separation distances in Fig. 6(a) and 7(a), respectively. As pointed out above, the near-field spectra are shown only for positive separation distances as the local field is quenched for overlapping dimers. As the spheres approach, Coulomb interactions lead to a strong charge concentration at the opposed metal surfaces and, thus, to very intense fields[120] in the gap and to hybridized plasmonic modes that strongly redshift. The charges are antisymmetric with respect to the plane perpendicular to the axis of the dimer at the center of the gap ($z = 0$). Dark symmetric modes are not excited in our planewave scheme of illumination. For moderate separations d, the lowest-energy Bonding Dimer Plasmon (BDP) dominates the response but, for small d, higher order modes are also efficiently excited. All modes have a clear signature in both the extinction and the near field. As the gap closes, $d \to 0^+$, both the redshift of the modes and the magnitude of the associated field enhancement diverge.

The response changes abruptly when conductive contact is established at $d \leq 0$. For overlapping particles, electron transfer between the particles becomes possible, and Charge Transfer Plasmon (CTP) modes[29–31,121] emerge. The lowest energy CTP appears at distinctively large wavelengths and corresponds to a dipolar-like oscillation characterized by non-zero net charges of opposite sign at each particle, an unphysical situation before contact. For small overlaps, the narrow wedges support very fast spatial oscillations of the charge distribution,[122]

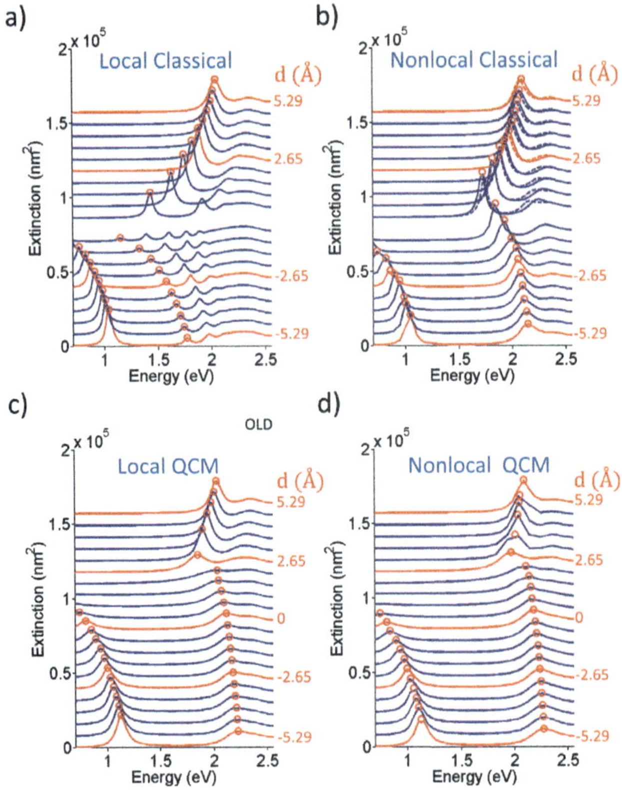

Fig. 6 Extinction cross-section of a Au dimer for different separation distances d within: (a) the local classical model, (b) the nonlocal classical model, (c) the local QCM and (d) the nonlocal QCM. The solid lines in (b) correspond to the hydrodynamical model and the dashed lines for $d > 0$ to the values obtained after rescaling the results from local classical calculations by $d_n = 2$ a.u $= 1.06$ Å. The system is composed of a dimer of 25 nm radius Au spherical particles in vacuum, illuminated by a plane-wave polarized along the dimer axis. The results are plotted for d steps of 1 a.u. (0.529 Å) and shifted vertically for clarity, with some representative distances indicated on the right. $d < 0$ correspond to overlapping particles. The classical local and nonlocal extinction spectra for $d = 0$ are not plotted as, in the absence of tunneling, the geometry is ill defined when contact is just established. The red open circles indicate the positions of the maxima of the modes that are further analyzed in Fig. 8.

leading to many high order modes. The resonant wavelength of these modes also diverges for $d \to 0^-$. As the overlap increases, the modes blueshift until eventually the spectra of a single sphere is recovered.

The divergence of the results as $d \to 0$, right before and after contact, is related to the description of the sharp interfaces between the metal and the surrounding medium within the local classical approach. This leads to extremely large and confined charges at the sharp wedges and narrow gaps producing the divergence. Rounding the edges of the neck of the contact between two particles would considerably diminish the expected number of resonances, their redshift and the field enhancement. While ref. 21 discusses the implications of geometrical rounding, nonlocal and quantum tunneling effects also result in effective

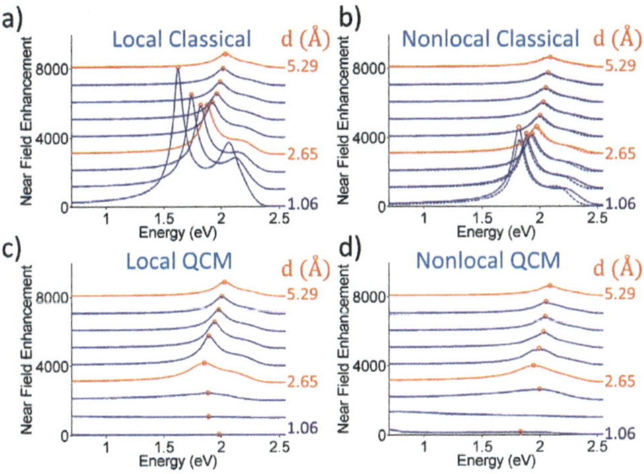

Fig. 7 Near field enhancement at the centre of the gap $|E_g/E_0|$ between two 25 nm radius Au spheres in vacuum as a function of the separation distance d within: (a) the local classical model, (b) the nonlocal classical model, (c) the local QCM and (d) the nonlocal QCM. The solid lines in (b) correspond to the hydrodynamical model and the dashed lines to the values obtained after rescaling the results from local classical calculations by $d_n = 2$ a.u = 1.06 Å. The system is composed of a dimer of 25 nm radius Au spherical particles in vacuum, illuminated by a plane-wave polarized along the dimer axis. The results are plotted for d steps of 1 a.u. (0.529 Å) and shifted vertically for clarity, with some representative distances indicated on the right. The red open circles indicate the position of the maxima of the modes that are further analyzed in Fig. 8.

rounding of the geometry of the junction as follows from the results below. Accounting for the nonlocal response within the hydrodynamical approach allows to remove the unphysical divergences of the optical response of narrow gaps[47,84] consistent with the results of Fig. 6(b) and 7(b).

Local and hydrodynamical nonlocal calculations indicate several similarities both in their far field [Fig. 6(a) and (b)] as well as in the near-field response [Fig. 7(a) and (b)]. As the particles approach, the modes redshift strongly and the corresponding field enhancement increases. Particle contact results in an abrupt change in the spectra with the appearance of Charge Transfer Plasmons that blueshift with the increasing overlap. However, important differences appear between the two models. Compared to the local results, the nonlocal description generally 'softens' the spectral trends,[85] as it captures the lack of sharp interfaces in real systems. As the distance approaches zero, either before or after contact, both the shifts of the modes and the number of resonances efficiently excited are smaller than for the local classical results. In a similar manner, the field enhancement predicted by the nonlocal approach can be very large just before contact but not as large as obtained in the local case, with no apparent divergence as the gap closes. In general, the differences between the two models are more significant for narrow gaps or small overlaps. For example, when d changes from 2.65 Å to 1.06 Å the local classical calculations predict that the low-energy BDP extinction resonance peak redshifts from 1.88 eV (660 nm) to 1.62 eV (764 nm), and that the near field enhancement increases from 2900 to 8100. This is to be compared with a wavelength change from 1.97 eV (628 nm) to 1.82 eV (683 nm), and a field enhancement increase from 1600 to 4600, obtained within the nonlocal hydrodynamical model for the same separation distances.

The distribution of surface charges is critical to understand the differences between the local and nonlocal classical responses. While in the local classical calculations the plasmon-induced screening charges are located at the geometrical surfaces of the metal, using the nonlocal hydrodynamic model results in a displacement of the screening charges into the metal, inwards from the metal surface.[80,86,87,123,124] Thus, when the distance between the geometrical surfaces is zero, the screening charges across the junction are separated by a distance d_n, where $d_n/2$ is the position of the centroid of the screening charge with respect to the geometrical surface (jellium edge) of the individual nanoparticles. Since the screened charges do not abruptly overlap when contact is established, the electrostatic interactions in the system do not diverge, in clear contrast to the classical model where infinite localization of the charges is produced. The effect of the spatial dispersion of the response for $d < 0$ is thus to round the sharp wedges of overlapping geometries.[85]

The effect of the shift of the centroid of the screening charge with respect to the geometrical surfaces of the gap can be further explored for $d > 0$ by using the concept of distance rescaling.[87] In Fig. 6(b) and 7(b), we show spectra corresponding to local classical calculations where the separation distance is rescaled to $d + d_n$ (with $d_n = 2$ a.u. $= 1.06$ Å) given by the actual separation between the screening charges, and not by the system geometry. Notably, this simple approach is in very good agreement with more complex hydrodynamical calculations, which supports the interpretation that the main role of nonlocality is to displace and spread the centroid of the charge density, effectively lifting unphysical divergences.

We move next to one of the main aspects of this discussion, the quantum effects due to electron tunneling across the junction, as captured using the QCM. Fig. 6(c) and 7(c) show the far- and near-field QCM results, respectively, for a local description of the plasmonic response. For $d \gtrsim 3.5$ Å, the tunneling starts being negligible and thus the classical behaviour is recovered. However, for shorter distances, as well as for all the overlapping cases considered, the electron tunneling strongly modifies the response, as expected from previous studies of related systems.[34,60,62] We thus denote the region of separation $d <$ 3–4 Å as the tunneling regime. Due to the tunneling current that emerges before contact and neutralizes the induced charges of opposite sign at the junction, the classical discontinuity of the optical response at contact, $d = 0$, is removed. The QCM calculations of the extinction cross-section [Fig. 6(c)] show a gradual transition, with a threshold distance $d_{th} \approx 0.35$ nm separating two different situations.[34,60] As the gap closes, still being larger than d_{th}, the hybridized plasmonic modes of the dimer redshift similarly to the results of classical calculations. At $d = d_{th}$, the redshifting modes gradually disappear and new modes corresponding to the Charge Transfer Plasmons (CTPs) emerge and blueshift with a further decrease of d.[33,125] The emergence of CTPs at small but positive separation distances, $d > 0$, points out that the conductive contact between nanoparticles is established prior to the direct geometrical overlap of their surfaces, owing to the electron tunneling through the potential barrier separating the nanoparticles.

The QCM extinction spectra exhibit a moderate number of modes for all separation distances, in contrast to the prediction from local classical calculations of dimers with narrow gaps and small overlaps. In the nonlocal classical case, we have discussed the relatively small number of modes as a consequence of the gradual variation of the electron distribution near the interfaces, which softens the plasmonic response. In the QCM, the softening is due to the charge transfer across the separating vacuum. The transfer probability, as codified by ε_g, changes gradually as we move transversally (plane xy) away from the center, from basically metallic to identical to the surrounding vacuum.

The results of the local QCM also show how the charge transfer between particles at tunneling distances has a dramatic effect on the near field enhancement at the center of the gap [Fig. 7(c)]. For sufficiently large gaps $d \gtrsim d_{th}$ the classical results are basically recovered, with increasing enhancement for narrowing gaps. However, for distances smaller than $d \approx d_{th}$, where the strongest classical enhancements are found, the quantum electron transfer screens the charges at the gap, quenching the near fields. This quenching can have important consequences in different optical techniques, such as surface-enhanced spectroscopies, where large field enhancements are often necessary.

Last, we analyze the situation where nonlocality and electron transfer across the gaps are included simultaneously in our QCM calculations with a hydrodynamical treatment of the metal response, as introduced in Section 3. For sufficiently large separation distances above d_{th}, the tunneling is negligible and the nonlocal screening is the main reason of the difference with classical local descriptions. We thus recover the classical nonlocal results both for the extinction, shown in Fig. 6(d), and for the near-field enhancement, Fig. 7(d). For separations below d_{th}, the nonlocality also modifies the details of the response, as observed when comparing the results of nonlocal QCM with those of the local QCM simulations. Nonetheless, the influence of nonlocality on the response remains relatively small when compared with the effect of quantum tunneling. The latter is thus the key aspect determining the optical response of the system. Generally, the nonlocal and local QCM calculations yield very similar results over the entire range of separations considered here.

Fig. 8(a) and (b) show the synthetic summary of the results obtained in this section using local classical, nonlocal classical, local QCM and nonlocal QCM calculations. The energy of the extinction plasmon resonance [Fig. 8(a)] and the resonance near-field enhancement [Fig. 8(b)] are shown as a function of the distance between nanoparticles, d, for the most relevant plasmon modes. In particular, Fig. 8(a) shows the spectral position of the extinction peaks for the lowest energy BDP mode that exists for positive distances ($d \gtrsim d_{th}$ when using the QCM, and $d > 0$ for the classical treatment) as well as for the two lowest energy CTP modes that emerge at $d \lesssim 1$–2 Å within the QCM and at $d < 0$ for classical treatments. Fig. 8(b) focuses on the range with $d \geq 0$ and displays the maximum field enhancement of the BDP, as well as of the second CTP′ charge transfer plasmon which emerges at ≈ 2.1 eV in the QCM calculations for $d \lesssim d_{th}$. The data points in Fig. 8(a) and (b) are extracted from Fig. 6 and 7, where they are marked with open circles. Along with the nonlocal results obtained within the hydrodynamical model, we also show the results obtained with the distance rescaling (dashed line), where the nonlocal classical results at distance d are taken as equal to the local classical values at $d + d_n$. As already discussed, we obtain a good agreement with the hydrodynamical approach for $d_n = 1.06$ Å.

Considering the plasmonic response of the dimer, three regions of the interparticle separation d can be defined, consistent with the introductory sketch of Fig. 1:

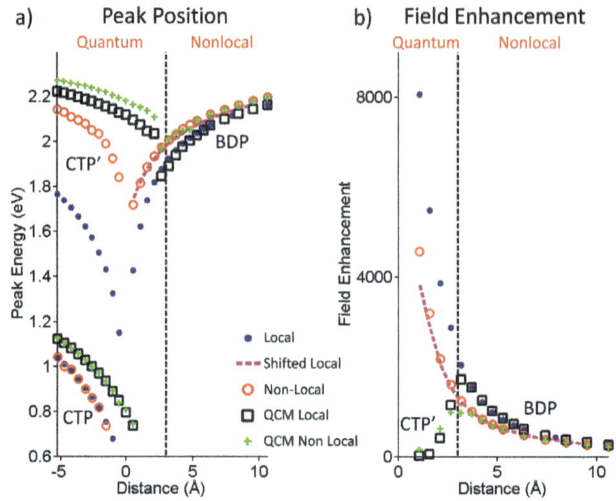

Fig. 8 Summary of the response predicted by the different models. (a) Energy of the extinction peaks of the main plasmon modes, and (b) maximum near field enhancement $|E_g/E_0|$ at the center of the gap as a function of separation distance, for local classical (blue dot), nonlocal classical (red open circle for hydrodynamical model and magenta dashed line for distance rescaling), local QCM (black open square) and nonlocal QCM (green +, calculated with the hydrodynamical formalism). The values correspond to those indicated by circles in Fig. 6 and 7. The vertical dashed line corresponds to $d = 3.0$ Å and approximately separates the distance ranges where either the quantum-induced charge transfer or nonlocal effects dominate the response, as indicated at the top of the figure. The labels of the plasmon peaks refer to the Bonding Dimer Plasmon (BDP) and to the two lowest-energy Charge Transfer Plasmons (CTP and CTP′).

• At large separations (in our case for $d \gtrsim d_{cl} = 2$ nm, not shown in Fig. 8) the local classical approach adequately captures the plasmonic response so that we identify this distance range as the *classical regime*.

• For $d_{th} \lesssim d \lesssim d_{cl}$ the differences between the local and nonlocal hydrodynamical results are clearly visible, and the electron tunneling through the potential barrier separating the nanoparticles is negligible. This is the *nonlocal* regime. According to the distance rescaling, including nonlocal effects gives a similar effect as modifying the distance in the local calculations by a factor $(1 + d_n/d)$, which allows to understand the increasing role of the nonlocality for narrowing gaps.

• The distance range $d < d_{th}$ corresponds to the *quantum regime*. At this Ångstrom-scale separations, the electron transfer across the junction dominates the optical response, and the differences between the nonlocal QCM and local QCM results remain relatively small. Nonlocality does modify to some extent the exact response, as can also be seen in Fig. 8. Nevertheless, the prediction of the local QCM calculations seems sufficient to understand most of the features of the plasmonic response.

To finish this section, we emphasize three particularly clear signatures of the tunneling regime for closely interacting spherical nanoparticles:[60,62] (i) the gradual transition in the dispersion of the gap modes at a small threshold separation distance d_{th}, where the modes stop redshifting and start a progressive blueshift as the particles become closer, (ii) the emergence of a CTP mode at low energies already for (small) positive distances, and (iii) the quenching of the near fields. The presence of much fewer modes for small overlaps and narrow gaps when comparing local QCM with local classical calculations is also striking but may not be the easiest path to identify quantum effects, as the number of modes is also comparatively small for nonlocal classical calculations or, after contact, for local classical results with rounded wedges.[21] We also notice that the signature of quantum effects on the far-field signal depends on the morphology of the gaps.[126] An additional consequence of quantum effects is that the significant quenching at the center of the gap expels the strong fields outside and thus establishes an intrinsic quantum limit to the ultimate field confinement.[6,46]

5 Extension of the QCM to different materials and temperatures

We have focused up to now on gold dimers at 0 K, but the QCM can be applied to more general materials and circumstances in plasmonics. We discuss in the following the effect of changing the temperature and of considering other metals.

5.1 Effect of temperature

The quantum conductivity σ_0 used in the implementation of the QCM discussed above has been obtained from eqn (8) assuming tunneling at 0 K temperature. More generally, eqn (7) should be applied to calculate the conductivity for non-zero temperatures.

Fig. 9(a) shows the static conductivity σ_0 of the vacuum gap between two flat Au surfaces. The results were obtained using eqn (7) as a function of the gap separation ℓ for the case of zero temperature (open red circles), 500 K (open blue circles) and 5000 K (open green circles). The temperature-independent electron transmission probability at the Fermi energy $T(E_F)$ is also shown for reference with a continuous line. The results were obtained as described in Subsections 2.1 and 2.2. Fig. 9(b) plots the distance and temperature dependent loss parameter γ_g used in the QCM description of the effective medium. A continuous line shows the results of the exponential approximation given by eqn (17), and the symbols show the results for γ_g directly deduced from σ_0 using eqn (16). The results for 0 K are identical to those presented in Fig. 3 and in Fig. 4.

Most notably, σ_0 and γ_g, which parameterize the electron transfer due to quantum tunneling, are very little sensitive to the temperature. The results for 0 K and 500 K are essentially identical, with a difference small enough not to be appreciated in the figure. Even for a temperature as high as 5000 K, which is considerably larger than the melting point of gold, the induced change in σ_0 and γ_g remains small. σ_0 is robust to temperature changes because the ensuing smearing of the Fermi distribution symmetrically opens the channels for electron tunneling in both directions. While the electron energy dependence of the barrier transmission $T(\Omega)$ breaks this symmetry, the characteristic electron energy scale of the change of $T(\Omega)$ is a fraction of eV. This requires very high temperatures for the corresponding electronic states to be populated. For practical purposes, the observed temperature dependence of γ_g should be negligible, inducing much

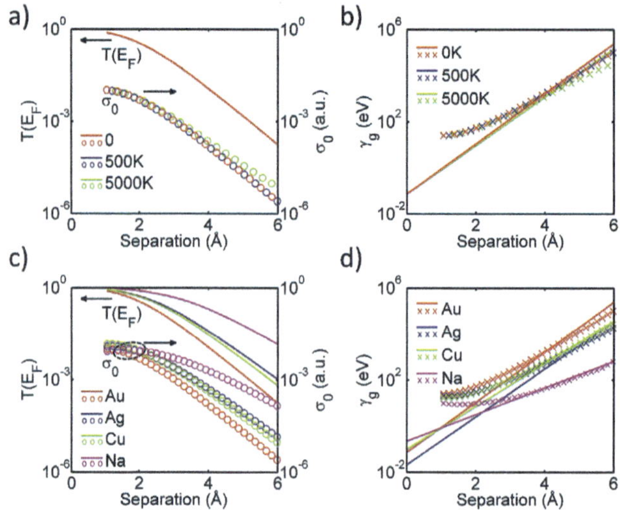

Fig. 9 Modeling for different materials and temperatures. (a and c) Tunneling probability at the Fermi Energy $T(E_F)$ (solid lines) and quasi-static conductance σ_0 (open circles) as a function of the gap separation distance obtained from quantum-mechanical calculations. In (a) different temperatures are considered (0 K in red, 500 K in blue, and 5000 K in green). $T(E_F)$ is independent of the temperature, therefore all three curves fall on the same position. The values represented here for Au at 0 K correspond to those in Fig. 3 and 4. In (c) different materials are considered at 0 K (Au in red, Ag in blue, Cu in green, and Na in magenta). (b and d) Tunneling loss parameter γ_g describing the effective medium in the gap as a function of the gap separation distance. Solid lines correspond to the exponential expression used for the QCM implementation, while crossed points are obtained from eqn (16) using the static conductivity σ_0 calculated quantum-mechanically. In (b) the same temperatures as in (a) are considered and overlap of the lines for 0 K and 500 K is observed. In (d) different materials are considered as in (c).

weaker changes than other temperature-dependent effects such as thermal expansion of the nanoparticles or changes in the bulk permittivity of gold.

5.2 Other materials

The QCM can be applied to other metals besides Au in a straightforward manner, following the description in Sections 2 and 3. Fig. 9(c) shows the evolution with ℓ of the tunneling probability at the Fermi energy $T(E_F)$ and of the quantum-mechanically calculated static conductivity σ_0, for Au, Ag, Cu and Na surfaces separated by vacuum. The results are generally similar in all these cases, particularly for the three noble metals. Overall, the lower the metal work function is (see Table 1 for the work functions of Na, Au, Ag and Cu), the larger is the conductivity and thus, the smaller is the loss parameter for a fixed value of gap separation (see Fig. 9(d)). For the four materials, and excluding the narrowest gaps, $T(E_F)$ and σ_0 follow an approximately exponential variation with ℓ. The exponential expression for γ_g used in the QCM (eqn (17)) is a good fit to the values directly deduced from σ_0 using eqn (16), with the exception of very short distances where this equation does not apply.

From a practical perspective, to set the permittivity ε_g of the effective medium at the gap, the QCM requires the knowledge of the permittivity of the metal surrounding the gap (ε_m),[113] the values of ω_p and γ_p determining the free-electron contribution to ε_m, and the characteristic lengths describing the distance dependence of the free-electron (ℓ_c) and the bound d-electron (ℓ_d) terms defining ε_g. ω_p and γ_p can be obtained on the basis of the ω dependence of ε_m at large wavelengths. ℓ_c and σ_0 can be obtained as discussed in Section 2. The d-electron contribution to the permittivity, ε_m^d, at $\ell = 0$ can then be obtained by subtracting the free-electron contribution from ε_m. For reference, Table 2 indicates the values used in our calculations for ω_p, γ_p, ℓ_c and ℓ_d in the case of Au, Ag, Cu and Na, all surrounded by vacuum. For pure Drude-metals such as Na, the d-electron contribution is absent and ℓ_d is not necessary. Notably, the decay length ℓ_c takes similar values for the three noble metals, but it is noticeably larger for Na. For systems separated by water or dielectrics,[53] we also expect to find significantly larger ℓ_c values.

In Fig. 10 we compare the spectra of extinction and near field enhancement $|E_g/E_0|$ at the center of the gap of Au, Ag and Cu spherical nanoparticle dimers in vacuum as a function of the separation d of the gap. The field is polarized along

Table 2 QCM parameters for Ag, Au, Cu and Na in vacuum: plasma frequency ω_p and losses γ_p of the Drude contribution to the dielectric function; characteristic lengths ℓ_c and ℓ_d of the exponential functions describing the increase of the loss parameter, and the decay of the d-electron contribution to the screening, respectively. ℓ_c and ℓ_d are given in Å, and γ_g and ω_p are given in eV. The value of ℓ_d is the one used in our calculations although other alternatives are possible (see text for discussion). ℓ_d is not given for Na because d-electrons do not play a role in the screening in this material

	ω_p (eV)	γ_p (eV)	ℓ_c (Å)	ℓ_d (Å)
Ag	9.175	0.0212	0.42	0.79
Au	9.065	0.0708	0.4	0.79
Cu	8.853	0.0954	0.47	0.79
Na	5.16	0.218	0.75	—

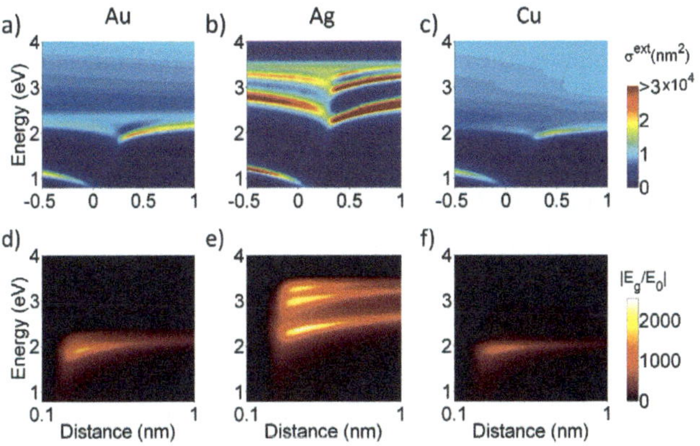

Fig. 10 Optical response for different materials. (a–c) Extinction cross-section σ^{ext}, and (d–f) near-field enhancement at the gap center $|E_g/E_0|$ for: (a and d) Au, (b and e) Ag and (c and f) Cu dimers in vacuum, as a function of the separation distance and energy. A radius $R = 25$ nm of the spherical particles and plane-wave illumination polarized along the dimer axis are used in all cases.

the dimer axis, and the radius of each of the two identical nanoparticles is $R = 25$ nm. The qualitative effect of tunneling is very similar for all three cases. As has been already discussed for Au, the extinction exhibits a gradual transition from large distances to the overlapping regime, and the CTP resonances emerge just before physical contact is established. The transition occurs around a threshold distance d_{th} of a few Ångströms which separates a region of redshifting (towards contact) and a region of blueshifting (progressive overlapping) of the plasmonic modes. The near field enhancement is maximized for all three materials for a gap separation of a few Ångstroms, near d_{th}, and it is quenched strongly for narrower gaps. d_{th} is similar for the three cases due to the similar values of ℓ_c obtained. Along with the qualitative similarities, the quantitative differences in the permittivity of different metals can strongly affect the details of the plasmonic response. Indeed, the number, strength, and spectral position of the resonances show a significant material dependence, mostly due to the material dependence of the losses parameter γ_p and the d-electron contributions to the metal permittivity ε_m^d.

6 Discussion and conclusions

In this paper, we have described in detail the Quantum Corrected Model, which allows accounting for the quantum tunneling with fully classical simulations of the optical properties of nanoparticle assemblies with narrow junctions. The main idea behind the QCM is to fill the interparticle junction with an effective material that results in the same local classical current density as the tunneling current density obtained from a quantum calculation. We describe the practical case of a noble metal, which shows a large contribution from d-electrons to the optical response. We provide a detailed description of the effective material for frequently used metals such as Au, Ag and Cu. We put particular emphasis in the extension of the QCM to incorporate the hydrodynamical treatment of nonlocal effects, and discuss the effect of non-zero temperatures on the optical properties. While this paper focuses on the linear response, the QCM can be extended to non-linear phenomena[34,127-130] by taking into account the dependence of tunneling across the gap on the strength of the local field.

We have also compared the respective influence of nonlocality and of quantum charge transfer on the plasmonic resonances of a dimer. Three regions of inter-particle separations could be identified. Above a few nanometers, local classical calculations adequately describe the optical properties of plasmonic dimers. In the intermediate range of distances, above ≈ 0.35 nm, classical nonlocality modifies the exact properties of the plasmonic modes. Nonetheless, nonlocal effects do not change the qualitative behaviour predicted by classical local models and they can be modeled by simple distance rescaling. For junction widths below ≈ 0.35 nm, the system is in the quantum tunneling regime. For these ultranarrow gaps, the charge transfer across the junction radically modifies the optical response of the dimer, leading to the quenching of the near fields and to a complete spectral redistribution of the plasmonic modes. These strong effects cannot be captured even qualitatively by current classical theories based on the local or nonlocal hydrodynamic descriptions of the metal. The different relative influence of nonlocal and tunneling effects depending on the separation distance should be observed in very general systems characterized by nanometer gaps, at least for spherical terminations,[126] and can thus be helpful in the interpretation of experimental measurements.

Acknowledgements

RE and JA acknowledge financial support from project FIS2013-41184P of MINECO and from project IT756-13 of the consolidated groups of the Department of Education of the Basque Government. PZ and FJGV acknowledge financial support from the European Research Council (ERC-2011-AdG Proposal No. 290981, Plasmonanoquanta). AGB and FJGV gratefully acknowledge the hospitality of Donostia International Physics Center. This work has been funded partly by the French Agence Nationale de la Recherche (ANR-2011-EMMA-003-01) and by Triangle de Physique.

References

1 K. L. Kelly, E. Coronado, L. L. Zhao and G. C. Schatz, *J. Phys. Chem. B*, 2003, **107**, 668–677.
2 R. Alvarez-Puebla, L. M. Liz-Marzán and F. J. Garcá de Abajo, *J. Phys. Chem. Lett.*, 2010, **1**, 2428–2434.
3 J. A. Schuller, E. S. Barnard, W. Cai, Y. C. Jun, J. S. White and M. L. Brongersma, *Nat. Mater.*, 2010, **9**, 193–204.
4 N. J. Halas, S. Lal, W.-S. Chang, S. Link and P. Nordlander, *Chem. Rev.*, 2011, **111**, 3913–3961.
5 A. J. Pasquale, B. M. Reinhard and L. Dal Negro, *ACS Nano*, 2011, **5**, 6578–6585.
6 H. Cang, A. Labno, C. Lu, X. Yin, M. Liu, C. Gladden, Y. Liu and X. Zhang, *Nature*, 2011, **469**, 385–388.
7 H. Xu, E. J. Bjerneld, M. Käll and L. Börjesson, *Phys. Rev. Lett.*, 1999, **83**, 4357–4360.
8 C. E. Talley, J. B. Jackson, C. Oubre, N. K. Grady, C. W. Hollars, S. M. Lane, T. R. Huser, P. Nordlander and N. J. Halas, *Nano Lett.*, 2005, **5**, 1569–1574.
9 J. Theiss, P. Pavaskar, P. M. Echternach, R. E. Muller and S. B. Cronin, *Nano Lett.*, 2010, **10**, 2749–2754.
10 M. Moskovits, *Rev. Mod. Phys.*, 1985, **57**, 783–826.
11 B. Fazio, C. D'Andrea, F. Bonaccorso, A. Irrera, G. Calogero, C. Vasi, P. G. Gucciardi, M. Allegrini, A. Toma, D. Chiappe, C. Martella and F. Buatier de Mongeot, *ACS Nano*, 2011, **5**, 5945–5956.
12 P. Mühlschlegel, H.-J. Eisler, O. J. F. Martin, B. Hecht and D. W. Pohl, *Science*, 2005, **308**, 1607–1609.
13 P. Bharadwaj, B. Deutsch and L. Novotny, *Adv. Opt. Photonics*, 2009, **1**, 438–483.
14 T. H. Taminiau, F. D. Stefani, F. B. Segerink and N. F. Van Hulst, *Nat. Photonics*, 2008, **2**, 234–237.
15 J. N. Farahani, H.-J. Eisler, D. W. Pohl, M. Pavius, P. Flückiger, P. Gasser and B. Hecht, *Nanotechnology*, 2007, **18**, 125506.
16 L. Rogobete, F. Kaminski, M. Agio and V. Sandoghdar, *Opt. Lett.*, 2007, **32**, 1623–1625.
17 R. Esteban, T. V. Teperik and J. J. Greffet, *Phys. Rev. Lett.*, 2010, **104**, 026802.

18 S. Kim, J. Jin, Y.-J. Kim, I.-Y. Park, Y. Kim and S.-W. Kim, *Nature*, 2008, **453**, 757–760.
19 A. García-Martín, D. R. Ward, D. Natelson and J. C. Cuevas, *Phys. Rev. B: Condens. Matter Mater. Phys.*, 2011, **83**, 193404.
20 E. Hao and G. C. Schatz, *J. Chem. Phys.*, 2004, **120**, 357–366.
21 I. Romero, J. Aizpurua, G. W. Bryant and F. J. Garcíade Abajo, *Opt. Express*, 2006, **14**, 9988–9999.
22 P. K. Jain and M. A. El-Sayed, *Chem. Phys. Lett.*, 2010, **487**, 153–164.
23 M. Danckwerts and L. Novotny, *Phys. Rev. Lett.*, 2007, **98**, 026104.
24 M. Kauranen and A. V. Zayats, *Nat. Photonics*, 2012, **6**, 737–748.
25 M. Scalora, M. A. Vincenti, D. de Ceglia and J. W. Haus, arXiv:1405.6999 [physics.optics], 2014.
26 P. J. Schuck, D. P. Fromm, A. Sundaramurthy, G. S. Kino and W. E. Moerner, *Phys. Rev. Lett.*, 2005, **94**, 017402.
27 A. I. Fernández-Domínguez, S. A. Maier and J. B. Pendry, *Phys. Rev. Lett.*, 2010, **105**, 266807.
28 L. Yang, H. Wang, B. Yan and B. M. Reinhard, *J. Phys. Chem. C*, 2010, **114**, 4901–4908.
29 T. Atay, J.-H. Song and A. V. Nurmikko, *Nano Lett.*, 2004, **4**, 1627–1631.
30 S. Marhaba, G. Bachelier, C. Bonnet, M. Broyer, E. Cottancin, N. Grillet, J. Lermé, J.-L. Vialle and M. Pellarin, *J. Phys. Chem. C*, 2009, **113**, 4349–4356.
31 S. S. Aćimović, M. P. Kreuzer, M. U. González and R. Quidant, *ACS Nano*, 2009, **3**, 1231–1237.
32 K.-H. Su, Q.-H. Wei, X. Zhang, J. J. Mock, D. R. Smith and S. Schultz, *Nano Lett.*, 2003, **3**, 1087–1090.
33 O. Pérez-González, N. Zabala, A. G. Borisov, N. J. Halas, P. Nordlander and J. Aizpurua, *Nano Lett.*, 2010, **10**, 3090–3095.
34 D. C. Marinica, A. K. Kazansky, P. Nordlander, J. Aizpurua and A. G. Borisov, *Nano Lett.*, 2012, **12**, 1333–1339.
35 Z. Jacob and V. M. Shalaev, *Science*, 2011, **334**, 463–464.
36 M. S. Tame, K. R. McEnery, Ş. K. Özdemir, J. Lee, S. A. Maier and M. S. Kim, *Nat. Phys.*, 2013, **9**, 329–340.
37 U. Kreibig and M. Vollmer, *Optical Properties of Metal Clusters*, Springer, 1995.
38 S. Berciaud, L. Cognet, P. Tamarat and B. Lounis, *Nano Lett.*, 2005, **5**, 515–518.
39 J. Tiggesbäumker, L. Köller, K.-H. Meiwes-Broer and A. Liebsch, *Phys. Rev. A: At., Mol., Opt. Phys.*, 1993, **48**, R1749–R1752.
40 S. Fedrigo, W. Harbich and J. Buttet, *Phys. Rev. B: Condens. Matter*, 1993, **47**, 10706–10715.
41 L. B. Scaffardi, N. Pellegri, O. de Sanctis and J. O. Tocho, *Nanotechnology*, 2005, **16**, 158–163.
42 E. Cottancin, G. Celep, J. Lermé, M. Pellarin, J. Huntzinger, J. Vialle and M. Broyer, *Theor. Chem. Acc.*, 2006, **116**, 514–523.
43 J. Lermé, B. Palpant, B. Prével, M. Pellarin, M. Treilleux, J. L. Vialle, A. Perez and M. Broyer, *Phys. Rev. Lett.*, 1998, **80**, 5105–5108.
44 V. V. Kresin, *Phys. Rev. B: Condens. Matter*, 1995, **51**, 1844–1849.
45 J. A. Scholl, A. L. Koh and J. A. Dionne, *Nature*, 2012, **483**, 421–427.
46 K. J. Savage, M. M. Hawkeye, R. Esteban, A. G. Borisov, J. Aizpurua and J. J. Baumberg, *Nature*, 2012, **491**, 574–577.

47 C. Ciracì, R. T. Hill, J. J. Mock, Y. Urzhumov, A. I. Fernández-Domínguez, S. A. Maier, J. B. Pendry, A. Chilkoti and D. R. Smith, *Science*, 2012, **337**, 1072–1074.

48 D.-K. Lim, K.-S. Jeon, J.-H. Hwang, H. Kim, S. Kwon, Y. D. Suh and J.-M. Nam, *Nat. Nanotechnol.*, 2011, **6**, 452–460.

49 R. W. Taylor, T.-C. Lee, O. A. Scherman, R. Esteban, J. Aizpurua, F. M. Huang, J. J. Baumberg and S. Mahajan, *ACS Nano*, 2011, **5**, 3878–3887.

50 D. R. Ward, N. K. Grady, C. S. Levin, N. J. Halas, Y. Wu, P. Nordlander and D. Natelson, *Nano Lett.*, 2007, **7**, 1396–1400.

51 J.-S. Huang, V. Callegari, P. Geisler, C. Brüning, J. Kern, J. C. Prangsma, X. Wu, T. Feichtner, J. Ziegler, P. Weinmann, M. Kamp, A. Forchel, P. Biagioni, U. Sennhauser and B. Hecht, *Nat. Commun.*, 2010, **1**, 150.

52 J. Kern, S. Großmann, N. V. Tarakina, T. Häckel, M. Emmerling, M. Kamp, J.-S. Huang, P. Biagioni, J. C. Prangsma and B. Hecht, *Nano Lett.*, 2012, **12**, 5504–5509.

53 S. F. Tan, L. Wu, J. K. Yang, P. Bai, M. Bosman and C. A. Nijhuis, *Science*, 2014, **343**, 1496–1499.

54 H. Cha, J. H. Yoon and S. Yoon, *ACS Nano*, 2014, **8**, 8554–8563.

55 Z. Wenqi and K. B. Crozier, *Nat. Comm.*, 2014, **5**, 5228.

56 J. Zuloaga, E. Prodan and P. Nordlander, *ACS Nano*, 2010, **4**, 5269–5276.

57 R. E. Stratmann, G. E. Scuseria and M. J. Frisch, *J. Chem. Phys.*, 1998, **109**, 8218–8224.

58 M. A. L. Marques and E. K. U. Gross, *Annu. Rev. Phys. Chem.*, 2004, **55**, 427–455.

59 L. Mao, Z. Li, B. Wu and H. Xu, *Appl. Phys. Lett.*, 2009, **94**, 243102.

60 J. Zuloaga, E. Prodan and P. Nordlander, *Nano Lett.*, 2009, **9**, 887–891.

61 P. Song, P. Nordlander and S. Gao, *J. Chem. Phys.*, 2011, **134**, 074701.

62 R. Esteban, A. G. Borisov, P. Nordlander and J. Aizpurua, *Nat. Commun.*, 2012, **3**, 825.

63 P. Zhang, J. Feist, A. Rubio, P. García-González and F. J. García-Vidal, *Phys. Rev. B: Condens. Matter Mater. Phys.*, 2014, **90**, 161407(R).

64 R. Ruppin, *J. Opt. Soc. Am.*, 1976, **66**, 449–453.

65 R. Fuchs and K. L. Kliewer, *Phys. Rev. B: Solid State*, 1971, **3**, 2270–2278.

66 P. J. Feibelman, *Phys. Rev. B: Solid State*, 1975, **12**, 1319–1336.

67 B. B. Dasgupta and R. Fuchs, *Phys. Rev. B*, 1981, **24**, 554–561.

68 W. L. Mochán, R. Fuchs and R. G. Barrera, *Phys. Rev. B*, 1983, **27**, 771–780.

69 R. Fuchs and F. Claro, *Phys. Rev. B*, 1987, **35**, 3722–3727.

70 R. Rojas, F. Claro and R. Fuchs, *Phys. Rev. B*, 1988, **37**, 6799–6807.

71 R. Ruppin, *J. Opt. Soc. Am. B*, 1989, **6**, 1559–1563.

72 A. Liebsch and W. L. Schaich, *Phys. Rev. B: Condens. Matter*, 1995, **52**, 14219–14234.

73 A. Liebsch, *Phys. Rev. B: Condens. Matter*, 1993, **48**, 11317–11328.

74 I. A. Larkin, M. I. Stockman, M. Achermann and V. I. Klimov, *Phys. Rev. B: Condens. Matter Mater. Phys.*, 2004, **69**, 121403.

75 P. Johansson, H. Xu and M. Käll, *Phys. Rev. B: Condens. Matter Mater. Phys.*, 2005, **72**, 035427.

76 J. Aizpurua and A. Rivacoba, *Phys. Rev. B: Condens. Matter Mater. Phys.*, 2008, **78**, 035404.

77 F. J. García de Abajo, *J. Phys. Chem. C*, 2008, **112**, 17983–17987.

78 J. M. McMahon, S. K. Gray and G. C. Schatz, *Phys. Rev. Lett.*, 2009, **103**, 097403.
79 J. M. McMahon, S. K. Gray and G. C. Schatz, *Nano Lett.*, 2010, **10**, 3473–3481.
80 C. David and F. J. García de Abajo, *J. Phys. Chem. C*, 2011, **115**, 19470–19475.
81 G. Toscano, S. Raza, A.-P. Jauho, N. A. Mortensen and M. Wubs, *Opt. Express*, 2012, **20**, 4176–4188.
82 S. Raza, N. Stenger, S. Kadkhodazadeh, S. V. Fischer, N. Kostesha, A.-P. Jauho, A. Burrows, M. Wubs and N. A. Mortensen, *Nanophotonics*, 2013, **2**, 131–138.
83 L. Stella, P. Zhang, F. J. García-Vidal, A. Rubio and P. García-González, *J. Phys. Chem. C*, 2013, **117**, 8941–8949.
84 N. A. Mortensen, S. Raza, M. Wubs, T. Søndergaard and S. I. Bozhevolnyi, *Nat. Commun.*, 2014, **5**, 3809.
85 A. I. Fernández-Domínguez, A. Wiener, F. J. García-Vidal, S. A. Maier and J. B. Pendry, *Phys. Rev. Lett.*, 2012, **108**, 106802.
86 T. V. Teperik, P. Nordlander, J. Aizpurua and A. G. Borisov, *Opt. Express*, 2013, **21**, 27306–27325.
87 T. V. Teperik, P. Nordlander, J. Aizpurua and A. G. Borisov, *Phys. Rev. Lett.*, 2013, **110**, 263901.
88 Y. Luo, A. I. Fernandez-Dominguez, A. Wiener, S. A. Maier and J. B. Pendry, *Phys. Rev. Lett.*, 2013, **111**, 093901.
89 G. Toscano, C. Rockstuhl, F. Evers, H. Xu, N. A. Mortensen and M. Wubs, arXiv:1408.5862 [physics.optics], 2014.
90 J. A. Scholl, A. García-Etxarri, A. L. Koh and J. A. Dionne, *Nano Lett.*, 2013, **13**, 564–569.
91 T. Dong, X. Ma and R. Mittra, *Appl. Phys. Lett.*, 2012, **101**, 233111.
92 E. Townsend and G. W. Bryant, *Nano Lett.*, 2012, **12**, 429–434.
93 W. Ekardt, *Phys. Rev. B*, 1985, **31**, 6360–6370.
94 W. A. de Heer, P. Milani and A. Châtelain, *Phys. Rev. Lett.*, 1989, **63**, 2834–2836.
95 P.-A. Hervieux and J.-Y. Bigot, *Phys. Rev. Lett.*, 2004, **92**, 197402.
96 E. Prodan, P. Nordlander and N. J. Halas, *Nano Lett.*, 2003, **3**, 1411–1415.
97 H. Winter, *Phys. Rep.*, 2002, **367**, 387–582.
98 N. D. Lang and P. Avouris, *Nano Lett.*, 2002, **2**, 1047–1050.
99 Th. Fennel, K.-H. Meiwes-Broer, J. Tiggesbäumker, P.-G. Reinhard, P. M. Dinh and E. Suraud, *Rev. Mod. Phys.*, 2010, **82**, 1793–1842.
100 F. Krausz and M. Ivanov, *Rev. Mod. Phys.*, 2009, **81**, 163–234.
101 O. I. Tolstikhin and T. Morishita, *Phys. Rev. A: At., Mol., Opt. Phys.*, 2012, **86**, 043417.
102 V. S. Popov, *Phys.-Usp.*, 2004, **47**, 855–885.
103 A. Zugarramurdi, N. Zabala, A. G. Borisov and E. V. Chulkov, *Phys. Rev. B: Condens. Matter Mater. Phys.*, 2011, **84**, 115422.
104 E. V. Chulkov, V. M. Silkin and P. M. Echenique, *Surf. Sci.*, 1999, **437**, 330–352.
105 J. M. Pitarke, F. Flores and P. M. Echenique, *Surf. Sci.*, 1990, **234**, 1–16.
106 J. M. Blanco, F. Flores and R. Pérez, *Prog. Surf. Sci.*, 2006, **81**, 403–443.
107 V. M. Silkin, J. Zhao, F. Guinea, E. V. Chulkov, P. M. Echenique and H. Petek, *Phys. Rev. B: Condens. Matter Mater. Phys.*, 2009, **80**, 121408.
108 D. Ferry and S. Goodnick, *Transport in nanostructures*, Crambridge University Press, 1999.
109 J. D. Jackson, *Classical Electrodynamics*, John Wiley, NY, 3rd edn, 1998.

110 J. W. Haus, D. de Ceglia, M. A. Vincenti and M. Scalora, *J. Opt. Soc. Am. B*, 2014, **31**, 259–269.
111 J. W. Haus, D. de Ceglia, M. A. Vincenti and M. Scalora, *J. Opt. Soc. Am. B*, 2014, **31**, A13–A19.
112 P. K. Tien and J. P. Gordon, *Phys. Rev.*, 1963, **129**, 647–651.
113 P. B. Johnson and R. W. Christy, *Phys. Rev. B: Solid State*, 1972, **6**, 4370–4379.
114 B. Hetényi, F. De Angelis, P. Giannozzi and R. Car, *J. Chem. Phys.*, 2001, **115**, 5791–5795.
115 F. J. García de Abajo and A. Howie, *Phys. Rev. B: Condens. Matter*, 2002, **65**, 115418.
116 A. Eguiluz, S. C. Ying and J. J. Quinn, *Phys. Rev. B: Solid State*, 1975, **11**, 2118–2121.
117 J. M. Pitarke, V. M. Silkin, E. V. Chulkov and P. M. Echenique, *Rep. Prog. Phys.*, 2007, **70**, 1–87.
118 P. Nordlander, C. Oubre, E. Prodan, K. Li and M. I. Stockman, *Nano Lett.*, 2004, **4**, 899–903.
119 W. Rechberger, A. Hohenau, A. Leitner, J. R. Krenn, B. Lamprecht and F. R. Aussenegg, *Opt. Commun.*, 2003, **220**, 137–141.
120 H. Xu, J. Aizpurua, M. Käll and P. Apell, *Phys. Rev. E: Stat. Phys., Plasmas, Fluids, Relat. Interdiscip. Top.*, 2000, **62**, 4318–4324.
121 J. B. Lassiter, J. Aizpurua, L. I. Hernandez, D. W. Brandl, I. Romero, S. Lal, J. H. Hafner, P. Nordlander and N. J. Halas, *Nano Lett.*, 2008, **8**, 1212–1218.
122 A. Aubry, D. Y. Lei, A. I. Fernández-Domínguez, Y. Sonnefraud, S. A. Maier and J. B. Pendry, *Nano Lett.*, 2010, **10**, 2574–2579.
123 A. I. Fernández-Domínguez, P. Zhang, Y. Luo, S. A. Maier, F. J. García-Vidal and J. B. Pendry, *Phys. Rev. B: Condens. Matter Mater. Phys.*, 2012, **86**, 241110.
124 R. C. Monreal, T. J. Antosiewicz and S. P. Apell, *New J. Phys.*, 2013, **15**, 083044.
125 M. Schnell, A. García Etxarri, A. J. Huber, K. Crozier, J. Aizpurua and R. Hillenbrand, *Nat. Photonics*, 2009, **3**, 287–291.
126 R. Esteban, G. Aguirregabiria, A. G. Borisov, Y. M. Wang, P. Nordlander, G. W. Bryant and J. Aizpurua, *ACS Photonics*, 2015, **2**, 295–305.
127 L. Wu, H. Duan, P. Bai, M. Bosman, J. K. W. Yang and E. Li, *ACS Nano*, 2013, **7**, 707–716.
128 D. R. Ward, F. Hüser, F. Pauly, J. C. Cuevas and D. Natelson, *Nat. Nanotechnol.*, 2010, **5**, 732–736.
129 A. Stolz, J. Berthelot, M.-M. Mennemanteuil, G. Colas des Francs, L. Markey, V. Meunier and A. Bouhelier, *Nano Lett.*, 2014, **14**, 2330–2338.
130 G. Hajisalem, M. S. Nezami, and R. Gordon, *Nano Lett.*, 2014, **14**, 6651–6654.

Faraday Discussions

PAPER

Unfolding the contents of sub-nm plasmonic gaps using normalising plasmon resonance spectroscopy†

Bart de Nijs,[a] Richard W. Bowman,[a] Lars O. Herrmann,[a] Felix Benz,[a] Steve J. Barrow,[b] Jan Mertens,[a] Daniel O. Sigle,[a] Rohit Chikkaraddy,[a] Anna Eiden,[c] Andrea Ferrari,[c] Oren A. Scherman[b] and Jeremy J. Baumberg*[a]

Received 6th October 2014, Accepted 14th November 2014
DOI: 10.1039/c4fd00195h

Plasmonic coupling of gold nanoparticles to a gold surface creates intense plasmonic hot spots with large electromagnetic field-enhancements within the cavity formed by the two metallic surfaces. The localised field in such structures is extremely sensitive to morphological fluctuations and subtle changes in the dielectric properties of the cavity contents. Here, we present an optical method that pins down the properties of the gap contents with high sensitivity, termed normalising plasmon resonance (NPR) spectroscopy. We use this on a variety of ultrathin molecular spacers such as filled and empty cucurbiturils, and graphene. Clear differences in the spectral positions and intensities of plasmonic modes observed in the scattering spectrum resolve thickness differences of 0.1 nm, and refractive index changes from molecular filling.

Introduction

When metal nanostructures are brought into close proximity, their electromagnetic (EM) fields couple resulting in intensely localised plasmonic hot spots confined to the gap between the structures.[1] This plasmonic coupling can result in large field-enhancements[2] making these systems promising surface enhanced Raman scattering (SERS) substrates.[3–5] For example, when a gold nanoparticle (Au NP) is placed just above a gold surface, currents induced in the Au NP couple to the induced image charges in the gold surface forming a virtual dimer with a plasmonic hot spot centred in the cavity between the nanoparticle and the substrate.[6–9]

[a]*NanoPhotonics Centre, Cavendish Laboratory, Department of Physics, University of Cambridge, JJ Thompson Ave, Cambridge, CB3 0HE, UK. E-mail: jjb12@cam.ac.uk; Tel: +44 01223 760945*

[b]*Department of Chemistry, University of Cambridge, Lensfield Road, Cambridge CB2 1EW, UK. E-mail: oas23@cam.ac.uk; Tel: +44 01223 331508*

[c]*Cambridge Graphene Centre, University of Cambridge, 9 JJ Thomson Avenue, Cambridge CB3 0FA, UK. E-mail: acf26@cam.ac.uk; Tel: +44 01223 748351*

† Electronic supplementary information (ESI) available. See DOI: 10.1039/c4fd00195h

This gold nanoparticle on mirror (NPoM) geometry has proved to be a powerful SERS substrate with enhancements allowing single molecule analysis.[9–11]

In this geometry two distinct plasmonic resonance modes are typically identified. Firstly, the characteristic *transverse* plasmonic resonance mode (T) for gold nanoparticles (around 533 nm for 80 nm Au NPs) denotes charge oscillations parallel to the gold surface and is therefore essentially unaffected by the thickness or dielectric properties of the cavity. Secondly, the coupled mode (C) results from the interaction of the EM fields of the nanoparticle and surface giving a red-shifted resonance.[12] This second mode is tightly confined within the small cavity formed between the two metallic interfaces with a resonant wavelength that increases strongly with decreasing gap spacing. Especially for sub-nm gaps, angstrom sensitivities in the spacer thickness are reached.[13–16] In addition, the plasmonic mode in the gap is highly sensitive to the dielectric properties of the spacer material which allows us to develop a new type of 'normalised' plasmon resonance (NPR) spectroscopy. That this mode results from the coupling can be shown by using a non-metallic substrate such as glass on which no C-mode is observed, as shown in Fig. S1 in the ESI.†

Results

We show how it is possible to develop an optical analysis utilising the high sensitivity of the C mode in order to probe the properties of sub-nm gaps by collecting scattering spectra of individual nanoparticles. The additional information is provided by comparing mode intensities using the uncoupled T mode as a reference. We demonstrate this on a prototypical plasmonic system but it is widely applicable across many nanostructures and devices.

To study the effect of the spacer material in coupled plasmonic structures, 80 nm Au NPs are deposited on a gold coated silicon wafer supporting a thin spacer layer. Other NP sizes can be equally well used, but below 20 nm diameter the spectral shifts can be too small for effective peak separation in the analysis below. Four different spacer layer systems are explored here: graphene (Fig. 1), monolayers of cucurbit[7]uril (CB[7], a rigid barrel shaped molecule), CB[7] with methyl viologen (MV^{2+}) as a guest, and CB[7] which is allowed to collect trace levels of molecules or ions from an aqueous solution over time.

In dark-field optical microscopy the 80 nm Au NPs on the gold surface in the NPoM geometry can be easily identified from their strong plasmonic scattering. A typical dark-field image shown in Fig. 2a resolves numerous Au NPs deposited on the gold surface with, in this case, CB[7] as a spacer layer with a spacing[3] of 0.9 nm. The two insets show two individual Au NPs, for which the dark field scattering spectra are presented in the subsequent Fig. 2b and c.

Spectral analysis when using microscopy objective lenses suffers from chromatic aberration, resulting in different focal points for each wavelength. In order to compare C mode intensities, the scattering spectra over a range of >5 μm focal distance are collected for each nanoparticle (Fig. 2b). This shows the strong effect of chromatic aberration for this lens (×100, NA0.8). Using Gaussian fits to the depth-dependent scattering for each wavelength of the focal-stack allows us to extract the focal positions (black dashed line). Using the scattering intensity along the focal line, we thus correct for axial chromatic aberration, producing the spectra shown in black in Fig. 2c. These can be compared to spectral cuts at focal

Fig. 1 Depiction of a gold nanoparticle on a gold surface with a 0.34 nm thick graphene spacer. The locations of the transverse mode (green) and coupled mode (red) are shown schematically.

heights of 0.35 μm (green) and 0.65 μm (red, top) or 1.05 μm (red, bottom), which produce different ratios of the two plasmons' strengths. The green curve shows the scattering spectrum at a focal height optimising the T mode at 533 nm, whilst

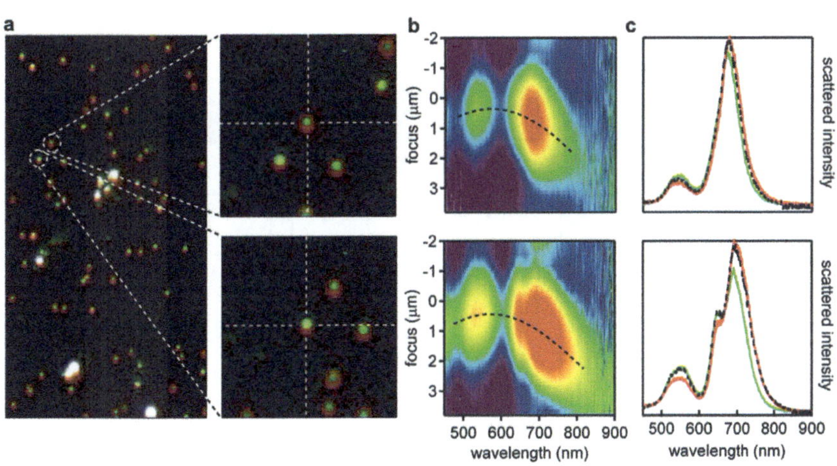

Fig. 2 Dark field images and spectra showing the NPoM scattering properties of Au NPs on Au separated by a 0.9 nm cucurbit[7]uril spacer. (a) Dark field optical microscopy image showing scattering from many NPoMs. The insets show two typical nanoparticles with the centre of collection spot indicated by white dashed lines. (b) Focal-stack of scattering spectra from the two particles from (a) showing the effect of chromatic aberration. (c) Scattering spectra at different focal heights, 0.35 μm (green) and 0.65 μm (red, top) or 1.05 μm (red, bottom), showing changes in peak ratios, with the aberration corrected spectrum in black.

the red curve shows the spectrum at the optimal focal height for the C modes between 600 nm and 800 nm. Scanning through the focal range not only eliminates the effects of chromatic aberration but also makes the method insensitive to changes in focal height when comparing nanoparticles.

To analyse the properties of the different spacers, chromatic aberration corrected spectra of a large (up to 1600) set of nanoparticles are recorded for each sample. A computer controlled stage is automatically scanned around to locate and measure each nanoparticle-on-mirror in turn, with reproducibility in spectra better than 1%. A typical example of the resulting spectra is shown in Fig. 3a where 700 superimposed corrected scattering spectra are shown for 80 nm Au NPs on a gold surface with a spacer of empty CB[7]. The majority of these spectra show a transverse peak around 533 nm and T modes visible between 700–800 nm.

To determine the averaged intensity of the plasmonic peak for each wavelength the spectral range from 450 nm to 900 nm is divided into 80 bins of 5.7 nm width, and the spectra sorted into these bins according to the position of the C mode. The resulting averaged spectra are shown in Fig. 3b together with the number of spectra found for each respective bin, shown in grey. Excluded from this analysis are spectra containing more than one coupled plasmon resonance peak, as previously reported for Au NPs on graphene,[8] which are likely due to faceting of the Au NP as suggested by Sigle *et al.*[17] This minimizes the influence of more non-spherical nanoparticles in the analysis.

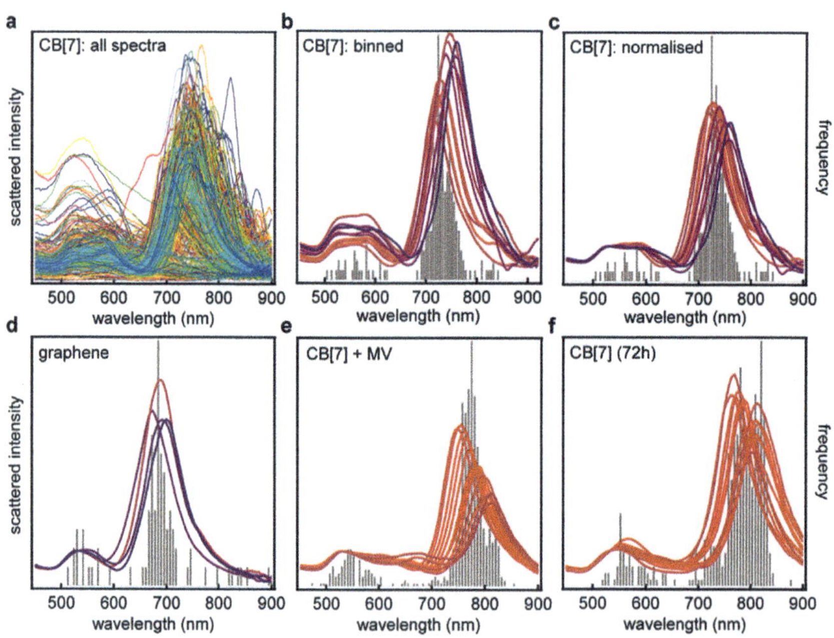

Fig. 3 Scattering spectra of 80 nm Au NPs corrected for chromatic aberration with statistical peak distributions shown in grey. (a) Superimposed spectra from 700 nanoparticles on Au with a CB[7] spacer showing two plasmon modes at 533 nm (T) and ~750 nm (C) (for the majority of the spectra). (b) Spectra from (a) sorted into bins and averaged. Bins with <1% of the spectra are not shown. (c–f) Spectra normalised to the maximum of the C mode at 533 are shown for spacers of (c) CB[7], (d) graphene, (e) CB[7] sequestering methyl viologen (MV^{2+}), and (f) CB[7] left in solution for 72 h.

Averaging the resulting spectra from each bin results in sharp peaks for the C mode with nearly identical shapes. In addition, the distribution of bin frequencies of the individual spectral peaks is a narrow Gaussian (Fig. 3c–f in grey) which for CB[7] alone has a FWHM of 15 nm. An increase in the absolute intensity of both the T and C mode is observed as the position of the C mode moves towards the infrared. This is likely produced by the polydispersity of the Au NPs, with larger nanoparticles red-shifted and scattering more strongly. Normalising all spectra to their T mode intensity at 533 nm results in a set of near-identical peaks with (for CB[7] alone) comparable intensity ratios of 1 : 5 between the T and C mode (Fig. 3c). While a third plasmon mode of quadrupolar nature is visible at 600 nm, it does not affect the analysis.

Using graphene as a spacer layer results in an even narrower FWHM of 13.4 nm with a distribution of coupled plasmon peak positions that is centred around 686 nm and with intensity ratio of 5.9 (Fig. 3d). This clearly demonstrates that changing the spacer layer strongly affects the characteristics of the C mode. Filling the CB[7] molecules with MV^{2+} prior to deposition on the gold surface results in a strong red shift of the C mode with a narrow FWHM of 14.5 nm. For this spacer the C peaks are centred around 773 nm and a decrease in the intensity ratio to between 4.5 to 3.2 is observed (Fig. 3e).

We find that leaving the empty CB[7] molecules bound to the gold surface in pure water over 3 days results in a significant cumulative shift in the coupled plasmon resonance peak. This is likely the result of CB[7] sequestering small molecules or ions from the water. This filling leads to two distinct populations in the distribution (Fig. 3f), where one group is spectrally located at 785 nm and the other at 805 nm. The two groups have intensity ratios of 5.3 and 4.5, respectively. We conclude that two or more types of molecules are sequestered by the CB[7] in this way which results in the observation of two populations. This shows the extreme sensitivity of the construct analysed in this way.

To better understand the coupled peak positions (λ_c) and intensity ratios (I_r), a series of numerical finite-difference time-domain (FDTD) simulations are performed for the NPoM geometry. Here, the same collection cone of the objective (θ_c) and the same angle of incidence for s- and p-polarized light (θ_i) as in our experiments are used (Fig. 4a). The thickness of the spacer (g) is varied in the simulations between 0.5 nm to 3 nm and the refractive index (n) is changed from

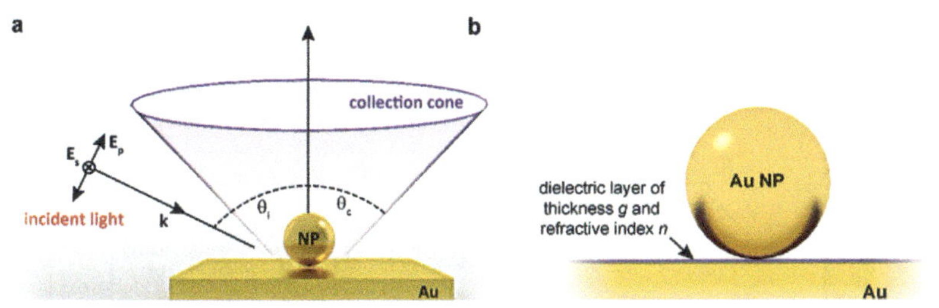

Fig. 4 The FDTD simulation geometry. (a) Particle illuminated with s- and p-polarized plane waves at angle θ_i and scattered light collected over a cone with half-angle θ_c. This replicates the experimental dark-field geometry. (b) Molecular or semiconductor monolayer in the gap modelled as an infinite dielectric sheet of thickness g (which determines the gap) and real refractive index n.

1.0 to 1.8 (Fig. 4b). The resulting resonance map for thickness and refractive index is plotted in Fig. 5 (lines). This highlights how the measured λ_c and I_r can be conformally mapped into the cavity contents. Extreme care has to be taken to ensure both the convergence of the simulations in such atomic-scale geometries, as well as the appropriate modelling of the experimental conditions.

Plotting the intensity ratios *versus* the coupled plasmon peak positions for each of the different spacer layers in one graph shows a clear distinction between them all (Fig. 5). Pleasingly, comparing the experimental results with the simulations shows a good agreement with expected spacer thicknesses for CB[7] (0.9 nm) and graphene (0.4 nm). In addition, the results from the simulations suggest that the filling of CB[7] molecules with MV^{2+} results in a larger gap spacing, which is hard to infer from any other technique. This is indeed not surprising as the MV^{2+} molecular length is greater than the 0.9 nm height of the CB[7] molecules. This shows that 0.1 nm accuracy appears to be available from this NPR spectroscopy.

In addition, the filling of the CB[7] molecules with MV^{2+} is found to result in an increase of the refractive index from <1.2 to between 1.4–1.7. By comparison, the resonance of initially empty NPoMs with CB[7] spacers that fill in solution over time shows a similarly strong increase in the refractive index from <1.2 to 1.4–1.6 with only a small variation in the gap spacing. This is likely due to mopping up of trace species in solution by the CBs, which can expel and replace the water molecules from inside, thus increasing the average electron density. Such

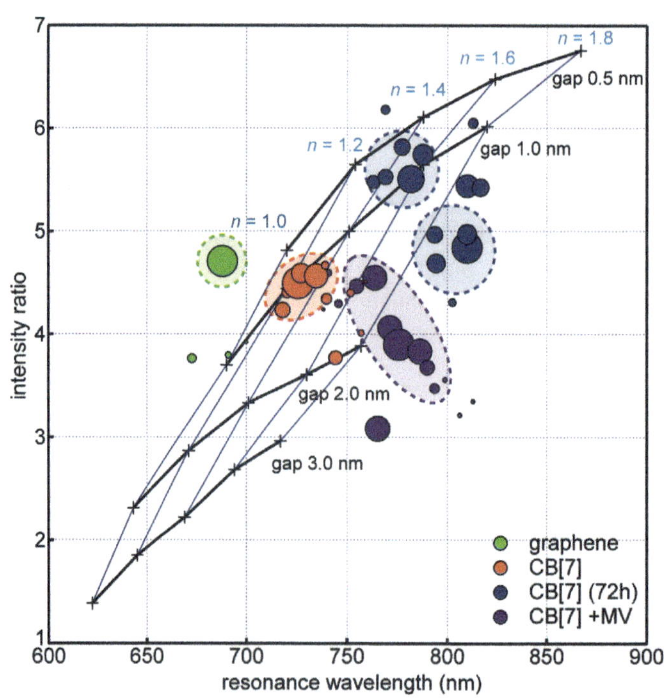

Fig. 5 Intensity ratios for each of the different spacers plotted against the resonance wavelength with results obtained from simulations using different spacer thickness and refractive indices.

changes have been noted from NMR studies on CBs,[18] and are here found in real time optical spectroscopy on 10 nm^3 volumes.

This NPR technique thus has great utility in studying nanoscale and molecular structures and their chemical modifications, in ambient conditions and in real time. It thus opens up a wide range of applications for utilising nano-optics for nano-chemistry.

Conclusions

We present a spectral protocol and analysis method to study the properties of sub-nm gaps between Au NPs and Au surfaces. This requires the correction of chromatic aberrations that are inevitably present in refractive lens based microscopy setups, but can be automated to deliver unbiased statistical samples. By comparing the intensities of the highly confined coupled plasmon with the unaffected transverse mode, a clear distinction between different sub-nm spacer layers is observed. Comparing the results from experiments with full simulations shows good agreement, recovering expected values for the different spacer layer thicknesses. Thus NPR spectroscopy reveals simple ways to monitor the contents of host–guest molecular systems, under real conditions. Our study thus opens up a new route for an all-spectroscopic analysis of the thickness and refractive index of sub-nm cavities, thin films, molecular spacers and other ultrathin layers for controlled separation in plasmonics, electronics, and molecular nanodevices.

Methods

Substrate preparation

For gold substrates, silicon wafers are coated using electron-beam-evaporation of a 70 nm thick gold layer on a 5 nm chrome adhesive layer. The surface roughness of the film is minimised by choosing a slow evaporation rate of 0.1 A s^{-1}. The gold coated substrate is submerged in a 1 mM solution of CB[7] in de-ionized water for 4 hours to deposit a layer of CB[7] molecules. For the MV^{2+} filled CB[7] sample a 1 : 1 molar ratio of CB[7] and MV is mixed and diluted to 1 mM and deposited on the gold surface as described for the empty CB[7] molecules. For the filling of the CB[7] molecules over time the gold substrate is submerged in a 1 mM CB[7] solution and left for 72 h. The graphene samples were prepared by chemical vapour deposition (CVD) on a copper substrate and transferred onto the gold film using a standard PMMA transfer technique as described in more detail in ref. 9.

Sample preparation

The 80 nm citrate capped gold nanoparticles (Au NPs) from BBI are used as received. For all CB[7] samples, a 5 μL drop of 80 nm Au NP suspension is deposited on the coated film where physisorption takes place, and is rinsed off with de-ionized water after 20 s (5 min. for graphene) to remove excess particles. The substrate is then blown dry using nitrogen. Using smaller Au NPs shifts the C mode close to the T mode and becomes weaker, as is shown in S2,†and using larger Au NPs shifts the C mode further towards the infrared, therefore 80 nm Au NPs were used since they scatter more strongly compared to the 60 nm Au NPs but still have a resonance in, or near to, the visible range.

Spectral acquisition

Spectra were collected using an Olympus BX51 optical microscope with an MPlanFLN 100×/0.80 dark-field objective fibre-coupled with a 50 μm VIS/NIR optical fibre to an Ocean Optics QE65pro spectrometer. To collect large data sets of Au NPoM scattering spectra, in-house particle tracking software was used. This software identifies particles based on a scattering intensity threshold and combined with a Prior Scientific *xyz*-motorized stage iteratively scans 9 spectral intensities in a 0.36 μm^2 map adjusting the *xy*-position accordingly to centre the collection spot on the Au NPoM. The final focus stack is then obtained by collecting 25 spectra over a focal range of 5.5 μm using a 1 second integration time.

Spectral analysis

The spectra are isolated by fitting a multi Gaussian fit over the range between 500 nm and 900 nm. The spectra are sorted and averaged in bin sizes of 5.7 nm. If more than one peak is found between 600 nm and 900 nm with an amplitude of more than half of the C mode, the spectra is considered to have a double-peaked coupled plasmon mode and is sorted out thereby minimizing the influence of non-spherical nanoparticles in the analysis. Finally, to eliminate angle dependent chromatic aberration, dark-field collection as a function of angle was normalised by scanning across a 100 μm diameter glass capillary which shows how infrared collection drops-off at high angles. The ratios, presented in Fig. 5, are corrected for this effect.

Simulations

Numerical simulations are carried out using Lumerical FDTD Solutions v8.9. The Au NP was modelled as a sphere on top of an infinite dielectric sheet of thickness *g*. The sheet thickness was varied between 0.5 and 3.0 nm. Underneath this sheet, a 200 nm thick gold layer was placed in order to replicate the experimental NP on mirror geometry. The dielectric function of gold was taken from Johnson and Christy[19] and the real refractive index of the dielectric sheet was varied between 1.0 and 1.8. The NP was illuminated with an s- and p-polarized plane wave from an angle of incidence of $\theta_i = 55°$. The scattered light was then collected within a cone of half-angle $\theta_c = 53°$ based on the numerical aperture of the objective. The resulting spectra were incoherently added in order to obtain the unpolarised scattering response that is measured experimentally.

Acknowledgements

We acknowledge financial support from EPSRC grants EP/G060649/1 and EP/I012060/1, and ERC grant LINASS 320503. Richard Bowman acknowledges financial support from Queens' College. Felix Benz acknowledges support from the Winton Programme for the Physics of Sustainability.

References

1 K. J. Savage, *et al.* Revealing the quantum regime in tunnelling plasmonics, *Nature*, 2012, **491**, 574–577.

2 S. Zou and G. C. Schatz, Silver nanoparticle array structures that produce giant enhancements in electromagnetic fields, *Chem. Phys. Lett.*, 2005, **403**, 62–67.
3 R. W. Taylor, *et al.* Precise Subnanometer Plasmonic Junctions for SERS within Gold Nanoparticle Assemblies Using Cucurbit[n]uril 'Glue', *ACS Nano*, 2011, **5**, 3878–3887.
4 R. W. Taylor, *et al.* In Situ SERS Monitoring of Photochemistry within a Nanojunction Reactor, *Nano Lett.*, 2013, **13**, 5985–5990.
5 K. Kneipp, H. Kneipp and J. Kneipp, Surface-Enhanced Raman Scattering in Local Optical Fields of Silver and Gold Nanoaggregates From Single-Molecule Raman Spectroscopy to Ultrasensitive Probing in Live Cells, *Acc. Chem. Res.*, 2006, **39**, 443–450.
6 P. K. Aravind, R. W. Rendell and H. Metiu, A new geometry for field enhancement in surface-enhanced spectroscopy, *Chem. Phys. Lett.*, 1982, **85**, 396–403.
7 P. K. Aravind and H. Metiu, The effects of the interaction between resonances in the electromagnetic response of a sphere-plane structure; applications to surface enhanced spectroscopy, *Surf. Sci.*, 1983, **124**, 506–528.
8 J. Mertens, *et al.* Controlling Subnanometer Gaps in Plasmonic Dimers Using Graphene, *Nano Lett.*, 2013, **13**, 5033–5038.
9 L. Li, T. Hutter, U. Steiner and S. Mahajan, Single molecule SERS and detection of biomolecules with a single gold nanoparticle on a mirror junction, *Analyst*, 2013, **138**, 4574–4578.
10 R. W. Taylor, *et al.* Watching individual molecules flex within lipid membranes using SERS, *Sci. Rep.*, 2014, **4**, 5940.
11 F. Benz, *et al.* Nanooptics of Molecular-Shunted Plasmonic Nanojunctions, *Nano Lett.*, 2015, **15**, 669–674.
12 J. J. Mock, *et al.* Distance-Dependent Plasmon Resonant Coupling between a Gold Nanoparticle and Gold Film, *Nano Lett.*, 2008, **8**, 2245–2252.
13 E. Hao and G. C. Schatz, Electromagnetic fields around silver nanoparticles and dimers, *J. Chem. Phys.*, 2004, **120**, 357–366.
14 J. Zuloaga, E. Prodan and P. Nordlander, Quantum Description of the Plasmon Resonances of a Nanoparticle Dimer, *Nano Lett.*, 2009, **9**, 887–891.
15 D. C. Marinica, A. K. Kazansky, P. Nordlander, J. Aizpurua and A. G. Borisov, Quantum Plasmonics: Nonlinear Effects in the Field Enhancement of a Plasmonic Nanoparticle Dimer, *Nano Lett.*, 2012, **12**, 1333–1339.
16 F. García de Abajo, J. Nonlocal Effects in the Plasmons of Strongly Interacting Nanoparticles, Dimers, and Waveguides, *J. Phys. Chem. C*, 2008, **112**, 17983–17987.
17 D. O. Sigle, *et al.* Monitoring morphological changes in 2D monolayer semiconductors using atom-thick plasmonic nanocavities, *ACS Nano*, 2015, **9**, 825–830.
18 K. A. Kellersberger, J. D. Anderson, S. M. Ward, K. E. Krakowiak and D. V. Dearden, Encapsulation of N_2, O_2, Methanol, or Acetonitrile by Decamethylcucurbit[5]uril(NH $4^+)_2$ Complexes in the Gas Phase: Influence of the Guest on 'Lid' Tightness, *J. Am. Chem. Soc.*, 2001, **123**, 11316–11317.
19 P. B. Johnson and R. W. Christy, Optical Constants of the Noble Metals, *Phys. Rev. B: Solid State*, 1972, **6**, 4370–4379.

Faraday Discussions

PAPER

High-density metallic nanogap arrays for the sensitive detection of single-walled carbon nanotube thin films

Hyeong-Ryeol Park, Seon Namgung, Xiaoshu Chen and Sang-Hyun Oh*

Received 24th November 2014, Accepted 5th December 2014
DOI: 10.1039/c4fd00233d

We have investigated the extraordinary optical transmission of terahertz waves through an array of nanogaps with varying dimensions and periodicities, and used this platform to demonstrate terahertz sensing of a thin film of single-walled carbon nanotubes. We have used atomic layer lithography to fabricate periodic arrays of nanogap loops that have a gap size of 2 nm and a loop length of 100 μm (aspect ratio of 50 000). These sub-mm-scale loops of nanogaps can sustain terahertz electromagnetic resonances along the contour. We have characterized the transmission of terahertz waves through the nanogap arrays and investigated the influence of inter-gap electromagnetic coupling as the array periodicity shrinks from 100 μm to 4 μm. While the gaps occupy only 0.1% of the surface area, we have measured an amplitude ($|E|$) transmittance of over 50% due to the strong and broadband field enhancement inside the nanogaps. The absolute transmission through the 2 nm gaps along the rectangular loops can be boosted up to 25%, while it is only 1% for annular gaps with the same perimeter. Furthermore, the extremely tight field confinement and strong field enhancement near the 2 nm gap lead to 43% extinction of THz waves in a 10 nm-thick film of single-walled carbon nanotubes over the gaps. On the other hand, THz extinction by the same nanotube film on a bare glass substrate is only 2%. These nanogaps pave the way toward developing sensitive terahertz detectors for biological and chemical targets.

1 Introduction

Light-matter interactions occurring in a nanometer-scale regime can reveal extreme confinement and enhancement of optical energies,[1–10] which have inspired diverse applications of plasmonic concepts such as quantum tunneling,[11–13] optical modulator,[14,15] nonlinear plasmonics,[16–18] optical rectification,[19] and biochemical sensing.[20] While the majority of the plasmonic experiments focus on the near-infrared and visible regions, enhanced light-matter

Department of Electrical and Computer Engineering, University of Minnesota, Minneapolis, MN 55455, USA. E-mail: sang@umn.edu

interactions in metallic nanogaps can also occur in the mid-infrared and terahertz (THz) regions.[21] However, the ability to observe and harness these phenomena for practical applications has been limited by the difficulty in mass-producing metallic nanostructures with nanometer-scale resolution.

We have recently developed a technique called atomic layer lithography to address this challenge and fabricate vertically oriented nanogaps with arbitrary geometries and atomic-scale resolution over a large area.[22–24] This technique uses atomic layer deposition (ALD)[25] as a top–down patterning method and converts its Angstrom-scale thickness control into lateral patterning resolution. As these metallic nanogaps have precisely controlled gap sizes, positions, and shapes, it is straightforward to engineer their plasmon-mediated optical transmission, reflection, and absorption in a broad spectral range from visible to THz regions.

Here we use this technique to produce a periodic array of 2 nm-wide gap loops with a perimeter of 100 μm, and exploit its strong THz resonances for sensing a thin film of single-walled carbon nanotubes (SWCNT). The resonance properties of our nanogap arrays strongly depend on the gap size as well as the transverse distance between the nanogap loops, which controls the coupling strength.[26] To characterize this inter-gap coupling with varying periods, the rectangular shape of the nanogap loop was chosen to shrink the horizontal period down to a deep subwavelength scale of $\lambda/100$. With close-packing of the nanogap loops, we show that the confinement of THz waves inside the 2 nm-wide gaps enables the detection of SWCNT films as thin as 10 nm with unprecedented performance.

2 Design and fabrication

As described in our previous work,[23] the atomic layer lithography technique consists of directional metal evaporation and a simple stripping process with adhesive tape (*e.g.* 3M Scotch® Magic Tape). The steps are summarized in Fig. 1a–d: (a) patterning of trenches in a metal film using a standard technique such as optical lithography; (b) conformal coating of the substrate with an Al_2O_3 layer *via* ALD; (c) filling of the trenches with metal *via* directional evaporation; and (d) removal of the excess metal atop the substrate using adhesive tape. To enhance the sensitivity for the detection of thin films, the remaining top Al_2O_3 layer was removed *via* reactive ion etching (RIE) with an inductively coupled plasma (Plasmalab System100 ICP180, Oxford Ltd.).

Using this method, we fabricated arrays of nanogaps with 100 μm-long contours of rectangular loops with varying periods (p) in 150 nm-thick gold films on glass substrates (Fig. 1e). The perimeter of the loop determines in-plane resonances at THz frequencies.[23,27] Compared to annular gaps with a perimeter of 100 μm (Fig. 1f), we can reduce the period of the rectangular loops with the same perimeter from 100 μm down to 4 μm (Fig. 1g–k). All of the nanogaps have the same gap size of 2 nm.

3 Results and discussion

To characterize the nanogap loops in the THz region, we performed THz time-domain spectroscopy in a frequency range from 0.2 to 1.0 THz. A single-cycle THz source was generated from a biased GaAs emitter (Tera-SED, Gigaoptics, GmbH) illuminated by a femtosecond Ti:sapphire laser pulse train with a center

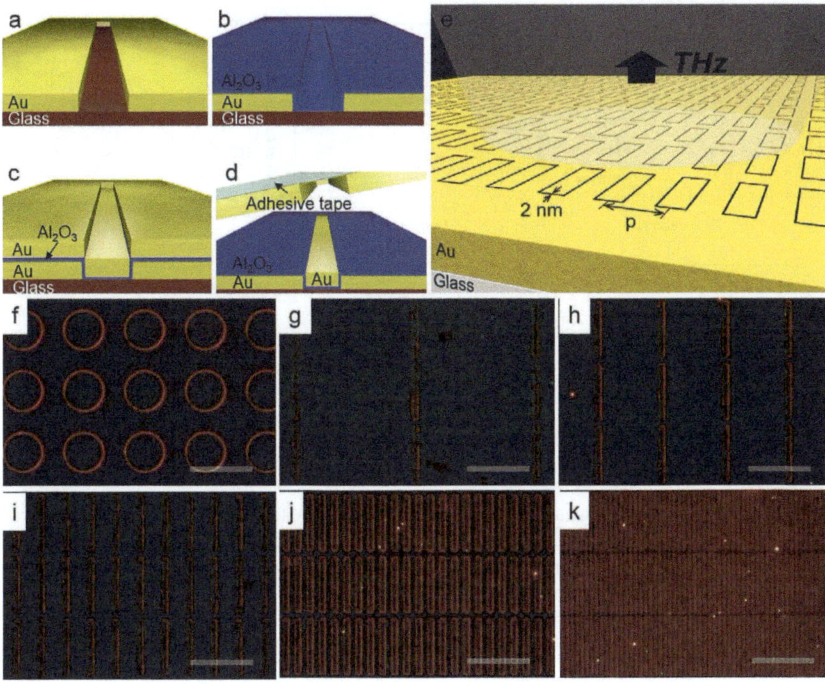

Fig. 1 (a) A trench is formed in a metal film by standard photolithography. (b) The structure is conformally coated with an Al_2O_3 layer using atomic layer deposition. (c) Another metal layer is embedded in the first metal structure using directional metal evaporation. (d) Excess metal atop the first metal layer is removed with Scotch tape. This is possible because of the poor adhesion between gold and Al_2O_3. (e) Illustration of an array of rectangular nanogap loops that have a gap size of 2 nm and transverse period (p), in a gold film on a glass substrate. Optical micrographs of: (f) annular gaps with a period of 50 μm and a perimeter of 100 μm, (g–j) nanogaps along 5×45 μm^2 rectangular patterns, and (k) nanogaps along 2×48 μm^2 rectangular patterns with diverse periods of (g) 100, (h) 50, (i) 20, (j) 10, and (k) 4 μm. All scale bars are 50 μm.

wavelength of 780 nm, a repetition rate of 80 MHz, and a pulse width of 90 fs (Mai Tai XF, Newport Corporation). We have only considered normal incidence, and the electric field of the incident wave was polarized along the short edge of the rectangle pattern. The p-polarized THz wave was illuminated from the substrate side of the nanogap loop sample and the transmitted THz pulses through the samples were detected with an electrooptic sampling method using a 1 mm-thick ZnTe crystal (INGCRYS Laser System Ltd.).

We have measured time traces of THz pulses transmitted through each nanogap loop array (Fig. 2a). The period of oscillation on the transmitted THz time signals decreases as the transverse period of the nanogaps shrinks from 100 μm to 4 μm. These distortions of transmitted THz pulses in the time domain correspond to blue-shifted resonance peaks in the spectral domain. By Fourier-transforming the temporal data, the transmitted (electric field) amplitude ($|E|$) spectra are obtained in the frequency domain (Fig. 2b). The measured spectra were normalized by a reference signal passing through a 3 mm × 3 mm aperture in a stainless steel plate. To estimate the field enhancement near the nanogap

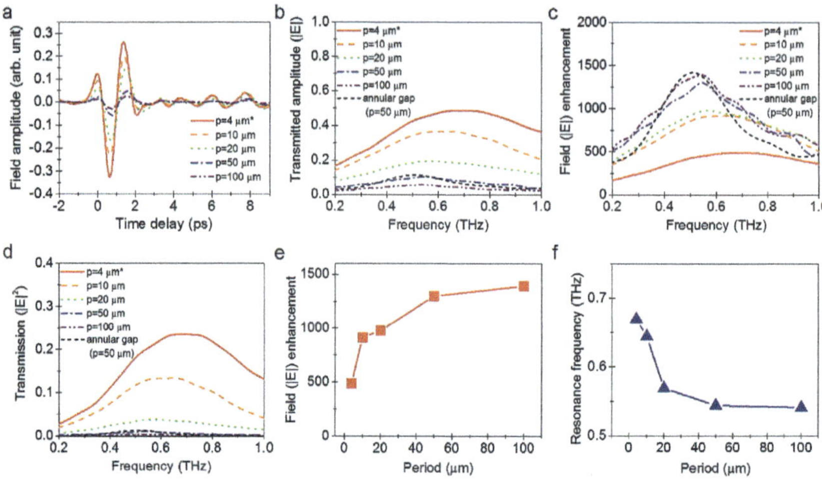

Fig. 2 (a) Time-domain signals for transmitted THz waves through arrays of nanogaps along 100 μm-long rectangular loops with periods, $p = 4$, 10, 20, 50, and 100 μm. The 4 μm-period sample has only the rectangle pattern of $2 \times 48\ \mu m^2$, and others have the rectangle pattern of $5 \times 45\ \mu m^2$. (b) Fourier-transformed THz transmitted amplitude ($|E|$) spectra, normalized by the reference signal through a bare glass substrate. The black dashed curve shows the transmitted amplitude spectrum for annular gaps with $p = 50$ μm and a perimeter of 100 μm, same as with the rectangular loops. (c) Electric field enhancement spectra in the same frequency domain as (b). (d) Transmission ($|E|^2$) spectra. (e) Field enhancement and (f) resonance frequency of the nanogap loops as a function of the array period.

using Kirchhoff's diffraction formalism,[4] we normalized the transmitted amplitude spectra by each aerial coverage ratio of the nanogaps and obtained the field ($|E|$) enhancement spectra shown in Fig. 2c.

For the 4 μm period, the transmitted field amplitude increases to 50% at the resonance frequency despite having an aerial coverage of only 0.1%. Furthermore, the maximum transmittance ($|E|^2$) for $p = 4$ μm reached up to 25% while the maximum transmission through annular gaps (black dashed line) with the same perimeter is only 1%, as shown in Fig. 2d. Fig. 2e and f show that, with the decreasing period, the field enhancement at resonance decreases, the linewidth is broader, and the resonance frequency shifts towards higher frequencies. These phenomena can be explained by short-range interactions between the adjacent gaps.[26] We emphasize that the 2 nm-wide gaps with the period of 4 μm lead to over 50% of the transmitted amplitude, still sustaining a high field enhancement of 500. These high-density metallic nanogaps will be useful to study surface-enhanced THz absorption.[20,28]

To demonstrate the utility of our high-density nanogap arrays, we measured the nanogap-enhanced THz absorption in a SWCNT film, which can be made as thin as 10 nm (Fig. 3a).[29] SWCNTs are relatively less metallic than multiwalled carbon nanotubes. Normally a thin film (~10 nm) of SWCNTs formed on a bare glass substrate interacts only weakly with an incident THz wave. Because the probing range of our sub-10 nm gap matches the thickness of the SWCNT films, we chose them as a model system to demonstrate nanogap-enhanced THz

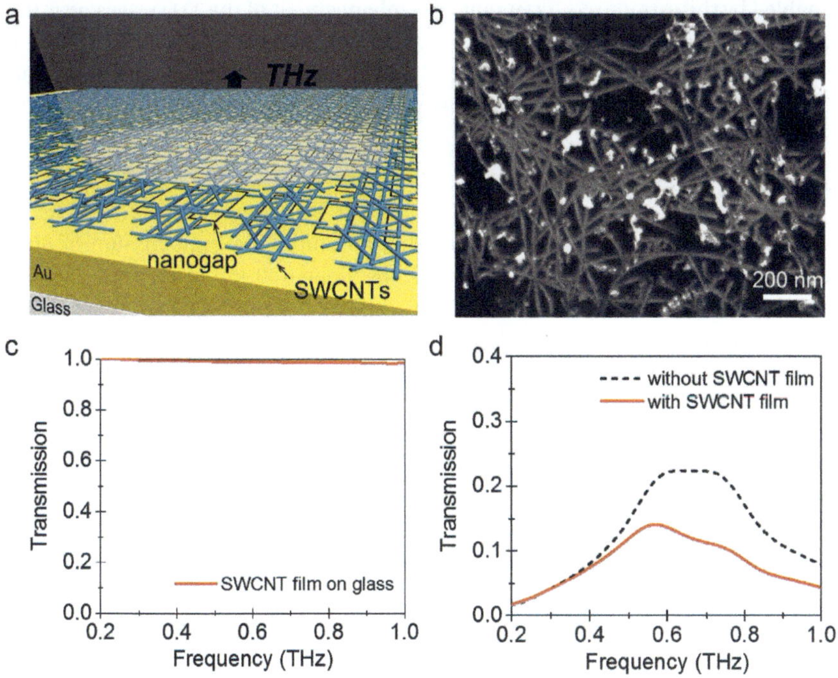

Fig. 3 (a) Schematic diagram of THz transmission ($|E|^2$) coupled to a single-walled carbon nanotube (SWCNT) film on 2 nm-wide gaps along a rectangular loop in a gold film. (b) Scanning electron micrograph of the SWCNT film on a bare glass substrate. (c) THz transmission spectrum through the SWCNT film on a glass substrate only. (d) Measured THz transmission spectra for the 2 nm-wide gap array with (red solid line) and without (black dashed line) the SWCNT film.

detection sensitivity. Furthermore, a molecular linker-free assembly method[29] allowed us to produce wafer-scale films of SWCNTs with sub-10 nm thickness on various substrates.

To prepare the SWCNT suspensions, commercial SWCNTs were dispersed in 1,2-dichlorobenzene with a concentration of 0.1 mg ml^{-1}. The SWCNTs were dip-coated on target substrates for 60 seconds, and rinsed thoroughly with 1,2-dichlorobenzene. Previous work[29] has shown that SWCNTs possess strong affinity for most bare surfaces, such as Au, SiO_2, Si, Al, and glass due to the natural polarization of the pristine surface, and the resulting SWCNT films had an effective thickness of 3–10 nm. Fig. 3b shows the top-view scanning electron micrograph of a SWCNT film on a bare glass substrate.

To experimentally measure the THz extinction in the SWCNT film, we performed THz time-domain spectroscopy on both the bare glass substrate and the 2 nm-wide gaps with a period of 4 μm with and without the SWCNT film. Surprisingly, while the SWCNT film on the bare substrate shows a tiny extinction of only 2% over a broad THz frequency range (Fig. 3c), the 2 nm-gaps with the SWCNT film allow for a 43% extinction at a resonance frequency of 0.65 THz. This strongly enhanced extinction is caused by the enhanced absorption in the SWCNT film and the large change of the local refractive index around the nanogaps enabled by the extreme confinement and enhancement of the THz waves near the nanogaps.[20,30,31] The highly conductive properties of the ultra-thin SWCNT films on the nanogaps also show promise as transparent conducting electrodes and lightweight shielding materials against electromagnetic interference.[32]

4 Conclusions

We have experimentally demonstrated that an amplitude transmission of 50% can be realized through periodic 2 nm-gaps along 100 μm-long loops with an aerial coverage ratio of as small as 0.1%. Atomic layer lithography enabled the wafer-scale production of tightly packed and well-defined arrays, wherein the separation between adjacent nanogap elements can be less than 5 μm. The ability to closely pack nanogap loops while maintaining a small aerial coverage is crucial in achieving a broad spectrum and high absolute transmission without sacrificing field enhancements. For practical sensing applications, we have shown that high density metallic nanogaps significantly enhance the THz extinction up to 43% in the SWCNT film, which is as thin as 10 nm, over the gaps, while the THz extinction through the same nanotube film on a bare glass substrate is only 2%. Our nanogap devices with high transmittance and high sensitivity could open up a new avenue for biochemical sensing, nonlinear optics, and active optical modulators.

Acknowledgements

This work was supported by the U.S. Department of Defense (DARPA Young Faculty Award N66001-11-1-4152; H.R.P., X.S.C., S.H.O.) and the NSF CAREER Award (H.R.P. and S.H.O.). Device fabrication was performed at the University of Minnesota, Nanofabrication Center, which receives support from the National Science Foundation (NSF) through the National Nanotechnology Infrastructure Network program, and the Characterization Facility, which has received capital equipment funding from NSF through the Materials Research Science and Engineering Center. X.S.C. acknowledges the support from the 3M Science and Technology Fellowship and the University of Minnesota Doctoral Dissertation Fellowship.

References

1 H. T. Miyazaki and Y. Kurokawa, *Phys. Rev. Lett.*, 2006, **96**, 097401.
2 H. A. Atwater, *Sci. Am.*, 2007, **296**, 56–63.
3 S. Lal, S. Link and N. J. Halas, *Nat. Photonics*, 2007, **1**, 641–648.
4 M. A. Seo, H. R. Park, S. M. Koo, D. J. Park, J. H. Kang, O. K. Suwal, S. S. Choi, P. C. M. Planken, G. S. Park, N. K. Park, Q. H. Park and D. S. Kim, *Nat. Photonics*, 2009, **3**, 152–156.
5 R. F. Oulton, V. J. Sorger, T. Zentgraf, R. M. Ma, C. Gladden, L. Dai, G. Bartal and X. Zhang, *Nature*, 2009, **461**, 629–632.
6 J. A. Schuller, E. S. Barnard, W. S. Cai, Y. C. Jun, J. S. White and M. L. Brongersma, *Nat. Mater.*, 2010, **9**, 193–204.
7 L. Novotny and N. van Hulst, *Nat. Photonics*, 2011, **5**, 83–90.

8 A. Garcia-Martin, D. R. Ward, D. Natelson and J. C. Cuevas, *Phys. Rev. B: Condens. Matter Mater. Phys.*, 2011, **83**, 193404.
9 C. Ciraci, R. T. Hill, J. J. Mock, Y. Urzhumov, A. I. Fernandez-Dominguez, S. A. Maier, J. B. Pendry, A. Chilkoti and D. R. Smith, *Science*, 2012, **337**, 1072–1074.
10 J. B. Pendry, A. Aubry, D. R. Smith and S. A. Maier, *Science*, 2012, **337**, 549–552.
11 R. Esteban, A. G. Borisov, P. Nordlander and J. Aizpurua, *Nat. Commun.*, 2012, **3**, 825.
12 K. J. Savage, M. M. Hawkeye, R. Esteban, A. G. Borisov, J. Aizpurua and J. J. Baumberg, *Nature*, 2012, **491**, 574–577.
13 P. Cristofolini, G. Christmann, S. I. Tsintzos, G. Deligeorgis, G. Konstantinidis, Z. Hatzopoulos, P. G. Savvidis and J. J. Baumberg, *Science*, 2012, **336**, 704–707.
14 M. Liu, X. B. Yin and X. Zhang, *Nano Lett.*, 2012, **12**, 1482–1485.
15 S. H. Lee, M. Choi, T. T. Kim, S. Lee, M. Liu, X. Yin, H. K. Choi, S. S. Lee, C. G. Choi, S. Y. Choi, X. Zhang and B. Min, *Nat. Mater.*, 2012, **11**, 936–941.
16 S. Palomba, H. Harutyunyan, J. Renger, R. Quidant, N. F. van Hulst and L. Novotny, *Philos. Trans. R. Soc., A*, 2011, **369**, 3497–3509.
17 M. Kauranen and A. V. Zayats, *Nat. Photonics*, 2012, **6**, 737–748.
18 J. B. Lassiter, X. Chen, X. Liu, C. Ciracì, T. B. Hoang, S. Larouche, S.-H. Oh, M. H. Mikkelsen and D. R. Smith, *ACS Photonics*, 2014, **1**, 1212–1217.
19 D. R. Ward, F. Huser, F. Pauly, J. C. Cuevas and D. Natelson, *Nat. Nanotechnol.*, 2010, **5**, 732–736.
20 H. R. Park, K. J. Ahn, S. Han, Y. M. Bahk, N. Park and D. S. Kim, *Nano Lett.*, 2013, **13**, 1782–1786.
21 D. M. Mittleman, *Nat. Photonics*, 2013, **7**, 666–669.
22 H. Im, K. C. Bantz, N. C. Lindquist, C. L. Haynes and S.-H. Oh, *Nano Lett.*, 2010, **10**, 2231–2236.
23 X. Chen, H. R. Park, M. Pelton, X. Piao, N. C. Lindquist, H. Im, Y. J. Kim, J. S. Ahn, K. J. Ahn, N. Park, D.-S. Kim and S.-H. Oh, *Nat. Commun.*, 2013, **4**, 2361.
24 X. Chen, H.-R. Park, N. C. Lindquist, J. Shaver, M. Pelton and S.-H. Oh, *Sci. Rep.*, 2014, **4**, 6722.
25 S. M. George, *Chem. Rev.*, 2010, **110**, 111–131.
26 Y. M. Bahk, H. R. Park, K. J. Ahn, H. S. Kim, Y. H. Ahn, D. S. Kim, J. Bravo-Abad, L. Martin-Moreno and F. J. Garcia-Vidal, *Phys. Rev. Lett.*, 2011, **106**, 013902.
27 J. Shu, C. Y. Qiu, V. Astley, D. Nickel, D. M. Mittleman and Q. F. Xu, *Opt. Express*, 2011, **19**, 26666–26671.
28 F. Le, D. W. Brandl, Y. A. Urzhumov, H. Wang, J. Kundu, N. J. Halas, J. Aizpurua and P. Nordlander, *ACS Nano*, 2008, **2**, 707–718.
29 M. Lee, J. Im, B. Y. Lee, S. Myung, J. Kang, L. Huang, Y. K. Kwon and S. Hong, *Nat. Nanotechnol.*, 2006, **1**, 66–71.
30 F. J. Garcia-Vidal, L. Martin-Moreno, T. W. Ebbesen and L. Kuipers, *Rev. Mod. Phys.*, 2010, **82**, 729–787.
31 W. Gao, J. Shu, K. Reichel, D. V. Nickel, X. He, G. Shi, R. Vajtai, P. M. Ajayan, J. Kono, D. M. Mittleman and Q. Xu, *Nano Lett.*, 2014, **14**, 1242–1248.
32 J. T. Hong, D. J. Park, J. Y. Moon, S. B. Choi, J. K. Park, F. Rotermund, J. Y. Park, S. Lee and Y. H. Ahn, *Appl. Phys. Express*, 2012, **5**, 015102.

Faraday Discussions

PAPER

The critical importance of gap modes in surface enhanced Raman scattering

Masayuki Futamata,* Maho Ishikura, Chiaki Iida and Saori Handa

Received 2nd October 2014, Accepted 21st November 2014
DOI: 10.1039/c4fd00188e

For efficient utilization of surface plasmons in surface enhanced Raman scattering (SERS), we investigated the gap modes in flocculates of metal nanoparticles (MNPs), and between MNPs and metal substrates under an external and an attenuated total reflection (ATR) geometry. First, the adsorbed state of thiol molecules and counter ions trapped in solutions were elucidated using flocculation-SERS, in which closely adjacent nanoparticles are formed by using interactions between MNPs and target species. Second, we obtained a pronounced enhancement of 10^5–10^8 at a nanogap between gold nanoparticles (AuNPs) and various metal substrates even with a large damping under an external geometry. Markedly larger enhancement was obtained for larger AuNPs, by a factor of 10^3 for particles with a radius (r) of 50 nm compared with those of $r = 15$ nm in this geometry. Finally, we attained an additional enhancement factor under an ATR geometry by a coupling of propagating surface plasmons with gap modes.

Introduction

Metal nanoparticles (MNPs) and metal nanostructures, which are precisely fabricated by top-down beam technologies and bottom-up chemical syntheses, have been utilized in surface enhanced Raman scattering (SERS). A marked enhancement in SERS, typically by a factor (G) of 10^4–10^6 on average, arises from specific electronic states in the chemical enhancement (CE) mechanism and localized surface plasmons (LSPs) of metal nanostructures in electromagnetic (EM) enhancement.[1] In particular, surface plasmons are crucial to the *in situ* characterization of individual molecules, because the EM mechanism provides a much larger enhancement ($G = 10^4$–10^5 for isolated, and $G = 10^8$–10^{10} for adjacent particles) irrespective of adsorbed species in contrast to the CE mechanism ($G = 10^2$–10^3) available only for specific molecules and vibrational modes. Even single molecule detection by Raman scattering has been reported at a nanogap of adjacent metal nanostructures (flocculates, intrinsically different from aggregates[2,3]) by using the coupling of LSP resonances, which is one of gap modes (Fig. 1a).

Graduate School of Science and Engineering, Saitama University, Saitama 338-8570, Japan. E-mail: futamata@chem.saitama-u.ac.jp

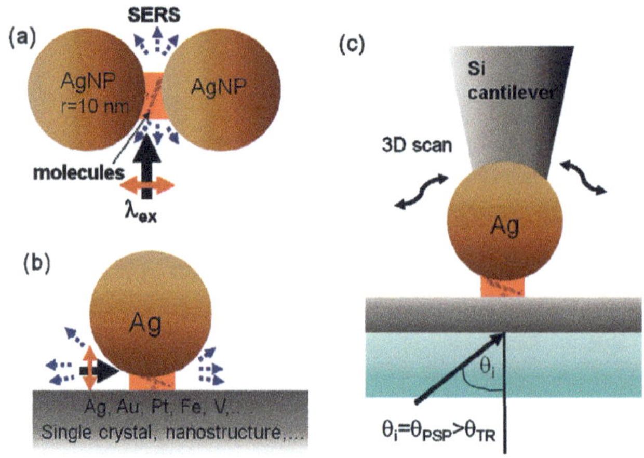

Fig. 1 Three distinct gap mode geometries: (a) in flocculation of MNPs, (b) under an external geometry, (c) in TERS under an ATR geometry.

To establish gap mode Raman spectroscopy using flocculated MNPs as an analytical tool, it is crucial to manage the conflicting properties of surface residuals like citrates, which cause electrostatic stabilization of MNPs in suspension, and steric hindrance for adsorption.[4–9] To overcome these issues, we exploited a self-assembled monolayer (SAM) of thiol molecules here, which to replace surface residuals and form flocculates of MNPs. In particular we focused on thiols having functional groups, such as carboxylic and amino groups at the solution side, to trap counter ions in solution. There are plenty of reports on SERS of various SAM films on MNPs,[11–15,28] however, their detailed adsorbed states are still under discussion. For instance, photochemical reactions,[11] and photo-induced charge transfer effects[12,13] have been reported for *p*-aminothiophenol (PATP) on metal surfaces. Also, distinct adsorption features of *p*-mercaptobenzoic acid (PMBA) *via* S–M bonds or COO^- groups are reported on Ag and Au surfaces,[15] which may affect pH sensing in living cells.[14] Thus, it is valid to investigate the effect of these PATP- and PMBA-SAM films on AgNPs using flocculation-SERS to characterize the detailed adsorbed state and interactions with those on neighboring AgNPs and counter ions.

As the second geometry for the excitation of gap modes, MNPs/adsorbates/metal substrates have recently been developed,[16–18] in which a coupling of dipoles in metal nanoparticles and image dipoles in metal substrates provides an enormously enhanced electric field at a nanogap (Fig. 1b). In this geometry, various metal substrates are available including single crystals,[18] and transition metal substrates with large damping.[16,17] MNPs are efficiently immobilized on metal substrates *via* adsorbed molecules through van der Waals and electrostatic interactions.[19] Metiu *et al.*[20] analyzed the enhanced electric field at a nanogap between MNPs and metal substrates in early '80s with respect to the electromagnetic enhancement mechanism of SERS. However, details of the gap modes have not been reported, for instance how the enhancement factor depends on the dielectric constants of substrates, and also on the size of the MNPs. Here we investigated these issues using FDTD calculations and SERS measurements.

In the third configuration, gap mode plasmons are excited in an ATR geometry, in which AgNPs are attached to adsorbates on a Ag film/prism. Propagating surface plasmons (PSPs) are excited at a resonance angle of incidence higher than the critical angle for total reflection, at which the wave vector and the energy of excitation light in accord with those of PSPs at an Ag film/(molecules) air interface (Fig. 1c). The ATR configuration is promising in tip-enhanced Raman scattering (TERS) to suppress plausible background signal from samples illuminated with a laser spot (1–10 μm). Indeed, only an evanescent wave at sample surfaces, where the tip probe (with a size of 10–30 nm) approaches in TERS, is transferred to propagating light under an ATR geometry. However, the coupling of PSPs and gap modes has not been evaluated in detail. For instance it is not obvious if gap modes under an ATR geometry provide larger enhancement compared with an external geometry. The excitation efficiency of PSPs is probably perturbed by the AgNPs attached to the Ag films/prism, and thereby related to the sensitivity in TERS. We evaluated the SERS enhancement factor under PSP and gap mode excitation, using reflectivity and SERS measurements as well as with theoretical calculations, while varying surface coverage of AgNPs (θ_{AgNP}).

Thus, we investigated gap mode Raman spectroscopy in three distinct geometries such as flocculates of AgNPs, an external geometry for AuNPs/adsorbates/metal substrates, and also an ATR configuration for AgNPs/adsorbates/Ag films/prisms to establish it as an analytical tool, while solving their present issues.

Experimental setup

AgNPs and AuNPs were prepared by chemical reduction using trisodium citrate.[4–8] Sulfuric acid, sodium hydroxide and various thiol molecules were purchased from Wako Pure Chemicals, and used without further purification.

(1) Flocculation SERS

p-Aminothiophenol (PATP) and *p*-mercaptobenzoic acid (PMBA) molecules were adsorbed onto AgNPs in aqueous solutions, whose pH values were adjusted between pH = 2–10 by the addition of H_2SO_4 and NaOH to protonate or deprotonate the thiol molecules. Here, the pK_a of an $-NH_3^+$ group in PATP and a -COOH group in PMBA in solution are 6.9 and 5.8, respectively.[8,21]

(2) Gap modes under an external configuration

Metal thin plates such as Sn, Zn, Pt, Pb, Cu and Al with sufficiently high purity (99.99–99.999 %) were purchased from Nilaco Co., and used as substrates in gap modes. Self-assembled monolayers of thiophenol (TP) were formed on these metals[22] by immersing them in ethanol solutions of TP (1 mM) for 1 h, and rinsed with pure methanol to remove excess and uncombined thiol molecules, yielding TP–SAM on the above metal substrates. Then, a small amount of AuNP suspension (~100 μL with a density of ~10^{12} particles per mL) was dropped onto the TP–SAM coated metal substrates, and left for 1 h to be immobilized through van der Waals interaction between TP, the metal substrates and the AuNPs.[19] Silicon wafers and cover glass slips were also used as substrates, which were immersed in *p*-styryl trimethoxy silane (PSTMS) in 2% methanol solutions for 10 min followed by through rinsing with pure methanol, and then by heating at 373 K for 10 min to polymerize the PSTMS. Thus, a monolayer of PSTMS was formed on Si *via* Si–O–Si covalent bonds.

(3) Coupling of PSPs and gap modes in an ATR geometry

TP–SAM films were formed on Ag films (45 nm thickness) evaporated on a Dove prism, BK-7. AgNPs, on which residual citrates were replaced by Cl^-, were immobilized on TP–SAM/Ag films/prism substrates using a drop coating method.

An FDTD calculation was used to evaluate the local electric field at a nanogap in three different geometries; (a) between AgNPs, (b) between AuNPs and various metal substrates under an external geometry, and (c) between AgNPs and Ag films on a Dove prism under an ATR geometry at various angles of incidence (40–60°) and at experimentally used excitation wavelengths of 632 nm (b) and 532 nm (c). Extinction spectra were measured to characterize the LSP resonance of isolated and flocculated AgNPs in solutions using a JASCO 540 UV-vis spectrophotometer. SERS spectra were measured for the sample solutions as those in extinction measurements, and also for AgNP/thiol–SAM/substrates under an external geometry (at an angle of incidence of 65°, with p-polarization) using a modified Renishaw micro-Raman spectrometer with a thin layer solution cuvette and a He–Ne laser (632.8 nm and <1 $\mu W\ \mu m^{-2}$). Also a Chromex polychromator (250is) equipped with a solid laser (532 nm, 20 mW) was used for SERS measurements of the AgNP/thiol/Ag film/prism samples under an ATR geometry. The surface coverage of the AgNPs (θ_{AgNP}) as well as the AuNPs, which was adjusted by dilution of the as-prepared AgNP solutions from 1/1 to 1/200 in experiments (3), were evaluated by scanning electron microscope (SEM) images using Hitachi S-4100, as exemplified in Fig. 9a–d. SERS enhancement factors at a nanogap in experiments (2) and (3) were estimated using θ_{AgNP} and by assuming a SERS active area on metal substrates with a diameter of 1/5 of the particles sizes, on the basis of the enhanced electric field distribution obtained by FDTD calculations.

Results and discussion

(1) Flocculation-SERS

AgNPs coated with PATP–SAM films feasibly formed flocculates when the pH was not controlled.[23] Because the pH in PATP solutions (10^{-4} M) used for the formation of SAM films is ~8.4, higher than the pK_a of 6.85 for $-S\text{-}\phi\text{-}NH_3^+ \leftrightarrow -S\text{-}\phi\text{-}NH_2$, neutral PATP ($-NH_2$) species are predominant on the AgNPs. Accordingly, the van der Waals attractive potential, ~200 kJ mol^{-1},[10] much larger than thermal energy 2.5 kJ mol^{-1}, caused the neutral PATP–SAM films and AgNPs to form flocculates. In contrast, AgNPs coated with PATP–SAM remained isolated at pH ~ 4, which is much lower than the pK_a for $-NH_3^+$ in PATP, owing to electrostatic repulsion between the protonated amino groups on neighboring AgNPs. This presumption was supported by the fact that flocculates of AgNPs (Fig. 2a) were formed by the addition of Na_2SO_4 (1 mM) or $HClO_4$ (5 mM). Accordingly, flocculates of AgNP–PATP$^+\cdots X^{n-}\cdots$PTAP$^+$–AgNP (here $X^{n+} = SO_4^{2-}$ or ClO_4^-) provided pronounced SERS spectra of trapped anions such as the ν_{1S} (SO_4^{2-}) band at 972 cm^{-1} and the ν_{1S} (ClO_4^-) band at 931 cm^{-1}, in addition to the SERS spectra from the protonated PATP ($-NH_3^+$) ions. Concerning the plausible photochemical reaction of PATP on Ag surfaces, we investigated the SERS of PATP in flocculated

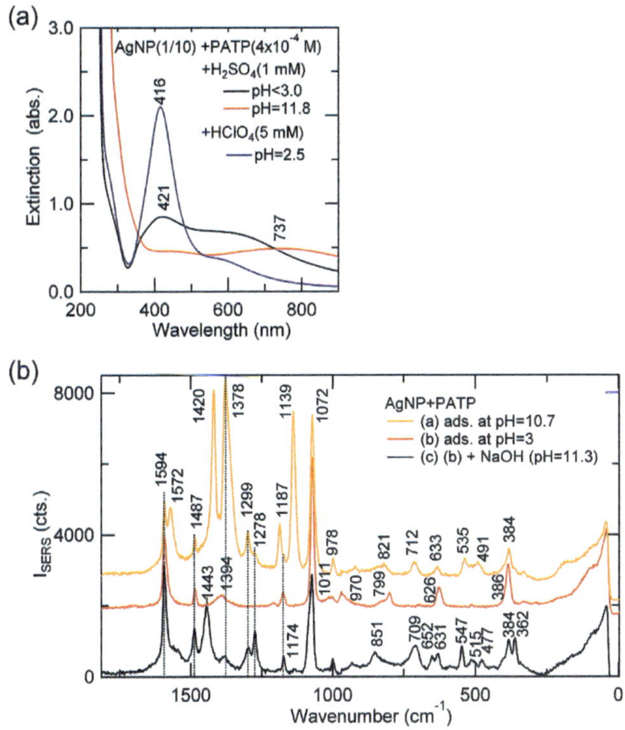

Fig. 2 Extinction spectra (a) and SERS spectra (b) of flocculates of PATP on AgNPs adsorbed at different pH (see the text for more detail).

AgNPs at different pH values. First, the SERS spectra of PATP were observed on AgNPs at basic pH (~10.7, Fig. 2b), which are significantly different from those of bulk powder of neutral PATP.[23] Prominent Raman bands were detected at 1572, 1420, 1378, 1187 and 1139 cm^{-1}, which are typically observed for PATP on Ag surfaces.[11,12] Distinct SERS spectra from PATPH$^+$ ions with protonated amino groups (–NH$_3$$^+$) appeared in acidic pH solutions (~3 ≪ pK_a ~ 5.8) at 1594, 1487, 1174, 1072, 970, 799, 626, and 386 cm^{-1} (Fig. 2b), which are similar to those for the bulk state. These observations are consistent with those in previous reports.[11,12] Note that the SERS spectra of PATPH$^+$ molecules at acidic pH significantly changed on the addition of NaOH solution, which increased the pH to 10–11, higher than pK_a. Indeed, new bands appeared at 1443, 1299, 1278, 1174, 851, 547, and 362 cm^{-1} in addition to the invariable bands at 1594, 1072, 631 and 384 cm^{-1} (Fig. 2b). These SERS spectra obtained after the addition of NaOH solution are also significantly different from those prepared at neutral pH (Fig. 2b). The observed spectral changes are tentatively attributed to deprotonated PATP molecules, although we need further effort to draw conclusions while adjusting the pH in solutions.

Next, PMBA with a carboxy group (–COOH), which was protonated in H_2SO_4 solution (pH = 1–3 ≪ pK_a = 5.8 of PMBA) before adsorption (first protonation), caused flocculation of AgNPs (Fig. 3a) similar to the case for neutral PATP. Interestingly, the SERS spectra of PMBA adsorbed in 1 × 10^{-6} M solutions, which

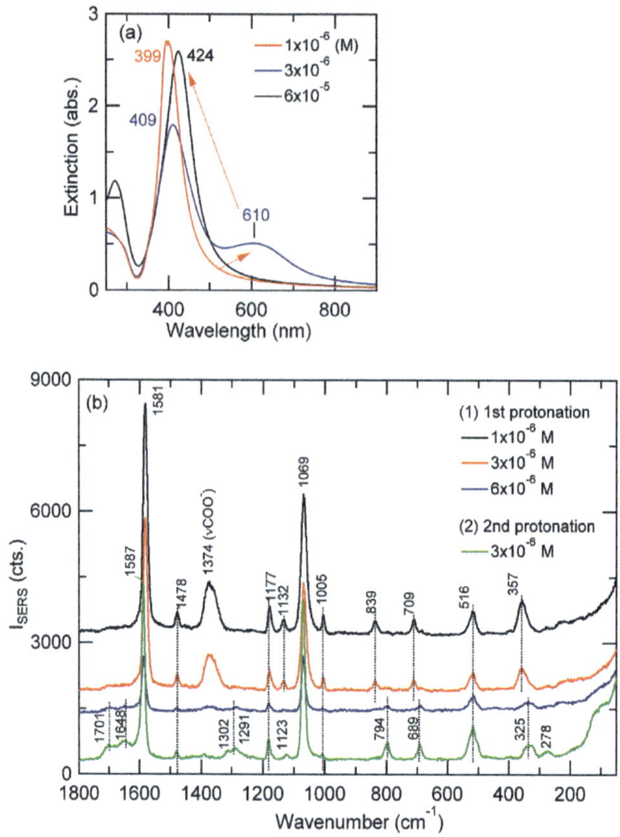

Fig. 3 Extinction spectra (a) and SERS spectra (b) of flocculates of PMBA on AgNPs adsorbed at different concentrations: 1×10^{-6} M, 3×10^{-6} M, and 6×10^{-6} M. Here, sulfuric acid was added to the solutions in Fig. 3a and those of (1) in Fig. 3b before adsorption of PMBA, while it was added to those of (2) in Fig. 3b after adsorption (see the text in detail).

yielded a surface coverage of <0.1 monolayer (ML) as determined by a subtractive absorption method,[8] are attributed to carboxylate anions ($-COO^-$). This assignment is validated by the observations of a pronounced broad peak at $\sim$1380 cm^{-1} from a symmetric stretching ν_{sCOO^-} mode of the carboxylate anions ($-COO^-$), and also by the absence of $\nu_{C=O}$ peaks at $\sim$1700 cm^{-1} from the carboxylic acid (–COOH, Fig. 3b). The Raman band intensity of the ν_{sCOO^-} band at $\sim$1380 cm^{-1} as well as those observed at 1132, 1005, 839, 709, and 357 cm^{-1}, which are assigned to vibrational modes of a deprotonated form of PMBA on Ag, decreased with increasing the concentration of PMBA from 1×10^{-6} (<0.1 ML) to 3×10^{-6} M ($\sim$0.1 ML) and 6×10^{-6} M ($\sim$0.25 ML). Concomitantly, the $\nu_{C=O}$ peaks at 1701 and 1648 cm^{-1} in addition to those at 1291, 794, 689 and 325 cm^{-1}, which are assigned to those of protonated PMBA on AgNPs on the basis of DFT calculations, appeared at a concentration of PMBA higher than 3×10^{-6} M. Here, the broad Raman band at $\sim$1380 cm^{-1}, observed at quite low coverage of PMBA, is attributed to carboxylate anions $-COO^-$ directly adsorbed on Ag surfaces with a flat or

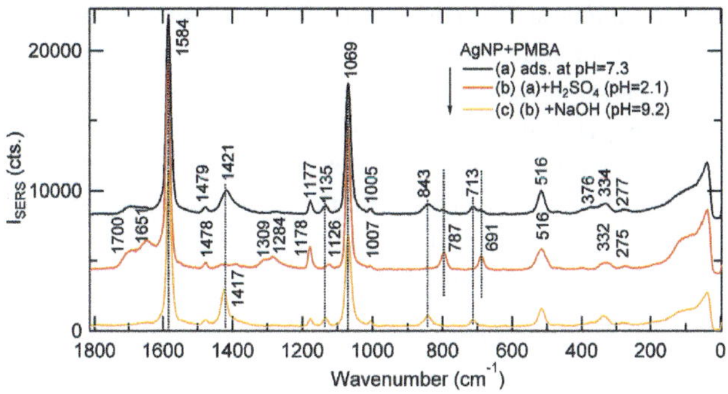

Fig. 4 Effect of pH tuning in solutions from neutral (a), *via* acidic (b), to neutral (c) values on SERS spectra of flocculates of PMBA adsorbed AgNPs: (a) deprotonated/protonated PMBA at pH=7.3, (b) protonated PMBA at pH=2.1, and (c) deprotonated PMBA at pH=9.2.

inclined orientation as suggested in the literature.[15] Associated with this attribution, the addition of $MgCl_2$ (0.3 mM) to the acidic solution (pH = 2) containing $PMBA^-$ anions adsorbed on AgNPs drastically decreased the intensity of the ν_{sCOO-} band at 1380 cm^{-1}, while increasing that of the $\nu_{C=O}$ bands at ~1700 and ~1650 cm^{-1}. It is likely that the divalent Mg^{2+}cations flipped the $PMBA^-$ anions with electrostatic interactions, after which $-COO^-$ anions facing the solution side resulted in protonation at pH $\ll$ pK_a. At pH ~ 12 ($\gg$pK_a), the corresponding sharp peak at 1430–1400 cm^{-1} from the carboxylate anions facing the solution side showed a significant shift (20–30 cm^{-1}) on addition of different metal ions.[8] Thus, a slightly higher coverage of PMBA between 0.1 and 0.25 ML changed its orientation from flat or inclined to vertical to the surface of the AgNPs, with the $-COO^-$ group facing the solution side. Such an orientation change was probably caused by the markedly lower surface pH in addition to the larger molecular interaction at higher coverage. Interestingly, PMBA-coated AgNPs were again isolated at a much higher coverage of ~1 ML at pH = 2 as evidenced by a single extinction peak at ~420 nm from isolated AgNPs (Fig. 2a and 3a). If the residual surface citrate ions on the AgNPs are completely replaced with PMBA, the AgNPs should immediately form flocculates or aggregates due to van der Waals attractive forces. Hence, the re-isolation at ~1 ML in acidic conditions, pH (2) $\ll$ pK_a (5.8), is probably caused by citrates partly remaining on the AgNPs, to which ion pair formation with hydrated protons is sterically hindered by the densely packed PMBA–SAM films. This is in contrast to the case at lower surface coverage (0.1–0.25 ML), in which hydrated protons are feasibly located nearby surface citrates to neutralize their charges. These presumptions were supported by the significantly negative zeta potential (*ca.* −20 mV) of the AgNPs observed at ~1 ML coverage of PMBA in acidic conditions, while the potential value is much smaller than that (*ca.* −45 mV) observed for the as-prepared AgNPs. In an alternative approach for the protonation, AgNPs adsorbed by $PMBA^-$ anions at pH = 5–6 in 3×10^{-6}–6×10^{-5} M solutions flocculated after the addition of H_2SO_4 (1–10 mM), which decreased the pH to 1–3 (second protonation). In this case, SERS spectra were obtained not from $PMBA^-$ anions with the $-COO^-$ groups facing the AgNPs, but only from protonated PMBA molecules (Fig. 3b). These distinct protonation and deprotonation features of PMBA on AgNPs are due to preferential adsorption *via* carboxylate groups at low surface coverage in acidic conditions and those *via* thiol groups at neutral pH. In addition, we confirmed that the carboxy groups in PMBA adsorbed in flocculates of AgNPs are almost reversibly protonated and deprotonated by tuning the pH of the solutions (Fig. 4).

Thus, we found the crucial role of van der Waals and electrostatic interactions in the flocculation of AgNPs, which facilitates the characterization of the adsorbed state of various species. Complicated adsorbed states of PATP– and PMBA–SAM films on AgNPs, and their interactions with counter ions, were elucidated at different pH values.

(2) Gap mode-SERS under an external geometry

2.1 FDTD calculations. A gap mode resonance occurred at ~550 nm for AuNPs ($r = 15$ nm), with a gap size of 1 nm, on various metal substrates such as Ag, Cu, Zn, Pt, Pb, Sn, and Al irrespective of their distinct optical properties.

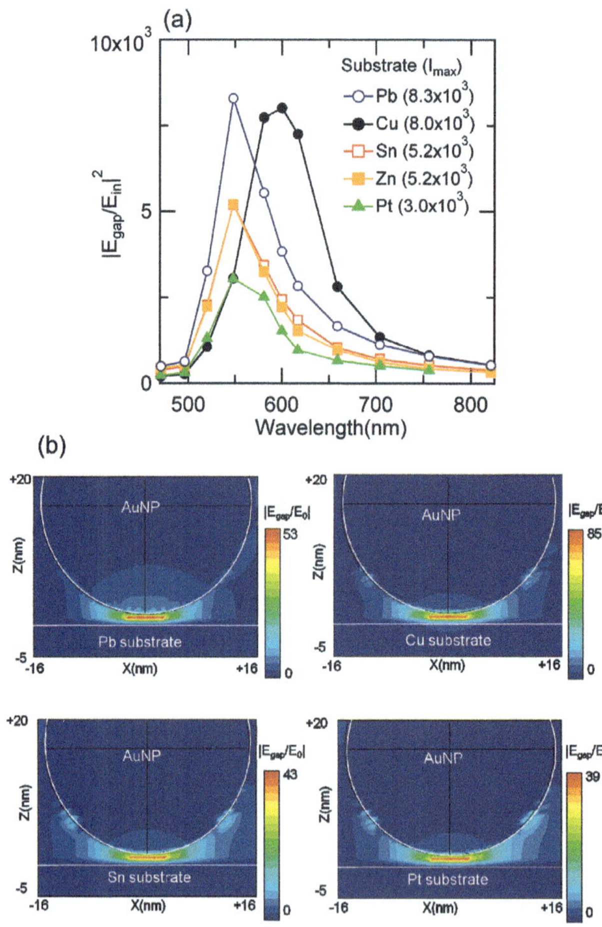

Fig. 5 A gap mode resonance and enhanced electric field intensity for AuNPs ($r = 15$ nm)/gap (1 nm)/metal (M) substrates, here M = Pb, Cu, Sn, Zn and Pt, calculated with an FDTD method: (a) at various excitation wavelengths, (b) spatial distribution at a nanogap.

Table 1 Enhancement factors for SERS signal intensities of TP under a gap mode resonance

Substrate materials	Observed ($\lambda_{ex} = 633$ nm)	Calculated ($\lambda = 617$ nm)[b]
Ag ($r_{AuNP} = 15$ nm)	1.6×10^6	1.3×10^7
Ag ($r_{AuNP} = 50$ nm)	1.3×10^8	1.0×10^{10}
Cu ($r_{AuNP} = 15$ nm)	3.3×10^5	5.3×10^7
Cu ($r_{AuNP} = 50$ nm)	3.4×10^7	1.2×10^{10}
Zn ($r_{AuNP} = 15$ nm)	1.3×10^6	2.3×10^6
Zn ($r_{AuNP} = 50$ nm)	9.8×10^6	1.3×10^9
Pb	1.5×10^5	8.1×10^6
Sn	1.1×10^6	3.4×10^6
Pt	1.2×10^6	2.3×10^6
Al	2.6×10^6	2.8×10^6
Si[a] ($r_{AuNP} = 50$ nm)	1.7×10^6	6.1×10^7
SiO_2 ($r_{AuNP} = 50$ nm)	$\ll 10^6$	5.1×10^4

[a] In calculation on Si substrate, naturally oxide layer (SiO_2, ~1 nm) was assumed. [b] $|E_{gap}/E_{in}|^4$.

Enormous enhancement factors of 2.5×10^3–1.0×10^4 were obtained in the electric field intensity at the nanogap, which were normalized to that in an incident light (here, amplitude $|E_{gap}/E_{in}|$, intensity $|E_{gap}/E_{in}|^2$, and Raman intensity $|E_{gap}/E_{in}|^4$ in Fig. 5a and b and Table 1). The larger sized AuNPs yielded the larger enhancement factors. For instance, the enhancement factor ($|E_{gap}/E_{in}|^2$) of 5.2×10^3 for $r = 15$ nm at a resonance wavelength of ~550 nm increased to 3.8×10^4 for $r = 50$ nm at resonance wavelengths of ~660 nm on Zn and Sn (Fig. 6a and Table 1), which roughly provides a Raman enhancement factor of ~1.4×10^9. Spherical gold particles with a size (r) larger than 50 nm yielded an almost identical enhancement to that for $r = 50$ nm, for example 1.3×10^5 (1.6×10^5) for $r = 100$ nm ($r = 50$ nm) on a Cu substrate, while resulting in wider resonance widths (FWHM) from 120 nm (for $r = 50$ nm) to 250–300 nm (for $r = 100$ nm) (Fig. 6a). Thus, gap mode Raman spectroscopy can be applied to various metal substrates even with large damping, which is given by the ratio of the real part (ε_r') to the imaginary part (ε_{im}'') of the optical constants ($\varepsilon_{metal} = \varepsilon_r' + i\varepsilon_{im}''$). For instance, the magnitude of the enhanced electric field on metal substrates under the resonance of propagating surface plasmons is proportional to $|\varepsilon_r'|^2/\varepsilon_{im}''$, which is 596 (Ag), 82.7 (Au), 53.9 (Cu), 15.0 (Zn), and 6.79 (Pt), 5.16 (Pb) at ~610 nm. It reminds us that only coin metals like Ag, Au and Cu are utilized as SERS active substrates on the basis of their high efficiency in exciting localized surface plasmons on roughened surfaces. Nevertheless, gap mode Raman spectroscopy is capable of elucidating catalytic reactions on transition metals like Pt,[10,11] Fe (ref. 19) or detailed adsorbed states on distinct atomically smooth surfaces of Au single crystals.[18]

The marked enhancement of the electric field at a nanogap is rationalized by a mechanism underlying in gap modes. Localized surface plasmons in the AuNPs excited by laser illumination form oscillating dipoles (p) or multipoles in the AuNPs at a resonance wavelength slightly modified by the presence of metal

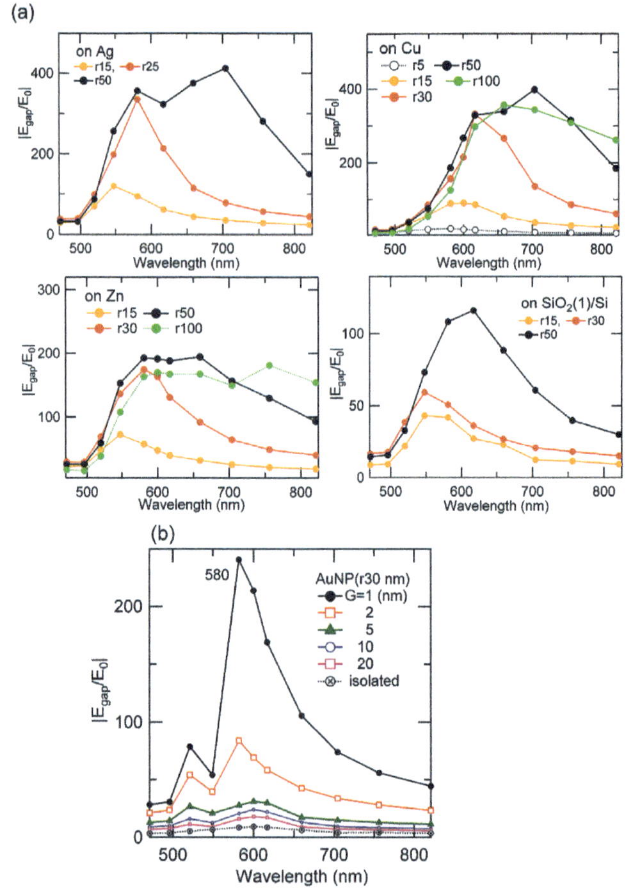

Fig. 6 Local electric field amplitudes ($|E_{gap}/E_{in}|$) at a nanogap for AuNP/gap (G)/metal (M) substrates under a gap mode resonance: (a) as a function of the size of the AuNPs (r = radius) for M = Ag, Cu, Zn and Si with $G = 1$ nm, and (b) as a function of the gap size for $r = 30$ nm for Al substrate.

substrates. These dipoles or multipoles induce image dipoles (p') or multipoles in metal substrates, which are located in close vicinity to the AuNPs, as given by eqn (1).[24]

$$p' = \frac{\varepsilon_{sub} - \varepsilon_{med}}{\varepsilon_{sub} + \varepsilon_{med}} p \tag{1}$$

Here, ε_{sub} and ε_{med} are the optical constants of the metal substrates and surrounding media, respectively. These dipoles and the image dipoles couple to form an enhanced electric field at a nanogap.[25] To confirm this mechanism, we calculated a gap mode resonance and enhanced electric field for an AuNP/gap/Al sample, while changing the gap size between 1 and 20 nm. As shown in Fig. 6b, an LSP resonance for an isolated AuNP at ~530 nm was only slightly modified to a gap mode resonance at ~560 nm with the decreasing gap size, in particular between 5 nm and 1 nm, proving the predominant role of the LSPs. Note that sufficiently large oscillating dipoles, which are similar to those in flocculates of

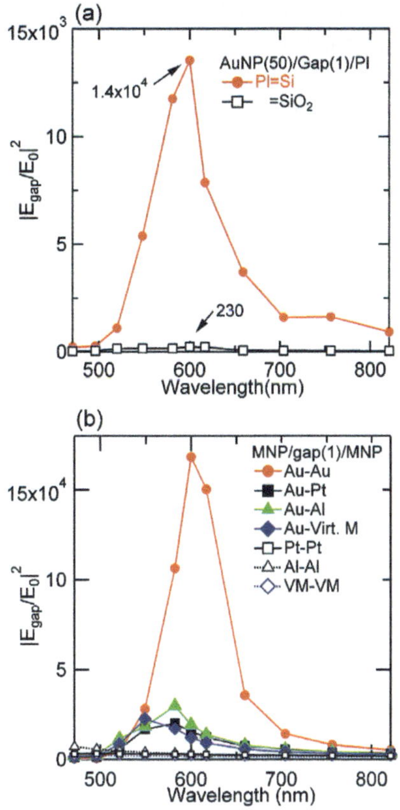

Fig. 7 A gap mode resonance and enhanced electric field calculated with an FDTD method: (a) for AuNP (r = 50 nm)/gap (1 nm)/dielectric (D) substrates, here D = Si, and SiO_2, and (b) for flocculates of MNPs, here MNP = Au, Pt, Al, and virtual metal (VM, $\varepsilon = -1 + 100i$).

AuNPs, are induced in various metal substrates, because optical constant of metals are in general much larger than those of the media. Indeed, optical constants of Pt, $\varepsilon_{\text{sub(Pt)}} = -11.3 + i18.7$ (ref. 26) are much larger than that of the media, air ($\varepsilon_{\text{med}} = 1$) in our experiments, yielding $|p'|^2 = 1.1 \times |p|^2$ as summarized in Table 1. Even virtual metals with a small real part and an extremely large imaginary part like $\varepsilon_{\text{sub}} = -1.0 + i100$, corresponding to $|p'|^2 = 1.0 \times |p|^2$, yielded enhancement factors in the electric field intensity of 10^3–10^4. Consequently, metal nanoparticles, here AuNPs, determine the gap mode resonance and enhancement factor at a nanogap *via* the LSP resonance (Table 1). Indeed, pronounced enhancement factors in Raman scattering for various metal substrates were experimentally confirmed as described in Section 2.2.

In addition, these results suggest that not only metals but also dielectric substrates with enough large optical constants, such as silicon ($\varepsilon_{\text{Si}} = 15.5 + i0.20$ at 600 nm[26]) are available in gap mode Raman spectroscopy. A silicon substrate indeed provided a pronounced enhancement factor of 1.4×10^4 (Si) in an electric field at a nanogap at a resonance wavelength of ~600 nm, compared with the more modest value (2.3×10^2) for BK-7 ($\varepsilon_{\text{BK-7}} = 2.10$ at 600 nm) as shown in

Fig. 7a. Accordingly, the enhancement factor in Raman scattering predicted for Si substrates is much larger than that for cover slips (BK-7). This calculation was experimentally confirmed in Section 2.2. Thus, the relevance of gap mode Raman spectroscopy was anticipated not only for largely damping metals but also for dielectric materials with sufficiently large optical constants.

We also corroborated that a markedly enhanced electric field is formed at nanogaps in flocculated MNPs by the coupling of dipoles and image dipoles in neighboring MNPs. In particular, a MNP with smaller damping like Ag and Au induces sufficiently large image dipoles in a closely adjacent MNP with larger damping such as Pt and Vi (virtual metal with $\varepsilon = -1 + i100$). Indeed, enhancement factors in electric field intensity for these heterogeneous pair of particles (with $r = 30$ nm and a gap size of 1 nm), such as 2.0×10^4 for AuNP–PtNP and 2.3×10^4 for AuNP–ViNP at their resonance wavelengths (550–600 nm), are comparable with those for AuNP–AuNP (1.7×10^5) and AgNP–AgNP (6.2×10^5). These enhancement factors are significantly larger than those anticipated for homogeneous pairs of MNPs with large damping like PtNP–PtNP (3.1×10^3) and ViNP–ViNP (1.8×10^3, Fig. 7b). Inherently the same results were obtained for MNPs on metal substrates. For instance, a rather modest enhancement in the electric field is formed at a nanogap between MNPs and metal substrates like AlNP–Al substrate (2.4×10^3) and PtNP–Pt substrate (1.8×10^3) both with large damping at ~600 nm.

2.2 Experimental observations. First, we found that all the metal substrates used here, Au, Ag, Pb, Pt, Sn, Cu, Al and Zn provided pronounced Raman spectra (Fig. 8a–e) and similar enhancement factors of 10^5–10^6 for TP molecules (Table 1),

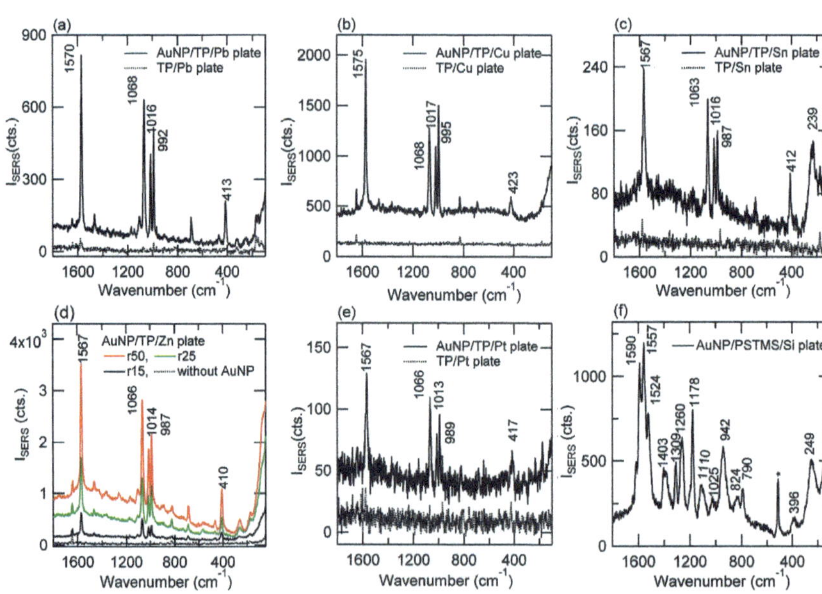

Fig. 8 Raman spectra of TP in AuNP/TP–SAM/metal substrates (M) under an external geometry: M = Pb (a), Cu (b), Sn (c), Zn (d),+ Pt (e) and Si (f).* +The effect of AuNP size, $r_{AuNP} = 15$, 25 and 50 nm, on enhancement was measured for Zn substrates. *PSTMS–SAM films were formed on Si instead of TP–SAM films in Fig. 8 and 9.

when a He–Ne laser illuminated AuNPs ($r \sim 15$ nm) immobilized on TP–SAM films on these metals. Essentially the same Raman spectra having identical peak positions and peak landscapes were observed for all of the substrates used. These are also similar to those observed for bulk liquid TP under the same geometry with the micro Raman equipment. These observations suggest a vertical or slightly tilted orientation of TP bound to these metal substrates *via* an S–Au bond, where in-plane modes are preferentially enhanced compared with out-of-plane modes. Since almost no Raman signals were detected without AuNPs on these samples, the gap modes between AuNPs and metal substrates provide a marked enhancement in the electric field at a nanogap. Here, the enhancement factors were estimated for a gap region approximated to a disc on metal substrates with a radius of 3 nm under AuNPs ($r = 15$ nm), where a remarkably enhanced electric field is centered.[4] In this estimation, SEM images of each sample of AuNP/TP/metal substrates were observed to evaluate the number of AuNPs immobilized on the metal substrates *via* TP–SAM molecules, as exemplified in Fig. 9a–d. Apparently, quite similar enhancement factors in Raman scattering were obtained for different metal species as anticipated by theoretical calculations, confirming the predominant role of the AuNPs to a gap mode resonance. These Raman enhancement factors observed for the immobilization of AuNPs are obviously not as large as those (10^8–10^9) for AgNPs.[21] This difference is caused by the distinct sizes of the AgNPs ($r = 30$ nm) and AuNPs ($r = 15$ nm) used. Indeed, the larger enhancement factors of 10^8–10^{10} were obtained for the larger AuNPs with $r = 30$–50 nm by FDTD calculations (*vide supra*, Fig. 6a and Table 1). We experimentally confirmed that the larger size AuNPs provided a larger enhancement, such as 9.8×10^6 for $r = 50$ nm compared with 1.3×10^6 for $r = 15$ nm on Zn substrates

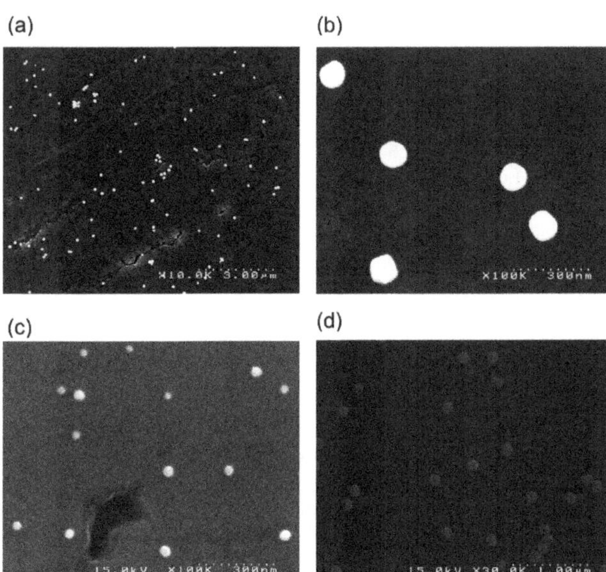

Fig. 9 SEM images of AuNPs immobilized on different substrates *via* TP–SAM films: (a) on Cu, (b) on Cu (with different magnification), (c) on Sn, and (d) on Si* substrates. *PSTMS was used on Si instead of TP.

(Fig. 8d). Essentially similar results for larger AuNPs were obtained on Cu substrates (Table 1). Potentially, single molecule detection can be achieved by using an appropriate size of AuNPs on various metal substrates.

Secondly, we applied gap modes to dielectric substrates such as a Si wafer and a glass cover slip (BK-7) to confirm the results predicted by the FDTD calculations, which also pertains to an enhancement mechanism of gap modes. As described in the Experimental section, PSTMS monolayers were formed on these substrates followed by the immobilization of AuNPs with van der Waals interaction. As shown in Fig. 8f, Si substrates provided a sufficiently large enhancement in the Raman scattering of PSTMS, whereas much weaker Raman signal intensity was obtained for the same samples on cover glass after the immobilization of AuNPs. The empirical enhancement factor for Raman scattering of PSTMS on Si under a gap mode resonance was estimated to be 1.7×10^6 using a Raman band at $1181\ cm^{-1}$. Thus, we found that gap mode Raman spectroscopy can be extended even to dielectric substrates with large optical constants.

(3) A coupling of PSPs and gap modes under an ATR configuration

The resonance angle for PSP excitation was observed to be ~43.6° at 532 nm with p-polarization for the TP–SAM/Ag film (45 nm)/BK-7 prism in air, which was in good agreement with the theoretical value of 43.2° (data not shown). The Raman enhancement factor for the TP band at $1069\ cm^{-1}$ (ν_{C-S}) was estimated to be 7×10^3, which corresponds well to the value of 3.3×10^4 predicted by the theoretical

Table 2 Resonance conditions and enhancement factor for PSPs coupled with gap modes

	θ_{AgNP} (%)	θ_{PSP} (°)	FWHM (°)	Enhancement (gap, G_{gap})	Enhancement (averaged, G_{av})
Calcd	0	43.2	0.4	—	3.3×10^4
	1	44.5	1	5.6×10^{11}[b]	—
	3.1	46.0	1.5	9.4×10^{11}[b] (4.9×10^{9}[c])	—
	5.6	47.5	2.8	3.6×10^{10}	—
	13	54.0	4	8.6×10^{9}[b]	—
Obsd	0	43.6	1.3	—	7×10^3
	0.7	44.3	1.6	1.3×10^{9}[a]	2.6×10^5
	1.4	44.0	1.4	1.8×10^{8}[a]	2.1×10^5
	2.8	45.0	2.2	1.2×10^{8}[a]	2.7×10^5
	4.6	44.3	2.9	3.7×10^{7}[a]	7.7×10^5
	10.6	44.0	4.1	2.4×10^{7}[a]	2.1×10^5
	17.6	45.0	5.3	6.8×10^{6}[a]	1.4×10^5
	25	47.7	>6	2.9×10^{6}[a]	2.1×10^5

[a] Enhancement in Raman scattering at a nanogap (G_{gap}) estimated from the observed enhancement factor averaged (G_{av}) for entire Ag films by using $G_{av} \times \{(r/r_{ac})^2/(\theta_{AgNP})\}$. Here, r and r_{ac} denote the averaged radii of AgNPs and active area under each AgNP, respectively. The Raman band of TP at $\sim 1069\ cm^{-1}$ (ν_{C-S}) was employed in the evaluation of the enhancement. The size of the active area under each AgNP roughly corresponds to $\sim r/5$ of the AgNP. [b] Enhancement in Raman scattering estimated by FDTD calculations by using $|E_{gap}/E_0|^4$. Here, E_{gap} and E_0 denote the electric field amplitude at a nanogap and incident light, respectively. [c] Enhancement in Raman scattering at a nanogap under an external geometry.

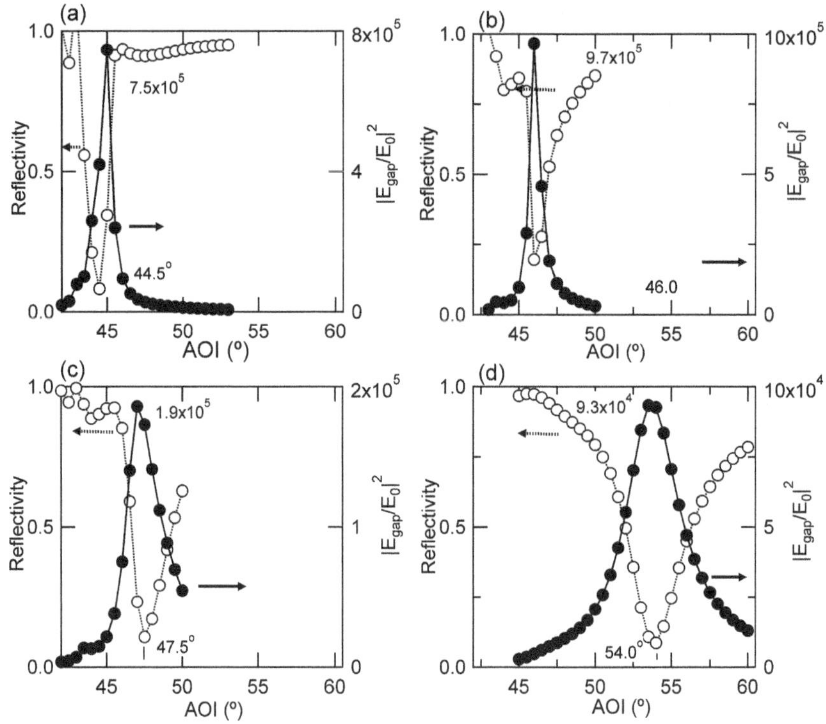

Fig. 10 Calculated ATR reflectivity at different coverage (θ) of AgNPs in a AgNPs/TP–SAM/Ag film (45 nm)/prism: (a) $\theta = 1$, (b) $\theta = 3.1$, (c) $\theta = 5.6$, and (d) $\theta = 13\%$.

calculations (Table 2). The observation that s-polarized light did not give any reflectivity dip or Raman enhancement in contrast to p-polarization enabled us to conclude that the above observations were caused by the excitation of PSPs. The immobilization of AgNPs onto TP–SAM/Ag film/prism samples with $\theta_{AgNP} \sim 0.7\%$ (Table 2, see also Fig. 10a–d, 11a–d and 12) slightly shifted the resonance angle to 44.3°, and provided a much larger Raman enhancement of 2.6×10^5, which resulted from the excitation of gap modes in addition to PSPs. The surface coverage of the AgNPs was empirically determined by SEM measurements as described in Section 2.2. The above enhancement factor is an averaged value over the entire Ag films illuminated by a laser spot. The actual enhancement factor for thiol molecules at a nanogap was estimated to be 1.3×10^9 on the basis of the surface coverage (~0.7%) of AgNP and the active area ($4\pi r^2 = 4\pi \times 4^2$–200 nm^2 under each AgNP) of the gap mode. The empirical enhancement in an ATR geometry (1.3×10^9), which corresponded to that (5.6×10^{11}) for the AgNP/gap (1 nm)/Ag (45 nm)/prism samples anticipated by the FDTD calculations, was significantly larger than that obtained in an external geometry (9×10^7). We also noted that a larger surface coverage of AgNPs gave a higher averaged Raman intensity, but a significantly smaller enhancement factor at the nanogap (Table 2). For instance, we obtained experimental (calculated) enhancement factors of 1.3×10^9 (5.6×10^{11}), 1.2×10^8 (9.4×10^{11}) and 2.4×10^7 (8.6×10^9) for surface coverages of AgNPs (θ_{AgNP}) of 1, 3 and 10%, respectively (Fig. 10a–d and 11a–d).

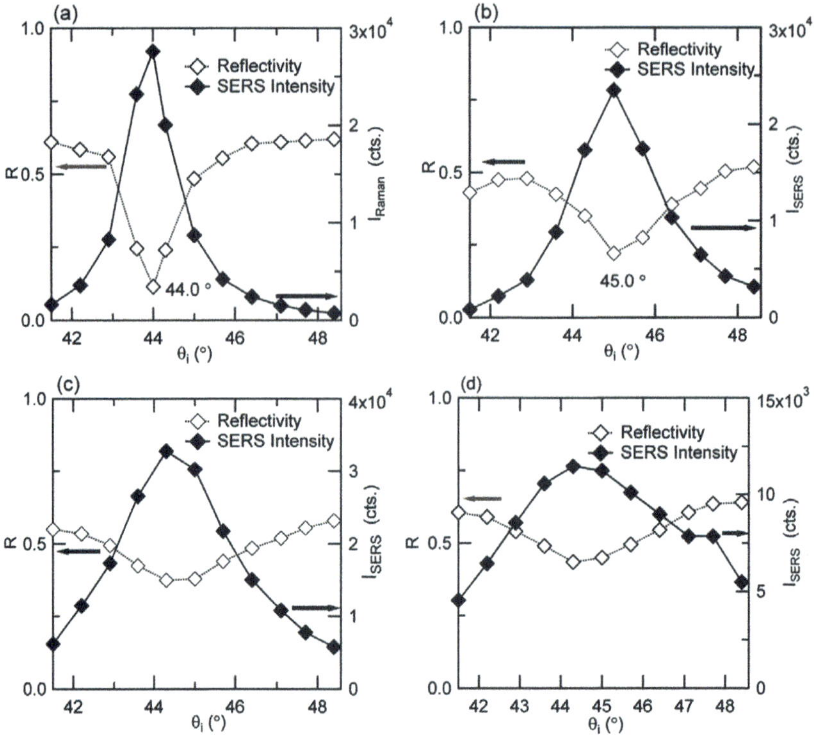

Fig. 11 Experimental ATR reflectivity at different coverage (θ) of AgNPs in a AgNPs/TP–SAM/Ag film (45 nm)/prism: (a) $\theta = 1.4$, (b) $\theta = 2.8$, (c) $\theta = 4.6$, and (d) $\theta = 10.6\%$.

Probably, the smaller enhancement at higher θ_{AgNP} results from the less efficient excitation of the PSPs. Indeed, a lager number of AgNPs increased the scattered light intensity above a prism, which decreased the evanescent field intensity on Ag films, as observed for roughened Ag films.[27] The lower excitation efficiency of the

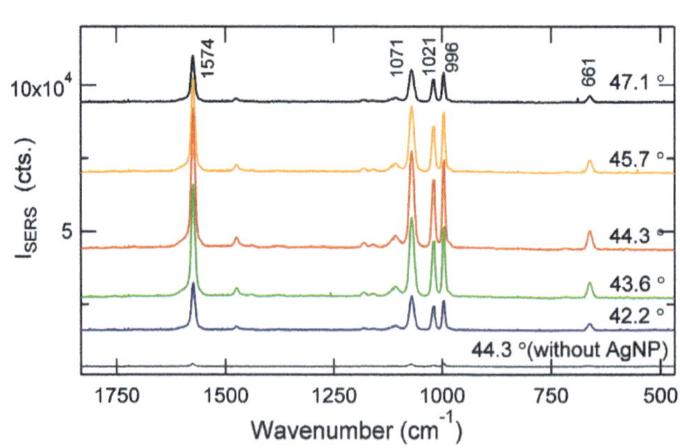

Fig. 12 Raman spectra at different angles of incidence for a AgNPs ($\theta = 4.6\%$)/TP–SAM/Ag film (45 nm)/prism.

PSPs was demonstrated by a shift in the resonance angle from 43.2° to higher angles of incidence, 44.5°, 46.0° and 54.0° with larger widths of 1, 1.5 and 4.0° for the surface coverage of AgNPs of 1, 3 and 10%, respectively. These observations indicate a larger damping of the PSP field at higher θ_{AgNP}, which corresponds well with the experimental observations (Table 2). In TERS measurements, only one AgNP attached on the tip of a cantilever does not perturb PSP resonances and thus provides the largest enhancement in electric field at a nanogap.

Gap modes combined with PSPs provided a Raman enhancement of 9×10^{11} at 532 nm (slight off-resonance), which is comparable with those of 5×10^{12} at the resonance wavelength of 470 nm. These values are significantly larger than those predicted for gap modes under an external geometry, such as 2×10^8 at 532 nm and 6×10^{10} at 470 nm. These differences were attributed to the fact that PSPs are efficiently excited at much wider wavelength than gap modes.[27] Indeed, only a slight shift of the resonance angle for PSPs was observed from 44.8° at 470 nm to 43.2° at 532 nm, which provided an equivalent evanescent field in proximity to the Ag films. This evanescent field is similarly enhanced by a factor of $\sim 10^2$ at these wavelengths relative to incident light intensity, which assures a significantly larger enhancement when coupled with gap modes in an ATR geometry compared to that in an external geometry.

Conclusions

Gap modes in flocculates of MNPs, and between MNPs and metal substrates under external and ATR configurations were investigated to establish them as an analytical tool. We succeeded in forming flocculates of AgNPs using electrostatic interactions between dissociated PMBA ($-COO^-$), protonated PATP ($-NH_3^+$) and counter ions (M^{n+}, X^-), as well as using van der Waals force between neutral PMBAs (–COOH) and PATP ($-NH_2$) on AgNPs. The detailed adsorbed states of PMBA and PATP as well as trapped counter ions, depending on the surface coverage and pH in solution, were elucidated using the enormous SERS enhancement. In gap modes under an external geometry, we found that various metal substrates with large damping including Pb, Sn, Cu, Zn, Al, and Pt are available. Note that these metal substrates, which were not effective in the early stage of SERS history using their roughened surfaces, provided markedly enhanced electric fields of 10^5–10^7 at a nanogap against AuNPs ($r = 15$ nm). These features are explained by a coupling of dipoles in the AuNPs and their image dipoles in the substrates. Hence, not only metal substrates but also dielectrics with large optical constants, such as Si with dielectric constants larger than 10 at 632 nm, relative to that of the surrounding media provided a pronounced enhancement ($\sim 10^6$) in SERS. The larger size of AuNPs provided the larger enhancement in the electric field at a nanogap, for instance an enhancement in the electric field from 5.2×10^3 (at $r_{AuNP} = 15$ nm) to 3.8×10^4 (at $r_{AuNP} = 50$ nm) on a Zn substrate, yielding 10^8–10^{10} of SERS enhancement on various metal substrates as experimentally supported. In gap modes under an ATR configuration, we corroborated that additional enhancement is obtained by a coupling of PSPs and gap modes by FDTD calculations and SERS measurements. Furthermore, a larger enhancement factor is provided at a lower surface coverage of AgNPs due to more efficient excitation of PSPs, suggesting the highest enhancement in tip-enhanced Raman scattering.

Acknowledgements

This research was supported by KAKENHI (25286014 and 25600022) by JSPS. M.F. appreciates Prof. Mitsuru Ishikawa (Josai University) for useful discussions. The authors thank to Shin-Etsu Chemical Co. Ltd for their kind supply of PSTMS.

References

1 A. Otto, I. Mrozek, H. Grabhorn and W. Akemann, *J. Phys.: Condens. Matter*, 1992, **4**, 1143–1212.
2 *Surface Enhanced Raman Scattering*, ed. K. Kneipp, M. Moskovic and H. Kneipp, Springer, Berlin, 2006.
3 E. C. Le Ru and P. G. Etchegoin, *Principles of Surface-Enhanced Raman Spectroscopy*, Elsevier, 2009.
4 M. Futamata, Y. Yu, T. Yanatori and T. Kokubun, *J. Phys. Chem. C*, 2010, **114**, 7502.
5 T. Yajima, Y. Yu and M. Futamata, *Phys. Chem. Chem. Phys.*, 2011, **13**, 12454.
6 M. Futamata, Y. Yu and T. Yajima, *J. Phys. Chem. C*, 2011, **115**, 5271.
7 T. Yajima, Y. Yu and M. Futamata, *J. Raman Spectrosc.*, 2013, **44**, 406.
8 Y. Yu, S. Handa, T. Yajima and M. Futamata, *Chem. Phys. Lett.*, 2013, **560**, 49.
9 S. Handa, Y. Yu and M. Futamata, *J. Surf. Sci. Soc. Jpn.*, 2013, **34**, 449–454.
10 J. N. Israelachvili, *Intermolecular and Surface Forces*, Academic Press, London, 1991.
11 L.-B. Zhao, M. Zhang, Y.-F. Huang, C. T. Williams, D.-Y. Wu, B. Ren and Z.-Q. Tian, *J. Phys. Chem. Lett.*, 2014, **5**, 1259–1266 and references therein.
12 K. Kim, K. L. Kim and K. S. Shin, *J. Phys. Chem. C*, 2013, **117**, 5975–5981 and references therein.
13 M. Osawa, N. Matsuda, K. Yoshii and I. Uchida, *J. Phys. Chem.*, 1994, **98**, 12702–12707.
14 J. Kneipp, H. Kneipp, B. Wittig and K. Kneipp, *J. Phys. Chem. C*, 2010, **114**, 7421–7426.
15 A. Michota and J. Bukowska, *J. Raman Spectrosc.*, 2003, **34**, 21.
16 Z. Liu, S.-Y. Ding, Z.-B. Chen, X. Wang, J.-H. Tian, J. R. Anema1, X.-S. Zhou, D.-Y. Wu, B.-W. Mao, X. Xu, B. Ren and Z.-Q. Tian, *Nat. Commun.*, 2011, **2**, 305.
17 K. Kim, H. B. Lee, J.-Y. Choi and K. S. Shin, *J. Phys. Chem. C*, 2012, **115**, 21047.
18 K. Ikeda, K. Takahashi, T. Masuda, H. Kobori, M. Kanehara, T. Teranishi and K. Uosaki, *J. Phys. Chem.*, 2012, **116**, 20806–20811.
19 H. Suzuki, H. Chiba and M. Futamata, *Vib. Spectrosc.*, 2014, **72**, 105–110.
20 P. K. Aravind and H. Metiu, *Surf. Sci.*, 1984, **124**, 506.
21 F. Raulin and J.-P. Lussiana, *Origins Life*, 1984, **14**, 157.
22 J. C. Love, L. A. Estroff, J. K. Kriebel, R. G. Nuzzo and G. M. Whitesides, *Chem. Rev.*, 2005, **105**, 1103–1169.
23 S. Handa, Y. Yu and M. Futamata, *Vib. Spectrosc.*, 2014, **72**, 128–134.
24 T. Okamoto and I. Yamaguchi, *Opt. Rev.*, 1999, **6**, 211–214.
25 J. D. Jackson, *Classical Electrodynamics*, John Wiley & Sons, Second edn, 1975, ch. 2.
26 D. Palik, *Handbook of Optical Constants of Solids*, Academic press, San Diego, 1998, pp. 333, 547 and 795.
27 H. Raether, *Surface Plasmons*, Springer, Berlin, 1988.
28 C. E. Talley, L. Jusinski, C. W. Hollars, S. M. Lane and T. Huser, *Anal. Chem.*, 2004, **76**, 7064–7068.

Faraday Discussions

PAPER

Selective TERS detection and imaging through controlled plasmonics†

Hao Wang,[a] Stacey L. Carrier,[a] Sheldon Park[b] and Zachary D. Schultz*[a]

Received 3rd October 2014, Accepted 24th November 2014
DOI: 10.1039/c4fd00190g

Enhanced Raman spectroscopy offers capabilities to detect molecules in the complex molecular environments and image chemical heterogeneity in a wide range of samples. It has been shown that plasmonic interactions between a TERS tip and a metal surface produce significant enhancements. In this report we show how SERS spectra from purified molecules can be used to selectively image proteins on surfaces and in cell membranes. The SERS response from the purified protein can be used to create a multivariate regression model that can be applied to nanoparticles that bind to protein receptors. Filtering the observed TERS spectra with the regression model can then selectively image the protein receptor. Experiments with mutant proteins suggest that key amino acids provide significant contributions to the observed TERS signal, which enables the differentiation of protein receptors. These results demonstrate the selectivity that can be obtained in TERS images through a controlled plasmonic interaction. This approach has further implications for identifying membrane receptors that bind specific molecules relevant to drug targeting and chemical signaling.

Introduction

The control of localized surface plasmon resonances has opened new avenues to high sensitivity spectroscopic detection.[1,2] The enhancement in Raman scattering, known as surface-enhanced Raman scattering (SERS), is strongly correlated with the plasmonic properties of nanostructures.[3] An understanding of the electric fields and observed enhancements has enabled high sensitivity spectroscopic and imaging applications.[4–6]

Tip-enhanced Raman scattering (TERS), utilizes a nanostructure on the apex of a scanning tunneling (STM), or atomic force (AFM), microscope tip to correlate Raman signals with nanometer spatial resolution.[7,8] In order to obtain high sensitivity TERS detection, the tip must be in close proximity to a metal surface and generate a gap plasmon resonance mode.[9] The interaction between the

[a]*University of Notre Dame Department of Chemistry and Biochemistry, Notre Dame, IN 46530, USA. E-mail: Schultz.41@nd.edu*
[b]*University of Buffalo, Department of Chemical and Biological Engineering, Buffalo, NY 14260, USA*

† Electronic supplementary information (ESI) available. See DOI: 10.1039/c4fd00190g

plasmon resonance on the tip and plasmon in the metal surface combine to generate significant enhancements in the Raman scattering. The enhancements observed in gap resonance modes are similar to the enhancements associated with gaps between nanoparticles, commonly known as hotspots, in SERS.[10]

This high sensitivity, high spatial resolution microscopy suggests possibilities to explore complex and chemically heterogeneous systems. The observed Raman scattering provides unique opportunities to identify various molecular species and distinguish minute compositional and structural alterations. Indeed TERS has been reported from cell membranes, however the signals arising from only the tip require longer acquisition times.[11–13] Examining intact cells to study membrane receptors or lipid domains is challenging. Obtaining gap-mode like enhancements on cell membranes could address these challenges.

It has been demonstrated that enhancements comparable to those in gap modes can be achieved when a TERS tip interacts with a nanoparticle.[14] Control of the polarization can increase the spatial resolution of this interaction to the size of the nanoparticle being detected.[15] Intriguingly, experiments with ligand-functionalized nanoparticles interacting with proteins on a surface give rise to Raman signals associated with the protein on the surface preferentially to the ligand on the nanoparticle probe.[16,17] Recently, peptide functionalized nanoparticles were used to selectively detect the Raman spectrum of a protein receptor within an intact cellular membrane.[18] This selective detection indicates that Raman scattering of molecules near the nanoparticle probe, but not in the gap junction, can be selectively enhanced in TERS imaging. Interestingly, the spectrum obtained from the tip coupling to the nanoparticle in intact cell membranes is significantly different than that observed from aggregated nanoparticles on the membrane, emphasizing the importance of a controlled interaction.[18]

Previous work suggests that plasmon modes in interacting dimers can generate substantial electric fields outside the gap junction.[19] Additionally, the scattering from a molecule outside of the gap junction was predicted to show increased scattering,[20] which has been observed experimentally using super-resolution SERS imaging approaches.[21] Interestingly, recent work suggests that when the nanostructures are within 1 nm of each other, quantum tunneling can alter the electric field in the gap junction and redistribute the electric field around the interacting nanostructures.[22,23]

The origin of this selective enhancement appears to arise from the controlled interaction of the TERS tip with the nanoparticle probe. It has been shown that the plasmon modes observed from interacting nanostructures are sensitive to the spacing and orientation of the nanostructures as well as the polarization of the incident and emitted electromagnetic radiation.[24–26] Furthermore, the observed SERS spectrum is strongly dependent upon nanoparticle spacing and orientation.[27,28]

In this report we use statistical analysis to correlate the enhanced Raman signals observed in SERS of aggregated nanoparticles with TERS detection of nanoparticle probes. The TERS experiments provide a controlled plasmonic interaction that is sensitive to molecules in and outside the gap junction. In experiments involving protein receptors binding to ligands, details of conformational changes associated with these events can be reflected in the vibrational spectrum. Our results demonstrate the utility of chemometric analysis in identifying specific molecules based on statistical similarity between each spectrum in a TERS map to the reference SERS spectrum of a pure component. In particular we observe that the SERS response from purified molecules can be used to identify that molecule in the complex chemical matrix of a cell membrane by TERS. Using mutant proteins, we are further able to validate this unprecedented selectivity. The improved image contrast enables us to examine the plasmonic enhancements responsible for the signals observed in TERS experiments.

Experimental methods

Reagents

Gold nanoparticles (GNP) and microscope slides functionalized with either biotin or streptavidin were purchased from Nanocs (New York, NY). Citrate stabilized GNP spheres (NanoXact™) were purchased from nanoComposix (San Diego, CA). Short peptides with the sequence of cRGDfC were synthesized by Peptide International (Louisville, KY). Purified human integrin $\alpha_v\beta_3$ protein was purchased from Chemicon, Temecula, CA (CC1021, EMD Millipore Corporation). Coverslips used in cell culture were poly-D-lysine coated to enhance cell attachment and spreading (BD BioCoat Cellware, BD Biosciences, San Jose, CA). Paraformaldehyde was purchased from Sigma-Aldrich (St. Louis, MO). All of these chemicals were used without any further purification. Ultrapure water (18.2 MΩ cm resistivity) was obtained from a Barnstead Nanopure filtration system.

Recombinant streptavidin mutants

Streptavidin mutants (M5, TMC, RC) and a wild type (WT) control were expressed, extracted, and purified from *Escherichia coli* T7 Express lysY competent cells with mutant plasmids and transformants on LB plates as previously reported.[29] All three mutants, as well as the wild type control, contain a FLAG tag (DYKDDDDK) and a six-His tag for affinity purification. These streptavidin mutants (M5, 2.14 mg mL^{-1}; TMC, 1.68 mg mL^{-1}; RC, 1.61 mg mL^{-1}; WT, 1.45 mg mL^{-1}) were dissolved in phosphate buffered saline (PBS) for further use. Details about the amino acid mutations for these mutants can be found in Table 1.

Preparation of TERS sample

The first set of TERS samples was prepared by drop depositing 20–50 μL biotinylated GNP colloid of appropriate concentration (0.05%) onto a streptavidin-coated slide or streptavidin derivatized GNP onto a biotin functionalized slide. The slide was rinsed with ultrapure water to remove physisorbed or weakly bound GNP on the functional slide.

TERS samples with recombinant streptavidin (M5, TMC, RC, WT) were made by depositing biotinylated GNPs onto recombinant streptavidin coated slides. The derivatization of glass slides with recombinant streptavidin was processed by first cleaning the slides for 1.5 hours in an acid bath, followed by rinsing with ultrapure H_2O and EtOH. After being dried, the slides were soaked in 1% (3-glycidyloxypropyl)trimethoxysilane ethanolic solution, then rinsed repeatedly in ultrapure H_2O and finally soaked in the respective recombinant protein solutions overnight.

Table 1 Amino acid sequence of streptavidin mutants

WT	MGSSHHHHHHSQDLASDPSKDSK
	AQVSAAEAGITGTWYNQLGSTFIVTAGAD
	GALTGTYESAVGNAESRYVLTGRYDSAPATD
	GSGTALGW**T**VAWKNNYRNAHSA
	TTWSGQYVGGAEARINTQW**L**LTSGTTEANA**W**KSTL**V**GHDTFTKVKPSAAS
	GSDYKDDDDK
M5	MGSSHHHHHHSQDLASDPSKDSKA
	QVSAAEAGITGTWYNQL
	GSTFIVTAGADGALT
	GTYESAVGNAESRY**T**LTGRYDSAPATDGS
	GTALGW**R**VAWKNNYRNAHSATTWS
	GQYVGGAEARINTQW**T**LTSGTTEANA**A**KSTL**R**GHDTFTKVKPSA
	ASGSDYKDDDDK
TMC	MGSSHHHHHHSQDL
	ASDPSKDSKAQVSAAEA
	GITGTWYNQLGSTFIVTAGAD
	GALTGTYESAVGNAESRY**T**LTGRYDSAPATDG
	SGTALGW**R**VAWKNNYRN
	AHSATTWSGQYV
	GGAEARINTQW**T**LTSGTTEANA**W**KSTL**C**GHDTFTKVKPSAA
	SGSDYKDDDDK
RC	MGSSHHHHHHSQDLA
	SDPSKDSKAQVSAAEAGITGTWY
	NQLGSTFIVTAGADGAL
	TGTYESAVGNAESRY**V**LTGRYDSAPAT
	DGSGTALGW**R**VAWKNNYRNAHSATT
	WSGQYVGGAEARINTQW**L**LTSGTTEANA**W**KSTL**C**GHDTFTKVKPSAA
	SGSDYKDDDDK

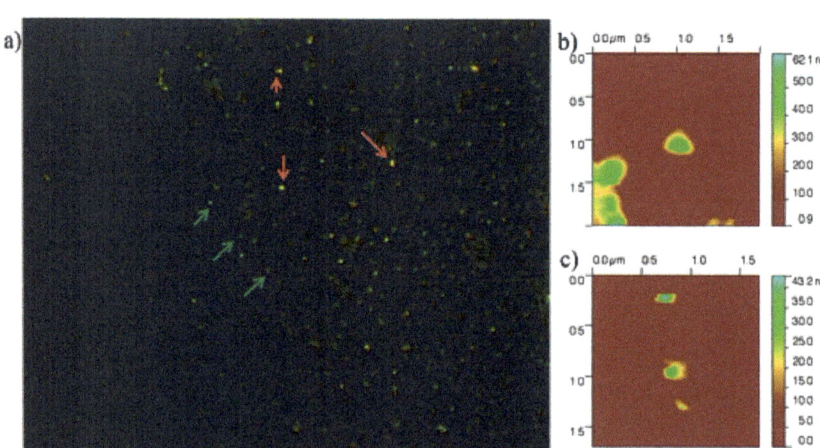

Fig. 1 (a) Representative dark-field image of 50 nm biotinylated GNPs deposited on a streptavidin functionalized glass slide. Red arrows point to bigger particles, multimers or clusters with bright yellow scattering and green arrows point to dim blue-greenish scattering spots for isolated single particles; (b) a representative AFM image of 80 nm biotin GNPs on a streptavidin slide surface is shown; (c) the AFM image shows 50 nm biotin GNPs on a streptavidin slide surface.

Fig. 1 shows the observed dark field scattering and representative AFM images observed from a sample prepared on glass slides. From these images, single and aggregated particles are readily discerned.

The other set of TERS samples was prepared by culturing cyclic RGD short peptide (cRGDfC) functionalized GNPs with colon cancer cells (SW620) plated on microscope coverslips. Rinsing and fixation of the cells with 4% (w/v) paraformaldehyde were carried out followed by a rinsing step with ultrapure H_2O to remove the residual fixatives. Fig. S1 (ESI†) shows the distribution of cRGDfC NPs on the cells and the observed AFM topography.

Preparation of SERS samples

Functionalized GNPs with biotin and streptavidin were pelleted through centrifugation. These pellets were re-dispersed in a microliter volume of ultrapure water and drop-coated on a clean microscope slide and then covered with a glass cover slip to avoid contamination during SERS measurements. Similarly, the RGD peptide modified GNPs were incubated with purified human integrin $\alpha_v\beta_3$ protein and then pelleted, re-dispersed and drop-coated onto a glass slide for SERS.

TERS spectroscopy

TERS was performed on a combined AFM-Raman instrument composed of a commercial atomic force microscope (Nanonics MV4000 AFM) and a homebuilt Raman spectrometer. The TERS signal was collected in a top-illumination geometry (normal incidence). A 632.8 nm HeNe laser (Melles Griot) was used for excitation. The intensity and polarization of the laser was controlled using a half-waveplate and polarizer. A liquid-crystal mode converter (ARCoptix, Neuchatel, NE, Switzerland) was used to generate a radial polarization, which has been shown to increase TERS contrast.[30,31] A long working distance dark-field objective (50×, NA = 0.5, LMPlanFLN, Olympus) was used to both focus the excitation and collect the scattered light from the TERS tip. The TERS tips were chemical-mechanical-polished (CMP) Au nanoparticle AFM tips (Nanonics Supertips Ltd.) with a diameter of *ca.* 100–200 nm. The laser power was adjusted to be 1 mW or less at the sample to avoid damage in all experiments. The collected Raman scattering was filtered through a dichroic beamsplitter (RazorEdge, Semrock, 633 nm) and a long pass filter (RazorEdge, Semrock, 633 nm), focused onto the 50 μm entrance slit of an imaging spectrograph (Horiba, Jobin Yvon), dispersed by a 600 g mm^{-1} grating (centered at 1600 cm^{-1}), and detected with a CCD camera (Horiba, Synapse).

TERS maps were obtained by scanning the sample stage under the TERS tip positioned in the laser focus. The acquisition time was 1 s per pixel. The tip position and status were checked after each mapping experiment to exclude cases when the tip either drifts out of the laser focus or is contaminated.

Raman spectroscopy

SERS spectra were collected using a home-built confocal Raman microscope with a 632.8 nm HeNe laser used as the excitation source. Sample illumination was achieved through a 50× (LMPlanFLN, Olympus) dark field objective. The laser power was kept at 1 mW. The SERS spectra for biotin and streptavidin were taken

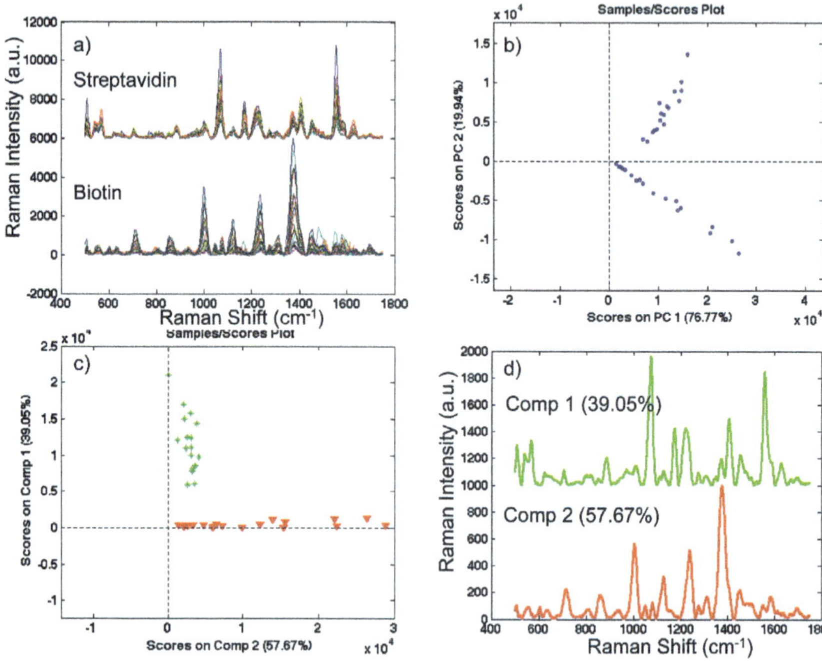

Fig. 2 (a) SERS spectra of pure streptavidin and biotin; (b) sample/scores plot along two principle components in the PCA analysis of the SERS data without any constraints; (c) sample/scores plot in MCR analysis of the SERS data with a non-negativity requirement and constrained number of components (2); and (d) MCR calculated pure components from the SERS data are shown. Green and red correspond to components 1 and 2, respectively.

from different spots on the prepared SERS nanoparticle film, with an acquisition time of 1 s per spectrum in all cases.

Chemometric analysis

Multivariate analysis was performed using PLS Toolbox (Eigenvector Research Inc., Wenatchee, WA, Version 7.0.2) working in a MATLAB R2011b (The Mathworks Inc, Natick, MA). All raw SERS and TERS spectra were preprocessed through a weighted least squares (WLS, Whittaker filter, 2nd order polynomial) automatic baseline subtraction to remove the offset differences due only to the background in each spectrum. These data were further used to decompose the pure components through analysis of the SERS data using multi-variant curve resolution (MCR). Classification analysis was performed using partial least-squares discriminate analysis (PLS-DA) on the obtained SERS and TERS datasets in order to determine the class of each spectrum in the TERS map. TERS maps and score maps were reconstructed in MATLAB according to single-peak intensities and PLS-DA scores, respectively.

Results and discussion

To identify the enhanced Raman signal of a protein bound to a ligand functionalized nanoparticle, we obtained the SERS spectra from biotinylated and streptavidin functionalized GNPs. Fig. 2a shows the SERS spectra of biotin and streptavidin in aggregated GNPs. SERS spectra from each molecule are shown to have a high degree of reproducibility, showing consistent frequencies. There is variation in the absolute intensities. These spectra provide a reference to distinguish the signals from biotin and streptavidin. Fig. 2b shows that the principle component analysis (PCA) of the spectra in Fig. 2a displays the clear, orthogonal separation expected of two components. Using constraints, associated with MCR analysis, we are able to determine orthogonal loadings that match the observed spectra as shown in Fig. 2c and d. The score on a specific loading can be used to infer concentration information with respect to each component, providing chemically interpretable results. In this controlled data set, multivariate analysis accurately captures each chemical component.

It has been shown that chemometric analysis can be used to identify spectrally similar features within the obtained images.[11] Vibrational spectroscopic maps are commonly produced by plotting the intensity of a single frequency over the measured region. This is straightforward and useful in a wide range of systems where a unique marker band can be identified for specific components. Raman tags, for example, have been exploited due to their strong Raman response and characteristic spectral features.[32] Apart from Raman tags, using a single marker

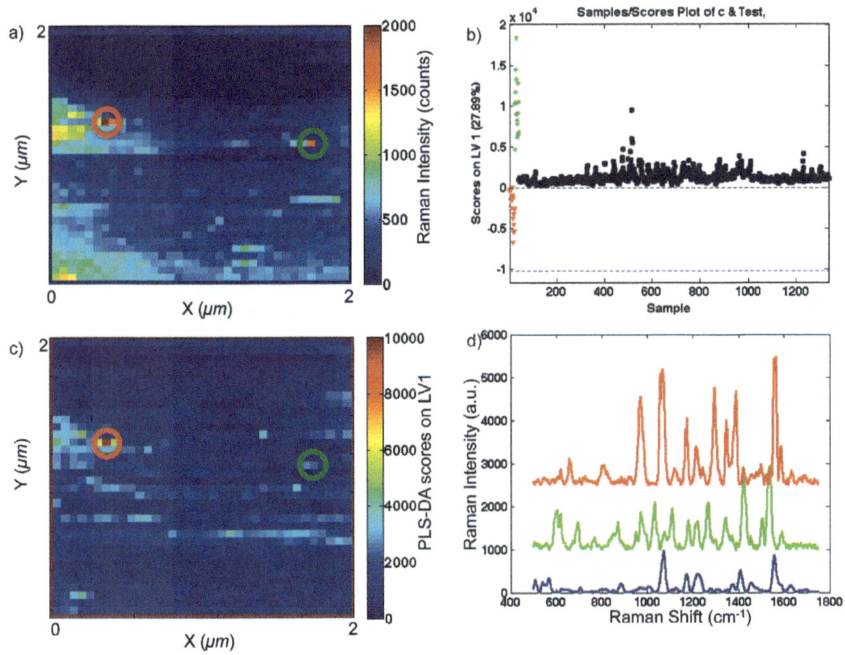

Fig. 3 (a) A 2 × 2 μm (36 × 36 pixel) TERS map at 960 cm^{-1} is shown from 50 nm streptavidin GNPs on a biotinylated slide. (b) Sample/scores are plotted against LV1 in PLS-DA analysis. The green and red data points are calibration SERS spectra of streptavidin and biotin, respectively. (c) The PLS-DA calculated scores on LV1 are mapped as an indicator of the full-spectral similarity of each TERS spectrum to LV1. (d) Selected TERS spectra are compared with PLS-DA calculated LV1 (blue); the green spectrum corresponds to the TERS pixel circled in green and the red spectrum corresponds to the TERS pixel circled in red.

band intensity is often problematic because the molecules share vibrational frequencies. Additionally, the intrinsic Raman signal of biomolecules is weak and depends on the conformation and structure of the molecule that can vary across a sample.

The reduction of data dimensionality shown in Fig. 2 suggests that SERS data sets can be used to selectively detect molecules in hyperspectral maps. We have previously reported the similarity between SERS spectra from aggregated GNPs and the spectra obtained from TERS detection of functionalized GNPs.[16,18] In order to take advantage of the multivariate regression of our data, we used partial least-squares discriminant analysis (PLS-DA) to determine the identity of components in our TERS imaging. PLS-DA calculates the spectral variance between a validation/prediction set and a calibration/training set, and assigns each sample to different classes.

Fig. 3 shows the use of plotting PLS-DA scores to improve image contrast for a specific chemical component. In Fig. 3a, the TERS map obtained from streptavidin functionalized GNPs detected by a TERS tip is plotted using a single intense marker band (960 cm^{-1}) to signify the detection of streptavidin GNPs. Using the PLS-DA model generated from the SERS spectra of biotin and streptavidin as the calibration set (39 spectra as in the MCR model), we can calculate the score of

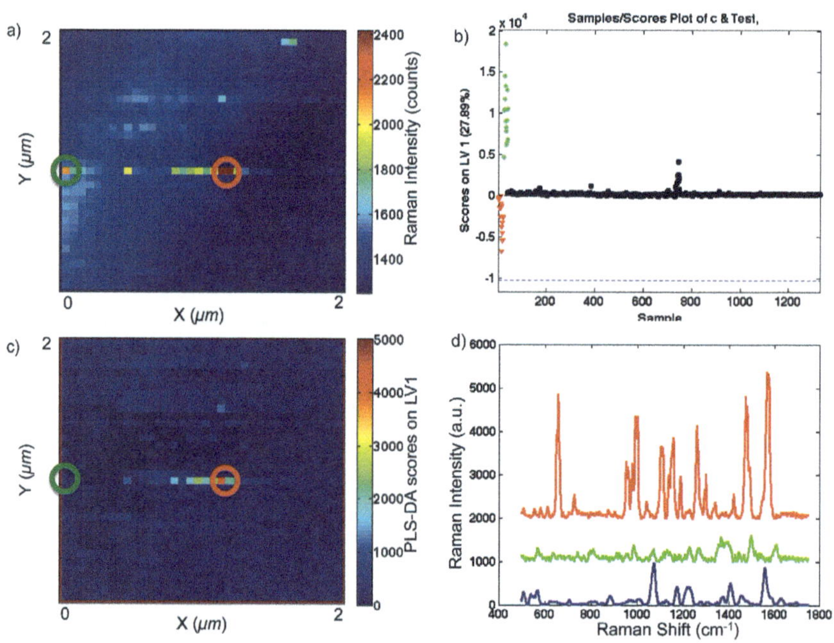

Fig. 4 (a) A 2 × 2 μm (36 × 36 pixel) TERS map at 960 cm^{-1} is shown from 80 nm biotinylated GNPs on a streptavidin slide. (b) Sample/scores are plotted against LV1 in PLS-DA analysis. The green and red data points are calibration SERS spectra of streptavidin and biotin, respectively. (c) The PLS-DA calculated scores on LV1 are mapped as an indicator of the full-spectral similarity of each TERS spectrum to LV1. (d) Selected TERS spectra are compared with PLS-DA calculated LV1 (blue). The green spectrum corresponds to the TERS pixel circled in green and the red spectrum corresponds to the TERS pixel circled in red.

each spectrum in the TERS map (1296 spectra) against the latent variable associated with streptavidin (LV1). Reconstructing the TERS map based on the scores of each TERS spectrum, we obtain an image with much better contrast (Fig. 3c). The multivariate analysis removes background signals not statistically similar to streptavidin but apparent in the single frequency intensity map.

A comparison in Fig. 3d between the selected pixels in the TERS map and the PLS-DA calculated LV1 illustrates how PLS-DA recognizes streptavidin by taking into account the full spectrum in calculating variances: the green spectrum shows intensity that appears to be streptavidin when plotting single band intensities, but does not show agreement with LV1 in the reconstructed scores map while the red spectrum is retained due to its increased correlation to the spectrum of the pure component.

To further assess PLS-DA score mapping to analyze TERS maps, we used the model to evaluate a sample with biotin functionalized GNPs on a streptavidin coated glass slide. Unlike the previous example where streptavidin resides between the GNPs and the TERS tip, in Fig. 4 the streptavidin is located underneath the biotinylated GNPs. Comparing the Raman map (Fig. 4a) through single band integration and the Raman map according to PLS-DA scores (Fig. 4c), again we obtain better contrast. Fig. 4d shows the spectra comparison where even the highest scoring pixel (red) shows only a partial match with the calculated LV1 associated with streptavidin (blue), and again the random intense pixel (green) shows poor correspondence with the LV1 and is removed from analysis of the protein.

Interestingly, the PLS-DA scores from the biotinylated GNPs on a streptavidin slide are lower than the PLS-DA scores using streptavidin functionalized GNPs on a biotinylated surface. The scores calculated by PLS-DA for the streptavidin–GNPs (Fig. 3) show a maximum around 10 000, while the score for the streptavidin detected beneath biotinylated GNPs only reaches 5000 (Fig. 4). The differences in PLS-DA scores suggest differences in enhancement related to the plasmonic environments experienced by the streptavidin protein and biotin ligands around the nanoparticles.

Large fields are expected in the gap region, which is likely where the protein resides in Fig. 3. However, in Fig. 4 where the protein is beneath the probe nanoparticle, the electric fields around the constructed dimer generate a lower level of scattering. One explanation for this is the redistribution of the electric fields arising from quantum tunneling effects.[23] The biotin coating may not keep the TERS tip from close (<1 nm) contact with the nanoparticle.[22] In this regime, it has been reported that the gap mode is quenched, but fields exist around the dimer construct. This quenching may explain the diminished signal from biotin observed in these experiments. Alternatively, the observed signals may arise from changes in the plasmon resonances as the distance between the functionalized GNP and the TERS tip change.[33]

In addition to changes in the electric field distribution, the gradient of the electric field can also explain spectroscopic differences. When the streptavidin is between GNP and the TERS tip (*e.g.* Fig. 3), the streptavidin signals is observed clearly and therefore scoring high on LV1. It is worth noting that streptavidin is a few nanometers in size and would inhibit the TERS tip from coming close enough to enter the quantum-tunneling regime. When biotinylated GNP resides are on top of the streptavidin, amino acids in the binding pocket of streptavidin and a small signal from biotin are observed.[16,17] It should be noted that the biotin is attached to the GNP through a flexible linker, which makes the exact distance between the GNP, the TERS tip, and the biotin–streptavidin binding uncertain. However, in this case the detected signal contains contributions from both biotin and streptavidin indicating protein resides near the GNP surface. In experiments with antibody functionalized GNPs, the binding partner could not be detected.[15]

We previously postulated that the electric field gradients noted above give rise to signals associated with amino acids in the protein's binding pocket.[17] To assess the selectivity of PLS-DA mapping for discerning minor changes in protein composition and structure, here we investigated streptavidin mutants using biotinylated GNPs and TERS detection. The mutants included in the study have the substitutions listed in Table 1. In addition to the mutations, these proteins have a His-tag (MGSSHHHHHHS) attached for purification but that can also interact with GNP probes. Of particular note, a tryptophan residue near the binding pocket is mutated in M5, which does not significantly alter its affinity for biotin.[29]

Fig. 5 is the TERS analysis obtained from biotinylated GNPs and the M5 mutant bound to a glass slide. While the TERS maps show signals from the nanoparticle probe, the PLS-DA indicates it is not statistically similar to the native

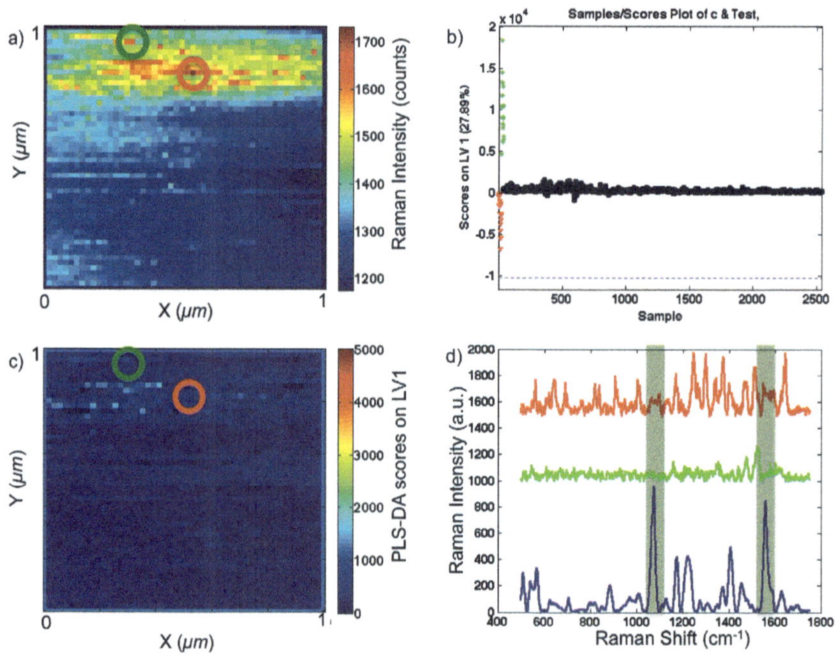

Fig. 5 (a) A 1 × 1 μm (50 × 50 pixel) TERS map at 960 cm^{-1} is shown from 50 nm biotinylated GNPs on a M5–streptavidin slide. (b) Sample/scores are plotted against LV1 in PLS-DA analysis. The green and red data points are calibration SERS spectra of streptavidin and biotin, respectively. (c) The PLS-DA calculated scores on LV1 are mapped as an indicator of the full-spectral similarity of each TERS spectrum to LV1. (d) Selected TERS spectra are compared with PLS-DA calculated LV1 (blue). The green spectrum corresponds to the TERS pixel circled in green and the red spectrum corresponds to the TERS pixel circled in red.

streptavidin (Fig. 5b and c). Using single Raman band integration (Fig. 5a), signals are observed, but the origin is difficult to discern. The TERS spectra shown in Fig. 5d, red and green as circled in Fig. 5a bear very little resemblance to the calculated LV1. Most prominently, a strong band at 1560 cm^{-1} is missing, which is consistent with the tryptophan mutation to alanine in the M5 mutant.

Fig. 6 shows the TERS mapping to the TMC mutant, which retains the tryptophan in the binding pocket and leads to a recovery of the PLS-DA score associated with streptavidin in the model (Fig. 6b). By comparing Fig. 6a and c, it is clear that the reconstructed Raman map with PLS-DA scores on LV1 exhibits improved accuracy in the class assignment of each TERS spectrum. The spectra plotted in Fig. 6d compares the high scoring and low scoring pixels, which can be easily confused as the same in Fig. 6a, and reveals the prowess of PLS-DA in spectral pattern recognition. Similar to the TMC mutant, the other two recombinant streptavidin proteins that retained the tryptophan residue, wild type and the RC-mutant, showed PLS-DA scores indicating the detection of streptavidin (see Fig. S2 and S3 in the ESI†). Worth noting is that the PLS-DA scores of the TMC, RC, and recombinant wild type are comparable to the native streptavidin in Fig. 4. Because M5 is known to retain its affinity for biotin, the lack of PLS-DA scores for streptavidin suggests the tryptophan near the binding pocket makes an important contribution to the TERS response from functionalized GNPs.

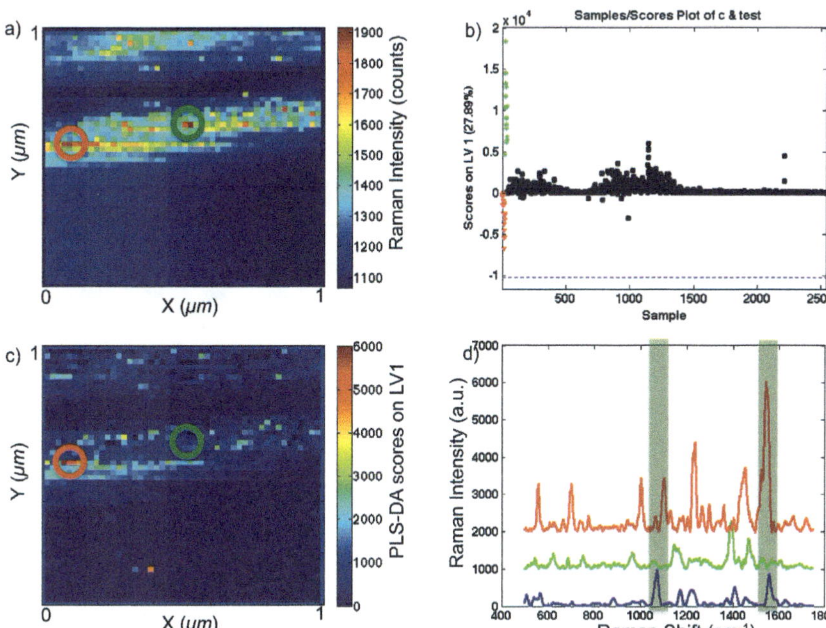

Fig. 6 (a) A 1 × 1 μm (50 × 50 pixel) TERS map at 960 cm^{-1} of 50 nm biotinylated GNPs on a TMC–streptavidin slide is shown. (b) Sample/scores are plotted against LV1 in the PLS-DA analysis. The green and red data points are calibration SERS spectra of streptavidin and biotin, respectively. (c) The PLS-DA calculated scores from LV1 are mapped as an indicator of the full-spectral similarity of each TERS spectrum to LV1. (d) Selected TERS spectra are compared with PLS-DA calculated LV1 (blue). The green spectrum corresponds to the TERS pixel circled in green and the red spectrum corresponds to the TERS pixel circled in red.

To assess the ability to selectively detect a specific biomolecule in a complex matrix, we applied this analysis to previously reported results from cyclo-arginine-glycine-aspartic acid-phenylalanine-cysteine (cRGD) peptides on GNPs bound to $\alpha_v\beta_3$ integrin receptors in intact cell membranes.[18] The PLS-DA model was produced from SERS spectra from the cRGD peptide and purified integrin detected in aggregated GNPs (Fig. S4 in the ESI†). This model was then applied to TERS spectra generated by TERS mapping of RGD functionalized GNP on fixed cells.

Fig. 7a shows the TERS map created by integrating the detected Raman band at 1003 cm^{-1}, which corresponds to a phenylalanine residue on the cRGD peptide. The PLS-DA scores on LV1 (Fig. 7b) show pixels with statistical similarity, which in this case correlates to the SERS spectrum of the purified integrin receptor. The PLS-DA map (Fig. 7c), again, improves the contrast and enables the location of the bound integrin on the cell membrane. The performance and utility of PLS-DA mapping is further illustrated through the analysis of additional TERS maps in the ESI (Fig. S5†).

The analysis done on streptavidin mutants shows how key amino acid differences can alter the observed signal. This suggests targeted nanoparticles have utility for identifying receptors that bind specific ligands. The observed TERS

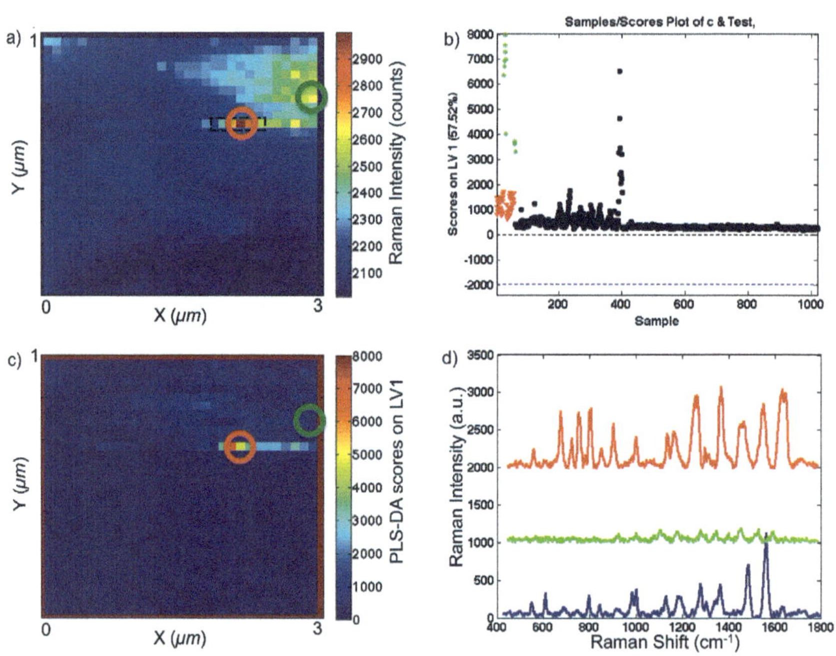

Fig. 7 (a) A 3 × 3 μm (31 × 31 pixel) TERS map at 1003 cm^{-1} of RGD GNPs on SW620 cells is shown. (b) The sample/scores are plotted against LV1 in the PLS-DA analysis. The green and red points are the calibration spectra from the integrin and RGD on GNPs, respectively. (c) Scores map shows the PLS-DA calculated scores on LV1 as an indicator of the full-spectral similarity of each TERS spectrum to LV1. (d) Selected TERS spectra are compared with the PLS-DA calculated LV1 (blue). The green spectrum corresponds to the TERS pixel circled in green and the red spectrum corresponds to the TERS pixel circled in red.

signal in these experiments can provide information that, for example, may differentiate binding to the $\alpha_v\beta_3$ integrin *versus* one of the other 22 integrins in the cell membrane. This would be a powerful tool for investigating drug targeting and signaling pathways.

Our experience with TERS detection of ligand-functionalized nanoparticles indicates the detected signal arises overwhelmingly from the targeted protein receptor. In the case where the protein is functionalized to the probe-nanoparticle, the electric field gradient does not extend beyond the protein and the enhancement observed appears to arise from the protein in the gap.[15] However, with small molecule ligands the observed signal appears to arise from the targeted protein. This selectivity suggests that only the receptor is within the enhancing region of the electric field around the nanostructures. Recently we showed that physical confinement near the enhancing surfaces is important for SERS detection.[34] This seems to also be involved in enhancing the TERS signal. Interestingly, the ligand-receptor interactions are non-covalent bonds which further implies that physical confinement is important but not a direct chemical interaction.

Additionally, the magnitude of the PLS-DA score may provide information relevant to receptor clustering. The score from multivariate analysis can be calibrated to provide information on whether one or more receptors are being detected. In the case of integrins, clustering is reported to play a significant role in its signal transduction mechanism.[35] Differences in the observed PLS-DA score from an isolated nanoparticle may provide insight into multiple receptors clustering on a single nanoparticle that can be compared with receptors binding to multiple nanoparticles in a cluster. Additional work may provide new insight into this important biological receptor.

Conclusions

The data presented illustrate the selectivity obtained from binding a functionalized nanoparticle to a protein receptor and detection with TERS. Using reference spectra obtained from SERS on purified biomolecules, the location of the biomolecule in a complex matrix can be imaged by TERS and background interference can be removed using PLS-DA. The regression analysis enables accurate identification and improved image contrast. Differences in the observed scores suggest different electromagnetic field environments may be responsible for the observed signals. Further analysis of the images may facilitate understanding of the plasmonic environments that generate the observed TERS signals. Using protein mutants, it is clear that the amino acids involved in binding, specifically tryptophan, contribute significantly to the observed TERS signal. The approach presented takes advantage of controlled plasmonic interactions to enable highly selective detection. This may further elucidate the chemistry of ligand binding to protein receptors and other chemical recognition events in complex matrices.

Acknowledgements

The authors acknowledge financial support for this work from NIH Award R21 GM107893, the Walther Cancer Foundation, a Cottrell Scholar Award to ZDS from

Research Corporation for Science Advancement, a Berry Family Graduate Fellowship to HW, and the university of Notre Dame.

References

1 Z. D. Schultz, J. M. Marr and H. Wang, *Nanophotonics*, 2014, **3**, 91–104.
2 M. D. Sonntag, J. M. Klingsporn, A. B. Zrimsek, B. Sharma, L. K. Ruvuna and R. P. Van Duyne, *Chem. Soc. Rev.*, 2014, **43**, 1230–1247.
3 M. Moskovits, *J. Chem. Phys.*, 1978, **69**, 4159–4161.
4 P. Negri, R. J. Flaherty, O. O. Dada and Z. D. Schultz, *Chem. Commun.*, 2014, **50**, 2707–2710.
5 K. Kneipp, Y. Wang, H. Kneipp, L. T. Perelman, I. Itzkan, R. Dasari and M. S. Feld, *Phys. Rev. Lett.*, 1997, **78**, 1667–1670.
6 R. Zhang, Y. Zhang, Z. C. Dong, S. Jiang, C. Zhang, L. G. Chen, L. Zhang, Y. Liao, J. Aizpurua, Y. Luo, J. L. Yang and J. G. Hou, *Nature*, 2013, **498**, 82–86.
7 R. M. Stockle, Y. D. Suh, V. Deckert and R. Zenobi, *Chem. Phys. Lett.*, 2000, **318**, 131–136.
8 N. Hayazawa, Y. Inouye, Z. Sekkat and S. Kawata, *Opt. Commun.*, 2000, **183**, 333–336.
9 J. Steidtner and B. Pettinger, *Phys. Rev. Lett.*, 2008, **100**, 236101.
10 K. L. Wustholz, A.-I. Henry, J. M. McMahon, R. G. Freeman, N. Valley, M. E. Piotti, M. J. Natan, G. C. Schatz and R. P. V. Duyne, *J. Am. Chem. Soc.*, 2010, **132**, 10903–10910.
11 M. Richter, M. Hedegaard, T. Deckert-Gaudig, P. Lampen and V. Deckert, *Small*, 2011, **7**, 209–214.
12 R. Bohme, M. Richter, D. Cialla, P. Rosch, V. Deckert and J. Popp, *J. Raman Spectrosc.*, 2009, **40**, 1452–1457.
13 U. Neugebauer, P. Rosch, M. Schmitt, J. Popp, C. Julien, A. Rasmussen, C. Budich and V. Deckert, *ChemPhysChem*, 2006, **7**, 1428–1430.
14 P. Olk, J. Renger, T. Hartling, M. T. Wenzel and L. M. Eng, *Nano Lett.*, 2007, **7**, 1736–1740.
15 K. D. Alexander and Z. D. Schultz, *Anal. Chem.*, 2012, **84**, 7408–7414.
16 H. Wang and Z. D. Schultz, *Analyst*, 2013, **138**, 3150–3157.
17 S. L. Carrier, C. M. Kownacki and Z. D. Schultz, *Chem. Commun.*, 2011, **47**, 2065–2067.
18 H. Wang and Z. D. Schultz, *ChemPhysChem*, 2014, **15**, 3944–3949.
19 E. Hao and G. C. Schatz, *J. Chem. Phys.*, 2004, **120**, 357–366.
20 L. K. Ausman and G. C. Schatz, *J. Chem. Phys.*, 2009, **131**, 10.
21 K. A. Willets, S. M. Stranahan and M. L. Weber, *J. Phys. Chem. Lett.*, 2012, **3**, 1286–1294.
22 R. Esteban, A. G. Borisov, P. Nordlander and J. Aizpurua, *Nat. Commun.*, 2012, **3**, 825.
23 T. V. Teperik, P. Nordlander, J. Aizpurua and A. G. Borisov, *Opt. Express*, 2013, **21**, 27306–27325.
24 L. Chuntonov and G. Haran, *Nano Lett.*, 2011, **11**, 2440–2445.
25 T. Shegai, Z. P. Li, T. Dadosh, Z. Y. Zhang, H. X. Xu and G. Haran, *Proc. Natl. Acad. Sci. U. S. A.*, 2008, **105**, 16448–16453.
26 N. J. Halas, S. Lal, W. S. Chang, S. Link and P. Nordlander, *Chem. Rev.*, 2011, **111**, 3913–3961.

27 J. Ye, F. F. Wen, H. Sobhani, J. B. Lassiter, P. Van Dorpe, P. Nordlander and N. J. Halas, *Nano Lett.*, 2012, **12**, 1660–1667.
28 K. D. Alexander, K. Skinner, S. P. Zhang, H. Wei and R. Lopez, *Nano Lett.*, 2010, **10**, 4488–4493.
29 K. H. Lim, H. Huang, A. Pralle and S. Park, *Biochemistry*, 2011, **50**, 8682–8691.
30 Z. D. Schultz, S. J. Stranick and I. W. Levin, *Anal. Chem.*, 2009, **81**, 9657–9663.
31 Z. D. Schultz, S. J. Stranick and I. W. Levin, *Appl. Spectrosc.*, 2008, **62**, 1173–1179.
32 D. Graham, R. Stevenson, D. G. Thompson, L. Barrett, C. Dalton and K. Faulds, *Faraday Discuss.*, 2011, **149**, 291–299.
33 J. D. Driskell, R. J. Lipert and M. D. Porter, *J. Phys. Chem. B*, 2006, **110**, 17444–17451.
34 S. M. Asiala and Z. D. Schultz, *Anal. Chem.*, 2014, **86**, 2625–2632.
35 E. Ruoslahti, *Annu. Rev. Cell Dev. Biol.*, 1996, **12**, 697–715.

Faraday Discussions

PAPER

Percolating plasmonic networks for light emission control

Michele Gaio,*[a] Marta Castro-Lopez,[ab] Jan Renger,†[b] Niek van Hulst[bc] and Riccardo Sapienza*[a]

Received 30th September 2014, Accepted 15th October 2014
DOI: 10.1039/c4fd00187g

Optical nanoantennas have revolutionised the way we manipulate single photons emitted by individual light sources in a nanostructured photonic environment. Complex plasmonic architectures allow for multiscale light control by shortening or stretching the light wavelength for a fixed operating frequency, meeting the size of the emitter and that of propagating modes. Here, we study self-assembled semi-continuous gold films and lithographic gold networks characterised by large local density of optical state (LDOS) fluctuations around the electrical percolation threshold, a regime where the surface is characterised by large metal clusters with fractal topology. We study the formation of plasmonic networks and their effect on light emission from embedded fluorescent probes in these systems. Through fluorescence dynamics experiments we discuss the role of global long-range interactions linked to the degree of percolation and to the network fractality, as well as the local near-field contributions coming from the local electro-magnetic fields and the topology. Our experiments indicate that local properties dominate the fluorescence modification.

1 Introduction

Quantum information processing at the single photon level is an increasingly attractive field aiming at coherent manipulation of information encoded into the photon degrees of freedom. Individual fluorescent emitters such as organic molecules or quantum dots with a single electron excitation are the ideal source of single photons, nevertheless, the intrinsic low efficiency of the interaction between light and matter limits photon absorption and emission and therefore poses great constraints to their practical use. Single emitters have dimensions much smaller than the wavelength of light, and therefore interact slowly and

[a]*Department of Physics, King's College London, Strand, London WCR 2LS, UK. E-mail: michele.gaio@kcl.ac.uk; riccardo.sapienza@kcl.ac.uk*
[b]*ICFO – Institut de Ciencies Fotoniques, Mediterranean Technology Park, 08860 Castelldefels (Barcelona), Spain*
[c]*ICREA – Institució Catalana de Recerca i Estudis Avançats, 08019 Barcelona, Spain*
† Present address: ETH Zurich, Photonics Laboratory, 8093 Zurich, Switzerland.

omni-directionally with radiation; these intrinsic fluorescence limits can be overcome when the source is placed in a structured photonic material.

The conventional quantum optics approach consists in surrounding the emitter with micrometer-sized high-finesse microcavities which increase photon–photon interaction by orders of magnitude.[1] Nanoscale modal engineering instead exploits light–matter coupling in the nanometer range where the optical modes can be strongly modified by nanostructured materials. For instance, photonic crystal structures in dielectric materials can enhance the emission of an emitter over large bandwidth,[2] or optical waveguides can achieve large coupling to molecules and transport the emitted light at distant locations.[3] In this context, plasmonic systems based on metallic structures with nano-sized features are particularly attractive because they act as optical antennas by squeezing propagating light into nanometer size volumes *via* surface plasmons. Stronger spatial localisation and field enhancement around these plasmonic antennas[4] leads to an increased optical mode density and can change the decay rate of an emitter over one thousand times.[5] This large Purcell factor, which is defined as the decay rate modification as compared to vacuum, is achieved because of the large dipole moment of the coupled system antenna–emitter and the large mode density increase at the emitter location. Both emission rates and directionality[6,7] can be designed for sources in the near field of the metal structure. The major drawback of metallic particles is the large optical absorption which is driving the search for alternative geometries and materials.

Different particle shapes with increasing complexity have been investigated to tune and strengthen the plasmonic response. Beyond simple spheres and rods, many other shapes have been fabricated, as for example nanoshells,[8] nanorings[9] and cylinders with crescent shaped cross sections.[10] Furthermore, coupling of many antennas into dimer, trimers, *etc.*..., can give rise to a novel collective response, as in metamaterials based on arrays of parallel nanorods.[11] When small metallic particles are coupled together, their optical modes hybridise due to near-field interactions and the resulting energy levels shift and split in analogy to molecular orbital formation.[12,13] Many particles can be also combined in chains, with plasmonic modes delocalized over the chain; the optical response of two-dimensional and three-dimensional chains have been interpreted as composed of the contributions from smaller functional one-dimensional chains embedded within them.[14] Planar deterministic plasmonic architectures have been investigated to engineer broad plasmonic resonances,[15] to design transparent metallic fractal electrodes,[16] and to control the optical wavefront with subwavelength layers.[17]

A more complex two-dimensional arrangement is constituted by metal semi-continuous films. Semi-continuous metal films were firstly investigated by means of scanning near-field optical microscopy (SNOM) showing light localisation into hotspots down to few nanometers and a strong spectral dependence on the probed site on the sample.[18] The localisation and coherence properties in these and other disordered films have been investigated numerically[19] showing a peculiar coexistence at the same frequency of both localised and delocalised modes (the so-called inhomogeneous localisation), with eigenmodes characterised by multiple hotspots coherently distributed over large scale. In particular, the near-field intensity fluctuations in random silver films have been studied experimentally and explained in terms of the aggregate size distribution using percolation theory.[20]

These systems have been extensively studied by scattering experiments aimed at identifying the plasmonic resonance. A different approach consists in studying the local density of optical states (LDOS), which is a fundamental quantity governing light–matter interactions. The LDOS is hard to probe by transmission and scattering experiments as it relates to all optical modes at a given frequency and spatial position. Instead, it can be obtained by fluorescence studies *via* the Purcell factor, as the spontaneous emission rate of optical emitters is proportional to the LDOS. In fact, the rate of spontaneous emission Γ is given by[21]

$$\Gamma = \frac{2\omega^2}{\hbar c^2}|\vec{p}|^2 \mathrm{Im}\left[\hat{\vec{n}}_p \cdot \vec{G}\left(\vec{r}_0, \vec{r}_0, \omega\right)\hat{\vec{n}}_p\right] = \frac{\pi\omega}{\hbar\varepsilon_0}|\vec{p}|^2 \cdot \rho_p(\vec{r}, \omega) \tag{1}$$

where $\vec{p}$ is the transition dipole defined between the two electronic states, $\vec{G}(\vec{r}, \vec{r}\,', \omega)$ is the dyadic Green function evaluated at the location of the emitting system ($\vec{r} = \vec{r}\,' = \vec{r}_0$), $\hat{\vec{n}}_\mathrm{p}$ is the unit vector in the direction of the dipole moment and $\rho_p(\vec{r})$ the partial local density of available optical states for a dipole in the direction $\vec{p}$. This partial LDOS $\rho_p(\vec{r},\omega)$ or the total LDOS $\rho(\vec{r},\omega)$ (integrated over all dipolar orientation) can thus be obtained by measuring the Purcell factor:

$$P = \Gamma/\Gamma_0, \tag{2}$$

as the ratio between the decay in the medium Γ and the reference decay rate in vacuum Γ_0 or, more commonly, as measured for instance inside an homogeneous substrate.

LDOS maps require scanning of the probe dipole and can be obtained by the fluorescence dynamics of a source placed at the tip of an atomic force microscope (AFM),[22,23] or by scanning the probe source generated by electron impact and then by recording the cathodoluminescence signal.[24] Scanning a sub-wavelength SNOM probe over the nanostructure surface gives information on the LDOS map.[25–27] These techniques are restricted to small areas of up to few μm^2, and cannot be easily applied to the large systems we are investigating here. In disordered media, the LDOS is best studied statistically. Several works have measured the fluorescence lifetime of molecules or quantum dots dispersed through the nanostructure. For example, in polymer films the LDOS fluctuations have been related to the local inhomogeneities of the film[28] and polymer segmental dynamics.[29] In 2D dielectric disordered media, the LDOS has been studied around Anderson localisation[30] while, in 3D structures, LDOS fluctuations have been measured in the diffusive regime,[31] and long long-tailed LDOS distributions have been measured and attributed to near-field and far-field interactions.[32]

In plasmonic systems, and in particular in semi-continuous metal films, the LDOS statistical distribution of complex plasmonic systems has been recently explored.[33] Fluorescence mapping on these systems has shown an increase in fluctuations of the LDOS, which has been linked to the presence of surface plasmons localised modes by relating the variance of the LDOS fluctuations to the inverse participation ratio R_IP which can be used as a qualitative measure of the area occupied by hotspots. Numerical computations have also allowed to discriminate between radiative and non-radiative contributions to the LDOS[34] and to highlight the dependence on the fluorescent distance between the source and the gold surface.[35] The overall spatial extent of the eigenmodes has been studied numerically estimating the averaged coherence length by computing the cross density of optical states which shows an overall reduction of the plasmonic mode extension around percolation.[36] In such systems, localised modes enhance the LDOS and therefore induce large Purcell factors, while delocalised ones can propagate the optical excitation to distant locations. Multiple coherent hotspots would potentially allow an efficient energy transfer between distant light sources. Understanding the nature of such optical modes is therefore of fundamental importance.

Here, we study how a large plasmonic network influences the emission properties of a source coupled to it. We fabricated plasmonic networks formed by (1) self-assembly and (2) electron beam lithography, and experimentally investigated the LDOS distribution by monitoring the change in the fluorescence decay rate of individual sources coupled to the networks to construct statistical distributions. We discuss the interplay between (i) long-range properties, such as the topology of the network at the percolation phase transition, *i.e.* the formation of a fractal cluster covering all the structure, and (ii) short-range local features on the order of few nanometers which are related to the inter particle gaps and strongly affect the local field enhancement. Our experiments indicate that the fluorescence modification due to local short-range properties dominates over the effect of the collective degree of percolation.

2 Self-assembled percolating networks

Semi-continuous metallic films can be obtained by thermal evaporation of noble metals like gold or silver onto a dielectric substrate and subsequent thermal annealing. We fabricated such sample on a glass and silicon substrate. The glass substrate is ideal for confocal microscopy, while the silicon conductive substrate is used here for the high resolution scanning electron microscopy images (SEM) shown in Fig. 1. We expect the network topology to be qualitatively similar in both cases, nevertheless, electrical conduction experiments (reported later in the manuscript) are used to pinpoint the percolation transition on the glass samples. The properties of the metal aggregates depend on the growth conditions and the substrate and can be controlled by tuning the amount of metal evaporated, which we describe by the equivalent thickness t. The equivalent thickness t is extrapolated by scaling from large thickness films as the thickness the sample would have had if it was a continuous film. For small t, the sample consists mostly of isolated metal particles of few nanometer size (Fig. 1a). As the amount of gold deposited increases, the clusters grow in size with irregular shapes, changing from elongated but isolated objects to more complex structures which touch at multiple points, increasing the level of connectivity of the system (Fig. 1b). It is worth noting that the clusters also grow in thickness, as for instance at $t = 8$ nm the covered area of the sample is smaller than for $t = 6$ nm. Eventually, as the filling fraction increases further, these clusters interconnect with each other at the electrical percolation threshold, when a continuous conducting path of metal is formed between the ends of the film (Fig. 1c). When more metal is deposited, the system starts looking like an irregular film with holes of smaller and smaller size (Fig. 1d). As shown in the SEM images, both local properties such as shape and

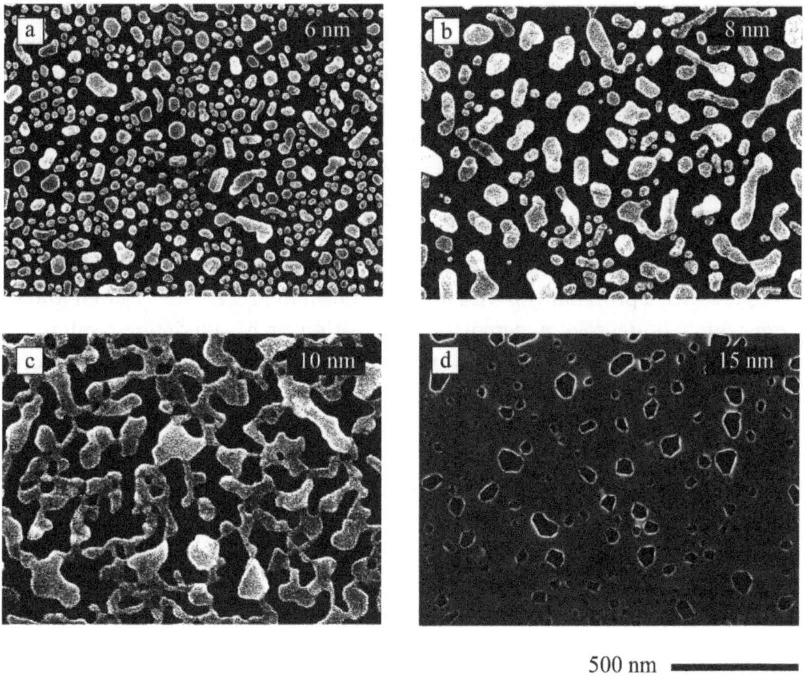

Fig. 1 Self-assembled gold films deposited on a silicon substrate. SEM images of films with different amounts of gold, labelled by means of the equivalent thickness. Panel (a) 6 nm of equivalent thickness, the structure is composed of mostly isolated particles; panel (b) 8 nm, larger isolated clusters form; panel (c) 10 nm, connected clusters develop, the system is above the percolation transition; panel (d) 15 nm, the sample is almost a continuous film with sparse holes. The size bar shown is the same for all images.

size of the clusters, the gaps and local connectivity, and the global network properties change across the different samples. Furthermore, it has been shown that the fractal dimension of the clusters in similar structures depends on the filling fraction.[33]

The semi-continuous gold networks were made on microscope cover-slip glass substrates by thermal evaporation of gold and no further thermal annealing. The samples have an equivalent thickness ranging from 2.5 nm to 50 nm. Due to the low contrast on the SEM images on non-conducting glass, we verified the crossing of the electrical phase transition by measuring the electrical resistance between the edges of the samples with a digital multimeter, as shown later in Section 5. The percolation transition occurs at $t \simeq 10$ nm.

3 Lithographic networks

Beyond thermal self-assembly, lithographic techniques allow samples of any given topology to be controlled and designed with high resolution.[37] Electron beam lithography (EBL) is a very flexible technique and allows the design and fabrication of structures beyond natural limits.[38,39] Using EBL, we fabricated plasmonic networks exhibiting a percolation transition and fractality similar to the self-assembled network, but with a mono disperse building block, a gold nanoantenna, and a well-defined degree of percolation. The fabricated networks have an area of $\sim 10 \times 10\ \mu m^2$ and inter-particle gaps of ~ 25 nm, while the building block is a gold nanoantenna of size 60 nm × 40 nm and 40 nm in height. The gold networks have been fabricated on ITO (10 nm)/SiO_2 substrates using poly(methyl methacrylate) (PMMA) as photo-resist and thermally evaporating 40 nm of gold prior lift-off. We have chosen bond percolating on a square lattice as this is one of the simplest geometries that combines strong near-field hotspots and collective percolation. In this system, each edge or bond of the square lattice is occupied with a probability p with a rod as shown in Fig. 2. This is a well-known system which undergoes a percolation transition at a well-defined value, *i.e.* $p = 0.5$.[40] We have created matrixes of up to 20 000 elements and filling probability $p = 0.1, 0.3, 0.4, 0.45, 0.48, 0.5, 0.52, 0.55, 0.6, 0.8$ and 1.

The nanorods, as sketched in Fig. 3a, are shaped in order to obtain a narrow gap between them of about 25 nm, a distance that guarantees the near-field coupling between their electromagnetic modes. In this design, the plasmon resonance of the single rod and the coupled system is expected to be in the visible range. The single rod has a plasmon resonance at around 650 nm. The resonance of two touching rods shifts to the infrared[41] while the effect of the hybridisation of the modes of two coupled but not touching rods is not as large;[12] in our case the two-rods resonance occurs at around 720 nm. Although there is no electrical percolation as the rods are not connected, at each vertex of the grid a maximum of four rods can interact and a collective plasmon mode can build up spanning many rods and spread across the system following the rods chain. The electrical field intensity map on the structure upon excitation with a dipole source obtained by finite difference time domain (FDTD) calculations is shown in Fig. 3b. Beside the field map visualisation, the FDTD simulation also allows to quantify the

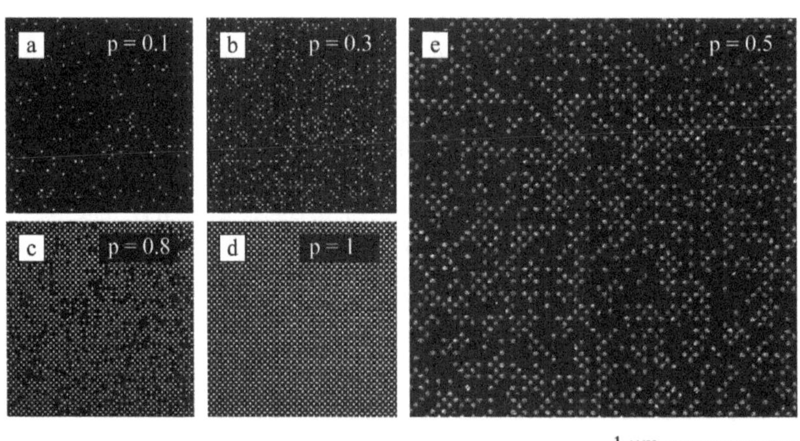

Fig. 2 SEM images of lithographic gold networks fabricated by EBL on glass substrates. As indicated on the image, panel (a) shows the structure for a value of the percolation parameter $p = 0.1$, panel (b) $p = 0.3$, panel (c) $p = 0.8$, panel (d) $p = 1$, and panel (e) $p = 0.5$. Each element of the network is a nanoantenna of size $60 \times 40 \times 40$ ($L \times W \times H$) nm. Each network is $10 \times 10\ \mu m^2$ and contains $\sim 10^4$ rods.

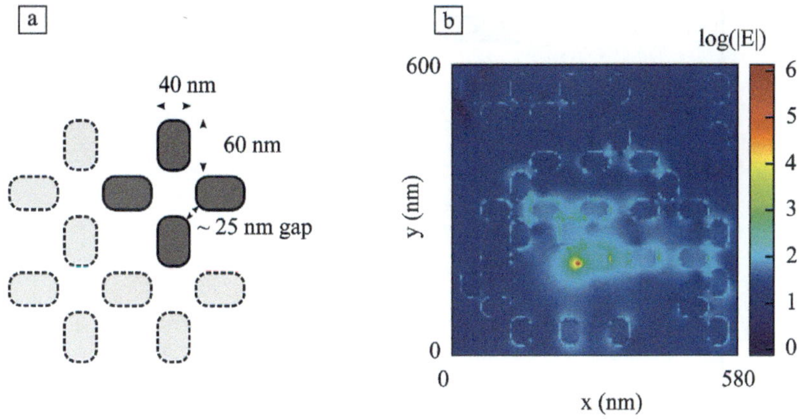

Fig. 3 Panel (a) sketch of the sample structure shown in Fig. 2: each nanoantenna has a size of 40 × 60 nm and the gap between the nanoantenna tips is ~25 nm. Panel (b) plot of the electric field intensity at $\lambda = 720$ nm obtained by FDTD simulations upon dipolar excitation at the location of highest intensity, 7 nm from the closest gold structure. Plasmonic waves of coupled nanoantennas are excited.

radiative and non-radiative contributions to the Purcell factor by considering the energy radiated into the far-field and the energy absorbed by the material. For a dipole oriented along the rod long axes and located 7 nm from the rod inside the gap region, the total Purcell factor calculated is ~100, dominated by non-radiative terms. In this case, the radiative Purcell factor is ~5.

3.1 Topological properties

The topology of the network evolves with the parameter p as it can be understood by direct visualization. Fig. 4 shows the results of Monte Carlo calculations on a grid with 100 × 100 nodes, thus containing up to 2×10^4 rods, which is the same size of the system we fabricated. Panels a–d show specific realisations of the system for different p-values below the percolation threshold (Fig. 4a $p = 0.4$ and Fig. 4b $p = 0.49$), at the threshold (Fig. 4c $p = 0.5$) and above the percolation threshold (Fig. Fig. 4d $p = 0.6$). The cumulative cluster mass distribution is defined as

$$C(N) = \int_0^N f(n)\, n\, \mathrm{d}n \tag{3}$$

with $f(n)$ being the cluster mass distribution probability, *i.e.* the number of clusters constituted by n elements normalised on the total number of elements. $C(N)$, which is shown below each realisation in Fig. 4, represents the fraction of the area which is filled with clusters of size up to N elements. The colour code in Fig. 4 is used to identify clusters of different size. For small p, the system consists of isolated particles or small clusters: at $p = 0.4$ about half of the system is composed by clusters with less than 10 elements, which are represented in blue. The size of the other clusters with more than 10 elements composing the system, which are represented in red, is limited at this filling probability to about 10^2 elements, which is the size of the largest cluster highlighted in green. As the

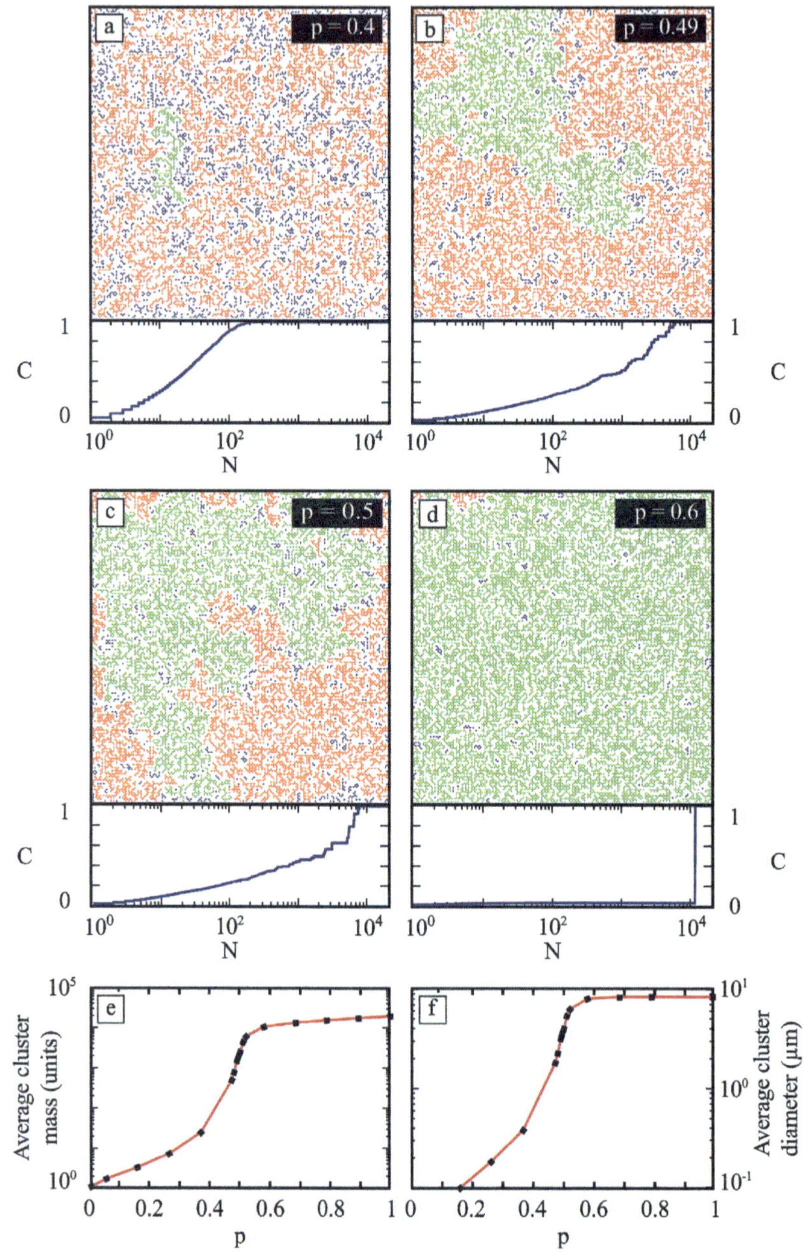

Fig. 4 Monte Carlo simulations of the nanoantenna network on a 100 × 100 grid. Realizations at different filling probabilities $p = 0.4$ (a), $p = 0.49$ (b), $p = 0.5$ (c) $p = 0.6$ (d). The different colours identify clusters of different sizes: red for clusters with mass bigger than 10 elements, blue for clusters with mass smaller or equal to 10 elements and green for the largest cluster. Below each structure, the cumulative cluster mass distribution $C(N)$ is shown. At $p = 0.4$ about half of the system is constituted by clusters with less than 10 elements and the maximum cluster size is limited to about 100 nanoantennas. For increasing values of p, the average cluster size quickly increases. Above the percolation transition, the system is dominated by the percolating cluster. Panel (e) shows the average cluster mass, and panel (f) the average cluster diameter.

p increases, larger clusters appears with a higher probability and the largest cluster starts to dominate the size distribution when the system approaches the percolation threshold, for instance at $p = 0.49$. At the phase transition ($p = 0.5$), the size of the largest cluster diverges reaching the system size. Above the threshold, the percolating cluster alone dominates and fills almost the whole system apart from few detached clusters of decreasing size.

In order to highlight the percolation transition, we computed the average cluster mass weighted with the cluster mass, which is defined as

$$m = \langle f(n)n \rangle. \tag{4}$$

As $f(n)n$ is the fraction of area filled by clusters with n elements, m corresponds to the average mass of the cluster that one would select by choosing a random point of the structure. As shown in Fig. 4e the average mass grows smoothly far from $p = 0.5$ both in the low and high p range, while it grows quickly approaching the phase transition. We estimated the cluster extension by means of the radius of gyration.

$$R_S = \frac{1}{n}\sum \left|\vec{r}_i - \vec{r}_{cm}\right|^2, \tag{5}$$

which is the root mean square distance of the elements from the centre of mass of the cluster $\vec{r}_{cm}$. The average cluster diameter weighted with the cluster mass is

$$d = 2\langle R_S f(n)n \rangle, \tag{6}$$

which is shown in Fig. 4f.

Two topological properties, the cluster diameter and cluster fractality, are expected to be important for the optical properties of the plasmon modes. A plasmon wave is bounded to the metal, and loses energy both by out-of-plane scattering and by ohmic absorption in the metal while propagating. These effects are reflected in the plasmon propagation length l_p which in continuous gold films is typically 5–20 μm.[42] The optical interference which determines the plasmon modes is limited to components within a length of the order of l_p. The comparison between the cluster diameter and the plasmon propagation length l_p is therefore a good indicator of how the plasmon excitation will spread in the system. If the cluster size is smaller than l_p then its shape and boundary will affect the hybridised plasmon mode and the optical response. Instead, if the cluster size grows larger than l_p, the plasmon response will saturate at a level that will not depend any more on the cluster boundary as no optical excitation will be able to probe it. In our networks in the region $p = 0.3$–0.5, the average cluster diameter grows quickly, approaching the sample size of 10 μm, and eventually crossing the plasmon propagation length (see Fig. 4f).

The second topological property, the fractality, is harder to be taken into account as it is a multiscale property which acts both on the local features and nanogaps as well as on the long-range shapes. At the phase transition the percolating cluster diverges in size, the structure is self-similar and the clusters are known to be fractal. Besides the average cluster diameter, which is an obvious topological measure, also the fractal dimension is expected to have very important relation with the optical properties of the plasmon modes. For example, in self-assembled networks, the degree of fractality of the system, calculated as the ratio of fractal to non-fractal (Euclidean) clusters, has been connected to the normalized variance of the LDOS.[33] For bond percolation networks, the fractal dimension D can be defined as the scaling of the cluster mass *versus* the area, *i.e.* $m \propto R_S{}^D$.[40] At the percolation threshold, the 2D bond percolation model used to design the lithographic networks is known to have a fractal dimension $D = 91/48$,[40] a typical value for 2D systems; similar values have been found also in semi-continuous films[33,43] even if using a different definition of the fractal dimension. Above the percolation threshold, the gold network evolves towards a continuous uniform film with Euclidean dimension $D = 2$ while, for very small values of p, the network consists of isolated antennas with no fractal properties and again Euclidean dimension $D = 2$. A crossover between a fractal and a Euclidean network is therefore expected in the region $p > 0.5$.

4 Fluorescence dynamics studies

As described in Section 1, the LDOS can be extracted from the decay rate of an emitter. Here, we performed fluorescent dynamics measurements *via* confocal microscopy of plasmonic networks with individual fluorescent point-like sources dispersed on the surface of the sample. We used dye-doped polymer beads (polystyrene–divinylbenzene) of nominal diameter 50 nm internally doped with red (absorption 542 nm, emission 612 nm) Firefly dye molecules (Thermo Scientific). Each bead contains more than $\sim 10^2$ molecules and has a well-defined, orientation-independent optical response and decay time that we measured to be $\tau = 5.5 \pm 0.3$ ns ($\langle \tau \rangle \pm \sigma$, where σ is the standard deviation of the distribution) on a glass substrate. The beads were randomly deposited on the structure by spin-coating with an average density of $\sim$1 bead per μm^2. A 532 nm laser (Nd:YAG second harmonic) with a 10 MHz repetition rate and 100 ps pulses width is used to excite the fluorescent source through an oil immersion objective (NA = 1.45). The same objective collects the light emitted which is directed to an avalanche photodiode (APD) for detection coupled to a time correlating single-photon counting (TCSPC) card for temporal analysis (TimeHarp 260, PicoQuant). The beads are identified by scanning the sample by means of a piezo stage and measuring the emitted photons with the APD for an integration time of 1 ms per pixel and a pixel size of 100 nm, as shown in the inset of Fig. 5. The lifetime of each bead was collected by increasing the acquisition time of the TCSPC to $\sim$10 s at a typical laser power of around 1 μW. Typical results are plotted in Fig. 5.

We fitted the data with a single exponential function in the region highlighted in Fig. 5. The small deviations from the single exponential decay are due to the averaging over the position and direction of the molecules inside the beads which experiment slightly different LDOS as they are at different distances from the gold surface. The large component with very short lifetime, which is more prominent in the self-assembled networks, is attributed to gold self-fluorescence. It shows as a peak at short time which is the result of the convolution of a very fast signal with lifetime <10 ps (as previously reported in ref. 44) with our instrumental response ($\sim$400 ps). After this first peak, the signal is due to the beads' fluorescence. The component we have analysed is in the range 2–18 ns (highlighted in Fig. 5) with a lifetime in the range 1–5 ns. For longer times, the signal merges with the background. The peak at $t = 18$ ns is due to after-pulsing in the APD, which is an

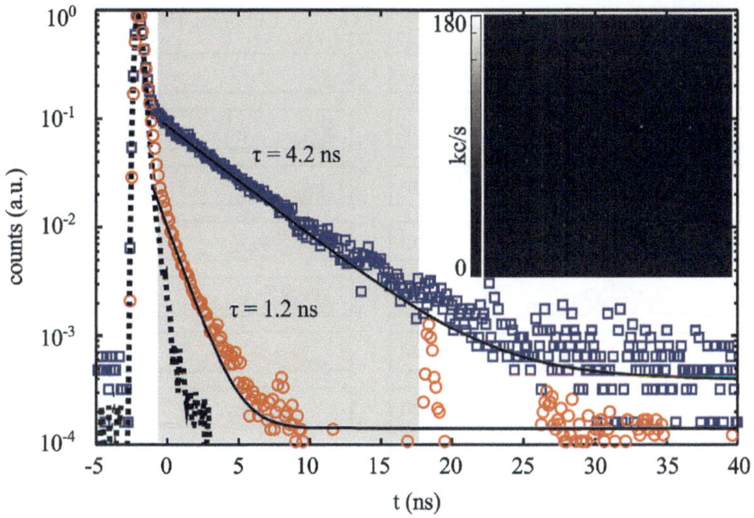

Fig. 5 Lifetime measurements from two emitters at two different positions on a self-assembled network, with high (red open circles) and low (blue open squares) decay rates. The initial peak, with very short lifetime, comes from the gold self-fluorescence which is resolved as the instrumental response (shown with the black dashed line). The continuous line through the data is the single exponential fit. The spurious signal at $t = 18$ ns is due to after-pulsing which is a common electronic artefact. In the inset, confocal scan on a 20 μm by 20 μm area of the sample which allows to identify the position of the sources.

electronic artefact of this type of electronics. The typical statistical error of the fit is <5%, much smaller than the typical lifetime variation observed.

5 Results and discussion

We performed an extensive decay rate statistics by collecting the lifetimes of up to 300 beads for each sample. In the case of the self-assembled networks, which extend over a few cm^2, a few different areas of the same sample were measured. In the case of lithographic structures, several realisations of the networks at the same p where measured on the same sample. The decay rate distributions are shown in Fig. 6. The topmost histograms are the references, *i.e.* the decay rate distribution for beads on a glass substrate, which have an average value $\Gamma_0 = 0.18\ ns^{-1}$. In both the self-assembled and the lithographic systems, the decay distributions change shape and broaden as compared to the reference. From the top to the bottom of Fig. 6a and b, we present the decay rate distribution resulting from the experiments on networks with increasing filling ratios, which we indicate by the equivalent thickness t for the self-assembled networks and by the filling probability p for the lithographic one. For the lithographic networks, measurements within the p-ranges of 0.4–0.45, 0.48–0.52 and 0.55–0.60 have been combined. The maximum mean Purcell factors are in the range 0.8–5 in both cases.

The variance of the distributions is plotted in Fig. 6c and d; in both cases the variance of the decay rate distribution increases with increasing values of t and p, as compared to the reference. For the self-assembled networks, the variance

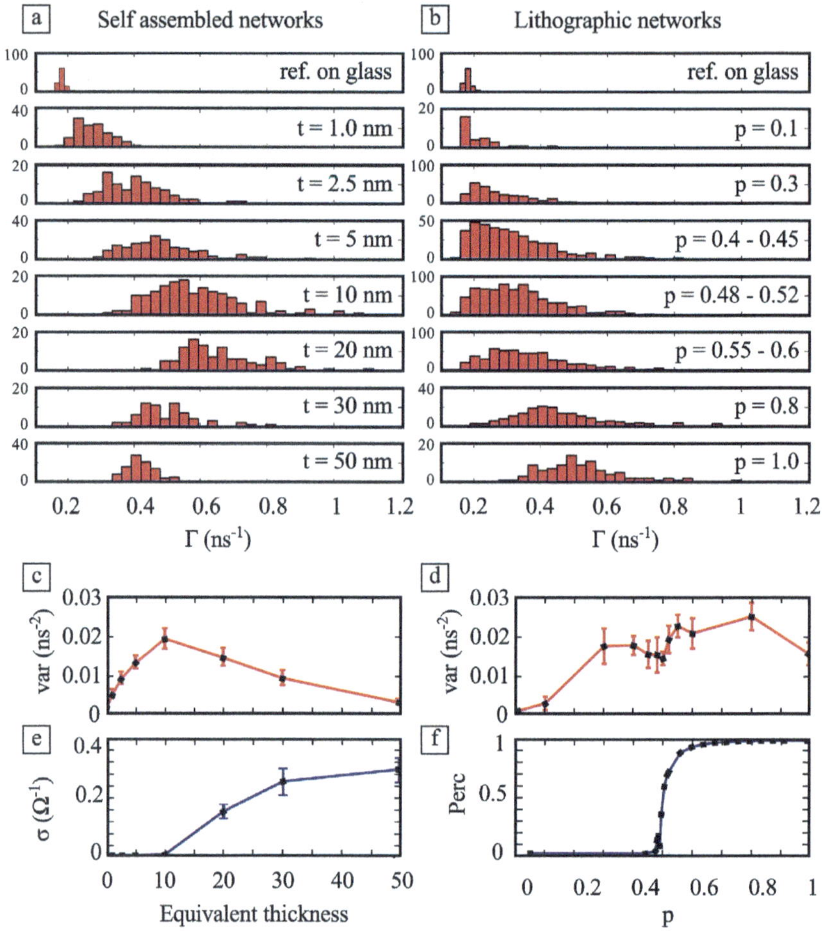

Fig. 6 Panel (a) shows the decay rate statistics of self-assembled networks as a function of the equivalent gold thickness while panel (b) shows the same for lithographic networks as a function of the percolation parameter p. A change of width and position of the decay rate distribution is visible for both cases. Panels (c) and (d) show the variance of the respective distributions. Panel (e) shows the measured conductibility, while panel (f) shows the percolation function, *i.e.* the probability that a randomly chosen nanoantenna belongs to the percolating cluster computed by Monte Carlo simulations. The maximum average Purcell enhancement (~3) and the maximum variance for self-assembled networks is between an equivalent thickness of 10 to 20 nm, which corresponds to the electrical percolation transition. The effect in the lithographic networks is less pronounced and occurs for large p values beyond the percolation transition. For small p values, the distributions are dominated by a significant amount of beads on glass, uncoupled to the structure which accumulates to low decay rate values. The maximum Purcell factor and the maximum FWHM of the distribution occurs towards the full samples ($p \sim 0.8$–1), far from the percolation transition.

increases reaching a maximum at around $t = 10$ nm, which corresponds also to the maximum Purcell enhancement, and then decreases to smaller values in the thicker samples. This maximum occurs together with a change in the topology of the network, which changes from disconnected clusters to a uniform plasmonic structure. Electrical conductivity measurements confirm the crossing of two electrical phases at around 10 nm as shown in Fig. 6f. For lower values of deposited gold, the conductivity is almost zero, while it grows abruptly once percolation is established just above 10 nm. This value of t corresponds to the peak of the variance in Fig. 6e. The behaviour of the self-assembled networks we report is consistent with works on similar samples as shown in ref. 33, although here the variance presents a smoother increase rather than an abrupt increase around the percolation value.

The decay rate distribution from the lithographic networks shown in Fig. 6b broadens as p increases but presents a different general trend. For small p values, the decay distribution is peaked at the same value as the reference, indicating that a significant fraction of the sources maintain an almost unchanged decay rate which we attribute to sources on glass, far and thus not coupled to the gold clusters. This is due to the low gold filling fraction of the system. On the contrary, the smaller grain size of the self-assembled networks guarantees a dense coverage of the sample surface down to the nanometer size at any filling ratio. In the lithographic networks, until $p < 0.5$, the distribution grows a long-tail toward larger Γ, but no real shift of the mean is visible. For $p > 0.6$, where the network is covered almost everywhere by the percolating cluster, as shown in Fig. 4d, the distribution shifts to larger values of the decay rate and maximum values are achieved towards the high density samples, towards $p = 1$. The marked difference with the self-assembled networks is a monotonic increase of the mean and variance of the distribution which do not seem to decrease again, nor to have a special behaviour around the percolation transition. This transition is expected at $p = 0.5$, as highlighted in Fig. 6f, by plotting the percolating function Perc(p), which is the probability of a rod to be part of the percolating cluster and is thus related to the conductivity of the system.[45] Below $p = 0.5$, there is no percolating cluster. At $p = 0.5$, where the percolating cluster begins to dominate the network, the probability that the antenna is part of the percolating cluster rapidly increases and for $p > 0.6$ it grows linearly as almost all rods which are added to the system are in the percolating cluster. Comparison of Fig. 6d and f indicates that the decay rate variance on the lithographic networks does not show a well-defined relation with the topological percolation transition. Only a minor dip in the variance is visible at $p = 0.5$, instead of the maximum of the self-assembled networks shown in Fig. 6c.

In Section 3 we have discussed the cluster size and the cluster fractality as two aspects which are expected to affect the plasmon response of the system. In the case of the lithographic networks, these quantities can be calculated, therefore an assessment of their role can be attempted. In the region of very small p-values, the spatial extension of the plasmonic modes l_p is expected to be larger than the average cluster size d which is <1 μm. For increasing values of p, we expect the cluster size d to become larger than l_p and therefore to saturate the optical response. Such an effect does not appear in our data. However, it is worth noting that the large number of uncoupled beads makes the analysis in the region below percolation difficult. Regarding the fractality of the system, for a 2D bond-percolating network, structural fractality occurs only in a small region close to $p = 0.5$, while the system is not fractal, as discussed in Section 3, both for very small p-values or in the region $p > 0.5$, when the percolating and large clusters dominate the system. Our data do not show an evident effect related to the fractality around $p = 0.5$. On the contrary, a general shift of the distribution happens at $p > 0.6$, when the percolating clusters span already all the system. At $p > 0.6$, the main structural difference is the local connectivity of the system: the sample is always dense but, as p increases, the average number of linked rods at each node (equal to $4p$) approaches its maximum value of 4. It has to be noted that, for self-assembled networks, the fractality of the system can extend down to the nanometer level, thus resulting in gaps between the metal particles of nanometer size; on the contrary, the lithographic networks always present a fixed gap of ~25 nm.

Another difference between the self-assembled and lithographic networks is related to the network shape: while in the former the components are physically connected, in the latter the rods are spaced by ~25 nm. As we discussed in Section 3.1, both networks are optically connected as the plasmons couple from rod to rod, while the self-assembled one is also electrically connected. This design was chosen to highlight the topology-related effects over those coming from the resonant nature of the elements composing the network. Our results seem to indicate that more attention and future investigations are required to fully understand the impact of the resonant response on the LDOS distribution, changing with the percolation parameter and extending to the IR.

These considerations suggest that the effect on the fluorescence modifications is dominated by local properties such as node connectivity and number of nanorods around each position, and are much less influenced by the degree of percolation, the fractality or the extension of the modes. It is tempting to extend the results of this observation to the self-assembled networks, but the lack of real control over the topological parameters and the change of the shape of the building blocks with t makes this far-fetched.

6 Conclusions

We have fabricated metal percolation networks by thermal self-assembly of semicontinuous films and by EBL fabrication of gold nanorod matrices. We have investigated the LDOS distribution for different filling ratios of the systems by measuring extensive statistics of decay rates and we have compared the broadening and shift of the distributions in the two cases. While both the average Purcell factor and the variance of the decay rate distribution evolve following the network topology, in the self-assembled films they peak close to the percolation threshold, while for the lithographic case the most significant effect is not observed in the vicinity of the percolation transition but at higher filling probabilities.

Acknowledgements

We wish to thank Chris Lorenz and Alexandre Cazé for fruitful discussions. This research was funded by the Engineering and Physical Sciences Research Council (EPSRC), the FP7 EU Project People, ERC Advanced Grant 247330-Nano Antennas, and MICINN program CONSOLIDER NanoLight CSD2007-046.

References

1 A. M. Armani, R. P. Kulkarni, S. E. Fraser, R. C. Flagan and K. J. Vahala, *Science*, 2007, **317**, 783–787.

2 P. Lodahl, A. F. van Driel, I. S. Nikolaev, A. Irman, K. Overgaag, D. Vanmaekelbergh and W. L. Vos, *Nature*, 2004, **430**, 654–657.

3 J. Hwang and E. A. Hinds, *New J. Phys.*, 2011, **13**, 085009.

4 L. Novotny and N. van Hulst, *Nat. Photonics*, 2011, **5**, 83–90.

5 A. Kinkhabwala, Z. Yu, S. Fan, Y. Avlasevich, K. Müllen and W. E. Moerner, *Nat. Photonics*, 2009, **3**, 654–657.

6 H. Aouani, O. Mahboub, N. Bonod, E. Devaux, E. Popov, H. Rigneault, T. W. Ebbesen and J. Wenger, *Nano Lett.*, 2011, **11**, 637–644.

7 A. G. Curto, G. Volpe, T. H. Taminiau, M. P. Kreuzer, R. Quidant and N. F. van Hulst, *Science*, 2010, **329**, 930–933.

8 R. D. Averitt, S. L. Westcott and N. J. Halas, *J. Opt. Soc. Am. B*, 1999, **16**, 1824–1832.

9 J. Aizpurua, P. Hanarp, D. S. Sutherland, M. Käll, G. W. Bryant and F. J. G. de Abajo, *Phys. Rev. Lett.*, 2003, **90**, 057401.

10 A. Aubry, D. Y. Lei, A. I. Fernández-Domínguez, Y. Sonnefraud, S. a. Maier and J. B. Pendry, *Nano Lett.*, 2010, **10**, 2574–2579.

11 A. V. Kabashin, P. Evans, S. Pastkovsky, W. Hendren, G. A. Wurtz, R. Atkinson, R. Pollard, V. A. Podolskiy and A. V. Zayats, *Nat. Mater.*, 2009, **8**, 867–871.

12 P. Nordlander, C. Oubre, E. Prodan, K. Li and M. I. Stockman, *Nano Lett.*, 2004, **4**, 899–903.

13 I. Romero, J. Aizpurua, G. W. Bryant and F. J. G. de Abajo, *Opt. Express*, 2006, **14**, 9988–9999.

14 R. Esteban, R. W. Taylor, J. J. Baumberg and J. Aizpurua, *Langmuir*, 2012, **28**, 8881–8890.

15 A. Gopinath, S. V. Boriskina, N.-N. Feng, B. M. Reinhard and L. Dal Negro, *Nano Lett.*, 2008, **8**, 2423–2431.

16 F. Afshinmanesh, A. G. Curto, K. M. Milaninia, N. F. van Hulst and M. L. Brongersma, *Nano Lett.*, 2014, **14**, 5068–5074.

17 N. Yu, P. Genevet, M. A. Kats, F. Aieta, J.-P. Tetienne, F. Capasso and Z. Gaburro, *Science*, 2011, **334**, 333–337.

18 S. Grésillon, L. Aigouy, A. C. Boccara, J. C. Rivoal, X. Quelin, C. Desmarest, P. Gadenne, V. A. Shubin, A. K. Sarychev and V. M. Shalaev, *Phys. Rev. Lett.*, 1999, **82**, 4520–4523.

19 M. I. Stockman, S. V. Faleev and D. J. Bergman, *Phys. Rev. Lett.*, 2001, **87**, 167401.

20 K. Seal, D. A. Genov, A. K. Sarychev, H. Noh, V. M. Shalaev, Z. C. Ying, X. Zhang and H. Cao, *Phys. Rev. Lett.*, 2006, **97**, 206103.

21 L. Novotny and B. Hecht, *Principles of Nano-Optics*, Cambridge University Press, 2012.

22 V. Krachmalnicoff, D. Cao, A. Cazé, E. Castanié, R. Pierrat, N. Bardou, S. Collin, R. Carminati and Y. D. Wilde, *Opt. Express*, 2013, **21**, 11536–11545.

23 M. Frimmer, Y. Chen and A. F. Koenderink, *Phys. Rev. Lett.*, 2011, **107**, 123602.

24 R. Sapienza, T. Coenen, J. Renger, M. Kuttge, N. F. van Hulst and A. Polman, *Nat. Mater.*, 2012, **11**, 781–787.

25 K. Joulain, R. Carminati, J.-P. Mulet and J.-J. Greffet, *Phys. Rev. B: Condens. Matter*, 2003, **68**, 245405.
26 S. Vignolini, F. Intonti, F. Riboli, D. S. Wiersma, L. Balet, L. H. Li, M. Francardi, A. Gerardino, A. Fiore and M. Gurioli, *Appl. Phys. Lett.*, 2009, **94**, 163102.
27 G. C. des Francs, C. Girard, J.-C. Weeber and A. Dereux, *Chem. Phys. Lett.*, 2001, **345**, 512–516.
28 R. A. L. Vallée, M. van Der Auweraer, F. C. De Schryver, D. Beljonne and M. Orrit, *ChemPhysChem*, 2005, **6**, 81–91.
29 R. A. L. Vallée, N. Tomczak, L. Kuipers, G. J. Vancso and N. F. van Hulst, *Phys. Rev. Lett.*, 2003, **91**, 038301.
30 P. D. García, S. Stobbe, I. Söllner and P. Lodahl, *Phys. Rev. Lett.*, 2012, **109**, 253902.
31 M. D. Birowosuto, S. E. Skipetrov, W. L. Vos and A. P. Mosk, *Phys. Rev. Lett.*, 2010, **105**, 013904.
32 R. Sapienza, P. Bondareff, R. Pierrat, B. Habert, R. Carminati and N. F. van Hulst, *Phys. Rev. Lett.*, 2011, **106**, 163902.
33 V. Krachmalnicoff, E. Castanié, Y. De Wilde and R. Carminati, *Phys. Rev. Lett.*, 2010, **105**, 1–4.
34 A. Cazé, R. Pierrat and R. Carminati, *Phot. Nano. Fund. Appl.*, 2012, **10**, 339–344.
35 E. Castanié, V. Krachmalnicoff, A. Cazé, R. Pierrat, Y. D. Wilde and R. Carminati, *Opt. Lett.*, 2012, **37**, 3006–3008.
36 A. Cazé, R. Pierrat and R. Carminati, *Phys. Rev. Lett.*, 2013, **110**, 063903.
37 H. Duan, A. I. Fernández-Domínguez, M. Bosman, S. a. Maier and J. K. W. Yang, *Nano Lett.*, 2012, **12**, 1683–1689.
38 A. Gopinath, S. V. Boriskina, B. M. Reinhard and L. D. Negro, *Opt. Express*, 2009, **17**, 3741–3753.
39 D. Brinks, M. Castro-Lopez, R. Hildner and N. F. van Hulst, *Proc. Natl. Acad. Sci. U. S. A.*, 2013, **110**, 18386–18390.
40 A. Aharony and D. Stauffer, *Introduction To Percolation Theory*, Taylor & Francis, 2003.
41 T. H. Taminiau, F. D. Stefani and N. F. van Hulst, *Nano Lett.*, 2011, **11**, 1020–1024.
42 A. Kolomenski, A. Kolomenskii, J. Noel, S. Peng and H. Schuessler, *Appl. Opt.*, 2009, **48**, 5683–5691.
43 A. Losquin, S. Camelio, D. Rossouw, M. Besbes, F. Pailloux, D. Babonneau, G. A. Botton, J.-J. Greffet, O. Stéphan and M. Kociak, *Phys. Rev. B: Condens. Matter Mater. Phys.*, 2013, **88**, 115427.
44 M. R. Beversluis, A. Bouhelier and L. Novotny, *Phys. Rev. B: Condens. Matter*, 2003, **68**, 115433.
45 S. Kirkpatrick, *Rev. Mod. Phys.*, 1973, **45**, 574–588.

Faraday Discussions

DISCUSSIONS

Surface plasmon enhanced spectroscopies and time and space resolved methods: general discussion

Jeremy Baumberg, Michael Nielsen, Sergey Bozhevolnyi, Viktor Podolskiy, Thomas Ebbesen, Kaiqiang Lin, Alexei A. Kornyshev, Jacob Khurgin, James Hutchison, Katarzyna Matczyszyn, Jino George, Emiliano Cortes, James T. Hugall, Adi Salomon, Paul Dawson, Olivier Martin, Santhosh Kotni, F. Javier García de Abajo, Michael Flatté, Martin Moskovits, Duncan Graham, Stefan Maier, Masayuki Futamata, Sang-Hyun Oh, Javier Aizpurua, Zachary Schultz and Riccardo Sapienza

DOI: 10.1039/c5fd90023a

F. Javier García de Abajo opened a general discussion of the two papers by Jeremy Baumberg and Javier Aizpurua: The plasmon frequencies and spatial distributions in sub-nanometer gaps are ultimately sensitive to the specific atomic structure near that region. It is thus conceivable that a single atom can make a difference in some systems. This is certainly the case in nanographenes, where an atomic defect has been observed to trap UV plasmons.[1] The presence of edge states in this atomically thin material also introduces a strong dependence on the specific atomic arrangement,[2,3] and it has been predicted that the removal of a row of atoms is sufficient to strongly scatter IR propagating plasmons.[4] I would anticipate a similar situation in atomically thin rows of noble metal atoms, where plasmons have been neatly observed through electron microscope spectroscopy,[5] and in which the removal of a single atom should prevent hopping between its neighbours. This extreme sensitivity to atomic detail is thus understandable when the system has atomic dimensions (graphene, atomic gold wires), while in gaps involving 3D materials (a gold tip near a gold surface) the situation might be different. The question arises, how does the dimensionality of the system affect the sensitivity of the plasmons to atomic detail? In connection to this, in noble-metal tip–tip or tip–planar-surface gaps, the atomic arrangement involves facets (*i.e.*, not just a single atom, but atomic planes, edges and kinks), so is there any dependence on the crystallographic orientation of the metals involved? Electronic surface states are part of this scenario, as their characteristics, and even their existence, depends on the material and orientation of the facets. Would the effect of these states be noticeable on the plasmons? I would appreciate having your opinion on these issues.

1 W. Zhow *et al.*, *Nat. Nanotech.*, 2012, 7, 161.
2 A. Manjavacas *et al.*, *Nanophotonics*, 2013, 2, 139.
3 S. Thongrattanasiri *et al.*, *Laser Photon. Rev.*, 2013, 7, 297.

4 I. Silveiro *et al.*, *Light Sci. Appl.*, 2015, **4**, e241.
5 T. Nagao *et al.*, *Phys. Rev. Lett.*, 2006, **97**, 116802.

Jeremy Baumberg answered: We believe that indeed there is strong sensitivity to single atoms, and rows of atoms, and current experiments progress towards this. We note that the crucial dimensionality is almost 2D since we have maybe 3–5 atom gaps, so that single Au atoms perturb the cavity modes significantly. We see such changes in real time and on single constructs. The questions raised are thus very pertinent. We have shown that facets are clearly important and modify the spectra strongly. We are not yet sure if electronic surface states have any effect.

Martin Moskovits addressed Javier Aizpurua: You used the term "full quantum treatment". Most quantum treatments of very complex subjects like this involves certain simplifications. A full quantum treatment would indeed be a time-dependent one, as the treatment described was, and have included, for example, a fully quantized electromagnetic field, the coordinates of every electron and perhaps even every nucleus, when dealing with the energetics of electrons that are in the process of thermalizing by interacting with phonons. We learned a tremendous amount from the inclusion of quantum tunneling as described by Aizpurua and Baumberg. One wonders, however, what new physical phenomena might arise when a "full quantum treatment" in the fuller sense is applied to such plasmonic systems in the fullness of time.

Javier Aizpurua replied: This comment is absolutely right. The use of the description of a "full" quantum treatment should be addressed with care, and in the right context. First of all, let me point out that one might consider the quantum nature of plasmons in a twofold way: there is the aspect of the quantum nature of the electrons and their correlations when forming the collective response of the electron gas that has been treated in Surface Physics for many years and now we are checking in nanoparticles and clusters to learn how to address nonlocalities and quantum aspects of the optical response; but there is also another important aspect of quantum plasmonics which is connected with a "full" quantization of the interaction of light and matter (quantum optics), in the spirit of what Professor Moskovits mentions, where the photon and plasmon fields can be quantized, and so can be the interaction between single photons and single plasmons whose quantum states follow Schrödinger's equation to describe the dynamics of the "full" coupled quantum-mechanical system. What we do is to focus in the first aspect of quantum effects in plasmonics, *i.e.* to describe the optical response of the electron gas in a nanoboundary in the framework of full time-dependent density functional theory to describe the time-evolution of the Kohn–Sham orbitals describing the dynamics of the electrons in the metal, under the perturbation of classical light. In that sense, the resolution of our problem is not "fully" quantum mechanical, but the optical response of the electron gas to a classical pulse of light is obtained in a "full" quantum mechanical scheme of response of the electron gas. Once we have this "full" quantum mechanical description of the optical response, we try to mimic it by establishing an effective classical model – QCM – that reproduces the properties of the "full" quantum mechanical response.

Jeremy Baumberg added: This is a very interesting question, and I think relates to what happens when light is localised on the atomic scale. In this case many things cannot be simply understood. For instance the dielectric model with a shape boundary between media would now have a fuzziness from the polarizability of the atom clouds which will be frequency dependent, as well a non-local susceptibility. How we treat molecules with partially delocalised electronic states must not be as simple as a dielectric constant. What are the effects of atomic faceting on the electronic states? All these concern both the quantum states of electrons at or around the surface, and in these tight gaps. In addition we look at the moment for effects related to quantized conductance, which can have an impact on the optical spectra. Hence there are very many unknowns at the moment in a treatment of light at this scale.

Alexei Kornyshev addressed Javier Aizpurua and Jeremy Baumberg: Nanoparticles are ordinarily functionalized with ligands that have charged terminals, to prevent their fusion. When assembled at the interface in a nanoplasmonic device, they may settle rather close to each other. What can generally be the role of ligands in 'plasmonic tunnelling', or quenching the hot spots?

Jeremy Baumberg responded: Indeed charged ligands are used to stabilise such nanoparticles, and thus they should play a role in both defining the geometry, and potentially offering conducting channels, as well as producing their own SERS signatures. It is possible that nanoparticles can assemble onto the mirror quite close, and we are currently studying such dimers on mirrors. The gold nanoparticles used are citrate capped and we confirm the vast majority are individually separated on the mirror. We do not believe they play a role in conduction compared to our recent work; switching out a single S-atom from conducting molecules resulting in a non-conducting molecule that causes drastic shifts in the plasmon resonance.[1] Surprisingly we see very little citrate SERS, suggesting the ligands are completely displaced from the gaps.

1 F. Benz *et al.*, Nanooptics of molecular-shunted plasmonic nanojunctions, *Nano Lett.*, 2015, **15**, 669.

Javier Aizpurua continued: The influence of ligands and other dielectrics in the gaps needs to be analyzed on a one-to-one basis. If those ligands are expelled from the gap, they will not influence the tunneling regime, but if they are located in the middle of the relevant tunneling or, more generally, conductive paths, they can have a very important influence in establishing the actual tunneling rates, due to the modification of tunneling barriers and effective separation distances. However, if a clear "classical" conductance path is established, with the conduction of several hundreds of G_o across the gap, then the evolution of modes is quite robust, as shown in ref. 1 and 2.

1 O. Pérez-González *et al.*, *Nano Lett.*, 2010, **10**, 3090–3095.
2 F. Benz *et al.*, *Nano Lett.*, 2014, **15**, 669–674.

Thomas Ebbesen asked: Atoms chemically bonded to the small metallic particles, such as S on an Au particle, should charge the particle (either withdrawing or injecting electrons). This could potentially modify the tunneling rate in the gap just as it does in Coulomb blockade experiments. Would this be a relevant issue in the structures you consider?

Javier Aizpurua responded: You are absolutely right. Our calculations are performed, so far, for a metal–insulator–metal cavity, where the metal edge is surrounded by a vacuum, but we have preliminary calculations where we locate a dielectric in the cavity, such as water, and we can see that the tunneling probability at optical frequencies is indeed enhanced, reaching effective tunneling for almost 1 nm separation distance. Along those lines, any atom or molecule, such as a S atom attached to the gold surface, has the potential to charge the particles and therefore modify the tunneling rates. To consider the actual molecular terminations and charging effects on the work functions and tunneling barriers is a very important issue that needs to be considered to properly account for the correct tunneling rate. Each case is particular and specific and needs to be treated on a one-by-one basis. We provide here general trends that will hold for other situations, but with larger or smaller rates, depending on the particular functionalization.

Michael Flatté commented: When describing the tunneling between nanoparticles in your theory you use the Bardeen transfer Hamiltonian, motivated in part by its use in scanning tunneling microscopy theory. This Hamiltonian, however, is a very poor approximation for the real system, as the transfer matrix between one material and the other is a very sensitive function of the quantum numbers of each of the two materials. It only works for STM because you look at the relative contribution at different sites, whereby the transfer Hamiltonian does not change much. In your case you are always looking at the transfer matrix between two points – those of closest approach of the two metallic objects. The value of the transfer matrix also perturbs the properties of each of the two metallic objects. Thus this does not seem like a well justified approximation.

Javier Aizpurua replied: In TDDFT we are calculating the tunneling current "exactly", without any approximations. Here "exactly" means that it is exact for our model system and up to the precision of the ALDA kernels. In our calculations aimed at establishing QCM parameters we do not use the Bardeen transfer Hamiltonian either. We apply the standard Landauer approach, where the transfer matrix is calculated "exactly" using a standard scattering theory approach once the model of the potential describing the junction is set. As to the electron current perturbing the interface potential, the variations of the electron density, and the corresponding eventual variation of the potential barrier due to the electron current at optical frequencies, we find that this is much smaller than the unperturbed density/potential. Thus this would be a second order correction. The situation might change eventually for a strong laser pulse (see for example ref. 1), but this is not the regime discussed in our paper (*Faraday Discuss.*, 2015, DOI: 10.1039/C4FD00196F). Actually, when implementing our approximation to the conductivity between two points in a metal–insulator–metal cavity, according to the method mentioned above, as is commonly used in STM, and then extending it to different frequencies within the QCM, one could be concerned about the validity of this approximation, especially when it comes to short separation distances. In order to discard those concerns, we compared the conductivity obtained under the static STM approach within the QCM, to that obtained in the full time-dependent density functional theory (TDDFT) for several relevant frequencies, where the effects that you mention (perturbation of the properties of each metallic object, *etc.*) can be naturally built-in. This comparison is presented in Fig. 5 of our paper, where one can see that the modulus and phase of the frequency-dependent conductivity obtained within the "exact" TDDFT is very well reproduced by this extended version of the static STM approach. We thus are quite comfortable about the use of this scheme to describe our frequency-dependent conductivity.

1 C. Marinica *et al.*, *Nano Lett.*, 2012, **12**, 1333.

Michael Flatté asked: For Fig. 5a of your paper (*Faraday Discuss.*, 2015, DOI: 10.1039/C4FD00196F), are the arbitrary units for QCM and TDDFT the same? That is, do the relative values of the calculation for TDDFT and QCM have meaning, or only the shape of the curve?

Javier Aizpurua replied: In Fig. 5a of our paper (*Faraday Discuss.*, 2015, DOI: 10.1039/C4FD00196F), (a.u.) stands for atomic units, and they stand for both the QCM and the TDDFT conductivity, therefore the comparison is direct and quantitatively meaningful.

F. Javier García de Abajo said: I find the term "full quantum" frequently used in your paper (*Faraday Discuss.*, 2015, DOI: 10.1039/C4FD00196F). Approximate schemes have been successfully developed by the *ab initio* community to reduce the properties of materials (including plasmons and the optical response in general) to solutions of the Schrödinger equation. However, the many-body nature of the system averts the possibility of exact solutions in most cases. Given a computational scheme (for example: time-dependent density-function theory (TDDFT), either in the local density approximation (LDA) or introducing gradient corrections to the functional; diagramatic expansions; or even configuration interaction), the level of approximation is commensurate with the computational capabilities. Some popular approximations involve, for instance, the assumption of a jellium model, the use of pseudopotentials to describe core electrons in atomic description, and very generally, the Born–Oppenheimer approximation (see however ref. 1 for a discussion of cases where this approximation fails). In this context, existing implementations of TDDFT are so far incomplete, as a practical recipe for an exact universal functional has not been formulated. Therefore, it is useful to know what are the approximations made, what type of functionals are used, *etc.* This issue is far from trivial, as there are numerous works questioning the validity of different methods depending on the kinds of systems that are simulated, based upon cross comparisons (*e.g.*, DFT *vs.* diagramatic expansions) and the predictability of experimental results (see for example ref. 2). I understand that TDDFT is used in this paper, but I miss a more detailed description of what the actual approximations are and what is meant by "full quantum". A related question pertains to the need of using a quantum mechanical description to simulate realistically plasmons in metallic gaps. For example, the spill-out of valence electrons (*i.e.*, their evanescent tails outside the metal, resulting from the wave nature of the electrons, which produce a smooth surface electron density profile, see ref. 3), affects the optical response, but a description of this response has been argued to be possible with classical methods (see ref. 4). It seems that the classical Drude model provides a reasonable level of description down to ~1 nm noble-metal gaps, so are there features of gap plasmons that reveal quantum properties below that distance? To be more precise, I am wondering whether it is possible to describe such features within the realm of classical theory (*i.e.*, beyond a phenomenological parametrisation of effective response functions)?

1 S. K. Min *et al.*, *Phys. Rev. Lett.*, 2014, **113**, 263004.
2 G. Onida *et al.*, *Rev. Mod. Phys.*, 2002, **74**, 601.
3 N. D. Lang and W. Kohn, *Phys. Rev. B*, 1970, **1**, 4555.
4 R. C. Monreal *et al.*, *New J. Phys.*, 2013, **15**, 083044.

Javier Aizpurua replied: As you point out, there are different levels of implementation of the "full" quantum mechanical calculations of the optical response. We used a local adiabatic approximation (ALDA) with exchange correlation kernel from Gunnarsson and Lundqvist.[1] The metal is described within the jellium model (JM), a well-established method to describe properly the plasmonic features in nanostructures. A list of references regarding the technical details of this implementation can be found at the end of our discussion paper (ref. 34 and 86 in *Faraday Discuss.*, 2015, DOI: 10.1039/C4FD00196F). The JM and ALDA we use are particularly well suited to describe the optical response of free electron metals like Na. Their validity has been thoroughly checked within the cluster community comparing the TDDFT results and experimental data. Part of this question is related to which extent TDDFT is an *ab initio* approach. This has been discussed in the quantum chemistry community some 10–20 years ago in the beginnings of DFT and TDDFT, and currently the commonly accepted appreciation is affirmative about this. Moreover, as Prof. J. García de Abajo has correctly pointed out, there are many exchange correlation kernels proposed in the literature, however, to our best knowledge plasmons are basically electrostatic in nature, and thus they show a main dependence on the Hartree term and are loosely sensitive to the precise Exch-Corr, at least for the free-electron metal systems that we address. We should also mention that we discuss the limitations of our techniques in the paper. In particular, an overestimation of the tunneling barrier within the local density approximation, and thus the necessity of the correction of the effective one-electron potentials by multiple images in the gap, as pointed out in Fig. 2. However, while these corrections to the potentials do change the specific distance at which resonant charge transfer becomes efficient, the qualitative conclusions of our works are robust. One could say that our uncertainty in the determination of the exact potentials is not larger than the experimental uncertainties and difficulties to produce large-scale junctions with perfectly controlled separations at the scale of 1 Angstrom. In this context, we understand by "full quantum", a methodology that traces the evolution in time of the Kohn–Sham orbitals, according to Schrödinger equations (also called time-dependent Kohn–Sham equations). Therefore, our theory naturally includes: tunneling and electron spill-out from the cluster boundaries, space-to-momentum uncertainty (resulting in the finite special extension of the induced electron densities), dynamic screening, plasmon interaction with binary electron–hole excitations, possible nonlinear effects, *etc.*; that is what we mean by "fully quantum". Of course, these models can be improved, to account for features that are missing in our JM approach, and therefore, we are currently working on improving the atomistic description of certain aspects of the plasmonic response. Regarding the possibility to describe quantum features within classical approaches, there have been several attempts during the last years, some of them pointed out by you in this question, but they all refer to partial aspects of the quantum effects, such as the quantum size effect or the charge density profile at the interfaces. We find no classical approach, apart from our QCM to account for the extreme quantum regime in gaps given by the tunneling at optical frequencies. Indeed, all these classical recipes, or our QCM have been elaborated based on the quantum mechanical calculations, as a guide to provide an understanding of the physics underlying the experimental observations. Particularly, regarding tunneling effects, there are nowadays many experimental data revealing the influence of tunneling in subnanometric gaps. These tunneling effects are purely quantum in nature and cannot be addressed with classical descriptions, to the best of our knowledge. This is why we developed our effective QCM presented here.

1 O. Gunnarsson and B. I. Lundqvist, *Phys. Rev. B*, 1976, **13**, 4274.

Jacob Khurgin addressed Jeremy Baumberg: Can one really talk about a refractive index for monolayers and single molecules? And is it necessary? What is wrong with bulk interface dipoles?

Jeremy Baumberg responded: All one needs to describe this system is a polarizability, and interface dipoles or refractive index are both ways to codify the same thing. Hence this is a pragmatic approach for now, especially since we observe changes in this notional refractive index when filling CB molecules. Effects of non-locality or discretisation are expected in higher level treatments.

Katarzyna Matczyszyn spoke to Javier Aizpurua: Do you observe any interesting impact of the peaks coming from nanostars or nanorods? Do you see the shape effects on the gaps coming from the plasmonic enhancement on the extremities of sharp shapes?

Javier Aizpurua responded: When it comes to quantum tunneling at optical frequencies, the relevant distances are at the Angstrom-scale in the transverse direction of the gap which is the most likely direction of the tunneling current. However, the morphology of the gap in the transverse direction determines the presence of new cavity modes of different nature that can modify totally the distribution of the "gap" modes that are affected by tunneling (see for example ref. 1). In this sense, having gaps formed by rods, or nanostar tips can of course modify the structure of modes and how tunneling affects these modes. The QCM can account precisely for these morphological issues, as we show in our paper (*Faraday Discuss.*, 2015, DOI: 10.1039/C4FD00196F).

1 R. Esteban *et al.*, *ACS Photonics*, 2015, **2**, 295–305.

Paul Dawson commented: Javier, you have constructed a sort of classical analogue for the quantum tunnelling regime where the tunnelling mechanism is described in terms of an effective medium in the tunnel gap. I am uncomfortable with this picture. Tunnelling is inherently a lossless process; it is a coupling between the electron wavefunctions in the media on each side of the gap resulting in a transfer of electrons across the gap and can be described, for example, by the use of a transfer Hamiltonian formalism. But no intrinsic losses are involved. If there are losses associated with tunnelling, these must occur in the real materials defining the tunnel gap; the tunnelling resistance is associated with what's going on inside the end of the tip and an equivalently small volume inside the sample underneath the tip. If we view tunnelling as a probablilistic process where electrons have a finite chance of crossing the barrier, then the losses are associated with the electrons (under an applied bias) that do not succeed in tunnelling but are scattered from the tip/sample surfaces. If the gap dimension is increased fewer electrons tunnel and more undergo internal scattering. This is the origin of resistance in tunnelling. You showed a simplified set of illustrative spheres depicting how the dielectric response function in the optical region is treated for the various regimes of interaction between adjacent metal bodies, ranging from the local, classical regime for larger separation distances, through a sphere with an outer shell described by a non-local function, through to the tunnelling and charge transfer regime where the tunnel gap is described by the insertion of an effective medium. But this medium is pure artefact and has enormous losses associated with it: the value of γg reaches into the 100s. Physically, I would much rather see a (further) modification to the dielectric function of the surface region of the spheres to describe the tunnelling resistance and have this as the input that feeds through to plasmon properties that deviate from the simple classical description at very small separations.

Javier Aizpurua replied: I will try to make you feel more comfortable about our quantum corrected model. The QCM is an effective model whose value is, merely, being able to describe effectively a very complex quantum mechanical process in a gap at optical frequencies. We do this by realizing that quantum tunneling is a pure resistive process involving the particles (electrodes) and the gap, as you point out, therefore our choice is to substitute the whole system by a resistive effective block in the gap. Of course, this does not mean that we have that particular material in the gap, but this nicely reproduces the optical response of the whole system in an effective way. The only ingredient that one needs in Maxwell equations is a formal relationship between the current and the field (conductivity). This is exactly what our model does: it imposes the correct tunneling current for the given field in the junction. If we wanted to build an effective model by changing the properties of the surrounding electrodes, as you suggest, then one would need to do full mechanical calculation to address the tunneling current. Otherwise, the tunneling effect would not be captured at all. It is to avoid this complex and very often undoable calculation that we bring the QCM in. We find our mathematical trick in the QCM as valid as, for instance, the description of a metamaterial effective response. The effective response does not mean that the whole metamaterial is homogeneous, but it reproduces very well the far-field response of the whole material. Of course it does not describe the fine details of the near-field within a unit cell of the metamaterial, but this is irrelevant if it is not needed, and still the effective response is very relevant. Our approach is doing precisely that: mimicking a complex process by a much simpler effective medium, which works very well to describe the tunneling regime at optical frequencies.

Emiliano Cortes turned the discussion to Jeremy Baumberg: What is the influence (if there is any) of the Au surface reconstruction during thiol adsorption in the plasmonic properties of nanogaps? Do the gold complexes (*i.e.* RS–Au–SR) play a role? The electronic structure of both the surface and the thiolated molecules might change significantly by considering/neglecting such reconstruction. Also, Au nanoparticles are usually described as a dodecahedron structure with both 111 and 100 facets. It is supposed that only 111 facets are reconstructed. Have you measured/calculated spectral broadening/shifts in plasmonic nanogaps due to different crystal faces in nanoparticles? Also the surface coverage and SAM lattices should be different in both faces presented in nanoparticles.

Jeremy Baumberg replied: Firstly, we have recently studied how such faceting influences the plasmonics modes, and will shortly publish detailed calculations. On one hand we have carefully studied the use of ultraflat gold mirrors and found no difference in the dark field scattering spectra. We therefore believe that the surface, in particular the gold mirror, experiences a change in morphology resulting in crystalline flat surfaces underneath the gold nanoparticles as a result of strong van der Waals forces and thiol–gold bonds. We have as of yet not yet attributed spectral broadening or shifts to differences between either 100 or 111 crystal structures on the nanoparticle, but this is in progress. Here we have tried to avoid such complications. The SAMs grown on the gold mirror are known to be extremely robust, and we can track their SERS to ensure that reconstruction or modifications are minor. However exploring how molecules attached to different Au crystal planes display different properties is relevant and underway.

Javier Aizpurua added: The description of the effects that you mention would require a full atomistic quantum description of the gold dimer with sufficiently large dimensions to develop well-defined plasmonic modes and facets with a considerable number of atoms. That kind of study is beyond current state-of-the-art computational capabilities, and it will probably take a while before computations for such large systems are feasible. At this point we are able to do full atomistic calculations of systems of about 1000 Na atoms. Our preliminary atomistic calculations, done in collaboration with the group of Daniel Sánchez Portal, in San Sebastian, point towards a two-fold influence of the reconstructions and faceting that you mention. On the one hand, the far field response seems quite robust against particular atomic arrangements associated with the crystallographic facets, the most pronounced effect being a consequence of the overall geometry of the faceting (size and shape). On the other hand, the near-field features at the gap are strongly dependent on the particular atomistic arrangement of the crystallographic facets and vertices. We are thus working to address these lattice restructuring and atomistic effects.

James Hugall followed on from this by asking: In the simulations Javier presents, he shows how the EM hot spot distribution on small plasmonic nanoparticles is highly sensitive to exact atomic position and roughness. In a material like gold, with high mobility (which must only be heightened at the nanoscale), one might expect the hot spot positions to be changing continuously on the nanoparticle surface. On larger nanostructures, one could believe that these atomic scale changes become less important, as the plasmonic mode becomes more distributed over a larger volume. SERS data of single molecules on nanoparticles generally show blinking behaviour, this is most likely due either to plasmonic hotspots moving on the nanoparticle or SERS molecules moving in and out of hotspots. Do you think the hotspot positions change rapidly on larger SERS particles or is it more likely the SERS molecules are moving around? Have you considered how fast hotspots may change and reconfigure on nanoparticles? Is it possible that the hotspots themselves induce or perhaps inhibit gold atom movement on the nanoparticles due to the optical gradients created?

Javier Aizpurua responded: You are referring to the atomistic calculations that we presented as additional material for discussion, complementing the description of the smooth surfaces given by the jellium model. The hotspots in our atomistic calculations are associated with edges and vertexes at the contact of atomic planes, and not with single adatoms. Therefore we believe that the distribution of these hot spots is not mobile and does not change position at the surface, once the gap is established. Therefore we favor an explanation based on the mobility of the molecules around those hot spots, rather than an eventual atom migration to explain the variability of the signal in spectroscopies such as SERS or TERS. Indeed, the atom motion due to the presence of hotspots will depend on the molecular absorption regime. For chemisorption the optical potentials should be about an order of magnitude smaller than chemical ones, however the situation might be different for systems showing physisorption.

Martin Moskovits addressed Jeremy Baumberg: Explaining the very large wavelength shift in plasmonic resonance between the system in which the monothiolated and the dithiolated molecule is resident in the gap by ascribing the difference to the fact the dithiolated molecule is "a conducting molecule" seems somewhat of a simplification, especially in light of the molecular conductance has been studied for a decade or more and the subject of at least a thousand papers. Surely, the conductance problem "through a molecule" is a quantum problem, so that for both molecules one has quantised conductance operating alongside tunneling.

Jeremy Baumberg replied: Indeed this literature is very important, and treated fully in our recent paper.[1] In this case it is well known how these molecules behave in conductivity, and it is this behaviour that we use in our simulations. Indeed it seems that a simple conductivity actually works very well in explaining the results. However, it is clearly more complicated, as the insulating molecule has a nominal gap which should allow quantum tunnelling. Perhaps the point is that the conductance measurements through single molecular layers are very confused and difficult to reproduce. By contrast, the optical measurements here are extremely clean and reproducible.

1 F. Benz *et al.*, Nanooptics of molecular-shunted plasmonic nanojunctions, *Nano Lett.*, 2015, **15**, 669.

Thomas Ebbesen said: The very large spectral changes you observe by just adding a S atom on the end of the spacer molecule suggest that the S–Au bond induces charge transfer that perturbs the plasmon resonance. Perhaps you could try injecting charge in the particles, like Paul Mulvaney does,[1,2] and see how it affects the spectrum?

1 C. Novo, A. M. Funston, A. K. Gooding and P. Mulvaney, *J. Am. Chem. Soc.*, 2009, **131**, 14664–14666.
2 R. Sardar, A. M. Funston, P. Mulvaney and R. W. Murray, *Langmuir*, 2009, **25**, 13840-13851.

Jeremy Baumberg responded: The potential to modify the charge chemically is indeed interesting. However the nanoparticles used here are >40 nm which are much too big to acquire electron density changes from a few hundred S–Au bonds. We note also that we would expect smaller changes from larger nanoparticles from this mechanism, the opposite to what we observe.

Alexei Kornyshev commented: One can create, in principle, an array of nanoparticles physisorbed near a metal surface, with some 2 nm surface-to-surface separation between the nanoparticles. Will the modes you are speaking about form bands in such periodic arrays?

Jeremy Baumberg answered: Indeed this is an experiment we have tried, and we expect interplay between the modes. Already we have previously shown that arrays of plasmonic particles with well controlled gaps form bands, and combinations of localised and delocalised states.[1,2] There will be different interactions between the transverse and longitudinal modes in such mat-on-mirror systems.

1 R. Esteban, R. W. Taylor, J. J. Baumberg and J. Aizpurua, How chain plasmons govern the optical response in strongly interacting self-assembled metallic clusters of nanoparticles, *Langmuir*, 2012, **28**, 8881.
2 M. G. Millyard, F. M. Huang, R. White, E. Spigone, J. Kivioja and J. J. Baumberg, Stretch-induced plasmonic anisotropy of self-assembled gold nanoparticle mats, *Appl. Phys. Lett.*, 2012, **100**, 073101.

James Hutchison asked: This question relates to your scattering spectra, particularly the spectral shifts accompanying encapsulation of methyl viologen by the cucurbituril in the nanogap. Your model accounts well for the shift in terms of a change in height of the cucurbituril film and a refractive index change, but does it also take into account the electrophilic nature of methyl viologen and its likely extraction of electrons from the metal nanoparticle? Such charging of nanoparticles has been shown to shift their scattering spectra.

Jeremy Baumberg responded: The change in the refractive index extracted in our experiments refers to the real part of the dielectric function. As suggested, there can also be changes in the molecular electronic levels from metal-controlled electron movements, however all our SERS is carefully ensured to be spectrally away from any electronic resonance so this will not have a direct effect. Indeed one might expect bond energies to change, and this is under study through SERS. Changes in conductivity modify the imaginary part of the dielectric function and can also cause alteration of the intensity ratio and spectral position, and this would also be good to include in our models.

Jino George remarked: How does the molecular packing and Au crystal packing affect the sub nano-gap effects? Do you have any kind of realization in this direction? Also, since you are below 1 nm, a small change in the metal packing and molecular packing will jump over approximations, and studies have been made before in surface chemistry that have direct proof of this. I invite you to read the elegant work of Professor Rasmita Raval and the team from Liverpool in this context.[1]

1 S. M. Barlow and R. Raval, *Surf. Sci. Rep.*, 2003, **50**, 201–342.

Jeremy Baumberg answered: This question also relates to the faceting of the nanoparticles, as a perfectly spherical nanoparticle would not have distinctive packing near the gap other than close-packing. As we have shown, the nanoparticle is faceted as a result of the crystal structure and this significantly influences the dark field scattering spectra.[1] Here, we aimed to minimize this effect by filtering out scattering spectra which showed more than one coupled mode (caused by faceting of the gold nanoparticles).

1 D. O. Sigle *et al.*, Monitoring morphological changes in 2D monolayer semiconductors using atom-thick plasmonic nanocavities, *ACS Nano*, 2015, **9**, 825–830.

Kaiqiang Lin remarked: The focal line is quite sensitive to the size of the nanoparticle on the substrate. So once the size of nanoparticles cannot be precisely determined, I think the LSPR peak of the gap mode is quite difficult to correct. How did you deal with that?

Jeremy Baumberg replied: The focal position seems not to be strongly modified by the size of the nanoparticles. In any case, for each nanoparticle we measure a full Z-stack of spectra and remove all the effects of aberration in the measurements.

James Hugall said: Your nanoparticle-on-mirror configurations seem to show very stable SERS and scattering signals over time. This appears distinct from what is normally observed in single to few molecule SERS experiments, where blinking is commonplace. This suggests that both molecular movement and any gold atom movement are supressed in your case. You hinted at how towards the edge of the nanoparticle, gold atoms move down towards the edges of the gap, finding a low energy configuration, and that this could also be seen in measurements. Could you comment on why in this case the system seems to find a stable low-energy configuration, whereas in many other few molecule SERS experiments this is not the case? Are optical forces driving this reconfiguration or something else?

Jeremy Baumberg answered: We believe our constructs are much more stable than typical hot spots produced by flocculating silver. Various reasons are possible, including the chemical instability of Ag, the stable nature of the molecular layers we use to fill the gaps, the metal surfaces underneath which passivate charge movement, and the prospect for atomic movement only at the edges of the facets underneath the nanoparticles. Blinking is occasionally observed but it is not the dominant effect, and we can track single molecules for many hours.[1] We believe optical forces are very strong[2] and further experiments are underway.

1 R. W. Taylor, F. Benz, D. O. Sigle, R. W. Bowman, P. Bao, J. S. Roth, G. R. Heath, S. D. Evans and J. J. Baumberg, Watching individual molecules flex within lipid membranes using SERS, *Sci. Rep.*, 2014, **4**, 5940.
2 L. O. Herrmann *et al.*, Threading plasmonic nanoparticle strings with light, *Nat. Commun.*, 2014, **5**, 4568.

Adi Salomon asked: A relatively large shift in the Raman spectra has been shown between two similar molecules; S–biphenyl and S–biphenyl–S. Are the shift results due to structural differences between the monolayers (such as tilt angle) or are they more related to energetic differences of the Au work function due to thiol adsorption (see ref. 1 and 2)?

1 A. Salomon *et al.*, *J. Phys. Chem. C*, 2008, **112**, 3969–3974.
2 A. Salomon *et al.*, *Adv. Mater.*, 2003, **15**, 1881–1890.

Jeremy Baumberg responded: We do not yet understand quantitatively the changes in the SERS spectra observed. This is because (as mentioned in another question), advanced DFT calculations inside such a dense electron environment are not yet possible. We know the tilt angle is similar (from ellipsometry) but have no way to probe underneath the nanoparticles. Many SERS lines are the same so we do not believe there are dramatic differences. However new SERS lines appear and these are still under discussion.

Thomas Ebbesen enquired: Could you have structural studies done by computational means (just like for proteins) to see if the structure of the molecules in the gap are modified, especially if a bond is formed like in the case of S?

Jeremy Baumberg replied: It is indeed crucial to develop computational models that would allow an understanding of how molecules react in this environment. However this is far beyond current capabilities which at most can connect molecules to a handful of gold atoms. To include all the image dipoles requires millions of gold atoms, and thus a different formalism is required. We believe our experimental probing using SERS proves that the molecules remain relatively unaffected.

Katarzyna Matczyszyn opened a general discussion of the paper by Sang-Hyun Oh: Regarding the possibility of the biosensing with the use of your system, what kind of molecules are you sensing? Did you try any interesting new molecules? What is the enhancement ratio, the limit of detection?

Sang-Hyun Oh answered: In our biosensing work with nanogap structures,[1] we demonstrated SERS detection of the biological analyte, adenine, which is a commonly detected molecule in intrinsic DNA sensing experiments. The limit of SERS detection in that case was 76 nM for adenine and measured Raman enhancement factor was near 10^8. We are also interested in using our nanogap systems to detect lipid molecules and antibodies.

1 H. Im *et al.*, *Adv. Mater.*, 2013, **25**, 2678.

Michael Nielsen commented: With such small nanogaps, ranging from 10 nm done to sub-1 nm, and high aspect ratios, how confident are you that you can remove the Al_2O_3 gap materials in order to backfill with molecules or other interesting materials? Have you proved to yourself that you have actually managed to get interesting molecules/materials into these nanogaps, rather than just sensing molecules on top of the gaps?

Sang-Hyun Oh responded: Our previous work showed that Al_2O_3 in the gap region can be removed in a buffered oxide etchant (BOE) solution.[1,2] The aspect ratio of our nanogap structure can be as high as 200 : 1 (1 nm gap width and 200 nm film thickness) thus it is very difficult to remove Al_2O_3 completely. Plasmonic hot spots are typically at the corners of the gap, so after removing just a bit of Al_2O_3 from the gap, it is possible to expose hot spots and backfill with molecules. We utilized very small (2 nm) gaps in the present manuscript to demonstrate tight squeezing of incoming radiation, but these sub-10-nm gaps are smaller than many biomolecules. For larger (>20 nm) gaps, it is possible to remove Al_2O_3 and backfill with molecules as we will show in another paper.

1 H. Im *et al.*, *Nano Lett.*, 2010, **10**, 2231.
2 H. Im *et al.*, *Adv. Mater.*, 2013, **25**, 2678.

F. Javier García de Abajo enquired: How are the size of the gap, the Ohmic losses arising from the material and the plasmon frequency interrelated in the gap systems that you are studying? How far can the plasmon frequency be pushed up (or down)?

Sang-Hyun Oh replied: Our previous work showed the behavior of gap plasmons in a metal slit.[1] At visible frequencies, the measured effective index of a gap plasmon mode in a 2 nm Ag slit was 17.8 at a free-space wavelength of 517 nm. When the slit forms a circular loop, we can also tune the ring resonance around each loop. We showed that the resonance frequency of such annular nanogap loops can be pushed up to mid-IR and even THz frequencies (for millimeter-scale loops). In this paper (*Faraday Discuss.*, 2015, DOI: 10.1039/C4FD00233D), we show how the ring resonance and transmittance change with the ring shape (circle *vs.* rectangle), packing density, and periodicity in the THz regime.

1 X. Chen *et al.*, *Nat. Commun.*, 2013, **4**, 2361.

Jeremy Baumberg asked: What is going on in the visible region to these narrow gap plasmon modes? The modes look very weak. Is this dominated by the losses?

Sang-Hyun Oh replied: This question refers to a set of transmission spectra measured from nanogap arrays with the gap size ranging from 1 to 10 nm and frequency from the visible to mid-infrared range (not shown in the paper). The resonance peaks in the visible region look relatively weaker than the peaks in the near-infrared (NIR) and mid-IR regime. These visible peaks are high-order Fabry–Pérot modes and the loss is high for gap plasmons in sub-10-nm gaps. In comparison, transmission spectra measured in the NIR and mid-IR regions show strong and well-defined peaks for gaps that are as much as 5000 times smaller than the free-space wavelength.

Olivier Martin remarked: For realistic sensing applications, two important issues arise: (1) The sensitivity of the device for an extremely small amount of analyte. The nanogaps used in your device are extremely small and very sensitive, provided that the analyte reaches there. How can this be achieved when only very small amounts of analyte are available; how can one assure that it will reach the sensing area? (2) Especially for point-of-care applications, it is important to reuse the sensor. Is it possible to wash away previous analyte and reuse such a sensor? What are the typical number of cycles that can be performed?

Sang-Hyun Oh answered: (1) This is an important problem in the biosensing community. When the active sensing area is very small and the concentration of analyte is low, often the sensitivity and detection time are limited by the transport of analytes. In that case, an active concentration scheme can be used. For example, optical trapping of analyte molecules toward the gap region, or other mechanisms based on electrokinetics may also be used (*e.g.* dielectrophoresis-based concentration; see ref. 1). That way, 'hot spots' for detection can also act as efficient trapping sites for analyte molecules. (2) While these sensors may be reused a few times after cleaning, we observe degraded performance after repeated experiments. Using atomic layer lithography method, we can now produce a large quantity of these sub-10-nm gap structures without using e-beam lithography, so our goal is to use a fresh new chip for each experiment.

1 A. Barik *et al.*, *Nano Lett.*, 2014, **14**, 2006.

James Hutchison commented: I have two questions relating to the situation where you remove the oxide layer to give 1–10 nm gaps in metal films. Firstly, can you easily flow liquids through these very small nanogaps, both for applications such as threaded DNA read-out, and in terms of rinsing/re-use of the nano-structures? Secondly, we heard from Professor Shalaev about the thermal deformation of gold at high temperatures (*Faraday Discuss.*, 2015, DOI: 10.1039/C4FD00208C). Have you seen or do you expect any deformation/sealing up of your very small gold nanogaps due to the high temperatures resulting from surface plasmon generation by intense laser light?

Sang-Hyun Oh replied: In response to your first question: It is possible to remove the initial oxide layer (Al_2O_3) in the gap and coat the exposed metal surfaces with a thinner layer of silica film. The gap region will then become hydrophilic and it will be possible to flow liquids through the gaps (*i.e.* to use the gaps as nanofluidic channels). In response to your second question: Yes, upon intense illumination with Ti : Sapphire laser light, it is possible to melt the gap region. In general, sub-10-nm metallic gaps can easily be damaged due to heat, laser illumination, or wet processing. For these operations (*e.g.* nonlinear spectroscopy, optical transmission measurements *etc.*) we usually leave Al_2O_3 inside the gap. The nanogaps filled with Al_2O_3 are robust, since the Al_2O_3 film prevents unwanted migration of metal atoms across the gap and also mechanically supports the gap structure.

Thomas Ebbesen added: In a follow-up to the previous question, it must be remembered that a liquid will only enter into the nanogaps if you have wetting (*i.e.* the contact angle is less than 90° in the Laplace–Young equation).

Sang-Hyun Oh replied: This is an important point. For THz spectroscopy, as shown in this paper (*Faraday Discuss.*, 2015, DOI: 10.1039/C4FD00233D), our focus was to tightly squeeze THz near fields using narrow (2 nm) gaps. For biomolecular sensing in a liquid environment, we use wider gaps (10–50 nm range) to ensure that proteins or other biomolecules can enter the gap region. First we remove Al_2O_3 in the gap region *via* wet etching, then to ensure proper wetting, the metal surface is coated with a thin layer of silica film *via* atomic layer deposition, which makes the surfaces hydrophilic.

Kaiqiang Lin addressed all three speakers: In experimental work, Professor Sang-Hyun Oh created a very large gap, which is formed by two large, flat surfaces. However, in theoretical calculation, as for SERS and sensing, we know that only a very small part of the gap can play a role, since the enhanced field is always localized in a limited area. So my question is: do you think it's good to fabricate a gap with very large surface? And do you have any idea what kind of gap could be promising for SERS and sensing?

Jeremy Baumberg responded: Indeed we are interested in understanding what facet shape or size would be the best for SERS. For providing careful comparison to theory, in our paper (*Faraday Discuss.*, 2015, DOI: 10.1039/C4FD00195H)we tried to minimize the influence from faceting by eliminating spectra with multiple resonances. This question is beyond the current scope, but addresses work that we focus on heavily. Currently we believe the plasmonic hot spot is more localised and intense as a result of the faceting, leading to even higher field enhancements useful for SERS, and allowing us to see single molecule SERS.[1]

1 F. Benz *et al.*, Nanooptics of molecular-shunted plasmonic nanojunctions, *Nano Lett.*, 2015, **15**, 669.

Sang-Hyun Oh also replied: In this question, I think that a “large gap” means a gap that is horizontally very long. In the present paper (*Faraday Discuss.*, 2015, DOI: 10.1039/C4FD00233D), we fabricated nanogaps that are long (a few hundred microns in loop perimeter) to utilize terahertz resonances. For visible spectroscopy applications, we can fabricate these rectangular loops (or circular loops) with sub-micron-scale diameter. The gap plasmon modes in such small loops can be tuned to have resonances at visible frequencies. One could also make a long and straight gap and utilize the vertical Fabry–Pérot resonance of gap plasmons for SERS and sensing. We are investigating both design options.

Javier Aizpurua then added: The generic signal in SERS depends both on the generation of large-enhancement hot spots as well as on a homogeneous distribution of these large enhancements in extended areas. Large gaps look promising for probing large assemblies of molecules. When it comes to probing a few or even single molecules, one might focus more on getting well-controlled cavity gaps where the signal is connected with the actual atomic-scale features of the field-enhancement. We are exploring this extremely localized regime to be exploited in single molecule events. However, large gap areas can be also very promising in other spectroscopies that rely on the active control of cavities, for example.

Alexei Kornyshev opened a general discussion of the paper by Riccardo Sapienza: When you speak about percolation in your systems, I do not think that you mean that your particles physically touch each other, but what is your connectivity rule?

Riccardo Sapienza responded: The particles are coupled in their near-field by a plasmon mode. Percolation is reached when the particles are optically connected.

F. Javier García de Abajo enquired: I guess each emitter is interacting with the local environment (*e.g.*, a specific gap), as well as with the sample regarded as a metamaterial (*i.e.*, we can think of a separation between local and average fields similar to that of the Clausius–Mossotti model). The metamaterial response undergoes a dramatic transition from optical dielectric (high, real epsilon) to optical metal (negative epsilon) when crossing the percolation limit (see my other question below). However, in both of those cases, the amplitude of the light reflected by the film is of the same order of magnitude as that of the incident light, so this only introduces an order-unity relative correction to the local field. It is possible that the local morphology influences to a larger extent the LDOS, and that the main effects arise from the abundance of narrow gaps that can substantially increase the decay rate. In this respect, I would expect that your random samples contain gaps of various shapes and separations, including a certain fraction of narrow gaps where the excited emitters undergo a strong reduction of their lifetime, in contrast to the more regular distribution of perhaps wider gaps in the lithographic samples. Is this picture plausible? Related to this, I find it interesting to know what is the relative contribution of dark channels to the decay rate (*e.g.*, creation of electron–hole pairs) and whether coupling to gap plasmons is important in this context.

Riccardo Sapienza replied: We agree that the source will probe the local and the global or effective properties. The large fluctuations of the LDOS we measure could be due to both these effects. Also, the natural network may present narrower gaps than the artificial one, leading to larger LDOS values. Nevertheless, this does not explain fully the different trends in the LDOS fluctuations, nor the role of percolation, unless a very narrow gap size appeared around percolation for reasons different than network connectivity. Non-radiative contributions are included in the calculations provided to answer a following question, but the dark channels mentioned there are not included. We expect them to contribute more and more for smaller gaps and distances, but also to be less important for large fluorescent probes, such as a 50-nm sized bead.

Jeremy Baumberg said: Can you clarify if the 50 nm dye $n = 1.7$ nanoparticles when placed in the gaps will shift the plasmon frequency, so that it no longer couples to the surrounding plasmons? In that case it is not surprising that the dye sees only the local response.

Riccardo Sapienza answered: The dye particles are embedded in a polymer film (PMMA) of similar refractive index (both ~1.6) to ensure index matching and minimal disruption of the plasmon resonance by the presence of the polymer bead. Therefore we believe the sources can couple well to the network.

Jeremy Baumberg said: What does the spectrum of the band of coupled resonances look like?

Riccardo Sapienza replied: The spectrum of the coupled resonances is shown in Fig. 1, with broad resonances around 600–900 nm.

Jeremy Baumberg asked: How do you explain the fact that when the fabricated system is "perfect" with no disorder, there is such a wide range of temporal decays of the dye seen?

Riccardo Sapienza responded: We attribute the distribution of decays for $p = 1$ to the variation of LDOS within the unit cell, which depends on position and distance to the gold rods. This is an additional proof of the importance of studies with point-sources even for the "perfect" case.

Sergey Bozhevolnyi spoke: I would like to comment and pose a question regarding the importance of local as opposed to global features observed near the percolation threshold. The history of studies, both theoretical and experimental, of electromagnetic properties of percolating metallic nanostructures is quite long. There were many papers published and research reported during the 90s, notably by V. Shalaev and M. Stockman, and including also a number of experimental studies using near-field optical microscopy on similar structures. I think you miss a lot of readily available optical information using only the lifetime measurements. Optical scattering or extinction (spatially resolved) spectra would give you an immediate answer as to whether electromagnetic processes in the system under consideration are governed by local or global interactions. Spectra of individual particles or even dimers and trimers would feature a finite number of distinct resonances, whereas systems near the percolation threshold and exhibiting strong interaction would feature very rich resonance spectra, becoming practically continuous. Did you try to look at optical spectra? Spectra can be measured with different spatial resolutions, ultimately using diffraction limited spots or even beyond the diffraction limit when using near-field techniques. Did you try to carry out any of these optical characterizations?

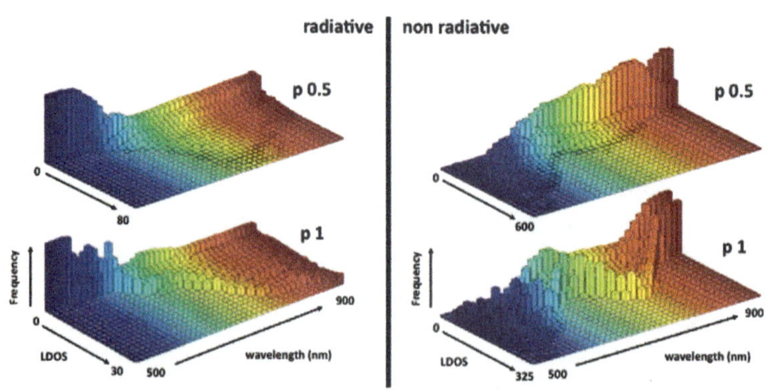

Fig. 1 LDOS calculations with FDTD method.

Riccardo Sapienza responded: Far-field spectra on plasmonic networks have been studied and indeed show rich resonant behavior that could be very beneficial when the metasurface is used as a complex plasmonic lens. We are here interested in localised and delocalised modes and in particular modes confined to ~10s of nm, therefore even a diffraction limit far-field study would only detect the average of the illuminated area, failing to probe them. Beyond far-field, we agree that near-field scattering experiments would have a better spatial resolution and would probe the full modal landscape providing a stronger link with the optical density of states, however the presence of the tip is surely disrupting the localised and delocalised modes of the network. Instead, we believe that fluorescence studies of the local density of state from 50 nm sources provide a more robust and clean local measurement.

Sergey Bozhevolnyi commented: With all respect to your opinion, I would not really agree with this line of reasoning – when conducting life-time experiments one does also measure the emission of radiation, *i.e.*, the radiation channel should be available, and that should also be reflected in the optical spectral measurements, which could be of different nature – scattering or extinction, for example. I would not discount the possibility to obtain additional and valuable information by conducting spectroscopic characterization. For example, gap surface plasmon modes that can result in very large Purcell factors are also difficult to excite with propagating waves, but have been proven to be possible.

Riccardo Sapienza responded: We agree that reciprocity dictates that if the fluorescence reaches the far-field the time-reversal wave is able to reach the near-field of the emitter, *i.e.* that a radiation channel is present for both directions. Nevertheless, lifetime studies provide the total LDOS, including the non-radiative contributions, which does not reach the far-field and therefore cannot be excited by far-field scattering experiments. This is valuable information that we have recorded in our experiments.

F. Javier García de Abajo commented: A medium consisting of metallic islands near percolation is known to behave as an effective dielectric of high index of refraction. This was shown by McKenzie and McPhedran for periodic arrangements of perfect-conductor spheres,[1] and experimentally corroborated, for example, by Pecharromán and Moya,[2] who measured dielectric functions as high as 10^5 in insulator–metal composites near percolation at low frequencies. Beyond percolation, the system then behaves as a metal from the optical viewpoint. I am intrigued to know how relevant this change of dielectric response is for the observations in this paper, particularly in the random films.

1 D. R. McKenzie and R. C. McPhedran, *Nature*, 1977, **265**, 128.
2 C. Pecharromán and J. S. Moya, *Adv. Mater.*, 2000, **12**, 294.

Riccardo Sapienza replied: A percolating network can be regarded as a dielectric with a large effective response or as a nanostructured medium with coupled scatterers. We are here focused on the fluctuating density of states, for which, in our opinion, the picture in term of far-field and near-field scattering and coupling is more appropriate. In our experiment we measure the local properties of an individual emitter, so we have not performed studies of the effective dielectric constant.

Jacob Khurgin enquired: Would it be possible to fabricate a 1-dimensional structure to study percolation?

Riccardo Sapienza responded: Yes, it would be possible, but its optics would also be quite different. In 1D the structure can be described as a chain and percolation does not appear as a phase transition.

Martin Moskovits asked: Adding to the significant literature pertinent to this important problem that others referred to, a great deal has been written, connecting SERS and percolating systems to microwave absorption – which might simplistically be viewed as a measure of thc "metallicity" of the evolving percolating materials. Qualitatively, the microwave absorption is expected to have a maximum at or near the percolation threshold. The SERS intensity of those substrates also seem to maximize just prior to percolation, where presumably the number of nanosized gaps would be very great and their gap sizes very small – a favorable condition for intense SERS. Such observations have been successfully interpreted in terms of an effective medium theory somewhat and reasonable modified for the application.[1] I wonder if one could make some connection between the results that you report the plasmonic effects of these substrates, viewed from the above perspective?

1 J. Obrzut, J. F. Douglas, O. Kirillov, F. Sharifi, and J. A. Liddle, Resonant microwave absorption in thermally deposited Au nanoparticle films near percolation coverage, *Langmuir*, 2013, **29**, 9010–9015.

Riccardo Sapienza answered: Thank you for the comment, indeed the large LDOS fluctuations are due to similar phenomena behind SERS enhancement, *i.e.* gap formation, but could also be due to delocalised modes. We were not aware of the microwave absorption enhancement experiments you mention, we thank you for suggesting this reference. A connection between these microwave measurements and the LDOS distributions should link average absorption and local optical properties, so we do not think this would be straightforward, but we leave this for further investigations.

Viktor Podolskiy remarked: There has been a lot of work on effective medium theories of percolating networks. Often, conventional effective medium theories do not work, and effective permittivity of percolating composite becomes scale-dependent. Scaling theory for transmission/reflection of percolating composites can be found in ref. 1. A rather comprehensive review of nonlinear properties of percolating materials can be found in ref. 2. The percolating composite can be thought as a set of coupled resonators, spanning resonance frequencies starting from a resonance of single particle (typically, near UV) to the resonance frequency of a (large) cluster (typically, IR). Therefore, spectral properties of your "natural" films are rather expected. However, it appears that spectral properties of your "controlled random" composites lack high-frequency resonances. Do you think this fact is related to the structure of "controlled random" composites that lack spherical particles and few-particle clusters?

1 Y. Yagil *et al.*, *Phys. Rev. B*, 1991, **43**, 11342.
2 A. K. Sarychev and V. M. Shalaev, *Phys. Rep.*, 2000, **335**, 275.

Riccardo Sapienza responded: We thank you for the comment, indeed our controlled random structure is not identical to the natural network as the building blocks are of different size. We have focused here on the topology and fractality (as also in ref. 1), and therefore on purpose we have used monodisperse building blocks. Nevertheless, other important effects could come from resonances in the clusters or from the gap sizes between particles and clusters, which would require additional investigations.

1 V. Krachmalnicoff *et al.*, *Phys. Rev. Lett.*, 2010, **105**, 183901.

Duncan Graham opened a general discussion of the paper by Zachary Schultz† by communicating: My question was in relation to the signals obtained from the protein underneath the nanoparticle and away from the tip. Is this limited to just being on the far side or will this work for the whole surface of the nanoparticle? Will different shapes of nanoparticles give the ability to perhaps tune where the maximum effect comes from in this situation?

Zachary Schultz communicated in response: We have modeled the basic response using COMSOL and demonstrated a polarization-dependent coupling that suggests the molecules enhanced are on the far side of the particle.[1] This is consistent with calculations by Schatz.[2] In the calculation, the location of the electric field enhancement depended on the spacing and plasmon resonance mode excited. A recent SERS paper by Ren and coworkers also showed that the excitation of different plasmon resonance will enhance molecules in different locations.[3] However, recent calculations by others, such as Aizpurua,[4] suggest that quantum tunneling can result in significant redistribution of the electric field. This is particularly interesting in light of our results on cellular membranes. We have some evidence that partial endocytosis of the nanoparticle probes can occur, but not correlated to the TERS experiment. This may indicate that the signals arise from the TERS tip contacting the probe and enhancing multiple receptors around the partially endocytosed nanoparticle probe. In regards to will different shape particles tune the maximum effect, this is very probable. Thus far we have only explored spherical particles, but rods may have an interesting effect and the tip probe dimer can be thought of as a rod. There is some asymmetry between the size of the TERS tip and the nanoparticle, which would correspond to odd-shaped particles. This may or may not be important. We do know that the observed signals between the nanoparticle TERS tip and a nanoparticle probe compared to aggregated nanoparticles on cell membranes are very different. Further experiments with different shaped particles should elucidate this effect.

1 K. D. Alexander and Z. D. Schultz, *Anal. Chem.*, 2012, **84**, 7408–7414.

† This paper was not presented at the meeting, but questions were submitted afterwards to Zachary Schultz.

2 E. Hao and G. C. Schatz, *J. Chem. Phys.*, 2004, **120**, 357–366.
3 X. Wang, M. H. Li, L. Y. Meng, K. Q. Lin, J. M. Feng, T. X. Huang, Z. L. Yang and B. Ren, *ACS Nano*, 2014, **8**, 528–536.
4 R. Esteban, A. G. Borisov, P. Nordlander and J. Aizpurua, *Nat. Commun.*, 2012, **3**, 825.

Stefan Maier communicated: Can you comment on the reproducibility of your TERS measurements, compared to standard SERS geometries?

Zachary Schultz communicated in response: Using a ligand that is specific for a protein, such as the c-RGD and $a_V\beta_3$ integrin or biotin and streptavidin, we see a reproducible signal, like that shown in Fig. 2a of our paper (*Faraday Discuss.*, 2015, DOI: 10.1039/C4FD00190G) for SERS of the biotin and streptavidin system. The statistical analysis we present shows that the signal in the TERS experiment is statistically similar to the SERS experiment of the purified protein. In the electronic supporting information of one of our other manuscripts,[1] we showed multiple examples of the observed signal that are very similar. However, when a non-specific ligand is used the signal is basically random. Some TERS spectra appear to show some variance, which is one reason we used multivariate analysis to interpret our results. The agreement in the TERS data is still quite strong. What appears to be important for a reproducible signal is controlling how the molecules interact with the plasmonic structure. The orientation and mobility of the protein appears to be highly confined in our TERS experiments, which we believe gives rise to the reproducibility. What is perhaps more remarkable is that the other molecules present do not perturb significantly the observed signal. This suggests that other molecules that are not preferentially confined/oriented are not significantly enhanced. To be clear, we do not believe this necessarily arises from "chemical enhancement" associated with forming a bond (biotin and streptavidin binding is a non-covalent interaction), but rather physisorption is sufficient for the selective enhancement. We have observed the importance of physisorption in SERS experiments as well (*e.g.* ref. 2). The SERS geometry we are comparing to is aggregated nanoparticles, perhaps this is a standard SERS geometry. I would argue "standard SERS geometries" is a bit ill-defined and that the variance across different SERS experiments would make for a lively session at a future discussion.

1 H. Wang and Z. D. Schultz, *ChemPhysChem*, 2014, **15**, 3944–3949.
2 S. M. Asiala and Z. D. Schultz, *Anal. Chem.*, 2014, **86**, 2625–2632.

James Hutchison opened a general discussion of the paper by Masayuki Futamata: In my experience, measuring SERS spectra of *para*-aminothiophenol (PATP in your paper, *Faraday Discuss.*, 2015, DOI: 10.1039/C4FD00188E) from junctions between silver nanoparticles, we often saw a lot of blinking of the SERS signal during imaging. I wonder if you observed these effects and if you have some insight into where they come from? Are they due to chemical effects like (de) protonation, or photochemical effects?

Masayuki Futamata replied: As reported in the literature, the origin of spectral changes observed for PATP molecules are still under discussion, probably relating to your observation. For instance, one group attributes it to photochemical dimerization of PATP on the basis of spectral resemblance of such species with those of dimer molecules (ref. 11 in our paper, *Faraday Discuss.*, 2015, DOI:

10.1039/C4FD00188E)), while another attributes to charge transfer resonance between metal and PATP molecules (ref. 12 and 13 in our paper). We also observed identical Raman spectra to the literature (ref. 11–13 in our paper) for flocculates formed in neutral pH (Fig. 2b(a) in our paper), which contain new peaks at 1420, 1378, 1139 cm^{-1} grown with the duration of laser illumination, and different from those of bulk state. If you prepare aggregates without controlling pH in solution, your spectra will probably show these peaks, the intensity of which can fluctuate with time due to photochemical reaction; this can be one reason for the blinking of your samples. In contrast to these spectra, we found a new spectra observed in flocculation of PATP-coated AgNPs, which were formed in acidic conditions in protonated PATP (PATPH+), and followed by addition of NaOH to deprotonate PATP molecules at the nanogap (spectra shown as the black and red lines in Fig. 2b(b) and 2b(c) in our paper, respectively). These spectra did not show any spectral changes with duration of illumination. Although more detailed measurements using deuterated water and *p*-dimethyl amino thiophenol are necessary to reach conclusive assignments of the spectra, we presume to attribute them to truly neutral PATP molecules at the nanogap. These rather complicated adsorption features (see also Fig. 3a–b in our paper) can be obtained by precise control of the experimental conditions in the flocculation SERS method.

Thomas Ebbesen said: How do you know if the pH in such gaps is the same as in the bulk?

Masayuki Futamata responded: We can monitor pH in nanogap from Raman spectra of PMBA (*p*-mercaptobenzoic acid) and PATP (*p*-aminothiophenol). Although more precise and detailed experiments are prerequisite for the determination of the surface pK_a of such adsorbed species, we found that surface pK_as of these molecules are not much different from those in bulk solutions, and protonation/deprotonation of functional groups in adsorbed thiols can be controlled by tuning the pH in solution.

Olivier Martin asked: In a paper called "Mode-selective surface-enhanced Raman spectroscopy using nanofabricated plasmonic dipole antennas" published in *J. Phys. Chem. C*,[1] we have shown that a molecule adsorbed on a plasmonic nanostructure sees its spectrum changed depending on where the plasmon resonance of the nanostructure lies. Specifically, by tuning the plasmonic nanostructure, we were able to enhance specific Raman modes of the molecule. Would it be possible to do the same with a tiny gap like you use, and wouldn't such a geometry provide additional degrees of freedom compared to the static experiment we performed?

1 W. Zhang *et al.*, *J. Phys. Chem. C*, 2009, **113**, 14672–14675.

Masayuki Futamata responded: Thank you for your comment. To control gap size in nanoarrays as you reported is also a quite efficient way to apply plasmon enhanced Raman scattering to various purposes. To our knowledge, unfortunately, to control the gap size with electron beam lithography (EBL) is not always feasible for people who are not familiar with such beam technology or not able to have sufficient machine time. We hope that EBL will be much more useful for us in the near future. In principle, the gap mode geometry, in which we have a self-assembled monolayer on metal substrates with a definite molecular size and orientation, yields a constant gap size, and thereby constant resonance conditions. In reality, this is not always assured due to the polycrystalline nature and surface roughness of metal substrates, and also due to inhomogeneous orientation of thiol molecules. Accordingly, if we observe extinction spectra or elastic scattering spectra for individual MNPs in the gap mode geometry, slightly distinct resonance features can be observed, which is in our future scope. Nevertheless, the gap mode resonance, expressed here by the electric field intensity at the nanogap, between AuNPs and metal substrates with large damping, does not vary with the gap size, since it is mostly determined by the resonance of localized surface plasmon of AuNPs. Indeed, the gap size did not vary the resonance features but the enhancement values, as anticipated by FDTD calculations, are shown for AuNPs on Al substrates in Fig. 6b in our paper (*Faraday Discuss.*, 2015, DOI: 10.1039/C4FD00188E).

Olivier Martin remarked: The investigated geometry is extremely interesting, as it provides access to very small gaps. To control these gaps further, would it be possible to use optical forces? How would you implement this?

Masayuki Futamata responded: Thank you for your comment. In principle, it is possible to utilize optical forces using a pulse laser, as partly reported in the literature, to form flocculates or aggregates of metal nanoparticles (MNPs) in solution; for instance please see the paper by M. Kall *et al.*[1] It is probable that, to restrict a number of MNPs to dimers or trimers, one should control the interaction between each MNP by modifying with various molecules with appropriate functional groups, in addition to controlling optical forces and the number of particles in solution. Because the flocculation and aggregation of MNPs are kinetic processes, it is possible to hold flocculates for a while to elucidate adsorbed species. Concerning the surface modification, we have used various interactions such as a covalent bond of S–Ag in thiol samples, a coordination bond of N:–Ag in DNA bases, electrostatic forces in AgNP/PMBA−…M+…−PMBA/AgNP, AgNP/PATP+…Xn−…+PATP/AgNP, and van der Waals interactions between PMBA/AgNPs and PATP/AgNPs. For this purpose, we tuned the surface charge of these molecules adsorbed on AgNPs and AuNPs by adjusting the pH in solution. Similarly, we immobilized AgNPs using van der Waals interactions against thiophenol molecules adsorbed on metal substrates. In general, it is also possible to use optical trapping to form a nanogap structure between MNPs and metal substrates, in which to suppress aggregation of MNPs is again critical.

1 M. Kall *et al.*, *Faraday Discuss.*, 2006, **132**, 35–44.

Jeremy Baumberg said: You tabulate very nicely the comparison of SERS in different systems, but it's not clear to me whether you actually measured the plasmonic resonance wavelengths? Do they vary from particle to particle, as that would completely change the results you obtain?

Masayuki Futamata replied: This is a very crucial point to utilize coupled localized surface plasmons (LSPs) in flocculates of silver nanoparticles (AgNPs), and the gap mode between gold nanoparticles (AuNPs) and metal substrates. It is because the resonance and enhancement in electric field at the nanogap depends sharply on the size and shape of each particle as well as the size of nanogap. First, most of our experiments have been performed not for individual dimer particles (a pair of particles), or single particle on the substrates, but for several or several tens of particles (Fig. 9a–d in our paper, *Faraday Discuss.*, 2015, DOI: 10.1039/C4FD00188E). So, we measured averaged resonance conditions for flocculates and the gap mode with extinction spectra for flocculates in solution, and elastic scattering spectra for gap mode geometry. In this approach we used an appropriate wavelength for the excitation of coupled LSP in flocculation, and the gap mode for many particles. For example, we used a He–Ne laser at 632.8 nm, which approximately accords with the additional LSP peak at 600–700 nm observed for PMBA- and PATP-coated AgNPs (Fig. 2a and 3a in our paper). Also, elastic scattering spectra in the gap mode for AuNPs, with a diameter of 40 nm and 100 nm, on various metal substrates provided broad resonance peak at 570–800 nm depending on the size of AuNPs (data not shown) in accordance with those obtained by FDTD calculations. From the viewpoint of spectroscopy in analytical applications, not a single but many (although still a countable number of) particles are feasible to obtain a sufficiently large signal intensity. In some exceptional cases, we used only a single AuNP ($r = 50$ nm) in gap-mode geometry to demonstrate the capability of huge SERS enhancement and related phenomena like blinking, in which the resonance wavelength depends on the size of the AuNP and nanogap. One single laser wavelength was available in the gap mode, since we use AuNPs with a size of 50 nm and also a definite size of nanogap determined by the molecular size and orientation of thiophenol self-assembled monolayer film.

Jeremy Baumberg asked: You use very different tuning conditions for the different nanoparticles on metals – sometimes the SERS pump laser is near resonance, sometimes it is way off. What assumptions do you then make to get calibrations of enhancements then that are comparable?

Masayuki Futamata responded: It is again crucial to utilize the coupled LSP and gap mode as we discussed in the preceding question. First, we utilized average resonance in flocculates of AgNPs, which means a single flocculate in many particles works to give a large enough SERS intensity, and we did not evaluate enhancement factors. The resonance wavelength in the gap mode using AuNPs with a definite radius of 15–50 nm (commercially available), with small size dispersion (Fig. 9a–d in our paper, *Faraday Discuss.*, 2015, DOI: 10.1039/C4FD00188E), does sharply depend on the size of the nanogap, as discussed in other presentations. To evaluate much more accurate resonance conditions and enhancements in the electric field at the nanogap under each AuNP is beyond our present scope. Instead, as mentioned above, we evaluate the averaged enhancement observed not for individual AuNPs, but for the countable number of AuNPs in the sampled area. From the viewpoint of analytical applications, it is more feasible to evaluate the averaged enhancement to show the capability to detect a small amount of adsorbed species. In addition, as reported in our manuscript, by reducing the surface coverage of AgNPs immobilized on an Ag film/Dove prism, a much larger enhancement in SERS intensity was obtained under attenuated total reflection (ATR) conditions. This is related to higher efficiency to excite a propagating surface plasmon (PSP) at a lower surface coverage of AgNPs. Thus, the largest enhancement was anticipated in ATR-TERS using PSP and the gap mode.

Jeremy Baumberg remarked: In some of your metals, the resonances are sharp, while others are very broad (and weaker). Does this mean that some of the SERS comparisons are not so valid?

Masayuki Futamata responded: In our experiments and calculations, the resonances of the gap mode for AuNPs on various metals with large damping are mainly determined by LSP resonance of AuNPs as shown by Fig. 5a and 6b in our manuscript (*Faraday Discuss.*, 2015, DOI: 10.1039/C4FD00188E). The enhancement in electric field at the nanogap is quite similar for different metal substrates (Table 1 in our paper). The LSP resonance wavelengths of AuNPs do not vary greatly with the size, but become broader with size, in particular at longer wavelengths (Fig. 6a in our paper). These features indicate the applicability of the gap mode to various metals with prominent SERS enhancement, which are quite different from those with much more complicated resonance features for small damping metal substrates, such as AgNPs on Ag substrates (Fig. 5 in our ref. 19).

Jeremy Baumberg asked: You still have citrate in your system; do you see SERS from these molecules, and if not, why not?

Masayuki Futamata replied: Citrate residuals on metal nanoparticles are indeed an experimentally important but embarrassing issue in various SERS experiments. First, in Raman spectral measurements reported here in flocculation-SERS, the surface of the AgNPs were covered by thiol molecules like PMBA (*p*-mercaptobenzoic acid), and PATP (*p*-aminothiophenol), which were formed by replacing citrate residuals on AgNPs. So, even if AgNPs have still some residual citrate, as evidenced by zeta-potential measurements, their amount is much smaller than that on as-prepared AgNPs without thiols. Second, even AgNPs just prepared by chemical reduction by citrates do not usually show Raman spectra of citrate. This is simply because the Raman scattering cross section of citrate is too small to be detected even under large enhancement. Hence, when we concentrate our AgNPs by centrifugation, we could observe the SERS band of citrates. We should also note that the quantity of residual citrate depends on the composition of reacting chemicals such as the molar ratio of $AgNO_3$ and trisodium citrate, as well as reaction conditions such as the temperature and duration of the reaction. Surface residuals like citrate are useful to isolate and stabilize metal nanoparticles in suspension using their electrostatic repulsion, but often cause difficulty in adsorption of neutral molecules like DNA bases due to their steric hindrance. Consequently, to control the adsorption of such residuals it is crucial to utilize metal nanoparticles in flocculates-SERS and gap mode Raman spectroscopy.

Emiliano Cortes remarked: Have you take into account that usually in Pt, Pd, *etc.* surfaces, thiol-based SAMs are formed by a mixed layer of sulphur and thiols? This complex interface might change the distance in the plasmonic gaps you are considering. How do you know that the interface can be described (in all cases) in the same way?

Masayuki Futamata responded: Thank you for your question. As you pointed out, the gap mode resonance depends sharply on the gap size. We measured gap-mode Raman for samples with average gap sizes to see if it was possible that such a mixed layer could be formed. In such a mixed layer, sulfur gives only an S–Ag stretching band, while the thiol molecules provide many other Raman bands. The sulfur layer should yield a significantly different intensity of the S–Ag band to those from thiol molecules. However, our Raman spectra did not show such a feature (Fig. 8 in our paper, *Faraday Discuss.*, 2015, DOI: 10.1039/C4FD00188E). To avoid such a mixed layer effect, we used a different Raman band from the S–Ag band to evaluate the enhancement factor, which resulted in no significant differences.

Santhosh Kotni communicated: The extinction spectra of gold nanoparticles clearly suggest to work with particle sizes above 50 nm to avoid losses due to absorption. Although the resonance condition can be achieved with particles bigger than 50 nm, which show stronger scattering than absorption, why do you prefer to work with such small particles of radius 15 nm?

Masayuki Futamata communicated in response: We have used various sizes of AuNPs to ensure the anticipated results by FDTD. Initially, we actually used a rather small size of AuNPs with $r = 15$ nm as you mentioned, which is related to the feasible preparation of such particles by the citrate reduction method (Frens' method, ref. 1). We agree with your comments on the size of AuNPs, since larger particles provide a larger enhancement in electric field at the nanogap as experimentally confirmed in our Table 1 (*Faraday Discuss.*, 2015, DOI: 10.1039/C4FD00188E).

1 G. Frens, *Nature*, 1973, **241**, 20–22.

Santhosh Kotni then communicated: In your Table 1 (*Faraday Discuss.*, 2015, DOI: 10.1039/C4FD00188E), the enhancement factors for different samples with different metals were compared. However, the number of molecules in each sample may not be same and that may lead to variation in the intensities. Please can you comment on whether the number of molecules in each sample are same or not? If they are same, how do you maintain the same number in each sample?

Masayuki Futamata communicated in response: First, a self-assembled monolayer (SAM) of thiophenol was formed on various metal substrates, which are known to form such SAM films with thiols (for instance see ref. 1). We have experimentally measured SERS enhancement as averaged values not for the individually sampled area, but under the countable number of AuNPs which corresponded well with those anticipated by FDTD calculations, suggesting that different metal substrates do not have a markedly different coverage and enhancement. The next plausible step is to improve the enhancement factor to be more quantitative, by measuring the enhancement in SERS intensity under each AuNP, which resulted from a different coverage of thiols and thus different gap sizes as you suggested, which is beyond our present scope.

1 J. Christopher Love *et al.*, *Chem. Rev.*, 2005, **105**, 1103–1169.

Faraday Discussions

PAPER

Ultra-strong coupling of molecular materials: spectroscopy and dynamics

Jino George, Shaojun Wang, Thibault Chervy, Antoine Canaguier-Durand, Gael Schaeffer, Jean-Marie Lehn, James A. Hutchison,* Cyriaque Genet and Thomas W. Ebbesen*

Received 7th October 2014, Accepted 20th October 2014
DOI: 10.1039/c4fd00197d

We report here a study of light–matter strong coupling involving three molecules with very different photo-physical properties. In particular we analyze their emission properties and show that the excitation spectra are very different from the static absorption of the coupled systems. Furthermore we report the emission quantum yields and excited state lifetimes, which are self-consistent. The above results raise a number of fundamental questions that are discussed and these demonstrate the need for further experiments and theoretical studies.

1. Introduction

One of the advantages of surface plasmon polaritons is their ability to concentrate the electromagnetic field in a very small volume, which in turn can enhance various phenomena and notably light–matter interactions. Therefore it is not surprising that plasmonic structures and micro-cavities have been used interchangeably to provide confined EM fields to induce light–matter strong coupling in molecular materials. Metallic cavities provide some of the strongest field confinements and have the additional feature that both the Fabry–Perot and plasmonic waveguide modes are present.[1,2] When molecular materials are placed in a high concentration in such confined fields, hybrid light–matter states, the so-called polaritonic states P+ and P−, are formed with Rabi splittings in the range of 0.1 to 1 eV, as illustrated in Fig. 1. When the Rabi splitting is a fraction of the transition energy ($\hbar\omega_M$), the perturbation of the coupled system is so strong that the energies of the other states of the system are modified. This is the so-called ultra-strong coupling regime where the ground state energy has been predicted to shift[3] and this has indeed been observed experimentally.[4,5] In addition, the hybrid light–matter eigen states are in principle delocalized over the optical mode and, upon excitation, quasi-bosonic exciton–polaritons are formed.

ISIS & icFRC, Université de Strasbourg & CNRS, 8 allée Gaspard Monge, 67000, Strasbourg, France. E-mail: hutchison@unistra.fr; ebbesen@unistra.fr

Such features of a strongly coupled system are quite unique and therefore have naturally generated much research activity from single atom–single photon studies to coupling in inorganic semiconductors and organic dyes.[6–18]

While early studies on organic materials focused on the J-aggregates of cyanine dyes,[18–32] due to their sharp and intense absorption peak, a greater variety of molecules are currently being investigated under strong coupling, depending on the scientific targets.[33–44] For instance, the composite bosonic nature of exciton–polaritons makes them of great interest for thresholdless lasing and other collective effects such as Bose–Einstein condensation. However, such reports remain elusive, being demonstrated only for a system incorporating an anthracene single crystal,[33] and more recently for fluorinated ladder polymers.[34,35] We have been more interested in how the molecular material properties of the coupled system can be modified relative to the bare molecules. Therefore we have chosen other molecules with specific functionalities, such as merocyanine, which is capable of (photo-)switching between two isomers, to demonstrate the effect of those functionalities on chemical reaction rates, work function, phase transitions, and electronic transport.[41–45] The observation of such effects belies the notion that only a very small percentage of molecules are coupled for each class of molecules. In fact, the stronger the Rabi splitting, the larger the fraction of coupled molecules.[5]

Molecular materials are typically highly disordered and inhomogeneously broadened and this has consequences on the strong coupling, both on the properties of the polaritonic states and on the whole system, as pointed out by Michetti and La Rocca.[16] In particular, the degree of delocalization of the polaritonic states inside a cavity, together with the fraction of coupled molecules, can be significantly affected. In addition, organic molecules used in strong coupling have typically 100 to 200 normal vibrational modes that also have strong consequences on their properties.[32,46,47]

With vacuum Rabi splittings ($\hbar\Omega_{RV}$) in the order of 1 eV, importantly, the coupled system is no longer in the Markovian regime since $\hbar\Omega_{RV} \gg k_BT$.[47] As a consequence, the lifetimes of the polaritonic states P+ and P− cannot be predicted from the lifetimes of the constituents of the system, namely the lifetime of the photon in the cavity and lifetime of the excited state. As we have pointed out elsewhere,[32] there are numerous examples of coupled systems that are longer lived than their constituents, for instance in the case of molecular excimers and exciplexes.

A large $\hbar\Omega_{RV}$ can be obtained by using a molecular transition with high absorption cross-sections or transition dipole moments (d) and metallic cavities with small mode volumes (v) since, in the absence of dissipation, the Rabi splitting energy ($\hbar\Omega_R$) between the two new hybrid light–matter states is given by:[6]

$$\hbar\Omega_R = 2Ed\sqrt{n_{ph}+1} = 2\sqrt{\frac{\hbar\omega}{2\varepsilon_0 v}}d\sqrt{n_{ph}+1} \tag{1}$$

where E is the electric field amplitude of the cavity, $\hbar\omega$ the cavity resonance or transition energy, ε_0 the vacuum permittivity, v the mode volume and n_{ph} the number of photons in the cavity mode. As eqn (1) shows, even when n_{ph} goes to zero, there remains a finite value for the Rabi splitting ($\hbar\Omega_{RV}$) due to the interaction with the vacuum field. This splitting is itself proportional to the square root of the number of molecules in the cavity $\sqrt{N}$, which implies that the $\hbar\Omega_{RV}$ is proportional to the square root of the molecular concentration $(\sqrt{\frac{N}{v}})$ in the optical field.

Crucial for all these studies is a better understanding of the properties of the polaritonic states (*e.g.* Fig. 1), their dynamics and how they are modified relative to the bare molecules. A greater variety of molecules need to be characterized to look for trends. We have undertaken such a study and, in this report, the photophysical properties of a subset of strongly coupled molecules with very different bare features are compared using both static and fs time-resolved pump–probe spectroscopy. The results raise fundamental questions that will be discussed.

In this paper the three different molecules that were investigated under strong coupling are as follows:

1.1 TDBC

The first one is the well-known 1,1′-diethyl-3,3′-bis(4-sulfobutyl)-5,5′,6,6′-tetrachlorobenzimidazolocarbocyanine (TDBC) in its J-aggregated form. J-aggregates are themselves formed by transition dipole–dipole strong coupling of neighboring molecules, where the symmetry of the aggregates is such that only the transition from the ground state to the lower coupled state is allowed, as illustrated in Fig. 2b. The Stoke shift is characteristically very small (<10 meV).

The Rabi splitting is such that P− lies below the Frenkel excitonic state involved in the coupling process (Fig. 2a). As a consequence, the energy level of the lower polariton branch lies amongst the higher vibrational states of the ground state manifold.[47] Thus, even if the polariton state is formed efficiently, it will relax very efficiently by dissipating energy into this bath of non-radiative states. This is despite the fact that polaritons, generated by excitation to P− or P+, are expected to interact (scatter) very weakly with phonons due to their low effective mass (being quasi-bosons). Recent measurements on TDBC (having ~180 normal vibration modes) show, indeed, that the non-radiative decay rate constant k_{nr}^{P-} of P− is smaller than the corresponding k_{nr} of the bare molecule.[32] Nevertheless, in both cases the non-radiative rate constants are in the order of 10^{12} s^{-1}. This dominates the decay kinetics of P−. It is also the source of the low P− fluorescence quantum yield, $\Phi_F^{P-} = \frac{k_r^{P-}}{k_{nr}^{P-} + k_r^{P-}}$, since the radiative rate

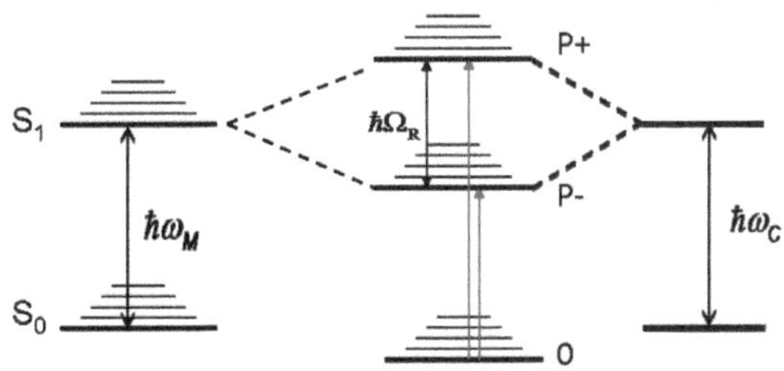

Fig. 1 A schematic diagram of the coupling between a molecular transition ($\hbar\omega_M$) and a cavity mode ($\hbar\omega_C$) giving rise to new hybrid light–matter eigen states P+ and P−. The dispersive nature of these states are not shown.

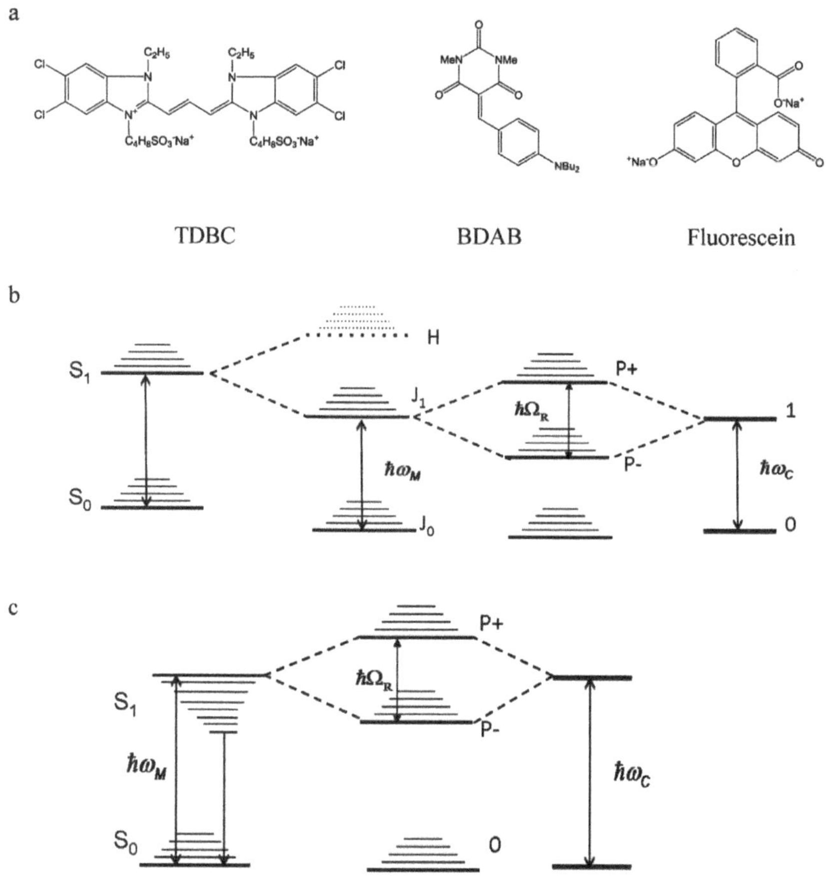

Fig. 2 Molecular structures of the three compounds (a) and the corresponding energy diagrams for TDBC (b) and BDAB (c) upon coupling. The TDBC dipole–dipole coupling gives rise to new states marked J and H, but for the J-aggregates only the J state is observed in the absorption spectrum as the transition to H is forbidden. Fluorescein is coupled schematically as in Fig. 1.

constant k_r^{P-} has an upper limit for a fully allowed transition in the order of $10^9 \sim 10^{10}$ s^{-1}. Vibrational relaxation *within* the excited states occurs on the 100 fs timescale and is characterized by small spectral shifts or sliding during the relaxation. This has also been seen for the P− of TDBC and it is important that this is not confused with other possible excited decay pathways when observing P− and P+ at very short time scales.[30]

1.2 BDAB

The second molecule is a so-called push–pull molecule, 5-(4-(dibutylamino)-benzylidene)-1,3-dimethylpyrimidine-2,4,6(1H,3H,5H)-trione (BDAB),[5] with an unusually large Stokes shift (~560 meV). The strong coupling energy level diagram for such molecular systems that undergo electronic energy re-distribution in the excited state, with the associated solvation shell re-distribution, *i.e.* systems featuring a large Stoke shift, is shown in Fig. 2c. The strong coupling in this case involves the transition from the molecular ground state to a higher vibrational state of the excited state manifold (the Franck–Condon transition). As a consequence, the lower P− polariton branch may lie above, or be isoenergetic with, the relaxed excited state of the bare molecule (see for instance ref. 37). This may have consequences on the dynamics and the decay processes of the coupled system, as has already been shown by Törmä and co-workers.[37]

1.3 Fluorescein

Fluorescein, as its name indicates, is a highly fluorescent dye in dilute solutions with a $\Phi_F \simeq 0.9$ and a Stoke shift of ~110 meV. It is a representative of standard organic dyes and will strongly couple to the cavity, as illustrated in Fig. 1.

2. Results

The strongly coupled samples were prepared by encasing polymer films containing the dyes between two silver films to form an optical cavity (see Experimental section for details). The thickness of the polymer film, and thus the final separation between the Ag mirrors in the cavity, was chosen to give an optical mode resonant with the absorption maximum of the dye/aggregate (at 590 nm for TDBC, 470 nm for BDAB, and 510 nm for fluorescein). The Q-factors of these cavities are low, typically ~10–20. Unless specified otherwise, the cavity thicknesses are tuned such that the Rabi splitting occurs at normal incidence ($k_{//} = 0$) where the photon–exciton mixing is 50 : 50.

2.1 Transition probability to P−

We start by comparing in Fig. 3a–c the optical absorption and emission (fluorescence) spectra of the Fabry–Perot cavities containing TDBC, fluorescein and BDAB at normal incidence with their bare spectra. All the absorption spectra are inhomogeneously broadened, even TDBC. The two new absorption peaks corresponding to P+ and P− (blue solid curves in Fig. 3) have a full width half maximum (FWHM) that is mainly determined by that of the cavity mode, as expected.[14] It should be noted that for fluorescein the cavity induced splitting is more complex, due to the double peaked absorption of the bare molecule, as seen before for Rhodamine.[37] The coupling strengths ($\hbar\Omega_{RV}$) for these samples are, respectively, 277, 554 and 720 meV for TDBC, BDAB and fluorescein and the corresponding fractions of the transition energy ($\hbar\Omega_{RV}/\hbar\omega_0$) are 13, 24 and 27%. These fractions, especially the latter two, are sufficiently high that the coupled systems are in the ultra-strong coupling regime, which has consequences on their properties as already discussed. Notice that the P− fluorescence peak of coupled BDAB is at higher energy than that of the bare molecule in agreement with the schematic illustration of Fig. 2c. Careful observation of the emission spectra of the coupled systems shows that at least some fluorescence of the bare molecules is always present in these cavities. In other words, some of the molecules are not coupled to the cavity due to the disorder (energetic and orientational) in the system.[5,16] This uncoupled emission does not disperse with angle and therefore can be easily distinguished from the P− emission. The apparent Stoke shift of the P− emission for all three molecules is extremely small or, depending on the conditions of the sample, it sometimes even appears to be slightly blue shifted

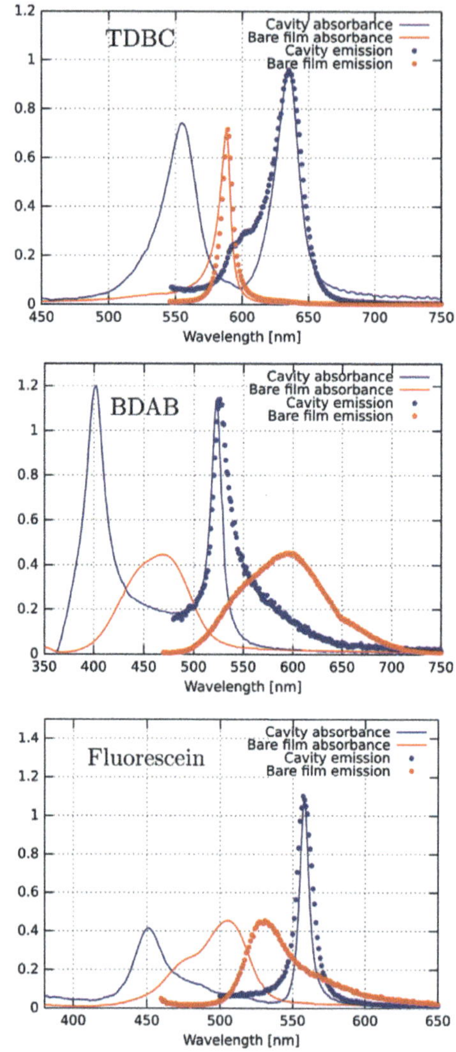

Fig. 3 Absorbance (solid lines) and emission (dotted lines) spectra of the bare molecules (red) and the corresponding coupled systems (blue). The excitation wavelengths were either 530 nm for TDBC or 450 nm for BDAB and fluorescein.

relative to absorption.[30] The blue shifted emission and its origin will be discussed again further on, after the presentation of more results.

Next, we measured the excitation spectra or action spectra associated with the P− emission. This is a standard technique used in molecular science to understand the origin of the emission and associated processes and we first introduced this for studying strong coupling in an earlier study.[30] It consists of measuring the emission or fluorescence intensity, $I_F(\lambda)$, at a given wavelength while scanning the excitation wavelength through the absorption spectrum of the sample. Since $I_F(\lambda) \propto \Phi_F\sigma_A(\lambda)I_0(\lambda)c = \Phi_F I_A(\lambda)$, where $\sigma_A(\lambda)$ is the absorption cross section of the molecule, c is the molecular concentration and $I_0(\lambda)$ is the incident light intensity, one would expect that when I_0 is normalized then $I_F(\lambda)$ will be proportional to the light absorbed, $I_A(\lambda)$. Therefore, by just plotting I_F at a given λ while scanning the normalized excitation, $I_0(\lambda)$, the shape obtained should reflect the absorbance of the states contributing to the emission process.

Fig. 4a, c and d show the excitation spectra of the three molecules recorded for their P− emission and these are compared to their absorbance spectra. The first striking feature is the relative contribution of P+ and P− to the emission of P− (here all the excitation spectra have been normalized to the P+ absorbance peak). It thus appears that the direct excitation of the P− peak contributes very little to the P− emission, while when exciting to P+ or the uncoupled molecules it rapidly relaxes to P−, where emission can occur. This relaxation to P− can occur *via* energy transfer between the coupled and the uncoupled molecular reservoirs as well as by rapid internal vibrational relaxation (IVR). The latter is a standard feature of molecules, with their very large number of normal modes and their overtones that provide a continuum of sublevels between the various states. It is interesting to note that the ratio of P−/P+ excitation peaks is the smallest for TDBC (~0.1) and the largest for BDAB (~0.5).

This raises the question of whether the direct excitation into the P− static absorption peak actually populates P− or not. To answer this question we also recorded the excitation spectrum of pump-probe experiments by looking at the

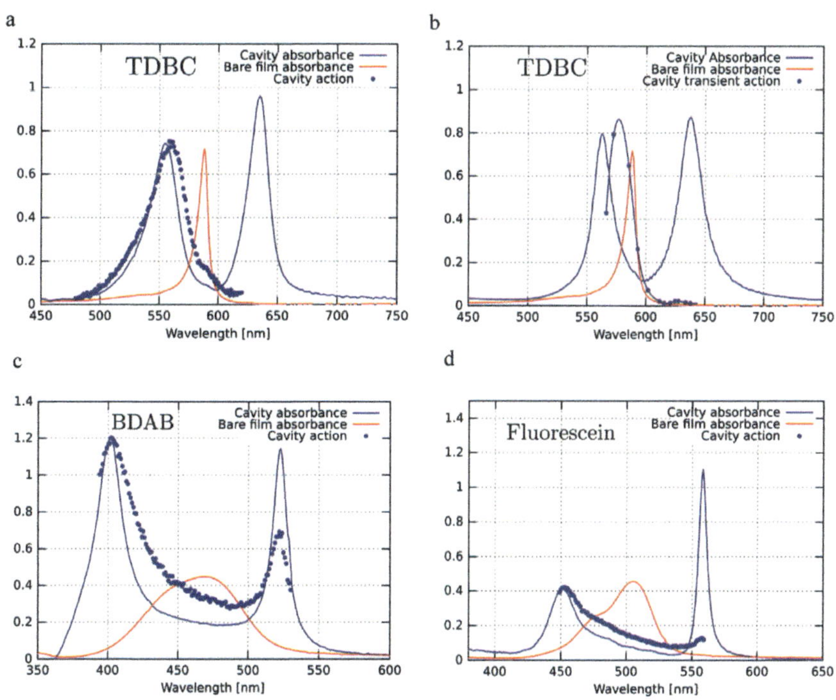

Fig. 4 Excitation or action spectra (blue dots) recorded at normal incidence for P− emission for (a) TDBC (at 680 nm), (c) BDAB (at 540 nm) and (d) Fluorescein (at 576 nm) and (b) for P− transient absorption of TDBC. These spectra are compared to the bare molecule absorbance spectra (solid red lines) and the strongly coupled system absorbance spectra (solid blue lines) of the same samples.

intensity of the transient absorption spectrum of P−, proportional to the population of P−, as a function of laser pump wavelength. This was possible to do for TDBC since our 150 fs-laser could be tuned to cover the static absorption peaks. The resulting spectrum is plotted in Fig. 4b and it has very similar features to the fluorescence excitation spectrum (Fig. 4a). Again, the direct excitation of P− appears not to be allowed. Interestingly, the $\hbar\Omega_{RV}$ deduced from the excitation spectra appears to be smaller than the static absorption spectrum indicates. This is probably the origin of the blue-shifted emission relative to the static absorption that is sometimes observed.

To try to understand the origin of these observations, the cavities were detuned and the fluorescence excitation spectra were recorded. For fluorescein, the detuning of the cavity to higher energies (*ca.* 60 meV) seems to enhance direct P− excitation, as shown in Fig. 5a (red curve). This could be explained by the higher exciton content of the P− branch. However, in the case of BDAB (Fig. 5b), the reverse tendency is seen, *i.e.* detuning to lower energies enhances the relative contribution of P− to the excitation spectrum. Taken together with the results of Fig. 4, it appears that the internal vibronic structure of the molecule plays an important role and also possibly the excited state rearrangement involved in the Stoke shift.

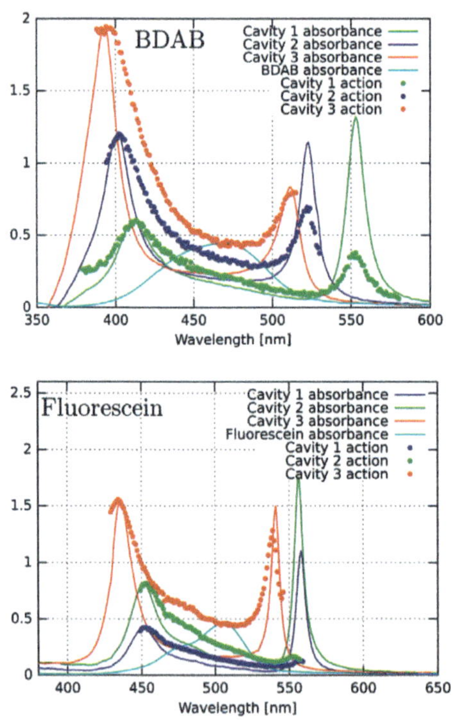

Fig. 5 Excitation spectra (dotted lines) for BDAB and fluorescein for detuned cavities (red and green curves) compared to the tuned cavities (blue curves) and compared to the corresponding absorbance spectra of the coupled systems and the bare molecules (light blue). The emission detection wavelengths for BDAB are 600 nm (red curve), 540 nm (blue curve) and 525 nm (green curve). For fluorescein these are 556 nm (red curve), 570 nm (green curve) and 576 nm (blue curve).

2.2 Lifetime of P−

The transient spectra of the three strongly coupled systems are shown in Fig. 6, together with their decay kinetics. The spectra do not evolve over the decay, indicating that they are dominated by a single transient species. The decays are not single exponentials due to the heterogeneities of the samples at the high concentrations that were used to induce strong coupling. The angle dependence of the decay was also measured for TDBC and was found to be invariant, indicating that the lifetime is not dependent on probed k-vector. This indicates that there is no bottle-neck along the P− branch shown in Fig. 6 for this sample. Such decay bottle-necks have been seen in other conditions and samples for strongly coupled TDBC.[24]

Table 1 below summarizes the dynamical data for the strongly coupled molecules. It first compares the observed half-lives measured by transient absorption with those of the 1/linewidth of the P− resonance and the experimental radiative lifetimes, as well as the predicted ones derived from absorbance peaks of P− using the Bowen and Wokes approximation.[48] It is notable that these two standard methods for estimating lifetimes give such very different values. The measured and calculated quantum yields for TDBC and fluorescein are in very good agreement, considering all the experimental uncertainty, while for BDAB the experimental value is an order of magnitude higher. The latter is not surprising

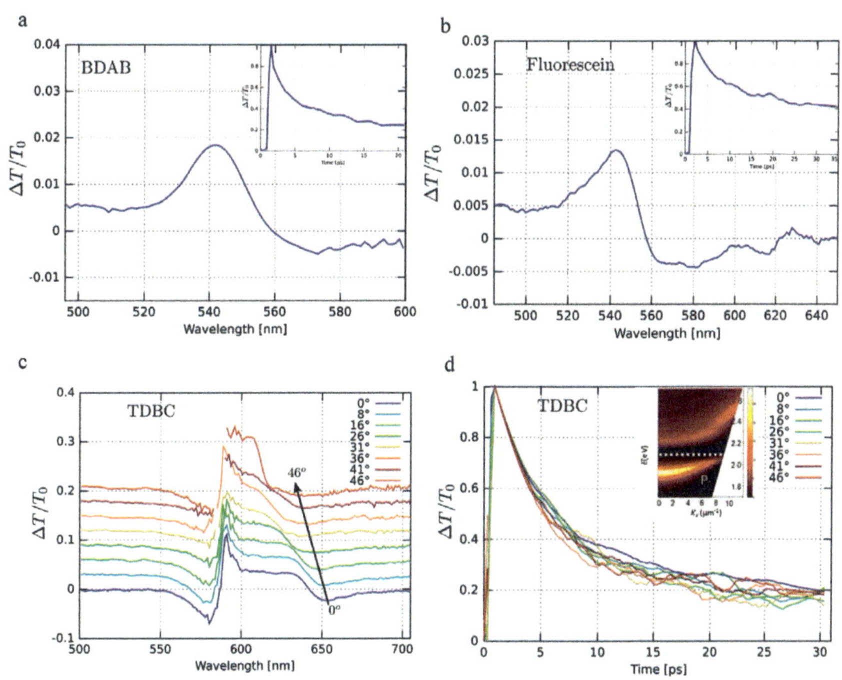

Fig. 6 (a), (b) and (c) Transient pump–probe spectra with the corresponding decays as inserts for BDAB and fluorescein. In the case of TDBC, the angular dependence of the transient spectra is shown in (c) and the corresponding decays in (d). The insert in (d) shows the dispersion of the static absorption of TDBC in the cavity. BDAB and fluorescein were pumped at 400 nm and TDBC at 585 nm.

Table 1

	TBDC	BDAB	Fluorescein
1st half life (transient absorption)	4.3 ps	3.5 ps	30 ps
Cavity lifetime (1/linewidth)	2 fs	5 fs	14 fs
Radiative lifetime (predicted value)	0.2 ns (0.4 ns)	1 ns (0.1 ns)	3 ns (1.5 ns)
Quantum yield (predicted value)	1×10^{-2} (2×10^{-2})	4×10^{-2} (3.5×10^{-3})	2×10^{-2} (1×10^{-2})

considering the fact that the emission is not from the Franck–Condon transition but due instead to an intra-molecular charge transfer state being at the origin of the large Stoke shift of the bare molecule. Thus, the Bowen and Wokes approximation is not the best model for such a molecule.

3. Discussion and conclusions

Conspicuous from the above excitation spectral experiments is the fact that the static absorption peak due to P– (in Fig. 3–5) does not readily lead to the generation of P– emission or even to the P– transient absorption. We have shown this excitation behaviour before for TDBC and it has since been confirmed by others.[24] This finding is surprising since one would expect that the static absorption peak of P– would lead directly to populating P– and subsequent emission. This is an important issue since it is a standard assumption that the emission in the spectral region of P– indeed comes from a corresponding state.

One could take a very naïve view to explain the above results from a purely classical interpretation. For instance, the shape of P– emission spectra can be very easily reproduced by taking the emission of the bare molecules filtered by the transmission properties of the cavity (as can be seen in Fig. 3). The excitation spectrum would therefore be dominated by the spectral regions where the bare molecule absorbs but enhanced by the modal properties of the cavity (as in Fig. 4). This is similar to the treatment of strong coupling by Zhu *et al.* where it is not necessary to assume the formation of hybrid light–matter states by vacuum Rabi splitting.[11] In this classical vision, the lifetimes of the P– measured by fluorescence or by transient absorption spectroscopy would be those of the bare molecule perturbed by the strong interaction with the confined electromagnetic environment of the cavity.

Nevertheless, the observation of novel and modified properties in strongly coupled systems argues strongly for the quantum interpretation and the existence of hybrid light–matter states. These include coherent emission over large distances,[23] condensation of polariton,[34,35] polariton lasing[33] and modification of such properties as chemical rates,[42] work-function[43] and conductivity,[45] amongst others. The significant P– emission quantum yields also argue against a simple cavity filter effect for the fluorescence. Therefore there is a need to understand why the direct excitation of P– is a very inefficient way of populating the state, despite the 50 : 50 photonic–exciton mixing for both P+ and P− in our experiments.

One possibility is that the transition probability is very weak despite the strong static absorption. This immediately raises the question of what is at the origin of the static absorption. One unique characteristic of these organic coupled systems is the large inhomogeneous broadening which, coupled to the cavity, could generate dissipation without populating P−. This would leave indirect excitation *via* higher energy states involving energy transfer and IVR as the most efficient means of then populating the P− state. However, as we discussed previously,[32] the recorded P− emission quantum efficiencies are only compatible with a lifetime in the picosecond regime, as measured for P− both in fluorescence and transient absorption. It is important to note that this is for a most favourable k_r corresponding to a highly allowed transition. If the transition to P− is only weakly allowed, then Φ_F^{P-} should be much smaller, unless the strongly coupled system boosts the emission for an otherwise weakly allowed transition.

Another conundrum for us in these systems is also related to lifetimes. From the FWHM of the P− and P+ absorption, principally determined by the Q-factor of the cavity,[14] one derives a lifetime of ~10 fs. If one integrates over the same absorbance peak to derive the radiative lifetime, one obtains a lifetime three orders of magnitude longer.[32] These represent two very different approaches, one optical and one molecular, to the same issue.

In the spirit of the Faraday discussions, we have shown here, through experiments, some of the open questions related to light–matter strong coupling involving molecular materials. Clearly more experiments and theory that can handle such complex materials and the particularities associated with individual molecular structures need to be developed to elucidate these strongly interacting systems. Only then will the full potential of using a confined electromagnetic environment to control material and molecular properties be realized.

4. Experimental

All the steady state measurements were performed in reflective cavities, in which the Ag mirror deposited on the glass was 200 nm thick, and the top mirror was 30 nm thick. For pump–probe measurements, experiments were performed in transmission with cavities in which both mirrors were 30 nm thick. The bottom Ag film was sputtered on borosilicate glass, followed by spin-coating of the molecules dispersed in a polymer to form a thin film, and then sputtering the second Ag film on top. TDBC films were spin-cast from aqueous solution (Milli-Q) containing 0.5% TDBC and 5% poly(vinyl) alcohol (PVA, M.W. 205 000) by weight. Fluorescein films were cast from aqueous solutions of 1% Fluorescein disodium salt and 2% PVA by weight. BDAB films were cast from toluene solutions of 0.5% BDAB and 1% poly(methyl methacrylate) (PMMA, M.W. 120 000) by weight. Toluene (Sigma Aldrich), PVA (FLUKA), PMMA (Sigma Aldrich), fluorescein disodium salt (Exciton) and TDBC (Few Chemicals) were obtained commercially and used without further purification. BDAB, 5-(4-(dibutylamino)benzylidene)-1,3-dimethylpyrimidine-2,4,6(1H,3H,5H)-trione, was synthesized as follows. 4-dibutylaminobenzaldehyde (196 mg, 840 μmol, 1.00 eq.) and 1,3-dimethyl-pyrimidine-2,4,6(1H,3H,5H)-trione (131 mg, 840 μmol, 1.00 eq.) were dissolved in 5 mL of ethanol. The resulting mixture was refluxed for 3 h. After cooling the precipitate was filtered off and washed with 10 mL of H_2O, 10 mL of ethanol and 10 mL of *n*-pentane. The yellow solid was dried in vacuum to give the product in quantitative yield (312 mg). 1H NMR (400 MHz, CDCl3, δ/ppm): 0.98 (t, $J = 7.3$ Hz, 6H), 1.45–1.32 (m, 4H), 1.69–1.57 (m, 4H), 3.44–3.37 (m, 10H), 6.66 (d, $J = 8.98$ Hz, 2H), 8.36–8.41 (d, $J = 9.23$ Hz, 3H). ESI-MS: *m/z*: 360.3 (100%). HRMS (ESI, *m/z*): calculated for $C_{21}H_{29}N_3NaO_3$ 394.210; found 394.210.

Steady-state transmission and reflection spectra were taken on a Shimadzu UV3101 spectrometer. Steady-state fluorescence spectra were taken with a Horiba Jobin Yvon-Spex Fluorolog-3 fluorimeter. Pump–probe experiments were carried out with narrow band (linewidth = 6 nm), 150 fs-pulse tunable SpectraPhysics laser system. The spectra were recorded using low-energy (<100 μJ cm^{-2}) pumping pulses to avoid spurious effects.

Acknowledgements

The authors acknowledge support from the ERC (Grants 227557), the International Center for Frontier Research in Chemistry (icFRC, Strasbourg), the ANR Equipex "Union" (ANR-10-EQPX-52-01), the Labex NIE projects (ANR-11-LABX-0058_NIE), CSC (ANR-10-LABX-0026_CSC) and USIAS within the Investissement d'Avenir program ANR-10-IDEX-0002-02.

References

1 S. Kéna-Cohen, S. A. Maier and D. D. C. Bradley, *Adv. Opt. Mater.*, 2013, **1**, 827.
2 E. N. Economou, *Phys. Rev.*, 1969, **182**, 539.
3 C. Ciuti, G. Bastard and I. Carusotto, *Phys. Rev. B: Condens. Matter Mater. Phys.*, 2005, **72**, 115303.
4 P. Forn-Díaz, J. Lisenfeld, D. Marcos, J. J. García-Ripoll, E. Solano, C. J. P. M. Harmans and J. E. Mooij, *Phys. Rev. Lett.*, 2010, **105**, 237001.
5 A. Canaguier-Durand, E. Devaux, J. George, Y. Pang, J. A. Hutchison, T. Schwartz, C. Genet, N. Wilhelms, J.-M. Lehn and T. W. Ebbesen, *Angew. Chem., Int. Ed.*, 2013, **52**, 10533.
6 S. Haroche and D. Kleppner, *Phys. Today*, 1989, **42**, 24.
7 R. Houdré, *Phys. Status Solidi B*, 2005, **242**, 2167.
8 P. Snoke and D. Littlewood, *Phys. Today*, 2010, **63**, 42.
9 Y. Kaluzny, P. Goy, M. Gross, J. M. Raimond and S. Haroche, *Phys. Rev. Lett.*, 1983, **51**, 1175.
10 G. Rempe, H. Walther and N. Klein, *Phys. Rev. Lett.*, 1987, **58**, 353.
11 Y. Zhu, D. J. Gauthier, S. E. Morin, Q. Wu, H. J. Carmichael and T. W. Mossberg, *Phys. Rev. Lett.*, 1990, **64**, 2499.
12 R. J. Thompson, G. Rempe and H. J. Kimble, *Phys. Rev. Lett.*, 1992, **68**, 1132.
13 C. Weisbuch, M. Nishioka, A. Ishikawa and Y. Arakawa, *Phys. Rev. Lett.*, 1992, **69**, 3314.
14 R. Houdré, R. P. Stanley and M. Ilegems, *Phys. Rev. A*, 1996, **53**, 2711.
15 F. Tassone, C. Piermarocchi, V. Savona, A. Quattropani and P. Schwendimann, *Phys. Rev. B: Condens. Matter Mater. Phys.*, 1997, **56**, 7554.
16 P. Michetti and G. C. La Rocca, *Phys. Rev. B: Condens. Matter Mater. Phys.*, 2005, **71**, 115320.

17 K. G. Lagoudakis, M. Wouters, M. Richard, A. Baas, I. Carusotto, R. Andre, L. E. S. I. Dang and B. Deveaud-Pledran, *Nat. Phys.*, 2008, **4**, 706.
18 D. G. Lidzey, D. D. C. Bradley, M. S. Skolnick, T. Virgili, S. Walker and D. M. Whittaker, *Nature*, 1998, **395**, 53.
19 P. A. Hobson, W. L. Barnes, D. G. Lidzey, G. A. Gehring, D. M. Whittaker, M. S. Skolnick and S. Walker, *Appl. Phys. Lett.*, 2002, **81**, 3519.
20 Y. Sugawara, T. A. Kelf, J. J. Baumberg, M. E Abdelsalam and P. N. Bartlett, *Phys. Rev. Lett.*, 2006, **97**, 266808.
21 N. T. Fofang, N. K. Grady, Z. Fan, A. O. Govorov and N. J. Halas, *Nano Lett.*, 2011, **11**, 1556.
22 J. R. Tischler, M. Scott Bradley, Q. Zhang, T. Atay, A. Nurmikko and V. Bulović, *Org. Electron.*, 2007, **8**, 94.
23 S. Aberra Guebrou, C. Symonds, E. Homeyer, J. C. Plenet, Y. N. Gartstein, V. M. Agranovich and J. Bellessa, *Phys. Rev. Lett.*, 2012, **108**, 066401.
24 D. M. Coles, R. T. Grant, D. G. Lidzey, C. Clark and P. G. Lagoudakis, *Phys. Rev. B: Condens. Matter Mater. Phys.*, 2013, **88**, 121303(R).
25 P. Vasa, W. Wang, R. Pomraenke, M. Lammers, M. Maiuri, C. Manzoni, G. Cerullo and C. Lienau, *Nat. Photonics*, 2013, **7**, 128.
26 G. Zengin, G. Johansson, P. Johansson, T. J. Antosiewicz, M. Käll and T. Shegai, *Sci. Rep.*, 2013, **3**, 3074.
27 Y.-W. Hao, H.-Y. Wang, Y. Jiang, Q.-D. Chen, K. Ueno, W.-Q. Wang, H. Misawa and H.-B. Sun, *Angew. Chem., Int. Ed.*, 2011, **123**, 7970.
28 J. Dintinger, S. Klein, F. Bustos, W. L. Barnes and T. W. Ebbesen, *Phys. Rev. B: Condens. Matter Mater. Phys.*, 2005, **71**, 035424.
29 A. Salomon, C. Genet and T. W. Ebbesen, *Angew. Chem., Int. Ed.*, 2009, **48**, 8748.
30 T. Schwartz, J. A. Hutchison, J. Léonard, C. Genet, S. Haacke and T. W. Ebbesen, *ChemPhysChem*, 2013, **14**, 125.
31 A. Salomon, S. Wang, J. A. Hutchison, C. Genet and T. W. Ebbesen, *ChemPhysChem*, 2013, **14**, 1882.
32 S. Wang, T. Chervy, J. George, J. A. Hutchison, C. Genet and T. W. Ebbesen, *J. Phys. Chem. Lett.*, 2014, **5**, 1433.
33 S. Kéna-Cohen and S. R. Forrest, *Nat. Photonics*, 2010, **4**, 371.
34 J. D. Plumhof, T. Stoferle, L. J. Mai, U. Scherf and R. F. Mahrt, *Nat. Mater.*, 2014, **13**, 248.
35 K. S. Daskalakis, S. A. Maier, R. Murray and S. Kéna-Cohen, *Nat. Mater.*, 2014, **13**, 271.
36 M. A. Koponen, U. Hohenester, T. K. Hakala and J. J. Toppari, *Phys. Rev. B: Condens. Matter Mater. Phys.*, 2013, **88**, 085425.
37 T. K. Hakala, J. J. Toppari, A. Kuzyk, M. Pettersson, H. Tikkanen, H. Kunttu and P. Törmä, *Phys. Rev. Lett.*, 2009, **103**, 053602.
38 W. Ni, Z. Yang, H. Chen, L. Li and J. Wang, *J. Am. Chem. Soc.*, 2008, **130**, 6692.
39 A. Berrier, R. Cools, C. Arnold, P. Offermans, M. Crego-Calama, S. H. Brongersma and J. Gómez-Rivas, *ACS Nano*, 2011, **5**, 6226.
40 F. Valmorra, M. Bröll, S. Schwaiger, N. Welzel, D. Heitmann and S. Mendach, *Appl. Phys. Lett.*, 2011, **99**, 051110.
41 T. Schwartz, J. A. Hutchison, C. Genet and T. W. Ebbesen, *Phys. Rev. Lett.*, 2011, **106**, 196405.

42 J. A. Hutchison, T. Schwartz, C. Genet, E. Devaux and T. W. Ebbesen, *Angew. Chem., Int. Ed.*, 2012, **51**, 1592.

43 J. A. Hutchison, A. Liscio, T. Schwartz, A. Canaguier-Durand, C. Genet, V. Palermo, P. Samorì and T. W. Ebbesen, *Adv. Mater.*, 2013, **25**, 2481.

44 S. J. Wang, A. Mika, J. A. Hutchison, C. Genet, A. Jouaiti, M. W. Hosseini and T. W. Ebbesen, *Nanoscale*, 2014, **6**, 7243.

45 E. Orgiu, J. George, J. A. Hutchison, E. Devaux, J.-F. Dayen, B. Doudin, F. Stellacci, C. Genet, P. Samorì and T. W. Ebbesen, 2014, arXiv:1409.1900 [cond-mat.mtrl-sci].

46 J. A. Ćwik, S. Reja, P. B. Littlewood and J. Keeling, *EPL*, 2014, **105**, 47009.

47 A. Canaguier-Durand, C. Genet, A. Lambrecht, T. W. Ebbesen and S. Reynaud, 2013, arXiv:1307.8378 [cond-mat.mes-hall].

48 J. B. Birks, *Photophysics of Aromatic Molecules*, Wiley-Interscience, 1970.

Faraday Discussions

PAPER

Relaxation dynamics of a quantum emitter resonantly coupled to a coherent state of a localized surface plasmon

Khachatur V. Nerkararyan[a] and Sergey I. Bozhevolnyi*[b]

Received 30th August 2014, Accepted 20th October 2014
DOI: 10.1039/c4fd00165f

We investigate the relaxation dynamics of a quantum dipole emitter (QDE), *e.g.*, a molecule or quantum dot, located near a metal nanoparticle (MNP) exhibiting a dipolar localized surface plasmon (LSP) resonance at the frequency of the QDE radiative transition. A generic three-level QDE, which is pumped with an external laser pulse and thereby brought into an optically active excited state, is considered to be weakly coupled to the resonant LSP described by a coherent state. It is shown that, under the condition of the QDE–MNP characteristic relaxation time being much shorter than that of the QDE in free space but much longer than the LSP lifetime, the QDE relaxation dynamics can be described analytically and feature, in general, non-exponential decay with complicated transient behaviour. The main physical consequence of this relaxation process is that the emission, being largely determined by the MNP, comes out with a substantial delay. It is also shown that energy dissipation in the QDE–MNP system is relatively weak with the probability of the photon emission being ~0.75, a number which, rather surprisingly, does not explicitly depend on the metal absorption characteristics. A large number of QDE–MNP system parameters in our analytical description open new possibilities for controlling quantum emitter dynamics.

Introduction

The interaction of quantum dipole emitters (QDEs), such as molecules or quantum dots, with metal nanoparticles (MNPs) at optical frequencies allows control over the flow of electromagnetic energy and lies at the core of an explosively growing field of quantum plasmonics.[1] The coupling of electromagnetic fields in dielectrics and free electron oscillations in metals, which results in surface plasmon excitations, enables the localization of light in subwavelength-sized volumes and, in general, command over light-matter interactions at the nanoscale. One of the very fundamental effects occurring as a result of these

[a]*Department of Physics, Yerevan State University, 375049 Yerevan, Armenia. E-mail: knerkar@ysu.am*
[b]*Department of Technology and Innovation, University of Southern Denmark, Niels Bohrs Allé 1, DK-5230 Odense M, Denmark. E-mail: seib@iti.sdu.dk*

interactions is the modification of QDE relaxation dynamics due to the presence of metal nanostructures, *e.g.* MNPs, strongly perturbing local electromagnetic fields. Recent advances in nano-optics, especially experiments with single molecules interacting with well-defined metal nanostructures,[2–4] often referred to as nanoantennas, serve as a strong impetus for further developments in this direction.[5,6] The most often discussed effect of QDE–MNP interactions is concerned with the modification (enhancement or quenching) of fluorescence yield determined by the balance between radiative and nonradiative decay rates, both enhanced near MNPs.[3,4,7–9] It is also expected the QDE–MNP interaction can enter the regime of strong coupling, where excitation energy is coherently transferred between the QDE and MNP in the form of Rabi oscillations.[1,10]

The strong coupling between a QDE and a resonator is a key element of the quantum-optics toolbox. In the absence of energy dissipation, two identical oscillators are strongly coupled even for small coupling constants. In practice, their coupling should be sufficiently strong so that the energy level splitting becomes larger than the width of the oscillator resonances or, alternatively, that the period of Rabi oscillations becomes smaller than the oscillator lifetimes. The QDE–MNP system is characterized by an enormous mismatch in the lifetimes of its components due to inevitable light absorption in metals, with the LSP lifetime being of the order of a few femtoseconds in the visible region.[11] It is clear that, for the strong QDE–MNP coupling regime to be realized, the normal mode splitting should be very large in order to exceed the LSP linewidth or, alternatively, for the Rabi frequency to exceed the LSP damping rate. Considering this requirement from the viewpoint of the QDE decay rate enhanced near a resonant MNP,[3,4,7–10] the MNP-enhanced QDE decay rate should be increased by *many orders of magnitude*, reaching the LSP damping rate.[11] The situation changes when considering the interaction of an ensemble of QDEs with plasmonic resonators due to the square-root scaling of the coupling constant with respect to the QDE concentration.[12] Several successful demonstrations of the strong-coupling regime in various resonant plasmonic configurations interacting with QDE ensembles have been reported during the last 10 years,[1] including recent real-time observations of ultrafast Rabi oscillations between J-aggregate excitons and surface plasmon polaritons supported by a periodic metal nanostructure.[13] However, for the QDE–MNP system, the strong-coupling regime requires very large QDE dipole moments ($\gg 1$ a.u.) and very small ($\ll 10$ nm) QDE–MNP separations,[10] and is yet to be experimentally realized.

Numerous investigations considering the modification of spontaneous emission in various QDE–MNP configurations have focused on the effect of dramatic enhancements of the QDE (radiative and nonradiative) decay rates in the vicinity of the MNP,[3,4,7–9,14,15] *always* implicitly assuming that the relaxation dynamics is purely exponential as obtained in the Weisskopf–Wigner treatment of an individual two-level atom.[16] At the same time, the occurrence of transient effects at early times was noted in the consideration of molecular dynamics modified by the presence of an MNP, stressing the following (at later times) exponential decay in the case of weak coupling but without further analysis of transient behaviour.[10] Here, we argue that, given an extremely large difference between the relaxation times of the excited QDE and the LSP mode of the MNP, the experimentally accessible regime of *weak* coupling, in which the QDE–MNP relaxation time is much shorter than that of the QDE in free space but much longer than the LSP lifetime, occurs *via coherent* interaction between an optically active excited QDE state and a coherent state of the resonant LSP. Recently, we considered a similar system of a generic three-level QDE located near an MNP exhibiting a dipolar resonance (that coincides with the frequency of the QE radiative transition) using a quasi-classical description, in which the QDE was represented by the coherent superposition of the excited and ground states obeying the time-dependent Schrödinger equation (for a two-level system), while the LSP field was considered to be induced in the MNP by the classical electromagnetic field created by the oscillating QDE dipole.[17] For this configuration under pulsed excitation, we have found an *intermediate* regime of relaxation (which we associated with the *self-stimulated* QDE transition[17]) that occurs in a coherent fashion (like in the strong-coupling regime) with a substantial delay of the emission followed by exponential relaxation dynamics akin to that of modified spontaneous emission in the weak-coupling regime.[3,4,7–9,14–16]

In this work, we develop further our treatment by elaborating the description of the LSP field and introducing the density-matrix formalism for essentially the same configuration as was treated previously.[17] The considered MNP–QDE configuration exhibits the following three main features:

(1) Resonant excitation of the LSP is realized as a result of free electron oscillations in the MNP that, for nm-size MNPs, can be regarded as classical current oscillations, since a large number of free electrons ($\sim$100 nm^{-3}) are involved and their energy spectrum can be considered continuous. This classical oscillating current is then represented by a quantum *coherent* state[18] of the LSP. Note that the coherent LSP state is fundamentally different from (often considered) LSP states with a definite number of quantized plasmons.[1]

(2) The LSP decay rate is larger than the QDE spontaneous emission (decay) rate by five orders of magnitude. Therefore, even for relatively strong QDE–MNP interactions, the relaxation of the QDE–MNP system is much slower than the LSP decay, a feature that allows one to disregard the LSP dynamics and consider the MNP response as being *instantaneous*.

(3) The magnitude of a dipole moment associated with the radiative QDE transition is one order of magnitude smaller than that of an LSP dipole moment induced by the QDE, a feature that allows one to consider the MNP acting as an *antenna* of the QDE–MNP system.

In our opinion, these distinct features of the considered MNP–QDE configuration open the possibility for the realization of *unique* (coherent) relaxation dynamics that exhibit characteristic signatures of both weak and strong coupling regimes.

Under these conditions and in the absence of external illumination, it becomes crucial to properly take into account self-action of the excited QDE, in which its dipole field generates an LSP mode that acts back on the QDE, thus providing a feedback in the QDE–MNP system. Considering the resonant QDE–MNP coupling to be strong enough to significantly decrease the excited QDE lifetime but weak enough so that the emission rate is still far smaller than the LSP dissipation rate, we find that the QDE–MNP emission comes with a delay producing a single emission pulse with characteristics that can be controlled by adjusting the system parameters, such as the QDE–MNP separation and MNP dimensions. We further consider energy dissipation in the QDE–MNP system due to the light absorption by the MNP and obtain a *general* result showing that the probability of the photon emission by the system is ~0.75, a number which, rather surprisingly, does *not* explicitly depend on the system parameters, including the metal absorption. We also discuss several publications, whose theoretical and experimental results might have already (albeit indirectly) supported our findings.

Theoretical framework

The QDE–MNP system under consideration is schematically presented in Fig. 1 and consists of a generic three-level QDE[9,10] and a spherical MNP. It is assumed that an external pump laser brings the QDE from the ground state g into the state s, where it decays nonradiatively into the optically active excited state e, and that the spherical MNP exhibits a dipolar LSP resonance at the frequency ω_0 of the radiative (dipole-allowed) transition e → g [Fig. 1(b)]. This allows us to separate the excitation dynamics, which are not influenced by the presence of the MNP, from the relaxation dynamics of state e, whose modification due to the QDE–MNP coupling is the main subject of this work. Note that the shape of an MNP is not important in this context and can be chosen specifically in order to produce a dipolar resonance at a given frequency,[10,19] for instance, to coincide with the QDE radiative transition frequency.

The Hamiltonian of the system can be represented as:

$$\hat{H} = \hat{H}_0 + \hat{H}_1, \tag{1}$$

$$\hat{H}_0 = \frac{1}{2}\hbar\omega_0\left(b_e^+ b_e - b_g^+ b_g\right) + \hbar\omega_p a^+ a, \tag{2}$$

$$\hat{H}_1 = q^*(b_e^+ b_g + b_g^+ b_e)a + q(b_e^+ b_g + b_g^+ b_e)a^+. \tag{3}$$

Here, ω_0 is the frequency of the radiative QDE transition, $\hbar$ is the reduced Planck constant, b_e^+ and b_e are the creation and annihilation operators of the excited QDE state e, b_g^+ and b_g are the creation and annihilation operators of the ground QDE state g, ω_p is the frequency of the resonant LSP excitation, a^+ and a are the creation and annihilation operators of the LSP, and q is the coupling constant characterizing the interaction between the QDE and LSP. By using the unitary transformation: $U_0 = \exp(-iH_0t/\hbar)$, we transform the system Hamiltonian into

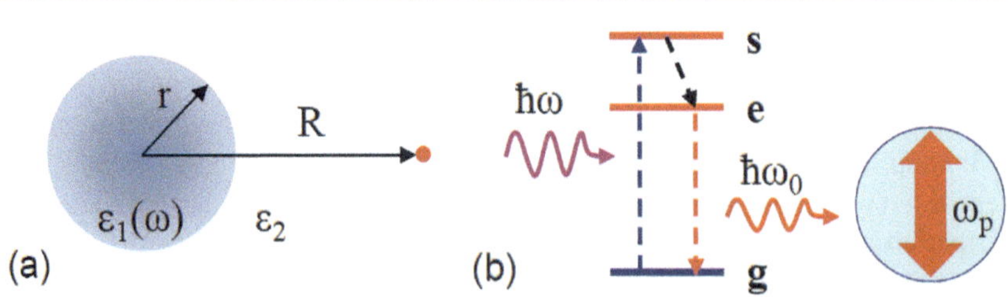

Fig. 1 Schematic of a system with a QDE placed near an MNP, indicating (a) system parameters and (b) QDE energetic levels along with an oscillating current associated with the LSP excitation.

one that, within the rotating wave approximation and under the condition of strict resonance ($\omega_0 = \omega_p$), has the following form:

$$\hat{H}' = q^* b_e^+ b_g a + q b_g^+ b_e a^+. \tag{4}$$

The first two of the QDE–MNP system features described in the previous section allow one to significantly simplify the solution procedure. Considering free electron oscillations in the resonantly excited MNP as classical current oscillations (due to the very large number of electrons involved and the continuity of their energy spectrum), we further make use of the concept of coherent states in quantum optics[18] for the description of this classical current. We also assume that, due to an extremely large difference in the decay rates of an LSP and an isolated QDE, it is possible and, indeed, highly probable that the relaxation of the QDE–MNP system is much slower than that of the LSP, but much faster than that of the isolated QDE. In such a situation, one can neglect the QDE relaxation due to its spontaneous emission and disregard the LSP dynamics, considering the MNP response as instantaneous. Under these conditions, the wave function of the full QDE–MNP system can be represented as follows:

$$|\psi\rangle = \{\beta_e(t)|e\rangle + \beta_g(t)|g\rangle\} e^{-\frac{1}{2}|\alpha(t)|^2} \sum_{n=0}^{\infty} \frac{\alpha^n(t)}{\sqrt{n!}} |n\rangle. \tag{5}$$

Here, $\beta_e(t)$ and $|e\rangle$ are the probability amplitude and the wave function of the excited QDE state, $\beta_g(t)$ and $|g\rangle$ are the probability amplitude and the wave function of the ground QDE state, $\alpha(t)$ is the eigenvalue of the operator a and $|n\rangle$ is the LSP wave function corresponding to the energy eigenvalue $n\hbar\omega_p$. Eqn (4) and (5) represent the starting point of our theoretical framework, allowing us to finally obtain relations for the probability density matrix elements of the QDE transitions and for the eigenvalue $\alpha(t)$ of the LSP operators:

$$\frac{\partial\langle b_e^+ b_e\rangle}{\partial t} = \frac{\partial \rho_{ee}}{\partial t} = \frac{i}{\hbar}\{q\alpha^* \rho_{ge} - q^* \alpha \rho_{eg}\} \quad \text{with } \rho_{gg} = 1 - \rho_{ee}, \tag{6}$$

$$\frac{\partial\langle b_e^+ b_g\rangle}{\partial t} = \frac{\partial \rho_{eg}}{\partial t} = -\frac{i}{\hbar} q\alpha^* \{\rho_{ee} - \rho_{gg}\} \quad \text{with } \rho_{ge} = \rho_{eg}^*, \tag{7}$$

$$\frac{\partial\langle a\rangle}{\partial t} = \frac{\partial \alpha}{\partial t} = -\frac{i}{\hbar} q \rho_{ge} - \Gamma\alpha, \tag{8}$$

where the LSP relaxation rate Γ is introduced. Here, we operate under the assumption of instantaneous MNP response: $\partial\alpha/\partial t \ll \Gamma\alpha$, hence

$$\alpha \approx -\frac{i}{\hbar\Gamma} q \rho_{ge}. \tag{9}$$

Finally, one can work out the following solution of the system of eqn (6)–(9):

$$\rho_{ee}(t) = \frac{1}{1 + e^{2\{\mu(t-\tau)-\varphi\}}}, \tag{10}$$

$$\rho_{eg}(t) = \rho_{ge}(t) = \frac{1}{2\cosh\{\mu(t-\tau)-\varphi\}}, \tag{11}$$

where

$$\mu = \frac{|q|^2}{\hbar^2 \Gamma} \quad \text{and} \quad \varphi = \ln\left[\frac{\rho_{ee}^{(\tau)}}{\rho_{gg}^{(\tau)}}\right], \tag{12}$$

and $\rho_{ee}^{(\tau)}$ and $\rho_{gg}^{(\tau)}$ are the probability density matrix elements at the initial moment $t = \tau$.

It is seen that the relaxation dynamics of the QDE–MNP configuration is characterized by the parameter μ that influences both the time delay in reaching the maximum QDE transition rate [eqn (11)] and the relaxation rate at later moments of time [eqn (10)]. This important system characteristic is related to the basic parameters of the considered configuration. Let us assume that the QDE dipole moment is collinear with the QDE–MNP axis and that the MNP center-to-QDE distance R is considerably larger than the MNP radius r [Fig. 1(a)], with all dimensions being much smaller than the wavelength λ of light, *i.e.*, that $\lambda \gg R \gg r$. Then, by making use of the electrostatic approximation we obtain for the coupling constant q, which represents the energy of the QDE–MNP (dipole–dipole) interaction, the following expression:

$$q = \frac{1}{4\pi\varepsilon_0\varepsilon_d}\frac{2\boldsymbol{d}_{eg}\cdot\boldsymbol{d}_p}{R^3}, \tag{13}$$

where ε_0 is the vacuum permittivity, ε_d is the relative permittivity of the dielectric environment, $\boldsymbol{d}_{eg}$ is the dipole moment of the QDE radiative transition, and $\boldsymbol{d}_p$ is the dipole moment associated with the LSP transition. In the vicinity of the LSP resonance, the polarizability δ of the MNP can be expressed as follows:[20]

$$\delta = \frac{1}{\hbar}\frac{|\boldsymbol{d}_p|^2}{\omega - \omega_p - i\Gamma} = 4\pi\varepsilon_0\varepsilon_d r^3 \frac{\varepsilon_m - \varepsilon_d}{\varepsilon_m + 2\varepsilon_d}, \tag{14}$$

where $\varepsilon_m = \varepsilon_m' + i\varepsilon_m''$ is the metal permittivity. Using the condition of strict resonance $\omega_0 = \omega_p$ and, alternatively, $|\varepsilon_m' + 2\varepsilon_d| \ll \varepsilon_m''$ one obtains:[21]

$$|\boldsymbol{d}_p|^2 = \frac{12\pi\hbar\varepsilon_0{\varepsilon_d}^2 r^3}{\dfrac{\partial\varepsilon_m'(\omega_p)}{\partial\omega_p}} \quad \text{and} \quad \Gamma = \frac{\varepsilon_m''(\omega_p)}{\dfrac{\partial\varepsilon_m'(\omega_p)}{\partial\omega_p}}. \tag{15}$$

Finally, the characteristic relaxation rate μ of the QDE–MNP system can be expressed *via* the system parameters by substituting eqn (13) and (15) into eqn (12):

$$\mu = \frac{3|\boldsymbol{d}_{eg}|^2 r^3}{\pi\hbar\varepsilon_0\varepsilon_m'' R^6}. \tag{16}$$

The QDE relaxation dynamics are strongly influenced by the excitation of the LSP mode that opens a very efficient relaxation channel. The QDE-induced dipole moment of the MNP, which plays the role of an antenna in the QDE–MNP system, can be represented in the following form:

$$\boldsymbol{p} = \frac{3i\varepsilon_d r^3}{\varepsilon_m'' R^3}\frac{\boldsymbol{d}_{eg}e^{-i\omega_0 t}}{\cosh\{\mu(t-\tau)-\varphi\}}. \tag{17}$$

The above expressions [eqn (10), (11), (16) and (17)] constitute the main theoretical outcome of our work, providing simple analytical formulas for the QDE–MNP relaxation and emission dynamics and demonstrating that the dynamics are in general quite complicated. Note that these expressions are also consistent with those obtained previously by us using the quasi-classical approach, in which the QDE was represented by the coherent superposition of the excited and ground states obeying the time-dependent Schrödinger equation (for a two-level system), while the LSP field was considered to be induced in the MNP by the classical electromagnetic field created by the oscillating QDE dipole.[17] In this respect, the above derivation can be considered as a justification of the previously used approach.

Results and discussion

One of the most important assumptions made in our theoretical framework described in the previous section is related to the strength of the QDE–MNP coupling, which should ensure considerably larger relaxation rates μ than that for the QDE in free space γ_0. Their ratio can be evaluated now with the help of eqn (16) and the Weisskopf–Wigner result[16] as follows:

$$\beta = \frac{\mu}{\gamma_0} = \frac{9}{\varepsilon''_{\mathrm{m}}\sqrt{\varepsilon_{\mathrm{d}}}}\left(\frac{\lambda_0}{2\pi R}\right)^3\left(\frac{r}{R}\right)^3, \tag{18}$$

with λ_0 being the vacuum wavelength corresponding to the QDE transition frequency ω_0. For a typical dielectric environment with $\varepsilon_{\mathrm{d}} = 2.25$ (*e.g.*, glass or polymer), the resonance condition (*i.e.*, $\varepsilon'_{\mathrm{m}} = -4.5$) is met, for gold, at a wavelength of $\sim$530 nm with $\varepsilon''_{\mathrm{m(g)}} = 2.35$ and, for silver, at $\sim$400 nm with $\varepsilon''_{\mathrm{m(s)}} = 0.22$.[22]

Considering an MNP with a radius of 5 nm and the QDE distance to the MNP center being 15 nm (in order to be within the electrostatic dipole description), one obtains the ratio $\beta \approx 17$ for gold and $\beta \approx 77$ for silver, thereby justifying the above assumption, $\mu \gg \gamma_0$. It is interesting that the effect is already pronounced at relatively large ($\sim$10 nm) distances between the QDEs and the MNP surface, which are in the range of distances explored in the recent experiments with 10 nm-size gold nanoparticles.[8] It is also apparent that even larger ratios can be achieved by exploiting the LSP shape dependence[10,19] and red-shifting the MNP resonance towards smaller metal absorption.[22]

The QDE relaxation described by eqn (10) and (11) begins at some moment in time when the QDE optically active state e, which is created by nonradiative decay of the excited state s, is partially relaxed into the ground state g, so that $\rho_{\mathrm{gg}}^{(\tau)} > 0$. This starting process can occur due to other inducements always found in an open system, for example, due to the free-space spontaneous emission, *i.e.*, without the influence of the QDE–MNP interaction, because the MNP dipole moment [eqn (17)] is still negligibly small. Let us now assume that the QDE relaxation at *initial* moments of time can be described *via* an exponential decay, so that $\rho_{\mathrm{ee}}^{r} = \exp(-\gamma t)$. It is reasonable to expect that this initial QDE relaxation is very similar to that of the free-space spontaneous emission, *i.e.*, that $\gamma \sim \gamma_0$. Applying the continuity condition at the transition between these two processes to both functions, $\rho_{\mathrm{ee}}^{r}(t)$ and $\rho_{\mathrm{ee}}(t)$, and their derivatives, one can determine the characteristic time $\tau = 1/2\mu$ and $\rho_{ee}^{(\tau)}/\rho_{gg}^{(\tau)} = \sqrt{2\mu/\gamma} \gg 1$. Note that the time τ does not depend on the initial QDE relaxation rate γ, after which the investigated process starts. Therefore, the present consideration allows one to analyze the whole relaxation process. The QDE relaxation dynamics depend strongly on the efficiency of the QDE–MNP interaction, which is characterized by the relaxation rate ratio μ/γ. Sharp step-like behavior, which is observed for very large ratios μ/γ, changes to more gradual population decay for smaller ratios μ/γ [Fig. 2(a)]. Emission of radiation from the QDE–MNP system is determined primarily by the square magnitude of the (induced) MNP dipole moment [eqn (17)], whose maximum is attained after a certain delay time $t_d = [1 + \ln(2\mu/\gamma)](2\mu)^{-1}$. For strong QDE–MNP interactions (very large μ/γ), the emission peak is narrow and occurs close to the initial moment of time [Fig. 2(b)]. Note that, in practice, $\mu/\gamma \sim \beta$ [eqn (18)] that can attain rather large values in realistic conditions as discussed above. It should also be noted that the delay time t_d is of the same order of magnitude as the width (at half maximum) of the emission peak. This interesting feature might be found useful when conducting and analyzing the corresponding experiments.

The described QDE relaxation process can be considered as the self-stimulated QDE transition from the excited coherent superposition state into the ground state because it is stimulated not by an external (to the QDE) monochromatic light but by the feedback field from the LSP, which is in turn excited in the MNP by the QDE dipole moment. It can be shown that this field represents in fact a π pulse, which ensures the QDE transition into the ground state.

We would like to mention that the occurrence of transient effects at early times was noted in the theoretical consideration of molecular dynamics modified by the presence of an MNP, stressing the following (at later times) exponential decay in the case of weak coupling but without further analysis of transient behavior.[10] However, a numerical analysis based on the SPP quantum-mechanical description indicated the occurrence of the emission peak with a certain delay in time in a fashion similar to our results (*cf.* Fig. 2 here and Fig. 2 in ref. 10). It is also relevant to mention the reported observation of non-Markovian dynamics of a quantum dot resonantly coupled to a micropillar cavity, resulting in a

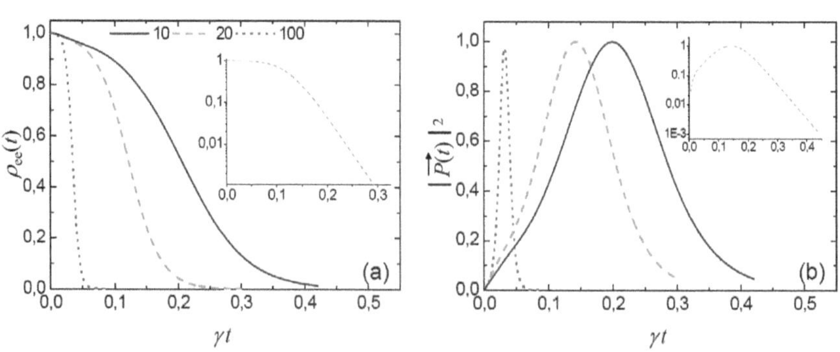

Fig. 2 Relaxation dynamics of the QDE–MNP configuration for different ratios of the established and initial relaxation rates μ/γ, showing (a) the population decay of the excited state e [eqn (10)] and (b) the squared magnitude of the MNP dipole moment [eqn (17)] normalized to its maximum value. Insets display the same dependencies on a logarithmic scale for the case of $\mu/\gamma = 20$.

nonexponential decay in time.[23] Finally, we believe that such a delay could have actually been present (but not elucidated) in the recent experiments with 10 nm-size gold nanoparticles connected by DNA to individual fluorophores (see Fig. 1(d) in ref. 8).

As already mentioned, the MNP serves as a radiative antenna in the QDE–MNP configuration. The energy dissipation in this configuration (that causes photon loss) occurs due to inevitable ohmic losses in the MNP. Then the probability of non-radiative decay of the system can be determined by relating the energy dissipated in the system[24] to the photon energy:

$$P = \frac{\frac{1}{2}\int_0^\infty \mathrm{Re}\left[\boldsymbol{p}\cdot\dot{\boldsymbol{E}}^*\right]\mathrm{d}t}{\hbar\omega_0}, \tag{19}$$

where $\boldsymbol{E}$ is the electric field induced at the MNP, which is characterized by the dipole moment $\boldsymbol{p}$ [eqn (17)], by the QDE dipole moment $\boldsymbol{d}_{\mathrm{eg}}$:

$$\dot{\boldsymbol{E}}^* = \frac{i\omega_0}{4\pi\varepsilon_0\varepsilon_\mathrm{d}R^3}\frac{\boldsymbol{d}^*_{\mathrm{eg}}\mathrm{e}^{i\omega t}}{\cosh\{\mu(t-\tau)-\varphi\}}. \tag{20}$$

In the discussed case, when $\rho_{\mathrm{ee}}^{(\tau)} \gg \rho_{\mathrm{gg}}^{(\tau)}$, after performing integration in eqn (19) one obtains a very simple formula for the loss probability: $P \cong 0.25[1-(3\gamma/8\mu)] \approx 0.25$. Under the condition of the resonant QDE–MNP coupling, the transition from the QDE optically active state e to the ground state g occurs during the time interval of $\sim 2t_\mathrm{d}$ (Fig. 2), implying the transition rate of $\sim 0.5/t_\mathrm{d}$. Assuming that the excitation of the upper state s by the pump [Fig. 1(b)] together with the subsequent relaxation to the QDE optically active state e takes considerably less time (also that the pump rate is smaller than any of the system relaxation rates), we can evaluate the fluorescence enhancement due to the QDE–MNP resonant coupling:

$$\eta = (1-P)\frac{1/2t_\mathrm{d}}{\gamma_0} \cong \frac{3}{4}\frac{\mu}{\gamma_0[1+\ln(2\mu/\gamma)]} \approx \frac{3}{4}\frac{\beta}{1+\ln(2\beta)}, \tag{21}$$

which is substantial [eqn (18)] and can easily reach several orders of magnitude.

The most interesting physical finding of our work, *viz.*, the spontaneous emission delay under pulsed excitation, is in fact quite general. Indeed, it is only required that a QDE is placed near a resonator that, at the frequency of QDE radiative transition, features a well-defined dipolar resonance with the damping rate, which is substantially larger than the QDE relaxation rate in free space. The former justifies the derivation of eqn (4) and ensures the $\pi/2$ phase delay in the QDE–MNP system (electromagnetic) feedback, while the latter is needed to realize the desirable weak-coupling regime. The appropriate interaction can be realized at practically any wavelength with non-spherical MNPs[19] or by using low absorbing dielectric (semiconductor) nanoparticles having large permittivity values and supporting strong Mie resonances that can be chosen propitiously by adjusting the particle shapes and sizes.[25] Another possibility would be to place a QDE near a metal surface, a configuration that is resonant if $\varepsilon_\mathrm{m}'(\omega_0) \approx -\varepsilon_\mathrm{d}$, with the strong-coupling regime requiring sub-nanometer QDE-surface distances.[26] Our approach can also be applied in this case, provided that $|\varepsilon_\mathrm{m}' + \varepsilon_\mathrm{d}| \ll \varepsilon_\mathrm{m}''$ and $2\varepsilon_\mathrm{d} \gg \varepsilon_\mathrm{m}''$, resulting in similar emission dynamics with the relaxation parameter μ given by

$$\mu_1 = \frac{|\boldsymbol{d}_{\mathrm{eg}}|^2}{8\pi\hbar\varepsilon_0\varepsilon''_{\mathrm{m}}R^3}, \tag{22}$$

with R being in this case the QDE-surface distance, and considering that the QDE dipole moment is perpendicular to the metal surface. Note, that the R^{-6} scaling in eqn (16) is transformed, for this configuration, into the R^{-3} scaling [eqn (22)], which is also expected to be the case for small QE–MNP separations with the dipolar MNP response to the homogeneous field becoming strongly multipolar and approaching that of a flat metal surface.[27] The condition $\varepsilon'_{\mathrm{m}}(\omega_0) \approx -\varepsilon_{\mathrm{d}}$ defines the physical situation, where the predicted relaxation dynamics can be observed. In the visible region of the spectrum, where the magnitude of the real part of the dielectric constant of metals, such as gold and silver, is large enough, semiconductor substrates can be used. In this case, excitons, which are typically localized in the vicinity of the interface, can be playing the role of individual QDEs. Indeed, very recent time-resolved spontaneous emission measurements in the configuration, consisting of a semiconductor (CdS) nanocrystal separated from a metal (silver) surface by a 8 nm-thin dielectric (MgF_2) layer, revealed that the spontaneous emission reaches its maximum with a significant delay in time.[28] This delay (a few hundred ps) is significantly larger than the characteristic time of reaching the equilibrium state (a few ps), with the overall time dependence being quite similar to our results (*cf.* Fig. 2 here and Fig. 3 in ref. 28).

Conclusions

In summary, we have considered the relaxation dynamics of a generic QDE excited with short pump pulses and located near a MNP that exhibits a dipolar LSP resonance at the frequency of the QDE radiative transition. It has been shown that, under the condition of the QDE–MNP characteristic relaxation time being much shorter than that of the QDE in free space but much longer than the LSP lifetime, the QDE relaxation dynamics can be described analytically and feature, in general, non-exponential decay with complicated transient behaviour. Our theoretical analysis resulted in the following main conclusions: (i) the relaxation dynamics in the resonantly coupled QDE–MNP system exhibits step-like behaviour thereby deviating significantly from the generally accepted exponential decay,[3,4,7–9] (ii) the QDE–MNP radiation emission reaches its maximum with a significant delay in time, and (iii) energy dissipation in the QDE–MNP system is relatively weak with the probability of the photon emission being $\sim$0.75, a number which, rather surprisingly, does not explicitly depend on the metal absorption characteristics. A large number of system parameters in our analytical description opens new possibilities for controlling the QDE relaxation and emission dynamics. Given the variety of resonant plasmonic[19] and semiconductor[25] nanoparticles, the experimental observation of the predicted effect seems feasible,[28] while the possibility of tuning the delay time by changing the QDE–MNP separation can be exploited in many applications, *e.g.*, for practical implementation of a nanoscopic ruler,[29] or optimization of scanning single QDE fluorescence lifetime imaging,[30] as well as in fundamental studies within quantum plasmonics.[1]

Acknowledgements

The authors gratefully acknowledge financial support for this work from the European Research Council, Grant no. 341054 (PLAQNAP), as well as partial support (KVN) from the Danish Council for Independent Research (Contract no. 09-072949, ANAP).

References

1 M. S. Tame, K. R. McEnery, Ş. K. Özdemir, J. Lee, S. A. Maier and M. S. Kim, *Nat. Phys.*, 2013, **9**, 329–340.
2 J. N. Farahani, D. W. Pohl, H. J. Eisler and B. Hecht, *Phys. Rev. Lett.*, 2005, **95**, 017402.
3 P. Anger, P. Bharadwaj and L. Novotny, *Phys. Rev. Lett.*, 2006, **96**, 113002.
4 S. Kühn, U. Håkanson, L. Rogobete and V. Sandoghdar, *Phys. Rev. Lett.*, 2006, **97**, 017402.
5 P. Mühlschlegel, H.-J. Eisler, B. Hecht and D. W. Pohl, *Science*, 2005, **308**, 1607–1609.
6 M. Agio, *Nanoscale*, 2012, **4**, 692–706.
7 X. W. Chen, M. Agio and V. Sandoghdar, *Phys. Rev. Lett.*, 2012, **108**, 233001.
8 G. P. Acuna, M. Bucher, I. H. Stein, C. Steinhauer, A. Kuzyk, P. Holzmeister, R. Schreiber, A. Moroz, F. D. Stefani, T. Liedl, F. C. Simmel and P. Tinnefeld, *ACS Nano*, 2012, **6**, 3189–3195.
9 K. E. Dorfman, P. K. Jha, D. V. Voronine, P. Genevet, F. Capasso and M. O. Scully, *Phys. Rev. Lett.*, 2013, **111**, 043601.
10 A. Trügler and U. Hohenester, *Phys. Rev. B: Condens. Matter Mater. Phys.*, 2008, **77**, 115403–115409.
11 C. Sönnichsen, T. Franzl, T. Wilk, G. von Plessen and J. Feldmann, *Phys. Rev. Lett.*, 2002, **88**, 077402.
12 D. G. Lidzey, D. D. C. Bradley, M. S. Skolnick, T. Virgili, S. Walker and D. M. Whittaker, *Nature*, 1998, **395**, 53–55.
13 P. Vasa, W. Wang, R. Pomraenke, M. Lammers, M. Maiuri, C. Manzoni, G. Cerullo and C. Lienau, *Nat. Photonics*, 2013, **7**, 128–132.
14 C. Sauvan, J. P. Hugonin, I. S. Maksymov and P. Lalanne, *Phys. Rev. Lett.*, 2013, **110**, 237401.
15 S. D'Agostino, F. D. Sala and L. C. Andreani, *Phys. Rev. B: Condens. Matter Mater. Phys.*, 2013, **87**, 205413.
16 V. F. Weisskof and E. P. Wigner, *Z. Phys.*, 1930, **63**, 54–73.
17 K. V. Nerkararyan and S. I. Bozhevolnyi, *Opt. Lett.*, 2014, **39**, 1617–1620.
18 C. C. Gerry and P. L. Knight, *Introductory Quantum Optics*, Cambridge University Press, Cambridge, 2005.
19 J. N. Anker, W. P. Hall, O. Lyandres, N. C. Shah, J. Zhao and R. P. Van Duyne, *Nat. Mater.*, 2008, **7**, 442–453.
20 V. B. Beresteskii, E. M. Lifshits, and L. P. Pitaevskii, *Quantum Electrodynamics*, Pergamon Press, Oxford, 1982.
21 M. I. Stockman, *Opt. Express*, 2011, **19**, 22029–22106.
22 P. B. Johnson and R. W. Christy, *Phys. Rev. B: Solid State*, 1972, **6**, 4370–4379.
23 K. H. Madsen, S. Ates, T. Lund-Hansen, A. Löffler, S. Reitzenstein, A. Forchel and P. Lodahl, *Phys. Rev. Lett.*, 2011, **106**, 233601.

24 L. D. Landau and E. M. Lifshitz, *Electrodynamics of Continuous Media*, Pergamon Press, Oxford, 1960.
25 A. B. Evlyukhin, S. M. Novikov, U. Zywietz, R. L. Eriksen, C. Reinhardt, S. I. Bozhevolnyi and B. N. Chichkov, *Nano Lett.*, 2012, **12**, 3749–3755.
26 M. M. Dvoynenko and J.-K. Wang, *Opt. Lett.*, 2013, **38**, 760–762.
27 A. Delga, J. Feist, J. Bravo-Abad and F. J. Garcia-Vidal, *Phys. Rev. Lett.*, 2014, **112**, 253601.
28 R.-M. Ma, S. Ota, Y. Li, S. Yang and X. Zhang, *Nat. Nanotechnol.*, 2014, **9**, 600–605.
29 J. Seelig, K. Leslie, A. Renn, S. Kühn, V. Jacobsen, M. van de Corput, C. Wyman and V. Sandoghdar, *Nano Lett.*, 2007, **7**, 685–689.
30 A. W. Schell, P. Engel, J. F. M. Werra, C. Wolff, K. Busch and O. Benson, *Nano Lett.*, 2014, **14**, 2623–2627.

Faraday Discussions

PAPER

Ultrafast dynamics of nanoplasmonic stopped-light lasing

Sebastian Wuestner, Tim Pickering, Joachim M. Hamm, A. Freddie Page, Andreas Pusch and Ortwin Hess*

Received 26th September 2014, Accepted 20th October 2014
DOI: 10.1039/c4fd00181h

We study the spatio-temporal dynamics of coherent amplification and lasing in planar gain-enhanced nanoplasmonic structures and show that a singularity in the density of optical states leads to a stopped-light feedback mechanism that allows for cavity-free photonic and surface-plasmon polariton nanolasing. We reveal that in the absence of cavity-induced feedback a phase-locked superposition of a quasi dispersion-free waveguide mode promotes the dynamic formation of a subwavelength lasing mode. Simulations on the basis of a full-time domain Maxwell–Bloch Langevin approach uncover a high spontaneous emission factor $\beta \approx 0.9$ and demonstrate that the stopped-light lasing/spasing mechanism is remarkably robust against interface roughness. Stopped-light surface-plasmon polariton lasing is shown to be stable for gain sections of a width of down to 200 nm but in wider gain structures of the order of 1 μm the dynamics is characterised by spatio-temporally oscillating lasing surface-plasmon polaritons with typical temporal and spatial periods of smaller than 5 fs and smaller than 100 nm. Stopped-light lasing thus provides opportunities for ultrafast nanolasing and the realization of ultra-thin lasing surfaces and offers a new route to ultrafast spasing and cavity-free active quantum plasmonics.

1 Introduction

Lasers have had tremendous impact on research and technology and have become an essential component of everyday life. Through innovative design concepts and new materials as well as advancements in fabrication techniques they have become ever more powerful, and can enter new frequency domains and emit shorter and shorter pulses. At the same time, it has also been possible to allow lasers to become smaller and smaller to recently culminate in an experimental demonstration of subwavelength nanolasing.[1–6] This miniaturisation was achieved through significant innovation in the two key ingredients that constitute lasers: light amplification and feedback. Semiconductor gain materials such as quantum wells and dots proved important in this shrinking process as did

The Blackett Laboratory, Department of Physics, Imperial College London, South Kensington Campus, London SW7 2AZ, UK. E-mail: o.hess@imperial.ac.uk; Fax: +44 (0)20 7594 2077; Tel: +44 (0)20 7594 7586

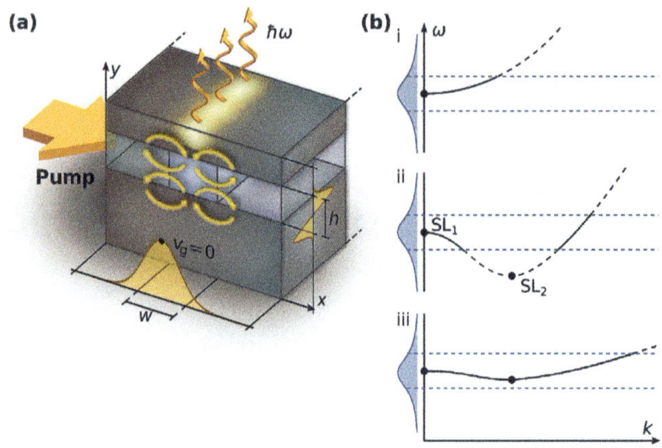

Fig. 1 (a) Subwavelength localised vortex-lasing in a stopped-light design based on a gain-enhanced plasmonic heterostructure. The stopped-light feature of the design allows for the trapping of photons in a closed vortex visualised by paired semi-circular yellow arrows. Exploiting leaky-modes, emission into free space takes place perpendicular to the surface. (b) By engineering the band structure of the metal–dielectric waveguide, several stopped-light points can be made to fall within the gain spectrum, resulting in a flat dispersion with an average slope of $(\omega_2 - \omega_1)/(k_2 - k_1)$.

innovative microcavity resonator designs,[7] for example, in the form of micro-disks,[8] micropillars,[9] defect mode photonic crystals[10,11] and photonic nanowires.

Recent plasmonic nanolasers employ plasmonic resonances for feedback,[12–15] allowing them to concentrate light into mode volumes that are no longer limited by diffraction. The use of localised surface plasmon resonances as cold-cavity modes, however, is only one route to lasing on subwavelength scales. Indeed, lasing does not require modes predefined by a resonator with a particular geometry but only a feedback mechanism (which traditionally is a resonator). However, let us consider the situation in a random laser. There, feedback arises from multiple random scattering events that lead to the spontaneous formation of closed optical paths within the disordered medium.[16,17] Recently, we have shown that cavity-free lasing is not restricted to micrometer sizes in disordered media (using random feedback) but can also be realised on subwavelength (nano-) scales using stopped light to provide local feedback.[18] Having established this principle and identified its main physical properties, we here study the spatio-temporal dynamics of a stopped-light (SL) laser as illustrated in Fig. 1 and uncover the ultrafast spatio-temporal dynamics of SL surface-plasmon polariton (SPP) lasing with <5 femtosecond dynamics on scales of less than 100 nm in a structure designed to support SPP modes.

2 Maxwell–Bloch Langevin theory

In the following we will briefly outline the Maxwell–Bloch Langevin approach which forms the bias for the time-resolved simulations discussed here. This involves, in particular, combining the Finite-Difference Time-Domain (FDTD) method with auxiliary differential equations that self-consistently describe both, the material dispersion of the plasmonic cladding and the nonlinear, spatially resolved polarization response of the gain media.

In the framework of the Maxwell–Bloch formalism we model the gain medium as a four-level system with pump ($0 \rightarrow 3$) and emission ($2 \rightarrow 1$) transitions.[19] These optical transitions are connected by and with (coherent) polarization densities $\mathbf{P}_a(\mathbf{r}, t)$ and $\mathbf{P}_e(\mathbf{r}, t)$ whose dynamics can be described on the basis of the differential equations

$$\frac{\partial^2 \mathbf{P}_{a,e}}{\partial t^2} + 2\gamma_{a,e}\frac{\partial \mathbf{P}_{a,e}}{\partial t} + {\omega_{a,e}}^2\mathbf{P}_{a,e} = -\sigma_{a,e}\Delta N_{a,e}\mathbf{E} - \kappa_{a,e}\left[\omega_{a,e}\mathrm{Im}(\mathbf{F}_{a,e}) + \frac{\partial \mathrm{Re}(\mathbf{F}_{a,e})}{\partial t}\right]. \quad (1)$$

These equations model the dynamics of spatio-temporal light-matter interaction *via* dipole coupling of the electric field to the electronic transitions. The resonance frequencies ω_i (and transition frequencies $\omega_{\mathbf{r},i} = \sqrt{\omega^2 - {\gamma_i}^2}$), the spectral half-widths γ_i, the phenomenological coupling constants $\sigma_i = 2\varepsilon_0 n_s' c\gamma_i\sigma_{0,i}$ (that are associated with the emission and absorption cross-sections $\sigma_{0,i}$), as well as the real part of the host refractive index n_s' reflect the particular choice of gain material.

The spatio-temporal dynamics of the occupation densities N_0 to N_3, which, in turn, determine the inversions $\Delta N_e(\mathbf{r}, t) = N_2(\mathbf{r}, t) - N_1(\mathbf{r}, t)$ and $\Delta N_a(\mathbf{r}, t) = N_3(\mathbf{r}, t) - N_0(\mathbf{r}, t)$ for emission and absorption, is given by the set of equations

$$\frac{\partial N_3}{\partial t} = \frac{1}{\hbar\omega_{r,a}}\left(\frac{\partial}{\partial t}\mathbf{P}_a + \gamma_a\mathbf{P}_a\right)\cdot\mathbf{E} - \frac{N_3}{\tau_{30}} - \frac{N_3}{\tau_{32}} + F_{33}, \quad (2)$$

$$\frac{\partial N_2}{\partial t} = \frac{N_3}{\tau_{32}} + \frac{1}{\hbar\omega_{r,e}}\left(\frac{\partial}{\partial t}\mathbf{P}_e + \gamma_e\mathbf{P}_e\right)\cdot\mathbf{E} - \frac{N_2}{\tau_{21}} + F_{22}, \quad (3)$$

$$\frac{\partial N_1}{\partial t} = \frac{N_2}{\tau_{21}} - \frac{1}{\hbar\omega_{r,e}}\left(\frac{\partial}{\partial t}\mathbf{P}_e + \gamma_e\mathbf{P}_e\right)\cdot\mathbf{E} - \frac{N_1}{\tau_{10}} + F_{11}, \quad (4)$$

$$\frac{\partial N_0}{\partial t} = \frac{N_1}{\tau_{10}} + \frac{N_3}{\tau_{30}} - \frac{1}{\hbar\omega_{r,a}}\left(\frac{\partial}{\partial t}\mathbf{P}_a + \gamma_a\mathbf{P}_a\right)\cdot\mathbf{E} + F_{00}. \quad (5)$$

The lifetimes τ_{jk} account for non-radiative relaxation processes between the levels. The term linked with the polarization density of the absorption transition, $(\hbar\omega_{r,a})^{-1}$ $(\partial\mathbf{P}_a/\partial t + \Gamma_a\mathbf{P}_a)\mathbf{E}$ represents the spatially and temporally inhomogeneous excitation. Quantum fluctuations and amplified spontaneous emission are represented *via* Langevin noise terms given by $\mathbf{F}_e(\mathbf{r}, t)$, $\mathbf{F}_a(\mathbf{r}, t)$, and $F_{00}(\mathbf{r}, t)$ to $F_{33}(\mathbf{r}, t)$.[18,19] Below the lasing threshold the noise acts as source for amplified spontaneous emission giving rise to the recorded background spectra, while above the lasing threshold, the noise acts as a seed aiding the transient build-up of the coherent lasing fields through feedback and stimulated emission processes in the gain medium.

3 Dispersionless stopped light in metal–dielectric waveguides

In our study of the characteristic spatio-temporal dynamics of stopped-light nanolasing we adopt, for specificity, the geometry as introduced in ref. 18 and illustrated in Fig. 2. It consists of a planar metal–dielectric–metal stack with a top metal layer of thickness $t = 500$ nm and a waveguide core of height $h = 290$ nm on

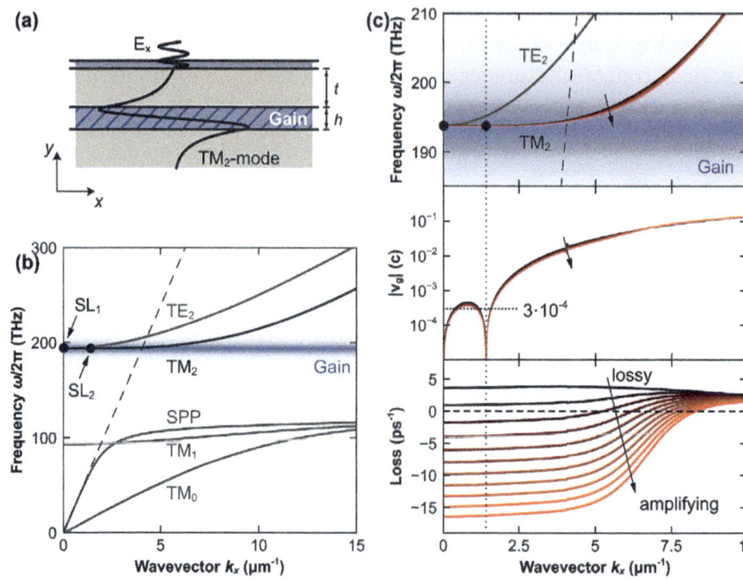

Fig. 2 (a) Field profile of the TM_2 mode at 1547 nm in a plasmonic heterostructure with $h = 290$ nm and $t = 500$ nm. (b) (Complex-frequency) dispersion of the structure, highlighting the wavevector dependence of the weakly-leaky TM_2 mode that includes two SL points as marked by black dots. The gain spectrum is highlighted in blue. (c) Impact of small-signal gain on the dispersion, group velocity and loss of the TM_2 mode from no gain (black) to a gain coefficient of 4180 cm^{-1} (red).

Table 1 Gain medium parameters for emission into the leaky TM_2 waveguide mode

Symbol	Description	Value
$\omega_e/2\pi$ (λ_e)	Emission frequency (wavelength)	193.41 THz (1550 nm)
γ_e	Emission spectral half-width	1/(40 fs) = 25 ps^{-1} (~32 nm)
$\sigma_{0,e}$	Emission cross-section	2.09×10^{-15} cm^2
N	Carrier density	2×10^{18} cm^{-3}
g_e	Maximum gain coefficient	4180 cm^{-1}

top of a thick metal substrate (Fig. 2a). The dielectric permittivities of the core and cladding materials are chosen to be characteristic of III–V semiconductors (such as InGaAsP) and transparent conductive oxides (indium tin oxide), respectively, giving rise to a (weakly) leaky TM_2 mode in the dispersion diagram (Fig. 2b) (Table 1).

By design this dispersion has two stopped-light singularities that lie close in frequency: SL_1 at $\omega_1/2\pi = 193.8$ THz ($\lambda_1 = 1546.9$ nm) and $k_1 = 0$ and SL_2 at $\omega_2/2\pi = 193.78$ THz ($\lambda_2 = 1547.06$ nm) and $k_2 = 1.42$ μm^{-1}. With both SL singularities being to the left of the (dashed) light-line, the SL lasing mode will couple to free-space radiation (perpendicular to the plane). The emission of light amounts to around 3–4% of the total energy lost per cycle and can be controlled by adjusting the thickness of the top metal cladding. We will start our discussion by initially highlighting the characteristics of this photonic (*i.e.*, leaky) design, before concentrating our discussion on analysing the spatio-temporal lasing/spasing dynamics of stopped surface-plasmon polaritons (*i.e.*, spasing[12]) where multiple SL singularities fall onto a bound plasmonic mode.

Fig. 2b shows that based on a passive mode theory we are to expect an extended band of vanishing dispersion. A small-signal gain analysis, where we assume that the gain medium fills the dielectric core of the structure except for a 10 nm buffer next to the metal (to account for gain quenching in close vicinity to the metal) confirms that the two SL singularities effectively pin down the dispersion between them and no significant change of the modal dispersion is observed with increasing inversion (Fig. 2c). The modal loss, on the other hand, displays a significant dependence on the inversion and changes sign at $\Delta N_{\mathrm{th}} \approx 0.13$ (Fig. 2c, bottom). This is the point at which the mode is undamped and experiences an almost uniform level of gain over a broad range of k-values (up to $k \approx 5\ \mu\mathrm{m}^{-1}$). As the loss rate includes both dissipative and radiative (leakage through the cladding layer) contributions, we can associate ΔN_{th} with the threshold inversion required for lasing.

So far, we have assumed that all interfaces and surfaces are ideally flat. However, it has been shown that even small structural imperfections can potentially have a significant (detrimental) impact on the propagation characteristics, particularly for light near SL points in slow light devices.[20] This is particularly challenging in photonic crystal waveguides where the disorder-increased light-scattering at interfaces not only imposes limitations on the achievable group index but can also result in Anderson localization of light.[21] In metal dielectric heterostructures as considered here, however, stopped light arises from a very different effect – negative energy flow inside the metallic cladding[22] and not from backscattering at geometric features as it is the case *e.g.* in photonic crystals.[20] This remarkable resilience of SL points to surface roughness has recently been observed in a passive metal–dielectric heterostructure.[22]

Fig. 3 illustrates the change of the centre of energy position of stopped-light wave packets without gain material where we inject wave packets at two distinct angles corresponding to wavevectors either side of the second SL point. We find that, below a critical rms roughness value, the energy velocity remains constant in time with a clear dependence on the injection angle and one can always find an optimum excitation angle for which the energy velocity along the waveguide direction is zero. It is apparent from the figure that the optimal angle depends on the level of surface roughness and additionally varies from sample to sample. This works up to an rms surface roughness of about 3 nm where the nature of pulse propagation is changed dramatically: the energy velocity is not constant any more and the pulse is equally likely to propagate in a forwards or backwards direction. In this regime, the propagation of energy is diffusive and correlates only weakly with the waveguide dispersion. As a result of the strong scattering at the surface inhomogeneities one observes a pulse breakup and the disappearance of the global SL point. In the active case with a gain material incorporated into the waveguides we found that SL lasing remains resilient to surface roughness with almost no variation in the average loss rate below rms values less than 1.5 nm.[18] However, above this rms value both the average loss rate and standard deviation steadily rose. The higher losses were seen to arise primarily from the scattering of the modal fields at the rough interfaces and an increased absorption of energy at

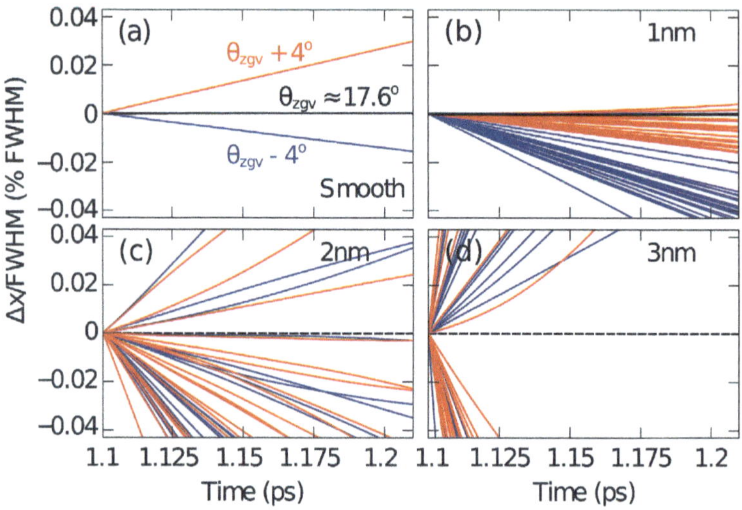

Fig. 3 Change in the centre of energy position of wavepackets excited in the heterostructure waveguides with smooth (a) and rough interfaces (1–3 nm rms height, (b–d)). Red lines indicate wavepackets incoupled at an angle of incidence $\theta = \theta_{ZGV} + 4$ degree, while blue lines correspond to an incidence angle $\theta = \theta_{ZGV} - 4$ degree. For cases (a and b) an optimal angle has been found corresponding to near zero propagation shown by the solid black lines. The dashed black lines in (c and d) mark the zero position.

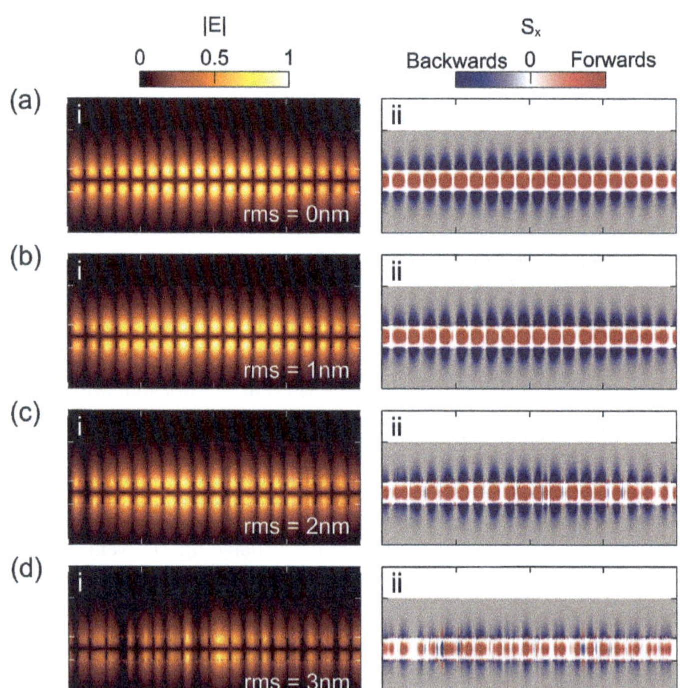

Fig. 4 Electric field amplitude (a.i–d.i) and the x-component of the Poynting vector (a.ii–d.ii) in the heterostructure waveguide with smooth interfaces (a) and rough interfaces with rms = 1 nm (b), 2 nm (c) and 3 nm (d). Higher electric field amplitudes are shown in lighter colours. Grey boxes highlight the positions of the ITO layers.

hotspots, surface irregularities over which the SL mode localises as observed in the mode profiles for different levels of rms roughness (Fig. 4).

4 Spatio-temporal dynamics of stopped-light lasing

Having established the foundations for stopped-light lasing by passive and small signal gain analysis we now perform first-principles time-domain simulations taking into account the full spatio-temporal nonlinearity of the gain material as well as the spatio-temporal interplay between light-field and inversion (Fig. 5). We model the gain material through four-level Maxwell–Bloch Langevin equations[14,18] (Section 2) with parameters that match those used in the semi-analytical mode calculations above. Initially the gain section shall be confined to a 400 nm wide strip, approximately four times smaller than the vacuum wavelength at the SL points. By design the gain section can be pumped optically using a propagating waveguide mode at a higher frequency. Subsequently we will establish a link between newly derived stopped-light laser rate equations to well-established rate-equation approaches for traditional lasers. The difference between the two will then reveal the particular features of stopped-light nanolasing and a comparison between the spatially resolved Maxwell–Bloch simulations and the stopped-light lasing rate equation model allows us, in turn, to extract the effective parameters of the system. We are particularly interested in the confinement factor and the effective mode volume of SL lasing.

4.1 High-β lasing through locking of spatial modes

At a pump level well above threshold, the transient field and inversion dynamics of a typical stopped-light laser can display ultrafast relaxation oscillations with a period of only around 10 ps,[18] gradually settling down to a quasi-steady-state. The SL-laser output energy follows a familiar S-shaped input–output curve (Fig. 5a) with a threshold pump rate $r_p \approx 1.43\ ps^{-1}$. Spectral narrowing of the emission line (comparing Fig. 5b and c) above the lasing threshold (Fig. 5c) is due to the

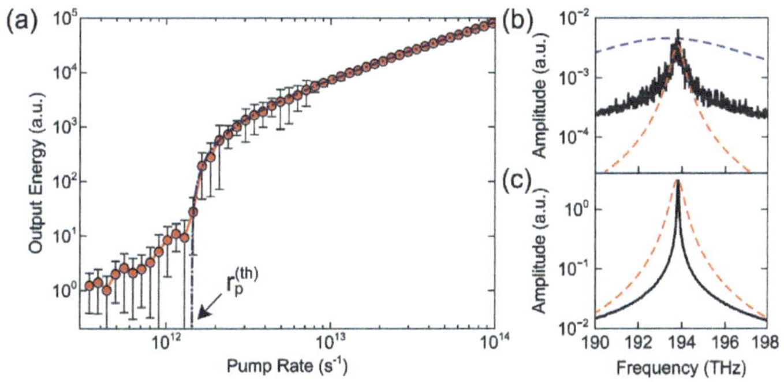

Fig. 5 (a) S-shaped input–output curve of the stopped-light laser with a threshold pump rate $r_p^{(th)} = 1.43\ ps^{-1}$. The ASE spectrum below threshold (b) can be compared to the lasing spectrum (c), indicating strong spectral narrowing. The spectral width of the ASE compares well with the (scaled) spectral density of states of the TM_2 mode at $k = 0$ in red, while the gain spectrum in blue is much wider.

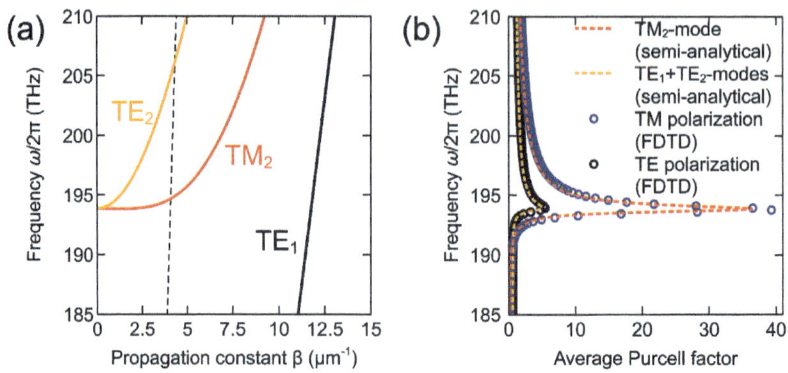

Fig. 6 (a) Dispersion of the TE_1 (black), TE_2 (orange) and TM_2 (red) waveguide modes and (b) the resulting Purcell factor for spontaneous emission into these modes spatially averaged over the position of emitters in the waveguide core. Two separate methods have been used to evaluate the Purcell factor: FDTD simulations (shown by black and blue circles) and semi-analytic calculations based on the waveguide mode dispersion (red and orange dashed lines).

build-up of phase coherence in the SL mode, while below threshold (Fig. 5b) we observe amplified spontaneous emission (ASE) with a spectrum that matches the spectral density of states of the TM_2 mode at $k = 0$.

Owing to a vanishing group velocity of the lasing TM_2 mode, the local density of optical states (LDOS) is strongly enhanced around the frequency of the SL singularities (Fig. 6). The TE_2 mode, which shares the SL singularity at $k = 0$ with the TM_2 mode (Fig. 2), offers a competing channel for spontaneous emission but its LDOS reaches only about 1/10th of the value of the TM_2 mode due to its stronger dispersion. Consequently, 90% of the spontaneous emission is channeled into the dispersionless TM-polarised lasing mode, which is confirmed by the spontaneous emission factor obtained from the threshold curve, $\beta = \gamma_c/(\gamma_c + \Gamma r_p) \approx 0.9$ (ref. 23) (total loss rate $\gamma_c \approx 3.71\ \mathrm{ps}^{-1}$). In contrast to microcavity lasers, where spontaneous emission channels can be suppressed *via* cavity design, the concentration of the spontaneous emission into the localised SL lasing mode is a result of the large density of k-modes that fall within the bandwidth of the gain. The observed high-β characteristics and picosecond relaxation oscillations of cavity-free SL lasing can potentially allow for the design of thresholdless plasmonic laser diodes that can be modulated with THz speeds.

4.2 Link to traditional laser models

Laser rate equations are a set of two coupled differential equations that approximate the dynamics of the photon and carrier number.[24] Applied to nanolasers[2,5] this simple model can reproduce the basic characteristics of a laser, such as its threshold behaviour, transient dynamics, and modulation speeds.

Here we start by first noting that Poynting's theorem, which describes the evolution of the electromagnetic energy density $U(\mathbf{r}, t)$, can be transformed into a rate equation

$$\frac{\partial S}{\partial t} = -\gamma_c S + R_{\mathrm{stim}}(N)S \tag{6}$$

for the photon number S at the lasing frequency ω_0 by volume integration of the total energy density,

$$S = (\hbar\omega_0)^{-1}\int_V d^3r U(\mathbf{r}, t). \tag{7}$$

This equation also defines the cavity loss rate γ_c and the stimulated emission rate R_{stim} (N). Fast phase-oscillations of the fields **E** and **H** at the lasing frequency are eliminated through a time averaging over one period and the integration volume V is taken to encompass the full SL lasing structure making the photon number a slowly-varying, effective variable of the system. In dispersive media, the energy density is given by

$$U(\mathbf{r}, t) = U_E(\mathbf{r}, t) + U_M(\mathbf{r}, t) = \frac{1}{2}\varepsilon_0 \frac{\partial\omega\varepsilon'(\mathbf{r}, \omega)}{\partial\omega}\bigg|_{\omega=\omega_0} \overline{\mathbf{E}^2(\mathbf{r}, t)} + \frac{1}{2}\mu_0\overline{\mathbf{H}^2(\mathbf{r}, t)}. \tag{8}$$

Here, $\varepsilon(\mathbf{r}, \omega)$ is the spatially-resolved relative permittivity which follows a Drude dispersion in the metal layers and is equal to ε_a in the active (gain) waveguide core layer (the prime denotes the real part of the complex quantity). It is important to account for the dispersive character of the permittivity to correctly describe the electric energy density U_E, in particular inside metals. In time-domain, this dispersive character is expressed by the dynamic polarization response $\mathbf{P}_f(\mathbf{r}, t)$ of the free electron plasma in the metal and its in-phase $\overline{\dot{\mathbf{P}}_f \cdot \mathbf{E}}$ contribution to the electric energy (see Box 1 in Hess *et al.*[14]).

In eqn (6), the cavity loss rate γ_c includes both the outcoupling of energy from the laser and the dissipation of energy inside the laser. The former is connected to the closed contour integral over the Poynting flux, while the second is given by the work that the fields perform on the free electron plasma of the metal. The rate $R_{stim}(N)$ arises from the stimulated emission of photons in the gain medium with an average carrier number N.

When describing semiconductor (microcavity) lasers, and more recently plasmonic nanolasers[2,25,26] using rate analysis, one finds that a confinement factor Γ must be introduced in the rate equations due to the imperfect spatial overlap of the mode profile with the gain section. This confinement factor expresses the fact that the mode volume V_{eff}, which connects the photon density s to the photon number $S = V_{eff}s$, is distinct from the active (gain) volume V_a, and it is defined as $\Gamma \equiv V_a/V_{eff}$.[27]

The confinement factor Γ enters the rate of stimulated emission

$$R_{stim}(N) = \Gamma v_g \zeta g(N), \tag{9}$$

as does a group velocity v_g (at this point the specific type of group velocity is not fixed), a factorization parameter ζ and the bulk gain coefficient $g(N) = g(\Delta N) = \sigma_a\Delta N$. The specific definition of the confinement factor and its physical interpretation then determines which group velocity must be used. We follow Chang and Chuang[27] and choose

$$\Gamma = \frac{\int_{V_a} d^3r \frac{1}{2}\varepsilon_0 \left[\partial(\omega\varepsilon_a')/\partial\omega + \varepsilon_a'\right]|\mathbf{E}|^2}{\int_V d^3rU} \approx \frac{2\int_{V_a} d^3rU_E}{\int_V d^3rU}, \tag{10}$$

which identifies v_g as the material group velocity of the gain material $v_g = v_{g,a} = c/n_g$ with $n_g = \partial(\omega n_a')/\partial\omega$. The approximation on the rhs of eqn (10) is valid for a weakly dispersive gain material with $v_{g,a} \approx v_{ph,a} = c/n_a'$. The factorization parameter ζ measures the degree of inhomogeneity of the inversion profile (spatial hole-burning) and is defined as

$$\zeta = \frac{\int_{V_a} d^3r\overline{\Delta N(\mathbf{r},t)}\cdot\overline{\mathbf{E}^2(\mathbf{r},t)}}{\Delta N(t)\cdot\int_{V_a} d^3r\overline{\mathbf{E}^2(\mathbf{r},t)}}. \tag{11}$$

with the average inversion density $\Delta N(t) = V_a^{-1}\int_{V_a} d^3r\overline{\Delta N(\mathbf{r},t)}$. Eqn (11) expresses a functional dependence of $\zeta(t)$ on the inversion and field intensity profiles and hence an implicit transient dynamics that stabilises when the laser reaches steady state. In lasers that exhibit negligible spatial hole-burning effects, $\zeta(t)$ is close to unity and can be approximated by a constant factor $\zeta \approx 1$. A time-constant ζ smaller than 1 can also be assumed when the spatial distribution of the inversion and the mode profiles vary only a little with time. In these cases, a linear relationship between the stimulated emission rate and the inversion density follows, $R_{stim} \propto \Delta N$, a functional dependence that is commonly adopted in rate equation analyses.[2]

For further comparison we calculate the stimulated emission rate R_{stim} from the dynamic change of the total energy, which we are able to extract in FDTD simulations using a rate retrieval method based on Poynting's theorem.[14,28] From eqn (9) and the knowledge of the confinement factor and average inversion, it is then possible to calculate the factorization parameter ζ. We also compare the confinement factor Γ of the rate equation analysis to the energy confinement

$$\Gamma_E = \int_{V_a} d^3rU \bigg/ \int_V d^3rU, \tag{12}$$

which is defined as the electromagnetic energy in the gain section divided by the total energy of the lasing mode. The two confinement factors will differ in strongly guiding or plasmonic systems because of the modal character of the fields, *i.e.* an unequal distribution of the electromagnetic energy into electric and magnetic components.

As a side note, recent publications on plasmonic nanolasers have suggested the use of a confinement factor defined to incorporate the average energy (or waveguide group) velocity v_E of the underlying waveguide system of the nanolaser.[25,26] This is possible because the stimulated emission rate R_{stim} and the bulk gain coefficient $g(N)$ in eqn (9) are invariant to the definition of the effective mode volume, while the confinement factor relates to the group velocity v_g used. The distinction into confinement factor and group velocity can therefore be made in terms of waveguide properties, describing the amplification of a wave packet when it travels along the active waveguide with energy velocity v_E. Comparing Γ' in $R_{stim}(\Delta N) = v_E\Gamma'\zeta g(N)$ (ref. 25 and 26) to Γ in eqn (10), we find the effective confinement factor $\Gamma' = \Gamma v_{g,a}/v_E$. The pre-factor $v_{g,a}/v_E$ points towards effective confinement factors Γ' that can become larger than unity in waveguiding systems with low energy velocity. This also implies a divergence of Γ' as $v_E \rightarrow 0$, *i.e.*, in the stopped-light regime. Clearly then, Γ' is not suitable for the description of the stationary lasing mode in the SL laser, in particular considering that the stimulated emission rate into the mode, R_{stim}, does not diverge.

5 Dynamics of stopped-light surface-plasmon polariton lasing

Our analysis of the lasing dynamics shows that the lasing mode localises in-plane to the subwavelength gain region even though there is no cavity that would encourage this behaviour. Moreover, possible gain-guiding effects would, in typical semiconductor gain materials, even promote an anti-guiding rather than a guiding. To explain the mechanism that governs the formation of the subwavelength modes we now analyse a structure that supports stopped-light lasing/spasing at higher wave numbers k, *i.e.* consider the coherent amplification of stopped surface-plasmon polaritons (a new form of spasing) using a plasmonic-mode design of the metal–dielectric stack structure that allows for a stopped-"light" singularity in the density of state and surface-plasmon polariton amplification.

In contrast to conventional single-mode lasing where the lasing frequency and mode profile are determined by the cavity, the SL lasing mode is characterised by a closed-loop energy vortex and forms dynamically by phase-locking the continuum of k-modes in the vicinity of the SL points. During the transient phase

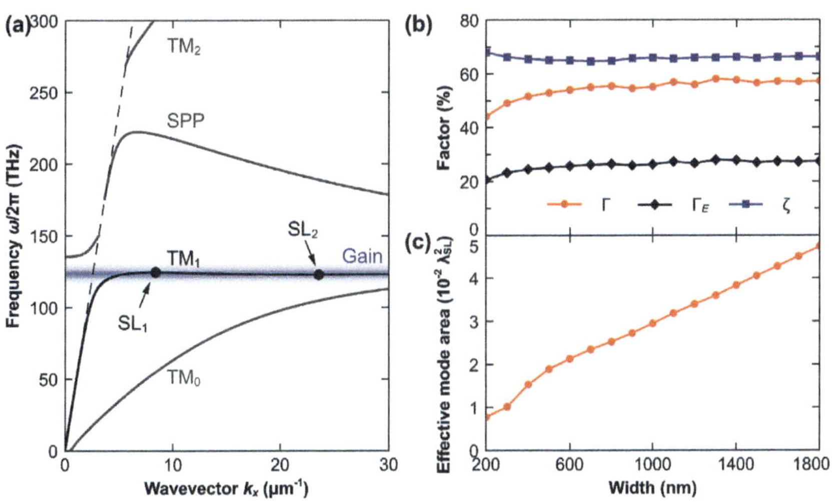

Fig. 7 (a) Complex-frequency dispersion of the plasmonic heterostructure with $h = 110$ nm and $t = 50$ nm (see Fig. 8a). Two SL points in the TM_1 mode with large wavevectors align with the gain spectrum. (b) Lasing properties of the mode for different gain widths. Shown are the confinement factor, the energy confinement and the factorization parameter ζ. (c) The effective mode area $A_{eff} = A_g/\Gamma$.

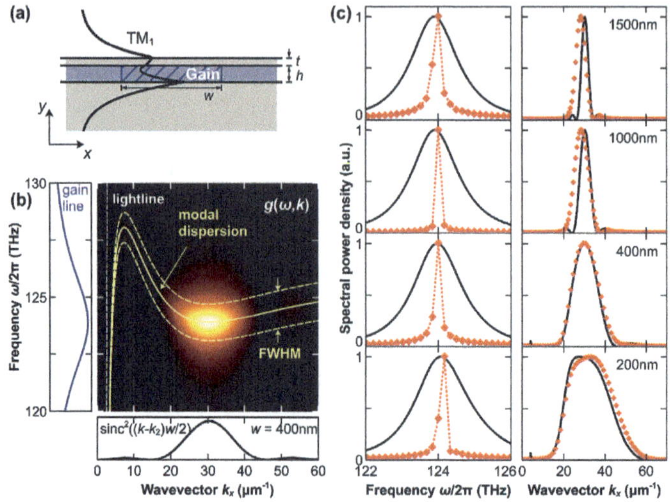

Fig. 8 (a) E_x field profile of the TM_1 waveguide mode in the plasmonic heterostructure waveguide. (b) Details of the TM_1 mode's complex frequency dispersion (yellow line) exhibiting two SL points with zero slope. The dispersion is overlaid over the semi-analytical results for the coupling strength $g(\omega, k)$. The gain spectrum in blue and the spatial sinc^2-function centred at the second SL point are given to the left and below the main graph. (c) Predicted lasing frequency on the left and wavevector spectrum on the right. Simulation results in red are compared to the semi-analytical result of $G(\omega) = \int dkg(\omega, k)$ in black.

Table 2 Gain medium parameters for emission into the second stopped light point of the bound TMs_1 waveguide mode

Symbol	Description	Value
$\omega_e/2\pi$ (λ_e)	Emission frequency (wavelength)	123.89 THz (2419.7 nm)
γ_e	Emission spectral half-width	1/(63.085 fs) $\approx$ 15.85 ps^{-1} (~49 nm)
$\sigma_{0,e}$	Emission cross-section	2.12 $\times$ 10^{-15} cm^2
N	Carrier density	2 $\times$ 10^{18} cm^{-3}
g_e	Maximum gain coefficient	4240 cm^{-1}

the lasing/spasing mode acquires a spectral composition that maximises the effective gain

$$G(\omega) \propto \int dk\mathscr{L}(\omega - \omega_0)\cdot D(\omega, k)\cdot \mathrm{sinc}^2((k-k')w/2) = \int dkg(\omega, k), \qquad (13)$$

where k' is centred on one of the SL singularities, by balancing the spatial confinement of the lasing mode to the gain section against its spectral overlap with the gain line $\mathscr{L}(\omega - \omega_0)$. The density of states

$$D(\omega, k) \propto \frac{\gamma(k)}{(\omega(k) - \omega)^2 + \gamma^2(k)} \qquad (14)$$

of the (TM_2) waveguide mode plays a critical role in this process as it links and controls the spatial and spectral overlap. The close agreement of the analytically predicted frequencies ω_{SLL} and k-spectra, $g(\omega_{SLL}, k)$, with those retrieved from full

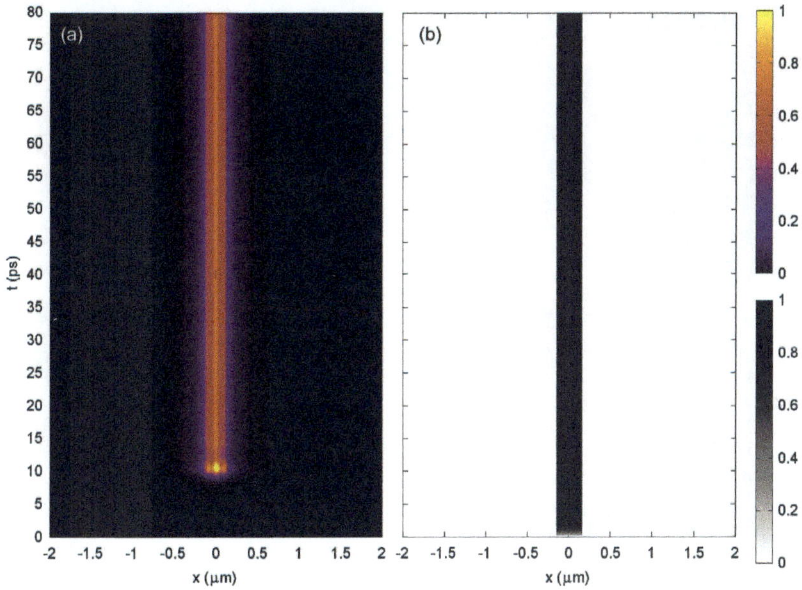

Fig. 9 Spatio-temporal dynamics of the stopped-light surface-plasmon polariton laser for a gain section width of 300 nm. Shown are the cycle-averaged electromagnetic energy density (a) and the inversion in the gain material (b). The cut along the x-axis is taken 30 nm above the metal substrate inside the waveguide core.

spatio-temporal simulations (Fig. 8c) confirms this principle: the frequency $\omega_{\rm SLL}$ of the SL lasing mode is selected by the maximization of $G(\omega)$, and its mode shape is determined by $g(\omega_{\rm SLL}, k)$. While the results further suggest that there is no fundamental limit for the spatial compression of the mode, a higher localization would of course require an increasingly flat dispersion to maintain a high spectral overlap.

With bound plasmonic modes one can exploit a wide range of wavevectors and achieve considerably smaller effective mode volumes than for photonic modes. We optimised the waveguide dispersion (Fig. 7 and 8) to exhibit two stopped light points in the plasmonic TM_1 mode at $\omega_1/2\pi = 128.14$ THz ($\lambda_1 \approx 2339.6$ nm) at $k_1 = 7.41\ \mu m^{-1}$ and $\omega_2/2\pi = 123.89$ THz ($\lambda_2 \approx 2419.7$ nm) at $k_2 = 30.34\ \mu m^{-1}$. There is no out-coupling to propagating free-space waves and the modal loss is approximately 3 times larger than the total loss of the TM_2 waveguide mode due to the higher field localization. Fig. 8b shows the predicted coupling strength $g(\omega, k)$ for a plasmonic stopped-light lasing structure. We choose to tune the emission line of the gain medium to the frequency of the second SL point ω_2 (Table 2) and consequently shift the peak of the sinc^2-function to k_2. The broad spectrum of supported wavevectors allows for localization of the SL lasing mode down to gain sections with a width of $w = 200$ nm (Fig. 7). For these small gain sections the maximum of the field profile aligns over the gain, effectively locking the amplitude and phase of the complementary modes with positive and negative phase velocities in space.

Fig. 9 demonstrates that the locking of the complementary modes in space takes place from the onset of the relaxation oscillations and remains constant in

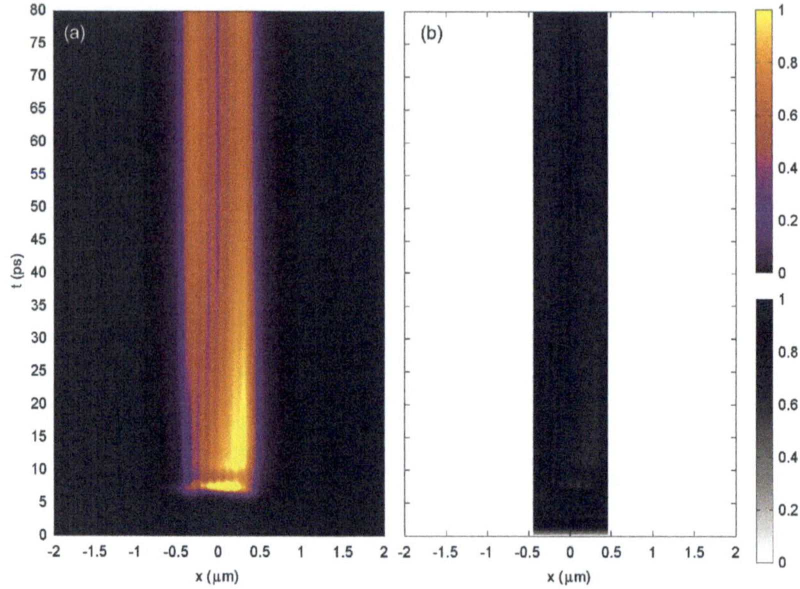

Fig. 10 Spatio-temporal dynamics of the stopped-light surface-plasmon polariton laser for a gain section width of 900 nm. Shown are the cycle-averaged electromagnetic energy density (a) and the inversion in the gain material (b). The cut along the x-axis is taken 30 nm above the metal substrate inside the waveguide core.

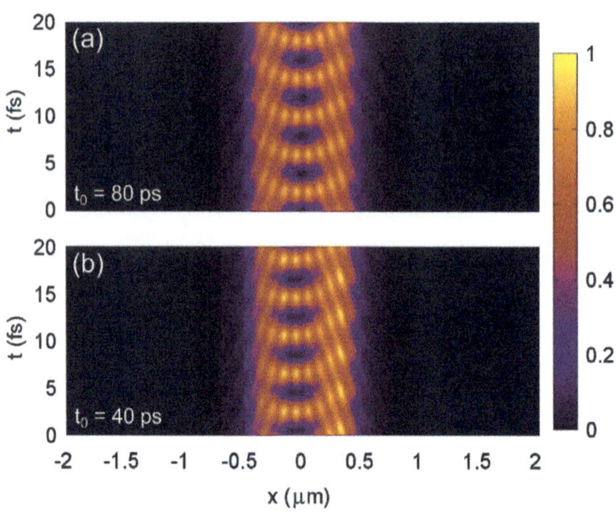

Fig. 11 Ultrafast spatio-temporal dynamics (non cycle-averaged) of the (surface-plasmon polariton) energy density on the femtosecond timescale. The window of 20 femtoseconds reveals the underlying ultrafast dynamics as shown in Fig. 10 after 80 ps (a) and 40 ps (b). The standing wave pattern in the centre is easily distinguishable from the propagating waves at the sides. The colour has been rescaled to the maximum energy density value and is not the same as in Fig. 10.

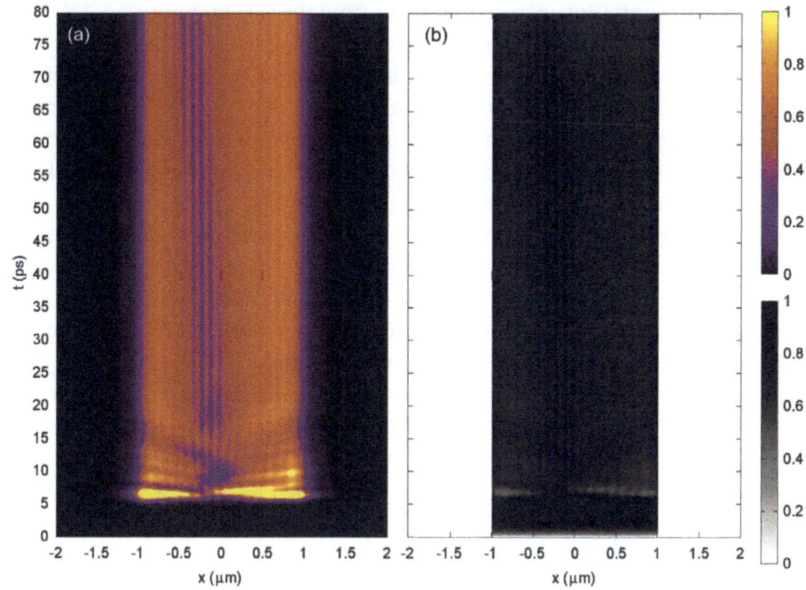

Fig. 12 Spatio-temporal dynamics of the stopped-light surface-plasmon polariton laser for a gain section width of 2000 nm. Shown are the cycle-averaged electromagnetic energy density (a) and the inversion in the gain material (b). The cut along the x-axis is taken 30 nm above the metal substrate inside the waveguide core.

time for small gain section widths. The figure shows the spatio-temporal dynamics of the cycle-averaged energy density (left) and gain material inversion (right) inside the waveguide core 30 nm above the metal substrate for a width of the $w = 300$ nm. After the initial build-up of inversion, a single relaxation oscillation leads into a steady-state regime with a spatially constant energy density profile. The profile is symmetric with respect to the centre of the gain section and is mirrored by the inversion where regions of smaller inversion are linked to corresponding regions of high field intensity due to depletion. As the spectrum of wavevectors is centred at a non-zero wavevector k_2 (see Fig. 8) the lasing mode becomes spatially modulated along x with nodes every $\pi/k_2 \approx 100$ nm.

For wider gain sections the positions of the nodal planes can move in time depending on the contributions to the lasing mode from the counterpropagating waveguide modes. Fig. 10 shows an exemplary time series for the case of a gain section of 900 nm width. Relaxation oscillations start earlier than for the smaller gain section and their spatial profile is no longer symmetric with respect to the centre of the section. Initially, the left-propagating waveguide mode has a higher amplitude leading to a high energy density with little spatial variation on the right and a standing wave pattern with lower amplitude on the left of the gain section. Again, this is mirrored in the (gain material) inversion shown in Fig. 10b.

The superposition of propagating and standing wave SPPs reveals itself in the spatio-temporal dynamics on the femtosecond timescale as shown in Fig. 11; the upper frame (Fig. 11a) illustrates the spatio-temporal dynamics of Fig. 10 in a time-interval of 20 femtoseconds starting at $t_0 = 80$ ps and the lower frame shows the corresponding frame starting at $t_0 = 40$ ps. In both cases, the phase fronts are seen to move inwards. But as time progresses from the earlier frame (Fig. 11b), the right-propagating waveguide mode becomes stronger, resulting in a more pronounced standing wave pattern that slowly moves towards the centre of the gain section. This behaviour is a direct consequence of the cavity-free property of SL lasing, where forward and backward modes become uncoupled due to the absence of back-reflections at geometric boundaries and form independently. At the end of the 80 picosecond time interval shown in Fig. 10, Fig. 11a shows that the profiles have become symmetric, just as was observed for small gain sections.

The richest ultrafast spatio-temporal dynamics, however, emerges in wide gain sections as epitomised in Fig. 12 for $w = 2000$ nm. Relaxation oscillations start independently at the edges of the gain section and several, clearly visible oscillations take place. During the relaxation oscillations a standing wave SPP component is formed close to the centre of the gain section which seems to perform an oscillatory motion in space. After approximately 20 ps, a quasi-steady state profile is reached, that now evolves on a much slower timescale than during the dynamics during the initial relaxation oscillations. In contrast to the cases of Fig. 9 and 10, the system now does not evolve into a more symmetric state within the timescales shown here. Instead, the standing wave pattern is seen to drift away from the centre towards the edge of the gain section. Judging the complexity of the spatio-temporal behaviour it is not clear at all if the dynamics will eventually lead to a symmetric steady state for very long times, as once an asymmetry is present in a nonlinear system it tends to prevail as a spatio-temporally varying structure.

6 Conclusions and outlook

Nanoplasmonic stopped-light lasing is a recently established principle[18] that may not only open the door to ultrafast cavity-free nanolasing, ultra-thin lasing surfaces and cavity-free quantum-electrodynamics but may also provide an entry point to quantum gain in quantum plasmonics[29] and quantum fluids of light.[30] Here, following these principles, we have studied its spatio-temporal dynamics, and, particularly, the ultrafast femto- and picosecond dynamics of surface plasmon polariton cavity-free nanolasing enabled by dispersion-less stopped-"light".

Engineered singularities in the density of optical states realised in a metal–dielectric–metal nano-waveguide structure are confirmed to lead to a stopped-light feedback mechanism that is the basis for the dynamic observed cavity-free photonic and surface-plasmon polariton nanolasing.

Having established a theoretical framework for characteristic laser parameters (such as the confinement factor) and deriving effective laser rate equations for cavity-free stopped-light nanolasing allowed us to explain the link with and expose the differences to traditional laser rate equations describing conventional lasers.

Extensive simulations on the basis of a full-time domain Maxwell–Bloch Langevin approach in combination with (semi-) analytic theory have uncovered that in the absence of cavity-induced feedback a phase-locked superposition of quasi dispersion-free waveguide mode promotes the dynamic formation of a subwavelength lasing mode. Our simulations have further demonstrated that this mechanism proves to be remarkably robust against interface roughness and the observed spontaneous emission factor $\beta \approx 0.9$ is a manifestation of a very high coupling of spontaneous emission into the stopped-light mode.

The ultrafast femto- and picosecond dynamics of stopped-light surface plasmon polariton lasing is shown to be stable for gain sections of widths of up to around 200 nm. In wider gain structures of the order of 1 μm, the initial dynamics is characterised by spatio-temporally oscillating amplified surface-plasmon polaritons on ultrafast timescales <5 femtoseconds and with spatial periods on the nanoscale <100 nm. Remarkably, this formation of a complex spatio-temporal SPP pattern has a surprising outward resemblance with the characteristic spatio-temporal light-field dynamics found in broad-area semiconductor lasers about 20 years ago.[31–33] But now the dynamics is much faster, happens on much smaller scales and is coherent.

Acknowledgements

This research was supported by the Leverhulme Trust and the Engineering and Physical Sciences Research Council (EPSRC) UK.

References

1 M. T. Hill, Y.-S. Oei, B. Smalbrugge, Y. Zhu, T. de Vries, P. J. van Veldhoven, F. W. M. van Otten, T. J. Eijkemans, J. P. Turkiewicz, H. de Waardt, E. J. Geluk, S.-H. Kwon, Y.-H. Lee, R. Ntzel and M. K. Smit, *Nat. Photonics*, 2007, **1**, 589–594.
2 R. F. Oulton, V. J. Sorger, T. Zentgraf, R.-M. Ma, C. Gladden, L. Dai, G. Bartal and X. Zhang, *Nature*, 2009, **461**, 629–632.
3 M. A. Noginov, G. Zhu, A. M. Belgrave, R. Bakker, V. M. Shalaev, E. E. Narimanov, S. Stout, E. Herz, T. Suteewong and U. Wiesner, *Nature*, 2009, **460**, 1110–1112.
4 R.-M. Ma, R. F. Oulton, V. J. Sorger, G. Bartal and X. Zhang, *Nat. Mater.*, 2011, **10**, 110–113.
5 M. Khajavikhan, A. Simic, M. Katz, J. H. Lee, B. Slutsky, A. Mizrahi, V. Lomakin and Y. Fainman, *Nature*, 2012, **482**, 204–207.
6 Y.-J. Lu, J. Kim, H.-Y. Chen, C. Wu, N. Dabidian, C. E. Sanders, C.-Y. Wang, M.-Y. Lu, B.-H. Li, X. Qiu, W.-H. Chang, L.-J. Chen, G. Shvets, C.-K. Shih and S. Gwo, *Science*, 2012, **337**, 450–453.
7 K. J. Vahala, *Nature*, 2003, **424**, 839–846.
8 S. L. McCall, A. F. J. Levi, R. E. Slusher, S. J. Pearton and R. A. Logan, *Appl. Phys. Lett.*, 1992, **60**, 289–291.
9 K. Iga, F. Koyama and S. Kinoshita, *IEEE J. Quantum Electron.*, 1988, **24**, 1845–1855.
10 O. Painter, R. K. Lee, A. Scherer, A. Yariv, J. D. O'Brien, P. D. Dapkus and I. Kim, *Science*, 1999, **284**, 1819–1821.
11 H. Altug, D. Englund and J. Vuckovic, *Nat. Phys.*, 2006, **2**, 484–488.
12 D. J. Bergman and M. I. Stockman, *Phys. Rev. Lett.*, 2003, **90**, 027402.
13 P. Berini and I. De Leon, *Nat. Photonics*, 2011, **6**, 16–24.
14 O. Hess, J. B. Pendry, S. A. Maier, R. F. Oulton, J. M. Hamm and K. L. Tsakmakidis, *Nat. Mater.*, 2012, **11**, 573–584.
15 J. B. Khurgin and G. Sun, *Nat. Photonics*, 2014, **18**, 10–14.

16 H. Cao, *Waves Random Media*, 2003, **13**, R1–R39.
17 D. S. Wiersma, *Nat. Phys.*, 2008, **4**, 359–367.
18 T. Pickering, J. M. Hamm, A. F. Page, S. Wuestner and O. Hess, *Nat. Commun.*, 2014, **5**, 4972.
19 A. Pusch, S. Wuestner, J. M. Hamm, K. L. Tsakmakidis and O. Hess, *ACS Nano*, 2012, **6**, 2420–2431.
20 T. Baba, *Nat. Photonics*, 2008, **2**, 465–473.
21 S. Mookherjea, J. S. Park, S.-H. Yang and P. R. Bandaru, *Nat. Photonics*, 2008, **2**, 90–93.
22 K. L. Tsakmakidis, T. W. Pickering, J. M. Hamm, A. F. Page and O. Hess, *Phys. Rev. Lett.*, 2014, **112**, 167401.
23 R.-M. Ma, R. F. Oulton, V. J. Sorger and X. Zhang, *Laser Photonics Rev.*, 2012, **7**, 1–21.
24 G. Björk and Y. Yamamoto, *IEEE J. Quantum Electron.*, 1991, **27**, 2386–2396.
25 C. Z. Ning, *Phys. Status Solidi B*, 2010, **247**, 774–788.
26 K. Ding and C. Z. Ning, *Light: Sci. Appl.*, 2012, **1**, e20.
27 S.-W. Chang and S. L. Chuang, *IEEE J. Quantum Electron.*, 2009, **45**, 1014–1023.
28 J. M. Hamm, S. Wuestner, K. L. Tsakmakidis and O. Hess, *Phys. Rev. Lett.*, 2011, **107**, 167405.
29 M. S. Tame, K. R. McEnery, K. Özdemir, J. Lee, S. A. Maier and M. S. Kim, *Nat. Phys.*, 2013, **9**, 329–340.
30 I. Carusotto and C. Ciuti, *Rev. Mod. Phys.*, 2013, **85**, 299–366.
31 O. Hess and T. Kuhn, *Phys. Rev. A: At., Mol., Opt. Phys.*, 1996, **54**, 3347–3359.
32 O. Hess and T. Kuhn, *Phys. Rev. A: At., Mol., Opt. Phys.*, 1996, **54**, 3360–3368.
33 E. Gehrig and O. Hess, *Spatio-Temporal Dynamics and Quantum Fluctuations in Semiconductor Lasers*, Springer, 2003.

DISCUSSIONS

Quantum plasmonics, gain and spasers: general discussion

Jeremy Baumberg, Mikhail Noginov, Javier Aizpurua, Kaiqiang Lin, Thomas Ebbesen, Alexei A Kornyshev, Riccardo Sapienza, Niek van Hulst, Santhosh Kotni, F. Javier García de Abajo, Pavel Ginzburg, Ortwin Hess, Mark Brongersma and Sergey Bozhevolnyi

DOI: 10.1039/c5fd90024g

Jeremy Baumberg opened discussion of the paper by Thomas Ebbesen: You were trying to understand the difference between the inverse of the linewidths and the cavity radiative lifetime. In the semiconductor community this comparison is well understood.

The inverse cavity linewidth is related to the coherent emission rate, which is that found for the re-radiating emitter when directly excited at the resonance energy. As photons excite the system, the excited wavefunction is well matched to directly re-radiate, and essentially the emission rate is the light-matter coupling, or the oscillator strength, and gives the inverse linewidth. In polaritons, this is a combination of the lifetime of photons in the cavity (or plasmon resonator) and the phase scattering time of the excited molecular state (or excitons in the semiconductor).

In semiconductors, this radiative re-emission is only possible from low wavevectors that can couple to light. On the other hand, the radiative recombination time assumes an incoherent population of excitons that are thermalised, and most of these are not well coupled to light at all, so do not radiate. The low wavevector excitons are emitted at the coherent rate, but then it takes time for other excitons to scatter down to take their place. In GaAs, coherent radiative rates are 20 ps, while incoherent recombination rates are 1 ns.

In the molecular systems here, the analogy is of mixed states, well coupled to the light field (fast coherent emission, giving the inverse of cavity linewidth), and slow recombination, given by continual scattering from excited states down into the lowest radiatively coupled state. This is the answer to your first question.

For the second question, my first query is whether the fraction of P− light that you see is dependent on the position of the detection wavelength. (It would be nice if you could include some data showing that this makes no difference in your paper.) The reason this is important is that the low energy tail states of P− are not necessarily well coupled to light.

Thomas Ebbesen responded: Thank you for your explanation of my first question! It is very clear and convincing. Regarding your query about whether the position of the detection wavelength affects the contribution we see from P− in the excitation spectra, it does not.

Jeremy Baumberg added: In inorganic semiconductor microcavity systems, other researchers have found rather different emissions when exciting at one phonon energy above the P− state, compared to exciting closer to P−. This is related to LO-phonon-assisted relaxation, and you could cite some of the papers from the Skolnick group, among others.[1–3]

1 S. Pau, G. Björk, H. Cao, F. Tassone, R. Huang, Y. Yamamoto and R. P. Stanley, *Phys. Rev. B*, 1997, **55**, R1942.
2 D. G. Lidzey, D. D. C. Bradley, M. S. Skolnick, T. Virgili, S. Walker and D. M. Whittaker, *Nature*, 1998, **395**, 53.
3 M. Maragkou, A. J. D. Grundy, T. Ostatnický and P. G. Lagoudakis, *Appl. Phys. Lett.*, 2010, **97**, 111110.

Thomas Ebbesen answered: Yes, indeed. The molecular systems are more complex, because there are 100 to 200 normal vibrational modes for each molecule, which create a continuum that favors intramolecular vibrational relaxation on sub-100 fs time scales.

Mikhail Noginov asked: Do the molecules coupled to the cavity oscillate coherently when they absorb or emit? Do you know all characteristic rates associated with relaxation from P+ to P− and then to the ground state?

Thomas Ebbesen replied: It is known from previous experiments, such as those by Bellessa's group in Lyon,[1] that the P− emission is coherent over micrometer distances in the coupled system. We also know the characteristic rates associated with the relaxation from P+ to P− (<100 fs) and then from P− to the ground state (picosecond time scale) for our systems.

1 S. Aberra Guebrou, C. Symonds, E. Homeyer, J. C. Plenet, Yu. N. Gartstein, V. M. Agranovich and J. Bellessa, *Phys. Rev. Lett.*, 2012, **108**, 066401.

Pavel Ginzburg commented: Thank you for presenting this very interesting contribution. One of the open questions that you formulated during your presentation was to fit a theoretical description to the experimental observations. The expression for vacuum Rabi splitting, being proportional to the square root of the number of molecules in the cavity (after eqn 1 in your paper, *Faraday Discuss.*, 2015, DOI: 10.1039/C4FD00197D), suggests the presence of strong collective effects, governing the quantum dynamics. Nevertheless, the theoretical description (given during the conference presentation) concentrated on the single particle formulation. My suggestion is to consider a description with coherent many-particles states (similar to Dicke's super/sub-radiant states, *e.g.* ref. 1), or any similar approach (*e.g.* ref. 2).

1 R. H. Dicke, *Phys. Rev.*, 1954, **93**, 99–110.
2 R. Bonifacio and L. A. Lugiato, *Phys. Rev. A*, 1975, **11**, 1507–1521.

Thomas Ebbesen responded: Thank you for this comment and the references, which I will look up. Indeed, the single particle formulation does not represent the conditions of our experiments. It was just the simplest way to illustrate the important parameters for strong coupling in the short time available. The Tavis–Cummings model is a more appropriate formulation.

Ortwin Hess asked: Do you observe any oscillatory dynamics in the ultra-strong coupled state?

Thomas Ebbesen answered: We have not tried to look for oscillations, as this would require extremely short fs laser pulses, shorter than the cavity lifetime.

Mark Brongersma enquired: In your very interesting strong coupling experiments, where a small cavity mode is coupled to the quantum emitters, have you analyzed the impact of the position of the emitters in the cavity? The excitation and emission processes may be dependent on the exact position and the strong coupling could be optimized. If strong coupling is observed in a cavity, does that necessarily imply that the chemistry of the molecules in the cavity is also changed? Do optics and chemistry probe the same properties of a molecule inside a cavity?

Thomas Ebbesen responded: Yes, we have published a paper in which we show the effect of the position of a thin dye layer inside the cavity.[1] The properties of the ensemble are modified by the position, as expected. Regarding whether the chemistry is changed or not under strong coupling, this is something we are exploring. We did find that the rate and yield of the photochemical reaction of a photochrome, where one of the two isomers is coupled to the cavity, was modified.[2] The question of whether the optics and the chemistry probe the same properties is a very good point and will probably only be answered after we have enough examples.

1 S. Wang, T. Chervy, J. George, J. A. Hutchison, C. Genet and T. W. Ebbesen, *J. Phys. Chem. Lett.*, 2014, **5**, 1433.
2 J. A. Hutchison, T. Schwartz, C. Genet, E. Devaux and T. W. Ebbesen, *Angew. Chemie Int. Ed.*, 2012, **51**, 1592.

Jeremy Baumberg remarked: I think it would be nice to include a diagram showing what you call action spectra, but are known in the semiconductor community as photoluminescence excitation (PLE) spectra, in your paper. It would be good to show the excitation and detection energies for each case of P+ or P− excitation, so that the clear issue shown is the rate of relaxation from excitation energy to the lowest state in the P− polariton manifold.

Thomas Ebbesen responded: Yes, the excitation and detection wavelengths are indicated in the figures in the paper (*Faraday Discuss.*, 2015, DOI: 10.1039/C4FD00197D).

Jeremy Baumberg said: A strong candidate for the answer to your second question then becomes clear. The question is why excitation in the higher P− manifold does not give much P− emission, while excitation in the P+ manifold gives strong P− emission. The answer is that, when higher P− states in the manifold are excited, they are very slow to relax to P− lower states, since they cannot emit higher energy phonons (the energy spacing is too small) and they have rather different "plane wave-like" wavefunctions, which do not overlap well with the tightly confined local phonon emission sites on each molecule. This has a direct analogue in inorganic microcavities when exciting low on the lower energy polariton branch, which can then scatter to even lower energies very slowly through acoustic phonons (of the order 100 ps). If non-radiative recombination is comparable, then the total emission at P− will be low. On the other hand, exciting at P+ gives many ways to relax through other states, such as singlets and much higher energy phonons which couple better to the P+ wavefunctions.

The interesting thing about this is that the slower scattering from P− higher energy states is a direct consequence of mixing many different parts of molecular wavefunctions with photons (or plasmons) into the combined P− polariton state. This polariton is large and delocalised compared to the phonons that are on each vibrational bond. The coupling is hence weak, and scattering is slow.

Thomas Ebbesen replied: Your explanation is very clear. However, unfortunately, it does not correspond to what we see experimentally: even the transient absorption spectra experiments give rise to little or no P− upon direct excitation (Fig. 4b, *Faraday Discuss.*, 2015, DOI: 10.1039/C4FD00197D). In other words, no signal (or a very weak one) is observed after the 150 fs pump and nothing appears over a longer ps time scale. It is as if we are not able to populate P− directly, independent of the subsequent scattering and phonon relaxation processes.

Kaiqiang Lin commented: I wonder whether light will still be necessary for the interaction between the cavity and the molecule?

Thomas Ebbesen replied: I did not go into such detail, but, even in the absence of light, strong coupling can occur between the cavity or surface plasmons and the molecules, due to vacuum fluctuations. In other words, the zero point fluctuations of the cavity mode and the molecular transition can lead to strong coupling.

Kaiqiang Lin queried: When we put one molecule into a cavity, do you think the real energy state of the molecule will be changed?

Thomas Ebbesen responded: It is very difficult to see strong coupling with one molecule in a cavity, simply because the Rabi splitting depends on the square root of the number of molecules coupled to the optical mode. In other words, the Rabi splitting for one molecule would be so tiny that you would need a high Q factor cavity to see it. High Q factor cavities are hard to obtain with metals.

Santhosh Kotni remarked: The reports on strong coupling between localized surface plasmons (LSP) and quantum emitters or J-aggregates show Rabi splitting at the absorption maximum of the emitter, signifying the coupling of LSP with the excitation mode of the emitter. Could you explain the possible conditions under which the emission mode couples to the surface plasmon and exhibits Rabi splitting at the emitter's emission maximum?

Thomas Ebbesen replied: The coupling of a surface plasmon mode with the emission mode of a quantum emitter is highly unlikely, because under normal conditions only a tiny fraction of the excited state is populated. Perhaps it could be observed under intense laser pumping, but that would be a difficult experiment.

Pavel Ginzburg opened discussion of the paper by Sergey Bozhevolnyi: Thank you for presenting this very interesting contribution. My question is regarding the electromagnetic field quantization for opened and lossy cavities and resonators. The theoretical formulation of the light-matter interaction dynamics in the plasmonic cavity is based on the Hamiltonian, which appears as eqn 1–3 in your paper (*Faraday Discuss.*, 2015, DOI: 10.1039/C4FD00165F). This formulation suggests the usage of creation/annihilation operators, which are attributed to the plasmonic cavity modes. In other words, the "mode decomposition method" was employed as a starting point for the subsequent derivations. This method has somewhat limited applicability, however, if losses and chromatic dispersion are present. Those effects imply that the modes' operators vanish with time commutation relations, for example (for more details, see *e.g.* ref. 1 and references therein). My question concerns the general limitations of the method that was used for the prediction of this new effect, which is itself, no doubt, very important and exciting.

1 P. Ginzburg and A. Zayats, *Opt. Express*, 2012, **20**, 6720–6727.

Sergey Bozhevolnyi replied: It is correctly pointed out in the question that the method we employed has general limitations, which are also related to the description of plasmonic modes. Specifically, we neglect the presence of all other plasmonic modes, apart from the fundamental dipolar mode, which has resonance at the exact frequency of the radiative transition of a quantum emitter. This mode is represented in our formalism by a coherent state, which usually appears when dealing with classical currents, as we argue is the case in our configuration with a relatively large nanoparticle. It is, however, difficult to assess the influence of this approximation on the final result of our analysis.

Niek van Hulst enquired: You mention that the molecule close to the nanoparticle is affected by self-stimulated emission. The emitted photon experiences a π-pulse on the particle and interferes destructively at the molecule. An interesting concept, yet what is the origin of the stimulated emitted photon – specifically, where does the first photon come from that makes this transition happen?

Sergey Bozhevolnyi responded: Strictly speaking, our mechanism sets in once the dipole moment of the considered radiative transition becomes non-zero, *i.e.* once the ground state becomes "slightly" populated. The field created by this dipole moment interacts with a particle, returning to the molecule as a π-pulse that brings the molecule very quickly to the ground state. There is, therefore, only one photon during the whole process, the emission of which is self-stimulated.

Niek van Hulst remarked: It is fascinating that a single quantum emitter close to a resonant metallic nanoparticle would first show a coherent build-up, displayed as a fs rise, and next the enhanced incoherent radiative decay. It seems this feature should be easy to detect experimentally, on a single molecule–nanoparticle system, simply using a pump–probe excitation with a few femtosecond time resolution. Would that be realistic?

Sergey Bozhevolnyi answered: The given description of the emission process should actually be rephrased, as the entire emission process is coherent, *i.e.* both during build-up and during decay (that is asymptotically approaching the exponential). In principle, it should be easy to detect this process experimentally given a high temporal resolution. It should, however, be kept in mind that the condition of strict resonance is very important, requiring careful selection of both the molecule and the nanoparticle.

F. Javier García de Abajo commented: I believe that the Hamiltonian that you are using with the rotating wave approximation (your eqn (1), (2) and (4) in *Faraday Discuss.*, 2015, DOI: 10.1039/C4FD00165F) is essentially the familiar Jaynes–Cummings (JC) model Hamiltonian,[1] which is well-known to admit closed-form solutions. You can find a tutorial description of these solutions in Appendix C, subsection 1 of our paper (*Faraday Discuss.*, 2015, DOI: 10.1039/C4FD00216D). The JC model can then be straightforwardly applied to find the state of the emitter, if the plasmon is in a Fock state. This situation is of practical interest to create a one-plasmon state; for example, starting with $n = 0$, so that the decay of the quantum dipole emitter (QDE) results in an $n = 1$ plasmon Fock state. If dissipation and radiative losses are neglected, one should observe perfect Rabi oscillations between the QDE and the plasmon, as we are then in the strong-coupling regime at the single quantum level. With dissipation and strong plasmon-radiation coupling, this type of system has been used to experimentally corroborate the preservation of quantum statistics in the decay of an emitter coupled to a plasmon.[2] You are then assuming in your eqn (5) (*Faraday Discuss.*, 2015, DOI: 10.1039/C4FD00165F) that the plasmon remains in a coherent state while the emitter undergoes a transition from the excited to the ground state, and I wonder how realistic this assumption is. The coherent plasmon state $|\alpha\rangle$ (rightmost part of eqn (5)) is an eigenstate of the plasmon annihilation operator (*i.e.* $a|\alpha\rangle = \alpha|\alpha\rangle$), but not of a^+. When the emitter decays, this transition is accompanied by a change in the plasmon state to $a^+|\alpha\rangle$, which I believe is not a coherent state. Can this be reconciled with the approximation of your eqn (5)?

1 E. T. Jaynes and F. W. Cummings, *Proc. IEEE*, 1963, **51**, 89.
2 A. V. Akimov, A. Mukherjee, C. L. Yu, D. E. Chang, A. S. Zibrov, P. R. Hemmer, H. Park and M. D. Lukin, *Nature*, 2007, **450**, 402.

Sergey Bozhevolnyi answered: Your consideration is based on the analogy of two resonantly coupled quantum emitters, where Fock states should indeed be used. The situation of resonant coupling of a quantum emitter and a localized plasmon is conceptually different. A localized surface plasmon is formed by classical oscillating currents, with many electrons (100 electrons per cubic nanometer) participating in the oscillations. The wave function of a multi-electron system with a continuous energy spectrum can always be described by a coherent state.[1] During the considered relaxation process, the plasmon coherent state does not change as such, only $\alpha(t)$ changes, describing the plasmon decay. No new state is thus created in this process. Finally, the experiment you mentioned is simply not relevant to our configuration, as it concerns propagating (not localized) surface plasmons.

1 C. C. Gerry and P. L. Knight, *Introductory Quantum Optics*, Cambridge University Press, Cambridge, 2005.

Alexei Kornyshev said: Apart from the fact that the plasmon resonance in a spherical particle takes place at a different frequency then at a flat interface, would there be any principal difference between your results and the case in which the emitter is close to a planar metal surface?

Sergey Bozhevolnyi replied: No, there would be no principal difference between these two cases. The corresponding discussion can actually be found in our paper, where the characteristic time constant for the emitter–surface interaction is also worked out (see eqn (22) in *Faraday Discuss.*, 2015, DOI: 10.1039/C4FD00165F).

Javier Aizpurua asked: Your Hamiltonian and your plasmonic state do not look very orthodox compared to other treatments in quantum nano-optics. How does this coherent state that describes the plasmon work in your formalism?

Sergey Bozhevolnyi answered: In our case, there are many electrons, which form the oscillating (in a metal nanoparticle) current induced by the field of a quantum emitter. This classical current is, in our formalism, represented by a coherent state (as detailed in the introduction of *Faraday Discuss.*, 2015, DOI: 10.1039/C4FD00165F).

Javier Aizpurua queried: Imagine that you have a decay of a single emitter, exciting just one single plasmon. How would your formalism describe such a situation? Even if you report a strong coupling regime, wouldn't you just be describing a Purcell factor affected by this coherent superposition, which defines the plasmon?

Sergey Bozhevolnyi replied: In our formalism, a coherent state is made up of wave functions corresponding to different numbers of plasmons, from zero to infinity (see eqn (5) in our paper). However, the average number is related to the particular situation, for example, it can represent only one plasmon. We do not really understand this question, because the Purcell factor can only be introduced in the weak-coupling approximation, whereas our formalism goes beyond the weak-coupling approximation.

Thomas Ebbesen asked: How would you experimentally distinguish the effect you are reporting from the Purcell effect? Would the lifetime be shorter or longer?

Sergey Bozhevolnyi replied: Actually, it would not be correct to compare these two regimes as they exclude each other (our process is a coherent emission, while the emission influenced by the Purcell factor in the weak-coupling regime is incoherent spontaneous emission). The difference that can be observed experimentally is a well-pronounced delay in the emission in our case, as opposed to exponentially decaying emission in the weak-coupling regime.

Jeremy Baumberg commented: It seems this paper is related to the phenomenon of superradiance, as discussed in the 1960s and 70s. Aligning inverted atomic emitters in a waveguide, once a photon starts at one end, it triggers stimulated emission all the way through and you get the same exact solution.

Would you say this is the same thing, with one emitter seeing itself through the plasmon field?

Sergey Bozhevolnyi answered: Typically, superradiance is understood as a phenomenon that occurs when a group of emitters (excited atoms or molecules) interact with a common light field. The latter could indeed be a waveguide mode, but I would have to study the literature more thoroughly before concluding that this would result in the same exact solution. It is an interesting idea to consider our situation from this point of view and definitely worth further careful consideration. One difference should be mentioned here, however: when an emitter sees itself through the plasmon field, its image has a definite phase shift of $\pi/2$ due to the plasmon resonance (which is a necessary condition for deriving our result).

Javier Aizpurua commented: It is quite puzzling that, in your results, the probability of emission is non-dependent on the losses of the plasmon, or on other details of the plasmonic cavity. How can this be possible? Do you understand why?

Sergey Bozhevolnyi answered: We believe that this very non-trivial result is the consequence of strict plasmon resonance, which, in turn, results in the particle polarizability being inversely proportional to losses in the metal (imaginary part of metal susceptibility). The larger polarizability is, therefore, the faster emission and the stronger absorption rate will be. These tendencies balance each other, producing the result of approximately 75% efficiency of emission.

Riccardo Sapienza opened discussion of the paper by Ortwin Hess: Tailoring the gain medium with low group velocity is a very powerful idea, but often limited by an associated increase in absorption and scattering, by unwanted impurities, for example. Could you comment on how this interplay of gain and loss would affect a realistic device? Is enhanced absorption included in the calculations?

Ortwin Hess responded: In our study of stopped-light lasing and condensation of surface-plasmons, we performed extensive numerical simulations on the basis of the Maxwell–Bloch theory, which also included a deliberate introduction of interface roughness and disorder. Of course, effective losses are increased with increased disorder, but we were able to confirm that the stopped-light effect itself is not compromised and lasing is still possible up to disorder levels beyond those of characteristic plasmonic systems. Indeed, plasmonic stopped-light singularities are much less affected by interface losses than those which could be realised in photonic crystals, which rely on Bragg-modes and are thus much more susceptible to disorder.

Mikhail Noginov asked: What is the characteristic delocalization length of the emitted photon? (How far can an electron go from the atom that emitted it?) How many of the other atoms are in this delocalization volume?

Ortwin Hess replied: The plasmic environment and the layer/stripe of quantum emitters, as well as the light field, are all considered on the basis of a patio-temporally dependent Maxwell–Bloch theory (typical resolution in space in this case ~1 nm, in time ~50 attoseconds). At every time step, there is a local interaction with the material (quantum emitters and plasmonic environment), but transported through space though Maxwell's equations. While electronic quantum coherence in space is thus not included on this level, non-locality is indeed dynamically considered, even if and while the light amplitude will be stationary at the stopped-light point. Indeed, any possible delocalisation of light will happen dynamically through the phase and the optical near-field as shown in the Poynting vector vortex with characteristic time scales corresponding to the optical frequency. Additionally, there is some light leakage due to losses, *etc.*

Thomas Ebbesen commented: You talked about Bose–Einstein condensation, but how would you see condensation occurring and appearing if the dispersion curve is nearly totally flat (for the stopped-light)?

Ortwin Hess answered: To trap non-equilibrium (pumped) surface-plasmon polaritons, the dispersion curve of the nanoplasmonic heterostructure is designed not to be totally flat, but to contain a suitable (second) stopped-light singularity. The resulting upward curvature of the dispersion creates a trap state, which is designed to be deep enough that it still successfully acts as a trap in the presence of thermal fluctuations.

F. Javier García de Abajo commented: The system considered in Fig. 8 and 9 (*Faraday Discuss.*, 2015, DOI: 10.1039/C4FD00181H) involves parallel wave vectors that are large compared with the light wave vector in the surrounding media, and thus these modes are to a large extent electrostatic, essentially involving collective electron motion mediated by Coulomb interactions. This is similar to plasmons in graphene, where retardation effects are not important. Perhaps "surface plasmons" is then an appropriate label to denote these modes, instead of surface-plasmon polaritons.

Ortwin Hess responded: Due to the stopped-light singularity in the surface-plasmon polariton dispersion at finite (high) wave vectors, the nature of the collective excitations is, indeed, less and less photon-like and more assumes the character of plasmons. At the same time, however, the trapping mechanism involves and corresponds to a highly non-equilibrium state, with continuous optical pumping providing compensation for the losses.

Pavel Ginzburg commented: Thank you for presenting this very interesting contribution, proposing employment of stopped-light for direct generation of coherent plasmonic fields. My question is on probable mode competition effects that could take place in effectively a two-dimensional structure, like the metal–insulator–metal waveguide under consideration. Modes with longitudinal field components along both x and z directions could be amplified and pass the threshold, if the symmetry of the structure is not broken (either purposefully or spontaneously, *e.g.* due to imperfections). Furthermore, in-plane confinement of the mode could lead to violation of the stopped-light conditions.

Ortwin Hess responded: In the case of stopped-light lasing (left of the light-line, low k-values), we have explicitly investigated (on the basis of finite-difference time-domain (FDTD) simulations which naturally take on board the possibility of any longitudinal or transverse "mode") the possible influence of other modes (such as TE) with respect to a potential amplification of spontaneously emitted light (which does go into all modes) at the stopped-light band, leading to stopped-light lasing (with photonic character). It turned out that, remarkably, most of the spontaneous emission was directed into the lasing mode, leading to an almost thresholdless behaviour with a β-factor of above 0.9. Additionally, the study of interfacial disorder also showed that, up to a characteristic level of disorder, there was insignificantly little coupling between co-propagating components of the wave. In the case of stopped-light trapping/condensation of surface-plasmon polaritons at higher k-values, we did not observe any other detrimental influences from other modes either, so far. Clearly, there are still many opportunities for further studies, which are currently ongoing. If additional in-plane confinement is considered, it would, of course, be necessary to adjust the particular mode that is targeted for a realisation of stopped-light singularities.

Jeremy Baumberg asked: Considering the coherent emission at finite k, how would symmetry breaking take place? It is unlikely to be in a ring, so it may choose a particular k-direction, or several coupled ones at once, or possible create some sort of vortex state. How would you be able to extend your calculations to find out what states emerge?

Ortwin Hess responded: In our dynamic simulations of stopped-light trapped, possibly condensed, surface-plasmon polaritons, we do indeed see the localisation of surface-plasmon polaritons with finite k. This preferred value of k is also reflected in the formation of coherent surface-plasmon polariton structures of a characteristic spatial extension, as seen in the space–time plots (*Faraday Discuss.*, 2015, DOI: 10.1039/C4FD00181H). While currently assuming an isotropically responding distribution of quantum emitters, we would expect, however, an induced anisotropy due to the symmetry breaking induced by the propagating (non-stopping) pump light, with a preference for the chosen propagation direction. If we were to extend our simulations with respect to an internal (*e.g.* molecular) anisotropy, we would anticipate additional contributions, which could either enhance or moderate this tendency, depending on their relative orientation.

Faraday Discussions

PAPER

Nonlinear quantum optics in the (ultra) strong light–matter coupling

Eduardo Sánchez-Burillo,[a] Juanjo García-Ripoll,[b] Luis Martín-Moreno[a] and David Zueco[ac]

Received 16th October 2014, Accepted 20th October 2014
DOI: 10.1039/c4fd00206g

The propagation of N photons in one dimensional waveguides coupled to M qubits is discussed, both in the strong and ultrastrong qubit–waveguide coupling. Special emphasis is placed on the characterisation of the nonlinear response and its linear limit for the scattered photons as a function of N, M, qubit inter distance and light–matter coupling. The quantum evolution is numerically solved *via* the matrix product states technique. The time evolutions for both the field and qubits are computed. The nonlinear character (as a function of N/M) depends on the computed observable. While perfect reflection is obtained for $N/M \cong 1$, photon–photon correlations are still resolved for ratios N/M = non-zero. Inter-qubit distance enhances the nonlinear response. Moving to the ultrastrong coupling regime, we observe that inelastic processes are robust against the number of qubits and that the qubit–qubit interaction mediated by the photons is qualitatively modified. The theory developed in this work models experiments in circuit QED, photonic crystals and dielectric waveguides.

1 Introduction

Typically, materials respond linearly to an electromagnetic (EM) field. Intense fields are usually demanded for accessing nonlinear responses.[1] Therefore, a long standing challenge in science and technology is to develop devices containing giant nonlinear properties that operate at small powers. The final goal is to shrink the required power to the few photon limit.[2,3] In doing so, the dipoles must interact more strongly with the driving photons than with the environment, which defines the strong light–matter coupling regime. Thus, quantum optical systems presenting strong light–matter interactions are excellent candidates for building nonlinear optical materials operating at tiny powers.

An ideal platform for having strong light–dipole coupling together with the possibility of generating and measuring few photon currents is waveguide QED.

[a]*Instituto de Ciencia de Materiales de Aragón y Departamento de Física de la Materia Condensada, CSIC-Universidad de Zaragoza, Zaragoza, E-50009, Spain*
[b]*Instituto de Física Fundamental, IFF-CSIC, Serrano 113-bis, Madrid E-28006, Spain*
[c]*Fundación ARAID, Paseo María Agustín 36, Zaragoza 50004, Spain*

There, the paradigmatic dipoles are two level systems (qubits) and the input and output fields travel through one dimensional waveguides. As there are only two propagation directions (left and right), interference effects are much larger than in 3D. Besides, the coupling to the qubits is enhanced by the reduced dimensionality (the Purcell effect). Different platforms can serve for the study: circuit-QED,[4–7] quantum dots interacting with photonic crystals,[8] dielectric waveguides,[9] molecules interacting with photons[10] or plasmonic devices.[11–15] With this kind of systems different nonlinear effects may be observed and used, such as photon–photon correlations,[16–30] non-classical light generation,[31] lasing[32] or effective interaction between noninteracting dipoles.[33–37]

Light–matter coupling, even when it is larger than the losses, is typically much smaller than the characteristic energy scales of qubits (dipoles) and photons. In this case, up to second order in this coupling, only processes where light and matter exchange excitations play a role. This is the Rotating Wave Approximation (RWA).[38] Quite recently, experiments have reached couplings large enough for this approximation to break down.[39–42] Then, the full dipole interaction must be taken into account and, in order to understand the experiments, it is indispensable to consider processes involving the spontaneous creation and annihilation of pairs of excitations. This is the ultrastrong coupling regime. Beyond the RWA picture novel physics has been predicted,[43–47] also from the scattering point of view.[48,49] Clearly, this regime has a great potential for nonlinear applications.

It is desirable to quantify the amount of nonlinearity for a given architecture with a given input driving, like in classical nonlinear optics, where materials are classified *via* their linear and nonlinear susceptibilities. Furthermore, some systems can behave as linear when looking at one quantity and nonlinear when measuring another. In general, the response is expected to be linear in the low polarisation limit, $N/M \ll 1$, with N the number of photons and M the number of qubits. In this work, we quantify more precisely this linear–nonlinear transition. In doing so, we must acknowledge that the description of qubits and currents containing few photons needs a quantum treatment. To compute this quantum evolution, we choose the Matrix Product State (MPS) technique adapted for photonic situations.[46,48] Within the MPS, the exact dynamics is computed for multiphoton input states passing through several qubits both in the strong and ultrastrong coupling regimes. We explore the dynamics by changing the ratio N/M and the light–matter coupling strength. Our aim here is to discuss the tradeoff between the enhancement of the coupling with the number of qubits (which, accordingly to the theory of Dicke states[50] scales as $g\sqrt{M}$) at the expense of decreasing the nonlinear response. Besides this, we also study the influence of the distance between the dipoles. Finally, we will explore the changes on this RWA phenomenology when moving to the ultrastrong coupling regime. As witnesses for the nonlinearities we will compute transmission and reflection probabilities, qubit populations and photon–photon correlations.

The rest of this discussions is organised as follows. Section 2 summarises the theoretical basis needed for understanding the results. Section 3 discusses the analytic properties for the scattering matrix in linear systems (*i.e.* when the scatterers are harmonic resonators), linking the notion of linear quantum system with linear optical response. The numerical results both in the RWA and beyond the RWA are presented in Sections 4 and 5 respectively. Our conclusions and two technical Appendices close the paper.

2 Theoretical background

In this section we summarise the quantum theory for scattering in waveguide QED. We also outline the MPS technique used in our numerical simulations.

2.1 Physical setup

This work deals with the dynamics of photons moving in a one dimensional waveguide interacting with a discrete number of dipoles. A pictorial representation is given in Fig. 1a. The waveguide is discretised and the quantum dipoles (scatterers) are point-like systems interacting with the photons accordingly to the dipole approximation, $H_{\text{int}} \propto \mathbf{E}\cdot\mathbf{p}$. The Hamiltonian is,

$$H_{\text{tot}} = \varepsilon\sum_x a_x^\dagger a_x - J\sum_x \left(a_x^\dagger a_{x+1} + hc\right) + \sum_{i=1}^{M}\left(h_i\left(c_i, c_i^\dagger\right) + g_i\left(c_i + c_i^\dagger\right)\left(a_{x_i} + a_{x_i}^\dagger\right)\right). \tag{1}$$

Here, the waveguide is modeled within the first two terms. It is an array of $N_{\text{cav}} = 2L + 1$ coupled cavities, running x from $-L$ to L, with bare frequencies ε and hopping between nearest neighbours J. The photon operators satisfy bosonic relations $[a_x, a_{x'}^\dagger] = \delta_{xx'}$. The free propagation is characterised by a cosine-shaped

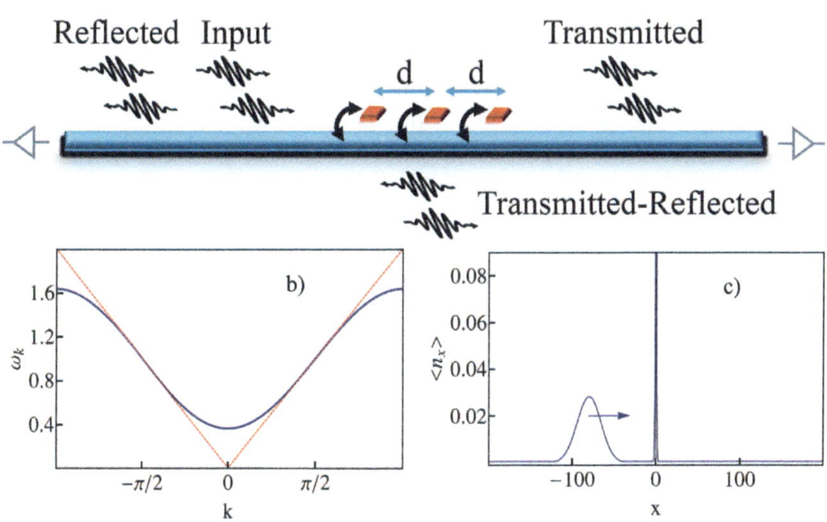

Fig. 1 (a) Schematic representation of the system: waveguide with input and output fields and the qubits (red squares) separated by a distance d. (b) Dispersion relation (blue curve) for $\varepsilon = 1$ and $J = 1/\pi$ and 400 cavities (these parameters will be used throughout the text). The dashed red line is the linearisation of ω_k around $k = \pi/2$. (c) Computed number of photons as a function of position in the chain (14) for an incident wavepacket ($N = 1$) impinging on one qubit ($\Delta = 1$) taking the full interaction Hamiltonian (1) with $g = 0.7$. There is a photonic cloud around the qubit position ($x = 0$), corresponding to the ground state. This ground state was computed with the MPS method, as explained in Section 2.3.

dispersion relation $\omega_k = \varepsilon - 2J\cos k$, as shown in Fig. 1(b). Under some circumstances and working in the middle of the band $k \cong \pi/2$, the linearised dispersion $\omega_k = \nu k$ with $\nu = 2J$ is well justified. Such a linearisation is not performed in our numerical investigations, but it can be important to have it in mind when comparing some of our findings with analytical treatments (mostly done with linear dispersion relations). The dipoles are characterised *via* the ladder operators $c_i^\dagger$ (c_i). In the numerical work presented in this paper, we will consider qubits,

$$c_i^\dagger = \sigma_i^+, \text{ with, } [\sigma_i^+, \sigma_j^-] = \delta_{ij}(2\sigma_i^+\sigma_i^- - 1) \tag{2}$$

and

$$h_i(c_i, c_i^\dagger) = \Delta_i c_i^\dagger c_i \tag{3}$$

with Δ_i the frequency for each qubit. Other nonlinear dipoles could be investigated without extra difficulty *e.g.* three level atoms[51,52] or Kerr-type punctual materials.[30] We can also consider linear dipoles, *i.e.*, harmonic oscillators satisfying bosonic commutation relations $[c_i, c_j^\dagger] = \delta_{ij}$, as we will do in Section 3.4 in order to explore the linear limit.

The last term in (1) results from the quantisation of the dipole interaction term $(H_{\text{int}})_i \propto \mathbf{E}(x_i)\cdot\mathbf{p}_i$, because $\mathbf{E}(x_i) \sim a_{xi} + \text{hc}$ (ref. 53) and $p_i \sim c_i^\dagger + \text{hc}$.[38] The coupling strength is given by the constants g_i, whose actual value will depend on the concrete physical realisation. If $g_i \ll \Delta_i$ we can approximate the interaction as,

$$H_{\text{int}} = g\sum_{i=1}^{M}\left(c_i + c_i^\dagger\right)\left(a_{x_i} + a_{x_i}^\dagger\right) \simeq g\sum_{i=1}^{M}\left(c_i^\dagger a_{x_i} + c_i a_{x_i}^\dagger\right) \equiv H_{\text{int}}^{\text{RWA}}. \tag{4}$$

i.e. we have neglected the counter-rotating terms $c_i^\dagger a_{x_i}^\dagger + \text{hc}$. This is the so-called Rotating Wave Approximation (RWA), which is valid up to $\mathcal{O}(g_i^2)$.[38] It is widely used since $g_i \ll \Delta_i$ typically holds in the experiments. Within the RWA, the number of excitations is conserved, $[H, \mathcal{N}] = 0$, with the number operator $\mathcal{N} \equiv \sum_x a_x^\dagger a_x + \sum_i c_i^\dagger c_i$. This symmetry highly reduces the complexity of solving the dynamics of (1).

If g_i is not a small parameter, the RWA is not justified and the full dipole interaction [the last term in (1)] must be taken into account. The regime where the RWA is not sufficient for describing the phenomenology is known as the ultrastrong coupling regime.[39–42] Here, the number of excitations is not conserved, making the problem much harder to solve.

A good example to observe the extra complexity is the computation of the ground state for (1), which is an essential step in the scattering process. Within the RWA, the ground state is trivial, $|\text{GS}\rangle_{\text{RWA}} = |0\rangle$, with $a_x|0\rangle = 0$ for all x and $c_j|0\rangle = 0$ for all j. Beyond the RWA, in the ultrastrong coupling regime, the $|\text{GS}\rangle$ is a correlated state with $\langle\text{GS}|\mathcal{N}|\text{GS}\rangle \neq 0$. In Fig. 1(c) the photon population in the waveguide per site, $\langle n_x\rangle = \langle\Psi|a_x^\dagger a_x|\Psi\rangle$ is plotted, being $|\Psi\rangle$ the ground state with a flying photon (eqn (5) for $N = 1$). There, only one qubit is coupled to the waveguide beyond the RWA ($g = 0.7[\Delta]$). Around the qubit (placed at $x = 0$) a non trivial structure appears. The peaked wavepacket around $x = -90$ is the flying photon.

2.2 Scattering

We are interested in computing the scattering characteristics for N-photon input states interacting with M qubits. The input state, our initial condition, is chosen to be the non-normalised quantum state.

$$|\Psi_{\text{in}}\rangle = \left(a_\phi^\dagger\right)^N |\text{GS}\rangle, \quad a_\phi^\dagger = \sum_x \phi_x^{\text{in}} a_x^\dagger, \tag{5}$$

where $|\text{GS}\rangle$ is the ground state of the system [see Fig. 1(c)] and ϕ_x^{in} is a Gaussian wavepacket centred in x_{in} with spatial width θ

$$\phi_x^{\text{in}} = \exp\left(-\frac{(x - x_{\text{in}})^2}{2\theta^2} + ik_{\text{in}} x \right). \tag{6}$$

Typically, we consider x_{in} located on the left hand side of the qubits with the wavepacket moving to the right toward the scatterers, as sketched in Fig. 1(a) and computed in Fig. 1(c).

In several scenarios, it will be convenient to work in momentum space,

$$a_k^\dagger = \frac{1}{\sqrt{L}} \sum_x \mathrm{e}^{ikx} a_x^\dagger. \tag{7}$$

In momentum space the wavepacket (6) is exponentially localised around k_{in}, with width $\sim \theta^{-1}$.

In our numerical simulations we evolve the initial state (5) unitarily,

$$|\Psi(t)\rangle = U(t,0)|\Psi_{\text{in}}\rangle = \mathrm{e}^{-iH_{\text{tot}}t}|\Psi_{\text{in}}\rangle. \tag{8}$$

We stop the simulation at a final time t_{out}, which must be sufficiently large for the photons to be moving freely along the waveguide, after having interacted with the scatterers. In doing so, our numerics account for stationary amplitudes encapsulated in the definition of the scattering matrix, S,

$$|\Psi_{\text{out}}\rangle = S|\Psi_{\text{in}}\rangle. \tag{9}$$

It is customary to specify the scattering matrices through their momentum components:

$$S_{p_1\ldots p_{N'},k_1\ldots k_N} = \langle GS|a_{p_1}\ldots a_{p_{N'}} S a_{k_1}^\dagger \ldots a_{k_N}^\dagger |\text{GS}\rangle. \tag{10}$$

Some comments are pertinent here. The $|\text{GS}\rangle$ appears in the definition of S. As discussed before, in the ultrastrong coupling regime the ground state differs from the vacuum state, and has a non-zero number of excitations, see Fig. 1(c). In the ultrastrong regime the number of excitations is not conserved. Therefore, in the above formula $N' \neq N$ in general. The components, $S_{p_1\ldots pN',k_1\ldots kN}$ can be numerically computed as projections of the evolved state as we will explain below.

2.3 Matrix Product States for scattering problems

As previously stated, beyond RWA even the ground state is non trivial. The problem becomes a many-body one and we must consider the full Hilbert space for any computation. If we truncate the number of particles per site to $n_{\max}$ and our scatterers are qubits, the dimension of the Hilbert space is $2^M(n_{\max}+1)^{N_{\rm cav}}$, which is exponential with both $N_{\rm cav}$ and M. Numerical brute force is impossible and analytical tools are really limited. Even within the RWA, if we work with N photons and M qubits the Hilbert space dimension, $(N_{\rm cav}+M)^N$, is also too large for multiphoton states.

In order to solve the problem, we use one celebrated method to deal with many-body 1D problems: Matrix Product States (MPS).[54–56] Let us summarise the idea behind this technique. A general state of a many-body system is usually written as

$$|\Psi\rangle = \sum_{i_1,\ldots,i_{N_{\rm cav}}} c_{i_1,\ldots,i_{N_{\rm cav}}} |i_1,\ldots,i_{N_{\rm cav}}\rangle, \tag{11}$$

where $\{|i_n\rangle\}_{i_n=1}^{d_n}$ is a basis of the local Hilbert space of the n-th body (or site, in our case), with d_n being the dimension of this local Hilbert space, $|i_i,\ldots,i_{N_{\rm cav}}\rangle = |i_1\rangle\ldots|i_{N_{\rm cav}}\rangle$, that is, the tensor product basis, and $c_{i_1,\ldots,i_{N_{\rm cav}}}\in\mathbb{C}$. However, it is possible to show that an equivalent parametrisation can be written as:

$$|\Psi\rangle = \sum_{i_1,\ldots,i_{N_{\rm cav}}} A_1{}^{i_1}\ldots A_{N_{\rm cav}}{}^{i_{N_{\rm cav}}} |i_1,\ldots,i_{N_{\rm cav}}\rangle, \tag{12}$$

where $A_n{}^{i_n}$ is a $D_n \times D_{n+1}$ matrix linked to the n-th site.† For the sake of simplicity, let us take $d_n = d$ and $D_n = D$ for all n. Then, $N_{\rm cav}D^2d$ coefficients are needed to describe any state. In principle, to represent $|\Psi\rangle$ exactly, D must be exponential in $N_{\rm cav}$. However, states in the low energy sector of well-behaved many-body Hamiltonians can be very accurately described by taking D polynomial in $N_{\rm cav}$.[57] Few photon dynamics belongs to this low energy sector and, as we will show, the number of MPS coefficients required to study scattering scales polynomially with $N_{\rm cav}$. This technique has been recently applied to propagation in bosonic chains interacting with impurities.[46,48]

Our simulations are as follows: (i) initialisation of the state as $|0\rangle$, (ii) computation of the ground state by means of imaginary time evolution, using the Suzuki–Trotter decomposition[58] adapted to the MPS representation,[59] (iii) generation of the input state (eqn (5)), (iv) time evolution of the wave function up to $t = t_{\rm out}$ (eqn (9)) again by means of Suzuki–Trotter decomposition and (v) computation of relevant quantities.

2.4 Observables and its computation with MPS

The time dependence of expected values,

$$\langle \mathcal{O}^t\rangle = \langle\Psi(t)|\mathcal{O}|\Psi(t)\rangle \tag{13}$$

where $\mathcal{O}$ is any Hermitian operator acting on waveguide, qubit variables or both, completely characterises the dynamics. Relevant values are the number of photons (in position and momentum spaces):

$$\langle n_x{}^t\rangle = \langle\Psi(t)|a_x^\dagger a_x|\Psi(t)\rangle \leftrightarrow \langle n_k{}^t\rangle = \frac{1}{L}\sum_{x_1x_2} e^{ik(x_1-x_2)}\left\langle\Psi(t)|a_{x_1}^\dagger a_{x_2}|\Psi(t)\right\rangle, \tag{14}$$

† We are taking open boundary conditions, so $D_{N_{\rm cav+1}} = D_1 = 1$.

or the qubit populations and correlations:

$$P_i^t = \langle \Psi(t)|\sigma_i^+\sigma_i^-|\Psi(t)\rangle,\ \langle\sigma_i^+\sigma_j^-\rangle = \langle \Psi(t)|\sigma_i^+\sigma_j^-|\Psi(t)\rangle. \tag{15}$$

Thus, for example, transmission and reflection coefficients can be obtained as

$$T_k = \langle n_k{}^{t_{\rm out}}\rangle / \langle n_k^0\rangle, \quad R_k = \langle n_{-k}{}^{t_{\rm out}}\rangle / \langle n_k^0\rangle. \tag{16}$$

The projectors

$$\phi^t_{x_1,\ldots,x_N} = \frac{1}{\sqrt{N!}}\langle {\rm GS}|a_{x_1}\ldots a_{x_N}|\Psi(t)\rangle, \tag{17}$$

are fundamental, since, setting $t = t_{\rm out}$ they are nothing but the Fourier transform of the scattering matrix $S_{p_1\ldots p_{N'},k_1\ldots k_N}$, eqn (10).

Let us start by considering an operator that can be expressed as product of local operators:

$$\mathscr{O} = o_1 o_2 \ldots o_{N_{\rm cav}} \tag{18}$$

e.g. $O = a^\dagger_{x_i}a_{x_j}$ or $O = \sigma_j^- a_{xj}$. For the states $|\Psi_A\rangle$ and $|\Psi_B\rangle$, characterised by the tensors $\{A_n{}^{i_n}\}_{n=1}^{N_{\rm cav}}$ and $\{B_n{}^{i_n}\}_{n=1}^{N_{\rm cav}}$ respectively [*cf.* eqn (12)] the matrix element $\langle\Psi_A|\mathscr{O}|\Psi_B\rangle$ is given by

$$\langle\Psi_A|\mathscr{O}|\Psi_B\rangle = \prod_{n=1}^{N_{\rm cav}} E_n(A_n, B_n, o_n), \tag{19}$$

with

$$E_n(A_n, B_n, o_n) = \sum_{i_n,j_n}\langle i_n|o_n|j_n\rangle\left(\left(A_n{}^{i_n}\right)^*\otimes B_n{}^{i_n}\right) \tag{20}$$

Since any operator can always be written as a sum of products of local operators, we can compute any expected value (13) and projector (17) without the explicit computation of the $c_{i_1,\ldots,i_{N_{\rm cav}}}$ coefficients in eqn (11).

Let us specify the concrete MPS formulas for the relevant observables. If we are interested in the photon number, $\langle n_x{}^t\rangle$ (14), we compute (19) with $|\Psi_A\rangle = |\Psi_B\rangle = |\Psi(t)\rangle$, $o_x = a_x^\dagger a_x$, and $o_n = \mathbb{I}_n$ for $n \neq x_i$, with $\mathbb{I}_n$ the identity operator. In the same way, for P_i^t defined in (15) we must replace $o_{xi} = \sigma_i^+\sigma_i^-$ and $o_n = \mathbb{I}_n$ for $n \neq x_i$. The momentum occupation number, $\langle n_k{}^t\rangle$ in (14) is computationally harder to obtain. We must compute every two-body operator $\langle a^\dagger_{x_1}a_{x_2}\rangle$ by taking $o_{x_1} = a^\dagger_{x_1}$, $o_{x2} = a_{x2}$ and $o_n = \mathbb{I}_n$ for $n \neq x_1, x_2$, with $|\Psi_A\rangle = |\Psi_B\rangle = |\Psi(t)\rangle$ and perform the sum in (14). We can also obtain two-body qubit correlators, $\sigma_i^+\sigma_j^-$, taking $o_{xi} = \sigma_i^+$, $o_{xj} = \sigma_j^-$ and $o_n = \mathbb{I}_n$ for $n \neq x_i, x_j$. Finally, for the projectors $\phi_{x_1,\ldots,x_N}{}^t$, eqn (17), one just takes $|\Psi_A\rangle = |{\rm GS}\rangle$, $|\Psi_B\rangle = |\Psi(t)\rangle$, $o_{xi} = a_{xi}$ with $i = 1,\ldots,N$ and $o_n = \mathbb{I}_n$ for $n \neq x_1,\ldots,x_N$.

3 Linear scattering

The main objective of this work is to realise how nonlinear the scattering of few photons through few qubits is. To accomplish this task we need first to know what linear scattering is. In this section we discuss in general what we understand for linear quantum optics and its realisation in waveguide QED. Finally, we establish under which conditions a collection of qubits behave as a linear optical medium.

3.1 Linear systems

In quantum physics, linear systems are those whose Heisenberg equations for the observables form a linear set. For the case of Hamiltonian (1) this happens whenever the scatterers are harmonic resonators (the waveguide itself is linear) both within RWA and non-RWA coupling regimes. In this case, the $c_j, c_j^\dagger$ operators in (1) are *bosonic* operators [*cf.* eqn (2)]:

$$[c_i, c_j^\dagger] = \delta_{ij} \tag{21}$$

and [*cf.* eqn (3)]:

$$h_i(c_i, c_i^\dagger) = \Delta_i c_i^\dagger c_i. \tag{22}$$

where Δ_i are the resonator frequencies.

3.2 Analytical properties for the *S*-matrix whit harmonic resonators as scatterers

Once we know what linear scattering means, we present our first result. An equivalent result was introduced by us in the ESI of ref. 48. We re-express it here in a more convenient way.

Theorem1 3.1. If the system is linear (in the sense of Section 3.1), the one-photon scattering matrix is given by:

$$\langle p|S|k\rangle = t_k \delta_{p,k} + r_k \delta_{p,-k}. \tag{23}$$

The apparent simplicity of theorem 3.1 requires some discussion. Firstly, photon creation is not possible. Second, eqn (23) fixes the actual form for the output states,

$$|\Psi_{\text{out}}\rangle = \sum_{k>0} \left(t_k \phi_k^{\text{in}} a_k^\dagger + r_k \phi_k^{\text{in}} a_{-k}^\dagger\right)|\text{GS}\rangle. \tag{24}$$

Therefore, the only scattering processes for one incoming photon occurring within linear models are the reflection and transmission of the photon without changing its input energy (momentum). Notice that this is a non trivial feature, since the Hamiltonian (1) is not number conserving: $\left[H, \sum_x a_x^\dagger a_x + \sum_j c_j^\dagger c_j\right] \neq 0$ and the ground state $|\text{GS}\rangle$ does not have a well defined number of excitations. However, the single photon scatters by reflecting and transmitting without changing the energy and without creating additional excitations in the system. This result mathematically relies on the Bogolioubov transformation and physically on the fact that (1) is a free model in quantum field theory language. The proof of this theorem is found in Appendix A.

The single photon result, Theorem 3.1 can be generalised to the multiphoton case:

Theorem1 3.2. If the system is linear the *N*-photon scattering matrix is given by:

$$\langle p_1, p_2, \ldots, p_N | S | k_1, k_2, \ldots, k_{N'} \rangle = \delta_{NN'} \sum_{n_1 \neq n_2 \neq \ldots \neq n_N} \langle p_1 | S | k_{n_1} \rangle \ldots \langle p_N | S | k_{n_N} \rangle. \tag{25}$$

The theorem says that, in linear systems the scattering matrix is a product of single photon scattering matrices. Consequently, no particle creation, Raman processes or photon–photon interactions are possible. The proof, detailed in Appendix B, is based on the single photon result, Theorem 3.1, together with the Wick's theorem. To better appreciate these two general results, let us apply them.

3.3 The classical limit: recovering the standard linear optics concept

Theorems 3.1 and 3.2 dictate the scattering in linear systems (*i.e.*, when the scatterers are harmonic resonators). So far, it is not completely clear whether the definition for linear systems in quantum mechanics given in Section 3.1 (together with the results in Section 3.2) corresponds to what linear optics means: the properties for the scattered currents are independent of the input intensity. Here we show that linear systems satisfy this intensity independence. Importantly, we comment on the classical limit for linear systems.

We consider a monochromatic coherent state as the N-photon input state,

$$|\Psi_{\text{in}}\rangle = |\alpha_k\rangle = \mathrm{e}^{-|\alpha_k|^2/2} \sum_{n=0}^{\infty} \frac{\alpha_k{}^n \left(a_k^\dagger\right)^n}{n!} |\text{GS}\rangle. \tag{26}$$

Applying the theorems 3.1 and 3.2 the output state can be written as,

$$|\Psi_{\text{out}}\rangle = \mathrm{e}^{-|\alpha_k|^2/2} \sum_{n=0}^{\infty} \frac{\alpha_k{}^n \left(t_k a_k^\dagger + r_k a_{-k}^\dagger\right)^n}{n!} |\text{GS}\rangle = |t_k \alpha_k\rangle \otimes |r_k \alpha_{-k}\rangle. \tag{27}$$

The second equality follows after some algebra.‡

Eqn (27) is a satisfactory result. The transmission and reflection coefficients t_k and r_k are independent of α_k. Recalling that $\langle \alpha_k | a_k^\dagger a_k | \alpha_k \rangle = |\alpha_k|^2$, this means independence from input intensity. We note that linear systems transform coherent states onto coherent states. Thus harmonic resonators neither change the statistics nor generate entanglement between the reflected and transmitted fields. Coherent states can be considered as classical inputs in the limit $\alpha_k \to \infty$, thus linear systems do not generate quantumness. Finally, the last expression for $|\Psi_{\text{out}}\rangle$ in (27) has a classical interpretation in terms of transmitted $|t_k\alpha_k|^2$ and reflected $|r_k\alpha_k|^2$ currents ($|t_k|^2 + |r_k|^2 = 1$).

3.4 From nonlinear to linear

Consider M qubits placed at the same point of the waveguide, $x_i = x_j$ for all i, j in Hamiltonian (1). For simplicity, assume that the couplings g_i are also the same. Introducing the operator,

‡ Note that: $\sum_{n=0}^{\infty} \frac{\alpha_k{}^n \left(a_k^\dagger(t_{\text{out}})\right)^n}{n!} = \sum_{n=0}^{\infty} \sum_{m=0}^{n} \frac{1}{m!(n-m)!} \left(t_k \alpha_k a_k^\dagger\right)^m \left(r_k \alpha_k a_{-k}^\dagger\right)^{n-m}$
$= \sum_{n=0}^{\infty} \frac{\left(t_k \alpha_k a_k^\dagger\right)^n}{n!} \sum_{n'=0}^{\infty} \frac{\left(r_k \alpha_k a_{-k}^\dagger\right)^{n'}}{n'!}.$

$$b \equiv \frac{1}{\sqrt{M}} \sum_{i=1}^{M} \sigma_i^-, \tag{28}$$

the total Hamiltonian, eqn (1), can be rewritten as

$$H_{\text{tot}} = \varepsilon \sum_x a_x^\dagger a_x - J \sum_x \left(a_x^\dagger a_{x+1} + \text{hc}\right) + \varDelta b^\dagger b + g\sqrt{M}\left(b^\dagger + b\right)\left(a_0^\dagger + a_0\right). \tag{29}$$

Thus, in terms of the Dicke states generated by (28) the effective coupling is $g\sqrt{M}$. Additionally, it is crucial to observe that[60]

$$\left[b, b^\dagger\right] = 1 - \frac{1}{M} \sum_j \sigma_j^+ \sigma_j^- \tag{30}$$

Therefore, in the limit $\left\langle \sum_j \sigma_j^+ \sigma_j^- \right\rangle / M \ll 1$ (a weak probe compared to the number of qubits) the operator b ($b^\dagger$) approximates an annihilation (creation) bosonic operator. Therefore, a large number of qubits is expected to behave as a harmonic resonator.

4 Nonlinear scattering: numerics in the RWA

We are interested in the nonlinear optical properties of a collection of qubits. The nonlinearities can be manifested in different observables. The theoretical results in Section 3.2 say nothing about the nonlinear regime. In the following, we will compute some natural quantities as the reflection and transmission probabilities or photon–photon correlation. We will evaluate how nonlinear the response is as a function of the number of incoming photons, number of qubits or inter qubit distance. Throughout this section the RWA is assumed. Physics beyond the RWA is discussed in the next section.

4.1 *N* photons *vs.* *M* qubits: total reflection spectrum

The combination of energy and number conservation implies that output states for one photon scattering through qubits in the RWA are also given by eqn (24) (like in linear systems). A well known result in this geometry is that a single monochromatic photon suffers perfect reflection, $|r_{k_{\text{in}}}|^2 = 1$, when its frequency $\omega_{\text{in}} = \varDelta$;[61–63] this effect has several useful applications.[13,64–67] For linear systems, following (25), the N-photon S matrix is a product of single photon S matrices. Thus, in linear systems, N-photon input states must also be perfectly reflected at resonance. On the other hand, a qubit cannot totally reflect more than one photon at the same time.[17] Then, for N-photon ($N > 1$) input states perfect reflection is not expected to occur with one qubit. If we want to overpass this saturation effect, we may increase the number of qubits. In the limit $N/M \ll 1$, with N photons and M the number of qubits the linear regime should be recovered, *i.e.* perfect reflection should happen [see Section 3.4].

Equipped with the MPS technique we can study the linear–nonlinear transition as a function of the ratio N/M. In doing so, we provide meaning to the inequality $N/M \ll 1$. In this subsection we consider that the M qubits are placed at

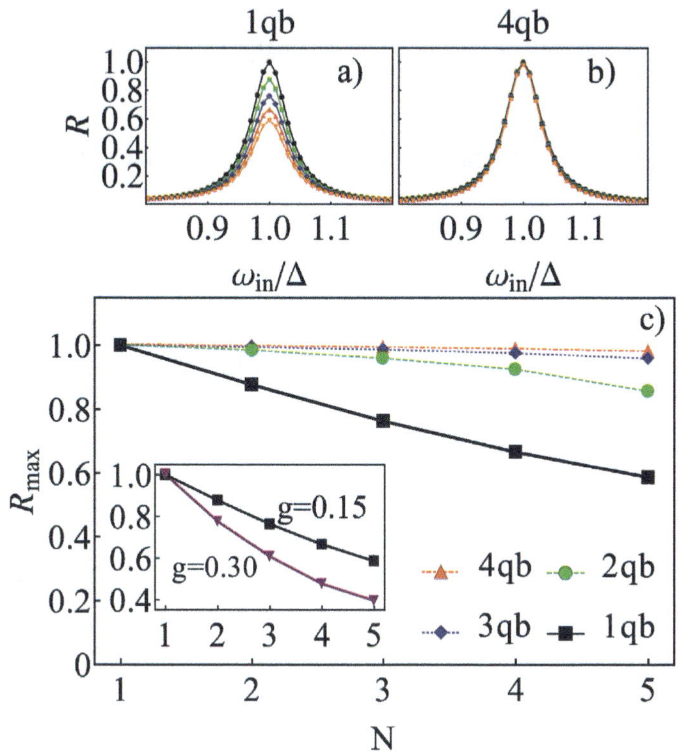

Fig. 2 ε, J and Δ are as in Fig. 1. The input state, eqn (5) and (6), with $k_{in} = \pi/2$ and $\theta = 2$. (a) R for $M = 1$ and $N = 1$–5 (black circles, green squares, blue diamonds, red triangles and orange inverted triangles respectively); (b) the same as in a) but with $M = 4$; (c) R_{max} for $g\sqrt{M} = 0.15$, with $M = 1$–5 qubits (black squares/solid line, green circles/dashed line, blue diamonds/dotted line and red triangles/dotted-dashed line respectively). In the inset we show R_{max} *vs.* N for $M = 1$, with $g = 0.15$ (black squares) and $g = 0.30$ (purple inverted triangles).

the same point, *i.e.*, their inter distance is zero. We compute $R_{\omega_{in}}$ from $N = 1$ to $N = 5$ photons input states given by (5) and (6) centred in the resonant value $k_{in} = \pi/2$ ($\omega_{k_{in}} = \Delta$). The scattering centre is composed by $M = 1$ to $M = 4$ qubits. We plot $R_{\omega_{in}}$ [*cf.* eqn (16)] for $M = 1$ and $M = 4$ qubits in Fig. 2(a and b) respectively. The spectral width scales with the one-photon effective coupling $g\sqrt{M}$ (29), which is maintained as constant in the calculations. As shown, the maximum reflection $R_{max} < 1$ as soon as $N > 1$. The effect is better observed in the $M = 1$ qubit case, as shown in Fig. 2(a). As argued before, by increasing the number of qubits we recover the full reflection $R_{max} \cong 1$. The dependence of R_{max} as a function of N/M is better represented in panel Fig. 2(c). R_{max} decreases much faster with N for $M = 1$ than for $M > 1$. For $M = 4$, R_{max} hardly gets modified by changing the number of photons in the considered range of N. Following refs 18 and 29 there is total reflection for $N = 2$ *vs.* $M = 2$ if the photon energies individually match with those of the qubits. We see a slight deviation from this result because we are taking a non-monochromatic input state and the component of the incident wavefunction for k_1, $k_2 \neq k_{in}$ is not negligible. It is remarkable that R_{max} does not only depend on N and M, but also on the coupling, see the inset in Fig. 2(c). Therefore the nonlinear characteristics do not only depend on the material (qubits) but on their coupling to the field too.

4.2 Two photons *vs.* M qubits: spatial photon–photon correlations in reflection

In this section we compute the photon–photon correlations created in the scattering process for two-photon input states, as a function of the number of qubits. For further comparison and understanding, we begin by discussing the linear case. In this case the scatterer is a harmonic oscillator $h = \Delta c^\dagger c$ with $[c, c^\dagger] = 1$ in eqn (1). It has already been discussed in this paper that no correlations are generated if the scatterer is linear [*cf.* Theorem 3.2]. The two point correlation factorises: $\left|\phi_{x_1,x_2}{}^{t_{\text{out}}}\right|^2 = \left|\phi_{x_1}{}^{t_{\text{out}}}\right|^2 \left|\phi_{x_2}{}^{t_{\text{out}}}\right|^2$ and, in particular, the two photons must be bunched both in reflection and transmission: $\left|\phi_{x,x}{}^{t_{\text{out}}}\right|^2 = \left|\phi_{x}{}^{t_{\text{out}}}\right|^4$. In the nonlinear case, however, the reflected field by one qubit is totally antibunched,[17] $\left|\phi_{x,x}{}^{t_{\text{out}}}\right|^2 = 0$. Thus, antibunching can be used as a witness for nonlinearities. With these antecedents, we provide answers to the following questions below. How does this depend on the number of qubits? Is it possible to interpolate between the highly nonlinear case of one qubit and the linear case of a harmonic oscillator by adding qubits? If so, how many qubits are needed for the system to be linear?

In Fig. 3d we plot the reflected $|\phi_{x_1,x_2}|^2$ against $x_1 - x_2$, fixing $(x_1 + x_2)/2$ such that the reflection component is maximum. The one-photon coupling, $g\sqrt{M}$ is kept constant for different M and equals the coupling g for the case of the harmonic oscillator. We remember that $|\phi_{x_1,x_2}|^2$ is proportional to the probability of having both photons separated by a distance $|x_1 - x_2|$. Setting $x_1 = x_2$ gives the probability of seeing both photons at the same point. For $M = 1$ the numerical results (Fig. 3d) show $|\phi_{xx}|^2 \simeq 0$, recovering the well known photon antibunching in reflection.[17,18] Surprisingly, photon antibunching and thus nonlinearity can still be resolved by increasing M, even for $M = 20$.§ The full contour for $|\phi_{x_1,x_2}|^2$ is shown in Fig. 3a–c for one and two qubits and a harmonic resonator respectively. Apart from the antibunched characteristics in the reflected signal, we can also appreciate that one qubit cannot reflect as much as two qubits, as we discussed in Section 4.1 [*cf.* Fig. 2]. The transmitted photons are always bunched. The components around $(x_1 + x_2)/2 \simeq 0$ and $|x_1 - x_2| \simeq 200$ correspond to one photon being transmitted and the other reflected.

Once the physics has been discussed, let us finish with a brief note on how to solve the two photon scattering against M qubits, for any M, placed at the same point and within the RWA. We start by remembering that the RWA implies the conservation for the number of excitations, *cf.* Section 2.1.

§ We note here that for the particular set of values $N = 2$ and $M = 2$ that, if $\omega_1 = \Delta_1$ and $\omega_2 = \Delta_2$ photons are bunched[18,29] as it would be linear scattering. However, our initial wavepacket is not monochromatic.

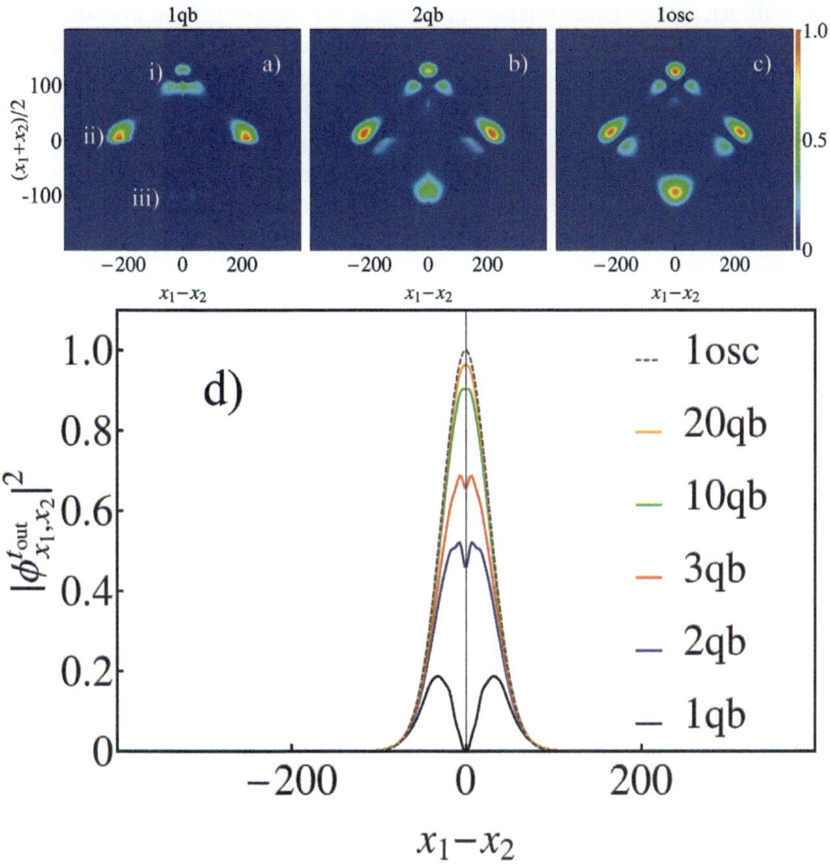

Fig. 3 $\left|\phi_{x_1,x_2}{}^{t_{out}}\right|^2$ for $g\sqrt{M} = 0.10$, with (a) $M = 1$ qubit, (b) $M = 2$ qubits and (c) a harmonic oscillator. $\left|\phi_{x_1,x_2}{}^{t_{out}}\right|^2$ is normalised in each panel such that its maximum is set to 1. In the contours, (i) corresponds to the transmitted–transmitted component, (ii) to the transmitted–reflected component and finally (iii) to the reflected–reflected component. (d) $\left|\phi_{x_1,x_2}{}^{t_{out}}\right|^2$ fixing $(x_1 + x_2)/2$ for $g\sqrt{M} = 0.10$, with $M = 1$, 2, 3, 10 and 20 qubits (solid, from bottom to top) and a harmonic oscillator (dashed). We normalise such that the correlator for the harmonic oscillator is 1 for $x_1 = x_2$. The input state is a 2-photon state with $k_{in} = \pi/2$ and $\theta = 20$.

Therefore, in the two excitation manifold and regarding the qubits, it will suffice to consider the following Dicke states: $\left\{|0\rangle, |1\rangle \equiv b^\dagger|0\rangle, |2\rangle \equiv \frac{(b^\dagger)^2|0\rangle}{\sqrt{2(1-1/M)}}\right\}$ [*cf.* eqn (28)]. Thus, if $N = 2$, the M qubits can be formally replaced by a three level system with states given as above. As expected [see Section 3.4], in the limit $M \rightarrow \infty$, $|2\rangle = \frac{(b^\dagger)^2|0\rangle}{\sqrt{2}}$ as in the harmonic oscillator case.

4.3 Three qubits with distance

We incorporate a new ingredient here, the inter-qubit distance. Interactions among qubits are mediated by the EM field decay with the distance in two and three dimensions.[68] One dimension is special: the field wavefront area does not grow with distance and, thus, the interaction does not decay but it is periodically modulated instead. The period depends on the qubits level splitting and the dispersion relation in the waveguide. This peculiarity has been theoretically investigated for qubits in a multitude of arrangements[34–37,69] and experimentally demonstrated quite recently.[7] For our purposes it is important to note that these works considered either classical or single photon input states. Two photon input states were considered too in refs 26 and 27.

Invoking the Markovian approximation the induced qubit–qubit interaction is considered instantaneous. This implies that the set of distances, $d \equiv x_{i'} - x_i$, related by

$$d' = d + \frac{\pi}{k_{\text{in}}} q, \quad q \in \mathbb{Z} \tag{31}$$

give the same scattering characteristics.

Let us consider $N = 2$ input photon states and three qubits separated by some nearest neighbour distance d. With the MPS tool, we solve the problem exactly, no matter the distance. Thus, we do not make any approximations, such as the small

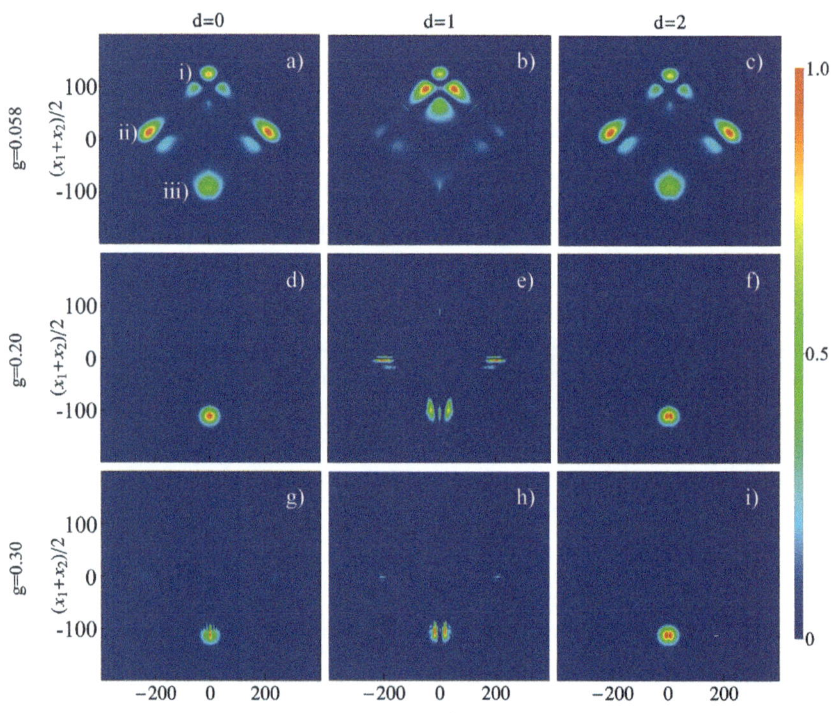

Fig. 4 $\left|\phi_{x_1,x_2}{}^{t_{\text{out}}}\right|^2$ for two incident photons and three qubits with inter-qubit distance d. (i–iii) mean the same as in Fig. 3. The incoming photons are characterised by $\theta = 20$ and $k_{\text{in}} = \pi/2$. We take $d = 0, 1, 2$ (left, middle and right columns respectively), so the first and the third cases should be equivalent because of (31). We consider the RWA Hamiltonian and take $g\sqrt{3} = 0.10$ in the first row, whereas we take the full Hamiltonian for the second and third rows, with $g = 0.20$ and $g = 0.30$ respectively.

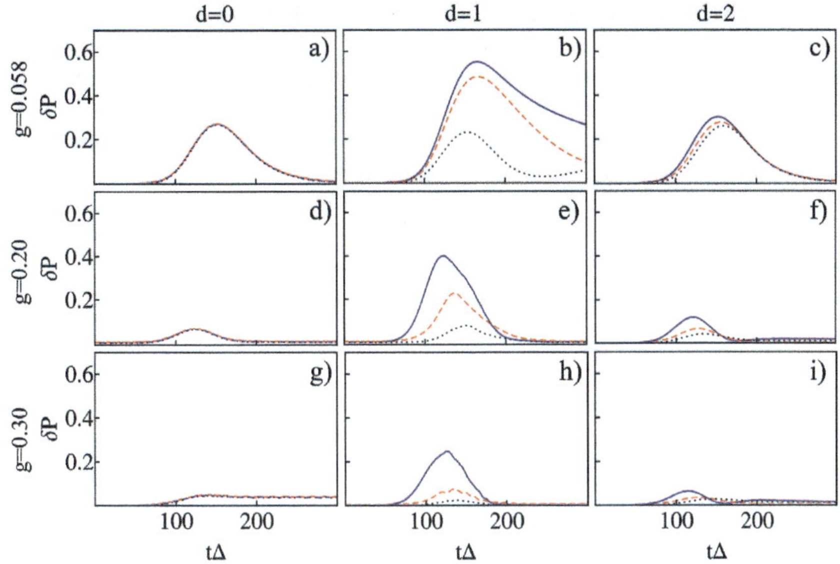

Fig. 5 Time evolution for the qubit population of the first (solid blue), the second (dashed red) and the third qubit (dotted black). The parameters are those of Fig. 4.

distance or Markovian approximations. We plot photon–photon correlations $\left|\phi_{x_1,x_2}{}^{t_{\text{out}}}\right|^2$ (17) and qubit populations $\delta P_i \equiv P_i - (P_i)_{\text{GS}}$ (15) for RWA couplings ($(P_i)_{\text{GS}} = 0$) in Fig. 4 and 5, panels (a–c) in both cases. Regarding the photons, the characteristics are the same at zero distance and $d = 2$, which is equivalent to zero distance, by means of (31) ($k_{\text{in}} = \pi/2$). On the other hand, the inherent non-Markovian properties of our exact simulation can be appreciated by comparing Fig. 5(a and c). If $d \neq 0$, the qubit at the left is excited first, then the central qubit and finally the one to the right.

The contours for the two-photon wave function $\left|\phi_{x_1,x_2}{}^{t_{\text{out}}}\right|^2$ are closer to the linear scattering result for $d = 0$ and $d = 2$ than those for the $d = 1$ case [*cf.* Fig. 3c and 4(a–c)]. This enhancement of nonlinear properties with the distance can be understood as follows. At distances $\frac{\pi}{k_{\text{in}}}q$, [equivalent to zero distance according to eqn (31)], only qubit states generated by the ladder operator b in eqn (28) are visited during the dynamics. For two photon input states, the nonlinearities die out as $1/M$ [*cf.* Section 4.2]. On the other hand, for other distances not fulfilling this condition of equivalence, more qubit states (satisfying the number conservation imposed by the RWA) can play a role. The fact that more qubit modes participate in the dynamics for $d = 1$ is apparent from Fig. 5(b). Clearly, more frequencies are involved in the evolution of P. Importantly enough, $d = 1$ corresponds to a distance where the coherent interaction between the qubits are maximised, while the correlations in the qubit decays are minimised. This is named as the subradiant case, as we can observe that the qubit decay (after excitation) is slower in this case.

5 Nonlinear scattering in the ultrastrong regime

We close this discussions by investigating stronger qubit–photon couplings, strong enough to break down the RWA approximation, eqn (4). The full dipole interaction must be considered. When moving to the ultrastrong regime, the problem becomes rather involved. The ground state, $|\mathrm{GS}\rangle$, for (1) is not trivial anymore but strongly correlated and the number of excitations is not conserved [recall Section 2.1]. This increase in the complexity brings a lot of new phenomena. We fix our attention on two characteristics. First, we discuss the appearance of inelastic scattering at high couplings.[48,70–72] Second, we revisit the qubit–qubit interactions mediated by the guide in the ultrastrong coupling regime.

5.1 Inelastic scattering with *M* qubits

So far, we have discussed elastic scattering. In linear systems, this is the only possible process. When the scatterers are qubits, however, inelastic processes may happen.[48,70–72] In an inelastic process, the photons can excite localised light–matter states. Those localised states, eigenstates for the total Hamiltonian, remain excited after the photons pass through. By energy conservation it is clear that the outgoing photons must have less energy. Within the RWA approximation,[70–72] at least two photons are needed for inelastic processes. When the actual full dipole interaction is considered (1), a non stimulated Raman scattering process occurs.[48] Indeed, it is quite remarkable that 100% efficiency can be achieved.[48]

We study the inelastic channel for one photon input states when interacting with *M* qubits placed at the same point. In Fig. 6 we plot the transmitted current in the inelastic channel,

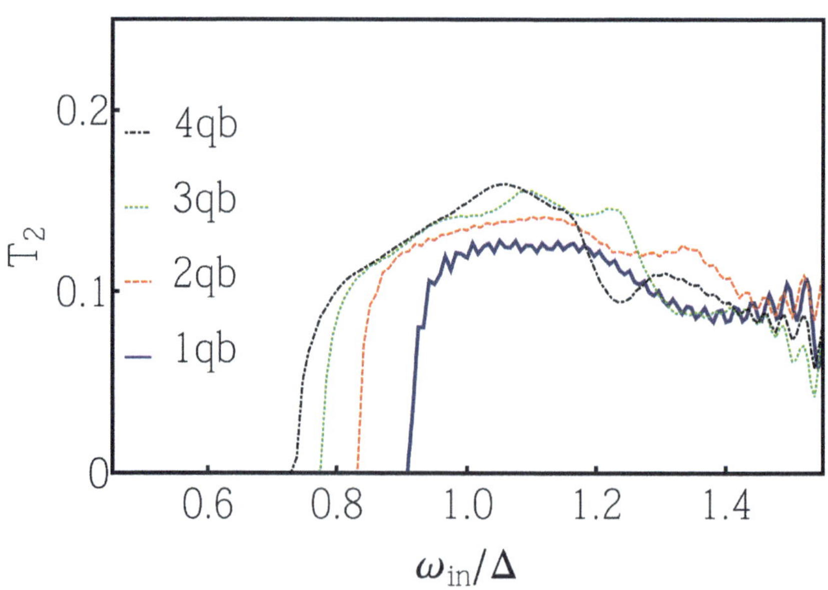

Fig. 6 Inelastic transmission probability for $g\sqrt{M} = 0.50$ and 1 qubit (solid blue), 2 qubits (dashed red), 3 qubits (dotted green) and 4 qubits (dashed-dotted black).

$$T_{2,k} = \frac{1}{2}\left(1 - |t_k|^2 - |r_k|^2\right) \quad (32)$$

In the figure we consider the qubits placed in the same point. This figure shows that inelastic scattering is another nonlinear characteristic which persists even for $N \ll M$ (recall that T_2 for a linear system is always zero, *cf.* Section 3.1).

5.2 Mixing distance and the ultrastrong regime

Interesting physics occurs with spatially separated qubits, when their interactions are mediated through the photons. At the same time the photons interact among themselves when they pass through those qubits. In the weak coupling regime, the output field is periodic in the qubit–qubit distance when taking the Markovian approximation (31) [see Section 4.3]. Below, we discuss how this scenario changes in the ultrastrong coupling regime.

In Fig. 4 the loss of this effective periodicity in the qubit distance is clearly shown by plotting $\left|\phi_{x_1,x_2}{}^{t_{\text{out}}}\right|^2$. At weak coupling (first row, already discussed in Section 4.3) the distances $d = 0$ and $d = 2$ satisfying (31) present identical output fields. The two lower rows show results at larger couplings. The condition (31) does not hold anymore, and the dipole–dipole periodic structure paradigm for one dimensional systems[7,34–36,69] is no longer true. Further confirmation is obtained when looking to the qubit dynamics. In the final column, we realise that, in the ultrastrong regime, signatures of subradiant dynamics are still perceived. The loss of periodicity is also appreciated here. If we fix our attention to the $d = 0$ case all qubits behave in the same manner, as they must, and they remain in an excited state, which is a signature of inelastic scattering. For the case $d = 2$, which in weak coupling is equivalent to $d = 0$ [*cf.* eqn (31)], the dynamics is completely different (see Fig. 5). At this set of parameters the qubits seem to evolve back to the ground state. Therefore no inelastic scattering occurs in this case.

6 Conclusions

The advent of artificial devices (optical cavities, superconducting circuits, dielectric and plasmonic guides, photonic crystals) opens the avenue for stronger and stronger light–matter coupling (at the single photon level). Recent impressive experimental advances are changing the paradigm and typical nonlinear effects can be observed at the few photon level. From the theoretical point of view, both light and matter must be described quantum mechanically. Going beyond one or two photons and one qubit is an analytically titanic task. Therefore, numerical tools are demanded to satisfy current experimental efforts in constructing devices responding nonlinearly at minimum powers. In this discussion we have shown that the MPS technique developed for one dimensional systems is a powerful tool in few photon nonlinear optics.

On the physical side, our task was to study the nonlinear response by increasing the number of qubits. It has a practical motif. The effective coupling scales with $g\sqrt{M}$ (M the number of qubits). Adding scatterers is a way to enhance the coupling but, at the same time, their nonlinear response is reduced. Some quantities, as the transmitted and reflected currents, already gave linear behaviour for ratios $N/M \cong 1$. However, we have found that for $N = 2$ input states, $M = 20$ qubits still generate photon–photon interactions.

We have also investigated the regime where light and matter are coupled ultrastrongly. New phenomena appear as Raman scattering or the breakdown of the periodicity in the qubit–qubit interaction, which should be observable by current technology, since some experimental setups already operate in the ultrastrong coupling regime. All together, this serves as a motivation for developing theories for its understanding and hopefully, will trigger further experimental studies for verifying the plethora of new phenomenology.

A Proof of theorem 3.1

If the scatterers are harmonic resonators the Hamiltonian (1) is linear [*cf.* Section 3.1] and it can be diagonalised with a Bogolioubov–Valatin (BV) transformation,

$$H = \sum \Lambda_l \alpha_l^\dagger \alpha_l, \tag{33}$$

with $[\alpha_l, \alpha_m^\dagger] = \delta_{lm}$ and $\Lambda_l > 0 \forall l$. The ground state is $\alpha_l|\text{GS}\rangle = 0 \forall l$. The BV transformation

$$\alpha_l = \sum_{i=-L}^{L} (\chi_{li}{}^a a_i + \eta_{li}{}^a a_i^\dagger) + \sum_{i=1}^{M} (\chi_{li}{}^c c_i + \eta_{li}{}^c c_i^\dagger) \tag{34}$$

is invertible.

Hamiltonian (33) commutes with $N_\alpha = \sum \alpha_l^\dagger \alpha_l$:

$$[H, N_\alpha] = 0. \tag{35}$$

The number of α-excitations, N_α, is a good quantum number. It also conserves parity, $P = \mathrm{e}^{i\pi \sum \alpha_l^\dagger \alpha_l}$:

$$[H, P] = 0. \tag{36}$$

Single photon input states, $N = 1$ in (5), are written in momentum space

$$|\Psi_{\text{in}}\rangle = \sum_{k>0} \tilde{\phi}_k^{\text{in}} a_k^\dagger |\text{GS}\rangle, \tag{37}$$

with $\tilde{\phi}_k^{\text{in}}$ the Fourier transform of ϕ_x^{in}. Using the transformation (34) and that $\alpha_l|\text{GS}\rangle = 0$, we can rewrite the state (37) in the α-representation

$$|\Psi_{\text{in}}\rangle = \sum_l \overline{\phi}_l^{\text{in}} \alpha_l^\dagger |\text{GS}\rangle. \tag{38}$$

Given the input state (38): $N_\alpha|\Psi_{\text{in}}\rangle = |\Psi_{\text{in}}\rangle$. Since N_α is a conserved quantity [*cf.* eqn (35)] the time evolution is restricted to the one α-excitation level (35). The output state is then

$$|\Psi_{\text{out}}\rangle = \sum_l \overline{\phi}_l^{\text{out}} \alpha_l^\dagger |\text{GS}\rangle, \tag{39}$$

with $\overline{\phi}_l^{\text{out}} \equiv \mathrm{e}^{-i\Lambda_l t_{\text{out}}} \overline{\phi}_l^{\text{in}}$. Using the transformation (34), the output state is rewritten

$$|\Psi_{out}\rangle = \sum_l \overline{\phi}_l^{\text{out}} \left(\sum_k \left(\left(\tilde{\chi}_{lk}{}^a\right)^* a_k^\dagger + \left(\tilde{\eta}_{lk}{}^a\right)^* a_k \right) + \sum_i \left(\left(\chi_{li}{}^c\right)^* c_i^\dagger + \left(\eta_{li}{}^c\right)^* c_i \right) \right) |\text{GS}\rangle, \tag{40}$$

with $\tilde{\chi}_{lk}{}^a$ and $\tilde{\eta}_{lk}{}^a$ the discrete Fourier transforms of $\tilde{\chi}_{li}{}^a$ and $\tilde{\eta}_{li}{}^a$ in the second index. The output state (40) removes the possibility of having multiphoton scattering states. Therefore, the scattering events can be elastic, with transmission and reflection amplitudes t_k and r_k and inelastic, with the scatterer relaxing to an excited state $|\text{EXC}\rangle$. In the latter, the photon emerges with a new momentum k_{new}, fulfilling energy conservation

$$\omega_{k_{\text{in}}} + E_{\text{GS}} = \omega_{k_{\text{new}}} + E_{\text{EXC}} \tag{41}$$

Therefore, the output state can be rewritten

$$|\Psi_{\text{out}}\rangle = \sum_{k>0} \tilde{\phi}_k^{\text{in}} \left(t_k a_k^\dagger + r_k a_{-k}^\dagger\right) |\text{GS}\rangle + \sum_k \tilde{\phi}_k^{\text{new}} a_k^\dagger |\text{EXC}\rangle, \tag{42}$$

with $\tilde{\phi}_k^{\text{new}}$ a wavepacket centred around k_{new}.

Let us fix our attention to the second term in the r.h.s. of (42), which is rewritten in terms of the α-operators with the help of the BV transformation (34):

$$\sum_k \tilde{\phi}_k^{\text{new}} a_k^\dagger |\text{EXC}\rangle = \sum_l \left(\overline{\phi}_l^{\text{new},p} \alpha_l^\dagger + \overline{\phi}_l^{\text{new},m} \alpha_l\right) |\text{EXC}\rangle \tag{43}$$

Using N_α and P conservation, eqn (35) and (36), $N_\alpha|\text{EXC}\rangle = 2n|\text{EXC}\rangle$ with $n \geq 1$. The first term in the r.h.s. of (43) has $2n + 1 \geq 3$ particles. Thus, $\bar{\phi}_l^{\text{new},p} = 0$. Finally $\alpha_l|\text{EXC}\rangle$ is an eigenstate of (33) with eigen-energy $E_{\text{EXC}} - \Lambda_l$ ($\Lambda_l > 0$). The latter must be equal to $\omega_{k_{\text{new}}} + E_{\text{EXC}}$ which is impossible. Therefore $\bar{\phi}_l^{\text{new},m} = 0$.

Putting this all together, the output state contains only the elastic channel,

$$|\Psi_{\text{out}}\rangle = \sum_{k>0} \tilde{\phi}_k^{\text{in}} \left(t_k a_k^\dagger + r_k a_{-k}^\dagger\right) |\text{GS}\rangle. \tag{44}$$

This ends the proof.

B Proof of theorem 3.2

The components for the scattering matrix (10) can be rewritten,

$$S_{p_1 \ldots p_{N'}, k_1 \ldots k_N} = \langle \text{GS}| a_{p_1} \ldots a_{p_{N'}} a_{k_1}^\dagger (t_{\text{out}}) \ldots a_{k_N}^\dagger (t_{\text{out}}) |\text{GS}\rangle \tag{45}$$

with

$$a_k^\dagger (t_{\text{out}}) = S^\dagger a_k^\dagger S = \left(t_k a_k^\dagger + r_k a_{-k}^\dagger\right). \tag{46}$$

In the last equality we have used (i) linearity: the Heisenberg evolution for the operators $a_{kn}(t)$ is independent of the input states, and (ii) the theorem 3.2. Eqn (46), together with the Wick theorem:

$$\langle \text{GS}| a_{p_1} \ldots a_{p_{N'}} a_{k_1}^\dagger \ldots a_{k_N}^\dagger |\text{GS}\rangle = \delta_{NN'} \sum_{m_1 \neq m_2 \neq \ldots \neq m_N} \delta_{p_1 k_{m_1}} \ldots \delta_{p_N k_{m_N}} \tag{47}$$

completes the proof.

Acknowledgements

We acknowledge support by the Spanish Ministerio de Economia y Competitividad within projects MAT2011-28581-C02, FIS2012-33022 and No. FIS2011-25167, the Gobierno de Aragon (FENOL group) and the European project PROMISCE.

References

1 R. W. Boyd, *Nonlinear Optics*, Academic Press, 2nd edn, 2003.
2 T. Peyronel, O. Firstenberg, Q.-Y. Liang, S. Hofferberth, A. V. Gorshkov, T. Pohl, M. D. Lukin and V. Vuletic, *Nature*, 2012, **488**, 57.
3 O. Firstenberg, T. Peyronel, Q.-Y. Liang, A. V. Gorshkov, M. D. Lukin and V. Vuleti, *Nature*, 2013, **502**, 71.
4 J. Q. You and F. Nori, *Nature*, 2011, **474**, 589–597.
5 O. Astafiev, A. M. Zagoskin, A. A. Abdumalikov, Y. A. Pashkin, T. Yamamoto, K. Inomata, Y. Nakamura and J. S. Tsai, *Science*, 2010, **327**, 840.
6 I. C. Hoi, C. M. Wilson, G. Johansson, J. Lindkvist, B. Peropadre, T. Palomaki and P. Delsing, *New J. Phys.*, 2013, **15**, 025011.
7 A. F. van Loo, A. Fedorov, K. Lalumière, B. C. Sanders, A. Blais and A. Wallraff, *Science*, 2013, **342**, 1494.
8 T. Lund-Hansen, S. Stobbe, B. Julsgaard, H. Thyrrestrup, T. Sünner, M. Kamp, A. Forchel and P. Lodahl, *Phys. Rev. Lett.*, 2008, **101**, 113903.
9 J. L. O'Brien, *Science*, 2007, **318**, 1567.
10 Y. Rezus, S. Walt, R. Lettow, A. Renn, G. Zumofen, S. Gotzinge and V. Sandoghdar, *Phys. Rev. Lett.*, 2012, **108**, 093601.
11 R. W. Boyd, *Plasmonics: Fundamentals and Applications*, Springer, 2nd edn, 2007.
12 D. E. Chang, A. S. Sørensen, P. R. Hemmer and M. Lukin, *Phys. Rev. Lett.*, 2006, **97**, 053002.
13 A. V. Akimov, A. Mukherjee, C. L. Yu, D. E. Chang, A. S. Zibrov, P. R. Hemmer, H. Park and M. D. Lukin, *Nat. Phys.*, 2007, **450**, 402.
14 A. L. Falk, F. H. L. Koppens, C. L. Yu, K. Kang, N. de Leon Snapp, A. V. Akimov, M.-H. Jo, M. D. Lukin and H. Park, *Nat. Phys.*, 2009, **10**, 475.
15 N. P. de Leon, M. D. Lukin and H. Park, *IEEE J. Sel. Top. Quantum Electron.*, 2012, **18**, 1781.
16 J.-T. Shen and S. Fan, *Phys. Rev. A*, 2007, **76**, 062709.
17 J.-T. Chen and S. Fan, *Phys. Rev. Lett.*, 2007, **98**, 153003.
18 T. Shi, S. Fan and C. P. Sun, *Phys. Rev. A*, 2011, **84**, 063803.
19 E. Rephaeli, Ş. Kocabaş and S. Fan, *Phys. Rev. A*, 2011, **84**, 063832.
20 S. Xu, E. Rephaeli and S. Fan, *Phys. Rev. Lett.*, 2013, **111**, 223602.
21 T. Shi and C. P. Sun, *Phys. Rev. B: Condens. Matter Mater. Phys.*, 2009, **79**, 205111.
22 V. I. Yudson and P. Reineker, *Phys. Rev. A*, 2008, **78**, 052713.
23 H. Zheng, D. J. Gauthier and H. U. Baranger, *Phys. Rev. A*, 2010, **82**, 063816.
24 H. Zheng, D. J. Gauthier and H. U. Baranger, *Phys. Rev. Lett.*, 2011, **107**, 223601.
25 H. Zheng, D. J. Gauthier and H. U. Baranger, *Phys. Rev. A*, 2012, **85**, 043832.

26 H. Zheng and H. U. Baranger, *Phys. Rev. Lett.*, 2013, **110**, 113601.
27 Y.-L. L. Fang, H. Zheng and H. U. Baranger, *EPJ Quantum Technol.*, 2014, **1**, 3.
28 D. Roy, *Phys. Rev. A*, 2011, **83**, 043823.
29 D. Roy, *Phys. Rev. A*, 2013, **87**, 063819.
30 J.-Q. Liao and C. K. Law, *Phys. Rev. A*, 2010, **82**, 053836.
31 B. Peropadre, J. Lindkvist, I. C. Hoi, C. M. Wilson, J. J. García-Ripoll, P. Delsing and G. Johansson, *New J. Phys.*, 2013, **15**, 035009.
32 E. Rephaeli and S. Fan, *Phys. Rev. Lett.*, 2012, **108**, 143602.
33 G.-Y. Chen, N. Lambert, C.-H. Chou, Y.-N. Chen and F. Nori, *Phys. Rev. B: Condens. Matter Mater. Phys.*, 2011, **84**, 045310.
34 A. González-Tudela, D. M. Cano, E. Moreno, L. Martín-Moreno, C. Tejedor and F. J. García-Vidal, *Phys. Rev. Lett.*, 2011, **106**, 020501.
35 D. Zueco, J. J. Mazo, E. Solano and J. J. García-Ripoll, *Phys. Rev. B: Condens. Matter Mater. Phys.*, 2012, **86**, 024503.
36 K. Lalumière, B. C. Sanders, A. F. van Loo, A. Fedorov, A. Wallraff and A. Blais, *Phys. Rev. A*, 2013, **88**, 043806.
37 C. González-Ballestero, E. Moreno and F. J. García-Vidal, *Phys. Rev. A*, 2014, **89**, 042328.
38 C. Cohen-Tannoudji, J. Dupont-Roc and G. Grynberg, *Atom-Photon Interactions: Basic Processes and Applications*, Wiley-Interscience, 1992, p. 680.
39 T. Niemczyk, F. Deppe, H. Huebl, E. Menzel, F. Hocke, M. Schwarz, J. J. García-Ripoll, D. Zueco, T. Hümmer, E. Solano, A. Marx and R. Gross, *Nat. Phys.*, 2010, **6**, 772.
40 P. Forn-Díaz, J. Lisenfeld, D. Marcos, J. J. García-Ripoll, E. Solano, C. J. P. M. Harmans and J. E. Mooij, *Phys. Rev. Lett.*, 2010, **105**, 237001.
41 T. Schwartz, J. A. Hutchinson, C. Genet and T. W. Ebbesen, *Phys. Rev. Lett.*, 2011, **106**, 196405.
42 G. Günter, A. A. Anappara, J. Hees, A. Sell, G. Biasol, L. Sorba, S. D. Liberato, C. Ciuti, A. Tredicucci, A. Leitenstorfer and R. Huber, *Nature*, 2009, **458**, 178.
43 S. De Liberato and C. Ciuti, *Phys. Rev. B: Condens. Matter Mater. Phys.*, 2008, **77**, 155321.
44 K. L. Hur, *Phys. Rev. B: Condens. Matter Mater. Phys.*, 2012, **85**, 140506.
45 G. Romero, D. Ballester, Y. M. Wang, V. Scarani and E. Solano, *Phys. Rev. Lett.*, 2012, **108**, 120501.
46 B. Peropadre, D. Zueco, D. Porras and J. J. García-Ripoll, *Phys. Rev. Lett.*, 2013, **111**, 243602.
47 U. Naether, J. J. García-Ripoll, J. J. Mazo and D. Zueco, *Phys. Rev. Lett.*, 2014, **112**, 074101.
48 E. Sánchez-Burillo, D. Zueco, J. J. García-Ripoll and L. Martín-Moreno, arXiv:1406.5779v1 [quant-ph], 2014.
49 Z. H. Wang, Y. Li, D. L. Zhou, C. P. Sun and P. Zhang, *Phys. Rev. A*, 2012, **86**, 023824.
50 R. H. Dicke, *Phys. Rev.*, 1954, **93**, 99.
51 D. Witthaut and A. S. Sørensen, *New J. Phys.*, 2010, **12**, 043052.
52 D. Roy, *Phys. Rev. Lett.*, 2011, **106**, 053601.
53 C. Cohen-Tannoudji, J. Dupont-Roc and G. Grynberg, *Photons & Atoms: Introduction to Quanutm Electrodynamics*, Wiley Vch, 1997, p. 486.
54 G. Vidal, *Phys. Rev. Lett.*, 2003, **91**, 147902.
55 G. Vidal, *Phys. Rev. Lett.*, 2004, **93**, 040502.

56 F. Verstraete, V. Murg and J. I. Cirac, *Adv. Phys.*, 2008, **57**, 143.
57 J. Eisert, M. Cramer and M. B. Plenio, *Rev. Mod. Phys.*, 2010, **82**, 277–306.
58 M. Suzuki, *J. Math. Phys.*, 1991, **32**, 400.
59 J. J. García-Ripoll, *New J. Phys.*, 2006, **8**, 305.
60 T. Hümmer, G. Reuther, P. Hänggi and D. Zueco, *Phys. Rev. A*, 2012, **85**, 052320.
61 J.-T. Shen and S. Fan, *Opt. Lett.*, 2005, **30**, 2001.
62 J.-T. Shen and S. Fan, *Phys. Rev. Lett.*, 2005, **95**, 213001.
63 L. Zhou, Z. R. Gong, Y. xi Liu, C. P. Sun and F. Nori, *Phys. Rev. Lett.*, 2008, **101**, 100501.
64 L. Zhou, S. Yang, Y. xi Liu, C. P. Sun and F. Nori, *Phys. Rev. A*, 2009, **80**, 062109.
65 J.-Q. Liao, Z. R. Gong, L. Zhou, Y. xi Liu, C. P. Sun and F. Nori, *Phys. Rev. A*, 2010, **81**, 042304.
66 L. Zhou, L.-P. Yang, Y. Li and C. P. Sun, *Phys. Rev. Lett.*, 2013, **111**, 103604.
67 J. Lu, L. Zhou, L.-M. Kuang and F. Nori, *Phys. Rev. A*, 2014, **89**, 013805.
68 R. H. Lehmberg, *Phys. Rev. A*, 1970, **2**, 883.
69 L. Zhou, H. Dong, Y. xi Liu, C. P. Sun and F. Nori, *Phys. Rev. A*, 2008, **78**, 063827.
70 P. Longo, P. Schmitteckert and K. Busch, *Phys. Rev. Lett.*, 2010, **104**, 023602.
71 P. Longo, P. Schmitteckert and K. Busch, *Phys. Rev. A*, 2011, **83**, 063828.
72 T. Shi and C. P. Sun, arXiv:0907.2776 [quant-ph], 2009.

Faraday Discussions

PAPER

Second harmonic generation on self-assembled tilted gold nanowires

Alessandro Belardini,*[a] Marco Centini,[a] Grigore Leahu,[a] Eugenio Fazio,[a] Concita Sibilia,[a] Joseph W. Haus*[b] and Andrew Sarangan[b]

Received 8th October 2014, Accepted 10th November 2014
DOI: 10.1039/c4fd00200h

Here we present measurements on a metasurface composed by tilted gold nanowires. The metasurface can induce an optical chiral response of the whole sample when the light impinges on the sample out of the normal incidence angle. In order to investigate the symmetry breaking induced by the geometry, we measured the second harmonic generation (SHG) signal generated by circular polarized pulsed light. This technique has demonstrated to be a powerful method in order to investigate the chiral morphology of nanostructures.

1 Introduction

Optical second harmonic (SH) generation is a very sensitive method for the study of symmetry properties of surfaces and has thus been widely used for the characterisation of chemical and biological samples.[1] In this frame, characterisation of surface chirality by SHG has attracted a large interest due to the great scientific importance of chiral molecules in life-science and pharmaceutical industries.[2] In order to enhance the SH conversion efficiency, a large variety of metal nanopatterned surfaces has been proposed as possible planar substrates, such as nano-holes,[3] nano-wires,[4,5] and nano-rods.[6] In these structures, the electric field intensity can increase up to several orders of magnitude due to the mutual effect of field localization near the sharp edges and localized surface plasmon excitation. Also, the possibility to obtain chirality (or its result, optical activity) with non-chiral elements has been studied in the past,[7] but it is worth to note that only very recently this possibility has awaken a new interest as demonstrated in ref. 8–11, where the phenomenon (which can be referred as extrinsic chirality) has been characterised from the linear and nonlinear points of view.

[a]*Università di Roma La Sapienza, Via Scarpa 16, 00161 Roma, Italy. E-mail: alessandro.belardini@uniroma1.it; Fax: +39 06 44280173; Tel: +39 06 49916800*
[b]*Department of Electrical and Computer Engineering, University of Dayton, Dayton, Ohio, USA. E-mail: jhaus1@udayton.edu; Tel: +1 937 985 2425*

The realization of large area nano-structured devices operating in the visible range, requires state of the art technologies and high fabrication costs. For this reason, self-assembly techniques are extremely appealing[12] since they allow the development of highly structured, large area samples without the need of slow and expensive serial operations.

In what follows, we present the SHG response of a sample composed by self-assembled tilted gold nanowires showing an extrinsic chiral behavior, better evidenced by a nonlinear optical response.

2 Samples and nonlinear optical measurements

Regular arrays of tilted nanowires (NWs) can be easily produced by grazing evaporating gold on a silicon substrate maintained at low temperature (100–300 K). In this case, self-ordered forests of Au NWs can be realized. The thin-film deposition method discussed in this study was carried out using a generic e-beam evaporation system (MDC evap-4000) made by Torr International.[13] This chamber was retrofitted with a custom-built substrate holder to support thin-film growth at cryogenic (~100 K) and room (300 K) temperatures concurrently. Liquid nitrogen was allowed to flow through the substrate holder in an open-loop configuration to achieve cryogenic substrate temperatures during the deposition process. Copper and teflon angled blocks with a preset angle of 88° were mounted on the substrate holder. Using mechanical clips, silicon (Si) substrates were attached to the angled surface of the block. Two type-K thermocouples were attached to the angle blocks to measure the substrate temperature during deposition. Further details of this experimental setup are discussed in ref. 13.

In this way, atoms of gold solidify in clusters by forming tilted nanowires (see Fig. 1).

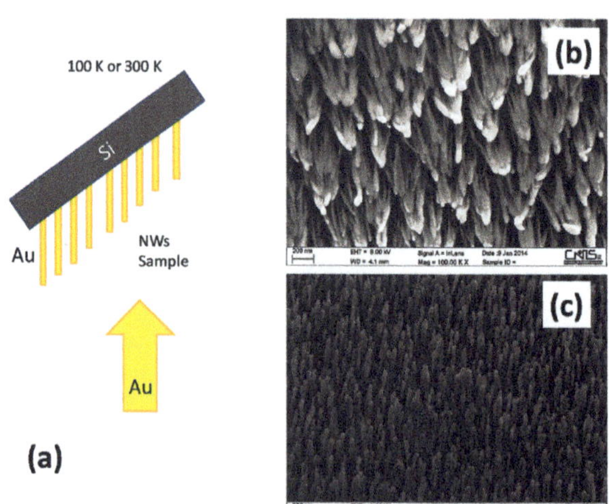

Fig. 1 (a) Sample fabrication technique; (b) SEM image of sample realised on a 100 K Si substrate (100 K sample); (c) SEM image of sample realised on a 300 K substrate (300 K sample).

The nanowires (NW) shown in Fig. 1 are homogenously distributed on a 2 cm × 2 cm surface. The length of the NWs ranges between 380 nm and 400 nm.

In the following, we present results for the sample shown in Fig. 1(c), *i.e.* the sample grown at 300 K. Investigations on different samples are in progress, with the aim of evaluating the influence of the temperature of sample realization and the thickness on the chiral response.

We have used SHG experiments (800–400 nm wavelength conversion) as a powerful method in the investigation of the surface morphology of gold nanowire arrays. In order to investigate the symmetry breaking induced by the geometry, we measured the SHG signal generated by generally elliptical polarized pulsed light. This technique[1] has demonstrated to be a powerful method in order to investigate the chiral morphology of self assembled nanostructures.[5,14]

In Fig. 2, the scheme of the light impinging on the sample with respect to the orientation of the wires is presented. The extrinsic chiral response of the metasurface depends indeed on the orientation of the wires with respect to the impinging light direction and polarization orientation. The input source is a 150 fs Ti:Sapphire laser at 800 nm in which the polarization (*p*-pol.) can be transformed by a $\lambda/4$ plate. By changing the orientation of the fast axis of the $\lambda/4$ plate with respect to the input linear polarization state (*p*-pol.) the polarization state of the light can become right-handed elliptical polarized (fast axis tilted in the range $-90° \div 0°$, where $-45°$ corresponds to right-handed circular pol. light), left handed elliptical polarized (fast axis tilted in the range $0° \div +90°$, where $+45°$ corresponds to left-handed circular pol. light), or linear *p*-polarised (fast axis at 0°). The incidence angle of the pump beam on the sample surface is fixed at $\alpha = 45°$.

The out-coming light is then filtered by an analyzer that can be set for either *p*- or *s*-polarization transmission and then is sent either to a photodiode (in case of linear measurements) or to a filtered photomultiplier (in case of nonlinear

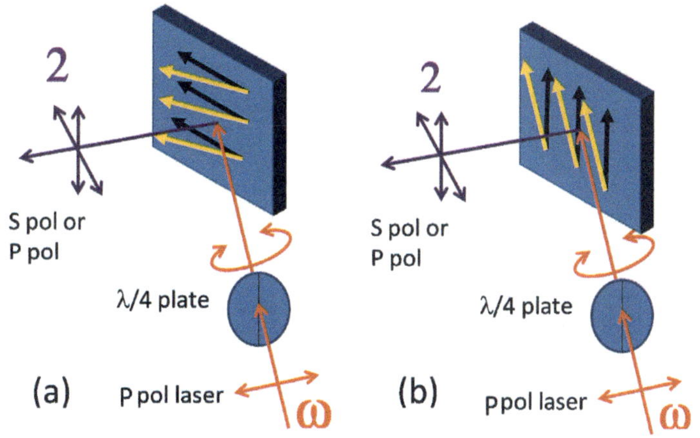

Fig. 2 Scheme of the input polarized beam with respect to the SHG beam generated in the reflection mode; (a) wires oriented mainly in the horizontal direction; (b) wires mainly oriented in the vertical direction.

measurements). The photomultiplier displays maximum sensitivity at the second harmonic wavelength of 400 nm; at the same time, the filter on the photomultiplier ensures that only light with wavelength shorter than 450 nm is detected.

Usually, the circular dichroism in the linear optic domain is defined by the difference between the transmitted intensity (at *s*- or *p*-polarization state) of the left-handed circular polarized light and the right-handed one divided by the average total intensity

$$\mathrm{CD} = 2\cdot({I_{\mathrm{circL}}}^{\omega} - {I_{\mathrm{circR}}}^{\omega})/({I_{\mathrm{circL}}}^{\omega} + {I_{\mathrm{circR}}}^{\omega}). \tag{1}$$

In a similar way, the circular dichroism in the nonlinear optical domain[1] is defined by the difference between the generated second harmonic field intensity (at *s*- or *p*-polarization state) obtained from the left-handed circular polarized light and the right-handed one divided by the average total generated intensity

$$\mathrm{SHG\_CD} = 2\cdot({I_{\mathrm{circL}}}^{2\omega} - {I_{\mathrm{circR}}}^{2\omega})/({I_{\mathrm{circL}}}^{2\omega} + {I_{\mathrm{circR}}}^{2\omega}). \tag{2}$$

In our experiments, we present the SHG signal as a function of the λ/4 fast axis orientation. The SHG signal produced by the 300 K sample is shown in Fig. 3, where a strong asymmetry between the right-handed and left-handed polarization states is present due to the orientation of the wires that are laid out of the plane of the incident light. In this case, the three non-coplanar directions of the incident light, the metasurface normal and the wires' main direction form a triad that is not mirror symmetric, originating the extrinsic chiral behavior. The blue line is the detected SHG signal in the *p* polarized state and the red one in the *s* polarized state. The detected SHG_CD value (obtained with the wave-plate fast axis oriented at +45° and −45°) results to be −0.19 in *s*-pol. state and −0.03 in *p*-pol. state.

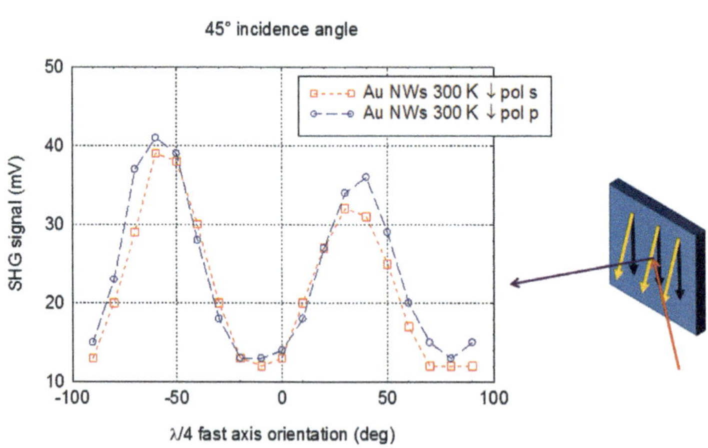

Fig. 3 SHG signal of the 300 K sample as a function of the λ/4 fast axis orientation. The blue line is the detected SHG signal in the *p* polarized state and the red curve is the SHG in the *s* polarized state.

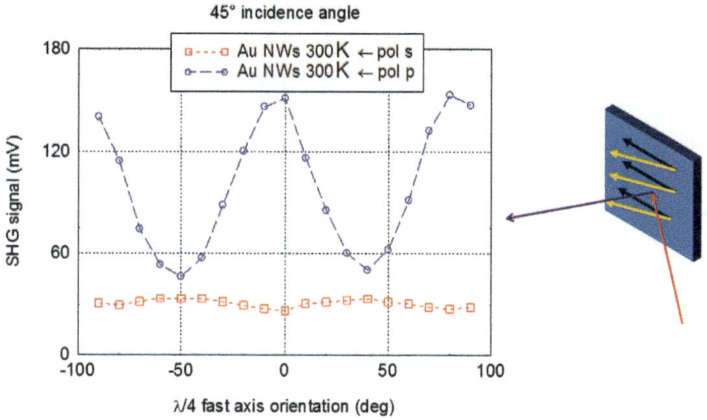

Fig. 4 SHG signal of the 300 K sample as a function of the λ/4 fast axis orientation. The blue line is the detected SHG signal in the *p* polarized state and the red curve is the SHG in the *s* polarized state.

More generally, since here we scan different polarization states that are not circular polarized but elliptically polarized, by continuously ranging from left-handed to right-handed, we introduce a phenomenological parameter that could define an elliptical dichroism in the second harmonic field as:

$$\mathrm{SHG\_ED} = 2\cdot({I_{\mathrm{ellipL}}}^{2\omega} - {I_{\mathrm{ellipR}}}^{2\omega})/({I_{\mathrm{ellipL}}}^{2\omega} + {I_{\mathrm{ellipR}}}^{2\omega}), \quad (3)$$

where the two left-handed and right-handed elliptic states are suitable states ranging 90° in the λ/4 wave-plate orientation, and where the starting state (ellipL) is arbitrarily selected. For example in what follows, we selected a starting polarization state (ellipL) by setting the λ/4 wave-plate orientation, which corresponds to the maximum SHG signal and then, adding 90°, we have the λ/4 wave-plate orientation which defines the state named ellipR.

In this case (from Fig. 3), the best SHG_ED value is −0.19 for the *p*-pol. state and −0.20 in the *s*-pol. state. The 'minus' sign indicates that the dominant symmetry is a right-handed symmetry (see formula (3)).

In Fig. 4, the SHG signal of the 300 K sample is shown as a function of the λ/4 fast axis orientation, for a different orientation of the wires with respect to the input pump beam. In this case, the wires are laid in the incidence plane, thus the three directions of the incident light, the metasurface normal and the wires' main direction are coplanar and the extrinsic chirality is not present. The measured SHG_ED is indeed close to zero as shown by the symmetric measurements in Fig. 4.

The phenomenological SHG_ED parameter evidences the asymmetry of the SHG response related to the NW orientation.

3 Conclusions

We report here the evidence of extrinsic chiral behavior of a self-assembled metasurface formed by a forest of tilted gold nanowires. The chiral behavior seems to be geometrically induced by the direction of the nanowires with respect to the direction of the input pump beam and the sample orientation.

The SHG experiments by means of a fs pump laser at 800 nm have been performed in the reflection mode, as a function of the either right-handed or left-handed elliptically polarized input light. The dependence of the SHG signal from the elliptical polarization state gives us the possibility of realizing a new kind of polarization filters in the visible and near IR domain.

More experiments are running on different types of samples, to evaluate the influence of growth conditions on the artificial chiral response.

Acknowledgements

This work was partially financed by MARINE project granted by the Italian Ministry of Defence. The Authors tank F. Mura and CNIS facilities at Sapienza University of Rome for the SEM images of the samples.

References

1 T. Verbiest, K. Clays and V. Rodriguez, *Second-Order Nonlinear Optical Characterization Techniques*, CRC Press, NY, USA, 2009.
2 T. Verbiest, M. Kauranen, A. Persoons, M. Ikonen, J. Kurkela and H. Lemmetyinen, *J. Am. Chem. Soc.*, 1994, **116**, 9203.
3 J. A. H. van Nieuwstadt, M. Sandtke, R. H. Harmsen, F. B. Segerink, J. C. Prangsma, S. Enoch and L. Kuipers, *Phys. Rev. Lett.*, 2006, **97**, 146102.
4 T. Kitahara, A. Sugawara, H. Sano and G. Mizutani, *J. Appl. Phys.*, 2004, **95**, 5002.
5 A. Belardini, M. C. Larciprete, M. Centini, E. Fazio, C. Sibilia, M. Bertolotti, A. Toma, D. Chiappe and F. Buatier de Mongeot, *Opt. Express*, 2009, **17**, 3603.
6 C. Hubert, L. Billot, P.-M. Adam, R. Bachelot, P. Royer, J. Grand, D. Gindre, K. D. Dorkenoo and A. Fort, *Appl. Phys. Lett.*, 2007, **90**, 181105.
7 R. Williams, *Phys. Rev. Lett.*, 1968, **21**, 342.
8 E. Plum1, X.-X. Liu, V. A. Fedotov, Y. Chen, D. P. Tsai and N. I. Zheludev, *Phys. Rev. Lett.*, 2009, **102**, 113902.
9 S. N. Volkov, K. Dolgaleva, R. W. Boyd, K. Jefimovs, J. Turunen, Y. Svirko, B. K. Canfield and M. Kauranen, *Phys. Rev.*, 2009, **A 79**, 043819.
10 S. V. Zhukovsky, C. Kremers and D. N. Chigrin, arXiv:1103.5008v1, 2011.
11 A. Belardini, M. C. Larciprete, M. Centini, E. Fazio, C. Sibilia, D. Chiappe, C. Martella, A. Toma, M. Giordano and F. Buatier de Mongeot, *Phys. Rev. Lett.*, 2011, **107**, 257401.
12 A. Parviz, D. Ryan and G. M. Whitesides, *IEEE Trans. Adv. Packag.*, 2003, **26**, 233.
13 M. Benson, P. Shah, M. Marciniak, A. Sarangan and A. Urbas, *J. Nanomater.*, 2014, **2014**, 694982.
14 A. Belardini, A. Benedetti, M. Centini, G. Leahu, F. Mura, S. Sennato, C. Sibilia, V. Robbiano, M. C. Giordano, C. Martella, D. Comoretto and F. B. de Mongeot, *Advanced Optical Materials*, 2013, **2**, 208.

Faraday Discussions

PAPER

Electrical control of Faraday rotation at a liquid–liquid interface†

Monica Marinescu,*[a] Alexei A. Kornyshev*[b] and Michael E. Flatté*[c]

Received 5th November 2014, Accepted 27th November 2014
DOI: 10.1039/c4fd00210e

A theory is developed for the Faraday rotation of light from a monolayer of charged magnetic nanoparticles at an electrified liquid–liquid interface. The polarization fields of neighboring nanoparticles enhance the Faraday rotation. At such interfaces, and for realistic sizes and charges of nanoparticles, their adsorption–desorption can be controlled with a voltage variation <1 V, providing electrovariable Faraday rotation. A calculation based on the Maxwell-Garnett theory predicts that the corresponding redistribution of 40 nm nanoparticles of yttrium iron garnet can switch a cavity with a quality factor larger than 10^4 for light of wavelength 500 nm at normal incidence.

Introduction

Recent theoretical and experimental advances in electrovariable nanooptics have focused attention on the electrically controllable optical response of nanoparticles (NPs) at liquid–liquid interfaces. Many of them are based on the interface between two immiscible electrolyte solutions (ITIES),[1–3] such as that formed between aqueous and organic oil salt solutions. Semiconductor NPs localized at the interface have been predicted to experience large electric fields, which can produce dramatic room-temperature Stark shifts in the optical transmission.[4] The reversible assembly of charged NPs at the ITIES has been predicted to occur as a result of applied potentials as low as ~1 V.[4] The reflection and transmission coefficients for light incidence on a monolayer of localised metallic NPs were calculated,[5,6] showing a dramatic frequency shift and linewidth broadening of the surface plasmon resonance. The optical properties of gold NPs adsorbed at an ITIES have also been measured.[7] The properties of magnetic NPs at such an interface, however, have not been explored to date, despite the extensive literature

[a]*Department of Mechanical Engineering, Faculty of Engineering, Imperial College, London, SW7 2AZ, UK. E-mail: m.marinescu@imperial.ac.uk*

[b]*Department of Chemistry, Faculty of Natural Sciences, Imperial College, London, SW7 2AZ, UK. E-mail: a.kornyshev@imperial.ac.uk*

[c]*Optical Science and Technology Center and Department of Physics and Astronomy, The University of Iowa, Iowa City, Iowa 52242, USA. E-mail: Michael_Flatte@mailaps.org*

† Electronic supplementary information (ESI) available. See DOI: 10.1039/c4fd00210e

on metallic and insulating magnetic NPs in solution, such as iron, cobalt, ferrite, and magnetite ferrofluids,[8] including measurements of Faraday rotation.[9,10] Theoretical[11] and experimental[12] studies have shown that Faraday rotation by magnetic NPs in a non-magnetic matrix depends on the volume fraction of NPs.

Here we predict that an ITIES with a ferrofluid component provides opportunities for electrically tuning the Faraday rotation, using modest ITIES-sustainable voltages, with possible applications in spectroscopy and for electrically switchable Faraday devices such as optical isolators. With a proper balance of forces between negatively charged NPs, controlled through the pH of the solution and the Debye screening length of the electrolytes, NPs are adsorbed at ITIES spontaneously[13,14] to block the energetically unfavorable water–oil interface.[5] When the aqueous phase is sufficiently positively polarized with respect to the oil phase, the NPs are dispersed in the bulk, whereas for negative voltages they are strongly driven into the ITIES. The applied static magnetic field causes the magnetization of the NPs to orient in the same direction. An increase in the Faraday rotation from localized NPs is expected, as a result of the nanoplasmonic interaction between neighboring particles.

Theory

We have calculated the Faraday rotation angle caused by magnetic NPs assembled at the ITIES according to the Maxwell-Garnett theory,[15,16] and compared it to the rotational angle caused by the same number of NPs dispersed in the bulk. Tuning

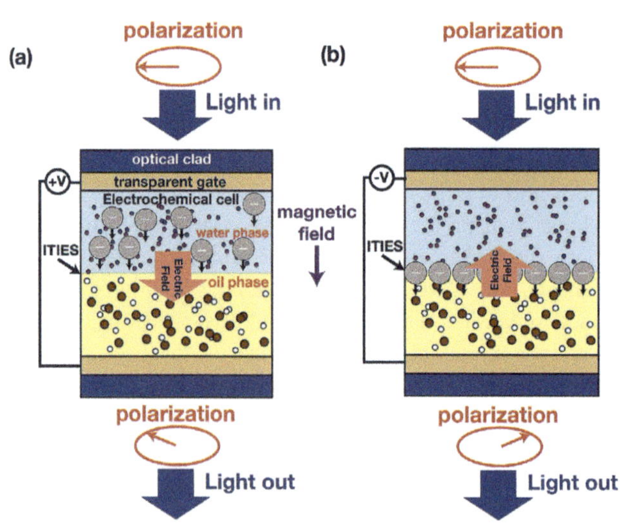

Fig. 1 When a voltage is applied to the interface of two immiscible electrolytic solutions, two back to back electrical double layers are formed, resulting in an electric field localized at the interface. That electric field interacts with the negatively charged nanoparticles to localize them at the interface or move them to the bulk of the aqueous phase. (a) The potential is such that the NPs remain in the bulk of the liquid, forming a dilute solution. (b) The potential is such that the NPs migrate to the liquid–liquid interface where, due to the interaction between the polarization fields, they produce a larger Faraday rotation. Not drawn to scale.

of the Faraday rotation angle is explored for an optical cavity (Fabry–Perot) geometry, in which the direction of light propagation and the magnetic field are perpendicular to the ITIES, and the electrodes are transparent, as shown in Fig. 1. For 40 nm radius NPs of yttrium iron garnet (YIG), tuning of the Faraday rotation angle by $\pi/2$ for 500 nm wavelength light is feasible for an optical cavity with a quality factor Q in excess of 10^4. The main steps in the mathematical derivation and the main model predictions are presented below.

Faraday rotation can be described as the result of imaginary, anti-symmetric, off-diagonal components of a dielectric permittivity tensor; for a magnetic field B_z:

$$\tilde{\varepsilon} = \begin{pmatrix} \varepsilon_x & iA & 0 \\ -iA & \varepsilon_y & 0 \\ 0 & 0 & \varepsilon_z \end{pmatrix}, \tag{1}$$

for which $|\varepsilon_x - \varepsilon_y| \ll A$, as is the case in typical magnetic materials. The polarization angle of linearly-polarized light propagating through a medium with the dielectric tensor of eqn (1) is rotated by an angle per unit length $\theta = (k_+ - k_-)/2$, which is proportional to the difference between the wave numbers of the left and right circularly polarized components: $k_\pm{}^2 = (\omega^2/c^2)(\varepsilon \pm A)^{-1}$. For $\varepsilon_x = \varepsilon_y = \varepsilon$

$$\theta = \frac{1}{2}\frac{\omega}{c}\frac{A}{\varepsilon^{1/2}}. \tag{2}$$

The polarizability of a sphere of radius R is given by the Clausius–Mossotti expression

$$\tilde{\alpha} = R^3\tilde{\varepsilon}_s\frac{\tilde{\varepsilon} - \tilde{\varepsilon}_s}{\tilde{\varepsilon} + 2\tilde{\varepsilon}_s}, \tag{3}$$

where $\tilde{\alpha}$ is the effective polarizability, $\tilde{\varepsilon}$ is the dielectric permittivity tensor of the sphere material, which has the form of eqn (1), and $\tilde{\varepsilon}_s = \varepsilon_s\mathbb{I}$ is the dielectric permittivity of the surrounding fluid, with $\mathbb{I}$ the unit matrix. We assume for simplicity that ε_s is the same for both electrolyte solutions. For a suspension of small particles homogeneously dispersed in the bulk of a liquid, the diagonal and off-diagonal elements of the effective permittivity tensor are[17]

$$\varepsilon_b = \varepsilon_s + \frac{3f\varepsilon_s(\varepsilon - \varepsilon_s)}{3\varepsilon_s + (1-f)(\varepsilon - \varepsilon_s)}, \tag{4a}$$

$$A_b = \frac{3f\varepsilon_s A}{3\varepsilon_s + (1-f)(\varepsilon - \varepsilon_s)}, \tag{4b}$$

where the volume fraction of NPs is denoted by f. In the limit of $f \ll 1$, the expressions coincide to those found for highly dilute solutions,[11] and eqn (4a) becomes the standard Maxwell-Garnett expression.[15,18] For a system with just enough NPs to cover the interface in a monolayer, $f = (4/3)(\pi R^3 n_s/l)$, where l is the depth of the bulk region (the dimension along the path of the incident light in Fig. 1) and n_s the number of particles per unit area of the interface. The corresponding Faraday rotation per unit length for a bulk region with dispersed nanoparticles (θ_b) is then calculated from eqn (2) with $\varepsilon \rightarrow \varepsilon_b$ and $A \rightarrow A_b$.

The dielectric permittivity of a $2R$-thick monolayer (or submonolayer) of NPs at the liquid–liquid interface where the Faraday rotation occurs is given by

$$\tilde{\varepsilon}_{\rm m} = \left(1 - n_{\rm s}\pi R^2\right)\varepsilon_{\rm s}\mathbb{I} + \frac{4\pi n_{\rm s}}{2R}\tilde{\alpha}_{\rm m}. \tag{5}$$

The first term in eqn (5) accounts for the unoccupied ITIES surface, while the second term provides the contribution of the NPs present at the interface. In the case where the two liquids on either side of the ITIES have significantly different permittivities, the expression in eqn (5) needs to be adjusted; here, for simplicity, we neglect the usually minor difference in the refractive indices of the two phases (this difference can be accounted for using the method of ref. 6). For a hexagonal lattice of NPs at the interface, $n_{\rm s} = \Theta/(2\sqrt{3}R^2)$, where $\Theta = N/N_{\rm max}$ is the coverage of the monolayer: the number N of NPs present at the interface relative to the maximum occupation number $N_{\rm max}$. The polarizability of the monolayer can be calculated as in eqn (8) of ref. 16.

$$\tilde{\alpha}_{\rm m} = \tilde{\alpha}(\mathbb{I} + \tilde{\alpha}\tilde{U})^{-1}, \tag{6a}$$

$$\tilde{U} = \frac{U_0\Theta^{3/2}}{a^3\varepsilon_{\rm s}}\begin{pmatrix} -1/2 & 0 & 0 \\ 0 & -1/2 & 0 \\ 0 & 0 & 1 \end{pmatrix}. \tag{6b}$$

Here $U_0 = 11.03$ defines the dipole–dipole interaction within the lattice and is constant for a given lattice type, here chosen as hexagonal;[19] a is the minimum lattice constant, corresponding to $N = N_{\rm max}$. Eqn (6) is valid for an average distance between neighboring particles of a few radii, as no multipole interactions are considered, and in the limit of long wavelengths of incident light compared to the lattice constant.

From eqn (1), (3) and (6), $\tilde{\alpha}_{\rm m}$ can be calculated:

$$\tilde{\alpha}_{\rm m} = \frac{\varepsilon_{\rm s}R^3}{\gamma}\begin{pmatrix} \dfrac{2}{2-C} & \dfrac{4iB}{(C-2)^2} & 0 \\ -\dfrac{4iB}{(C-2)^2} & \dfrac{2}{2-C} & 0 \\ 0 & 0 & \dfrac{1}{C+1} \end{pmatrix}, \quad \gamma = \frac{\varepsilon + 2\varepsilon_{\rm s}}{\varepsilon - \varepsilon_{\rm s}} \tag{7a}$$

$$B = \frac{3\varepsilon_{\rm s}A}{(\varepsilon - \varepsilon_{\rm s})(\varepsilon + 2\varepsilon_{\rm s})}, \quad C = \frac{R^3U_0\Theta^{3/2}}{\gamma a^3}. \tag{7b}$$

From eqn (5) and (7), $\varepsilon_{\rm m}$ can be calculated and thus, the corresponding Faraday rotation. In the limit of a vanishing dipole–dipole interaction, by identifying $2\pi n_{\rm s}R^2 \rightarrow 3f$ and taking the limit of a dilute solution $(1 - n_{\rm s}\pi R^2) \rightarrow 1$, the results for the bulk, eqn (4), are retrieved.

Results

Fig. 2 displays the net rotation in radians caused by NPs dispersed in the system for a transparent magnetic material commonly used in Faraday isolators, yttrium iron garnet (YIG). Full coverage of the interface is not crucial in obtaining a

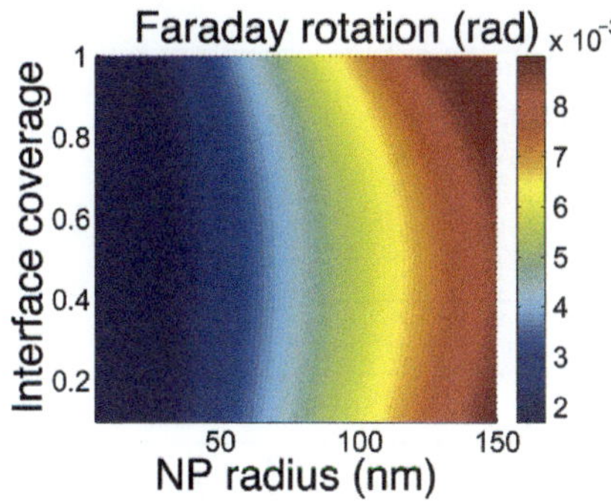

Fig. 2 Faraday rotation caused by NPs in the bulk and at the interface depends on the interfacial coverage and the NP size. The total number of NPs in the system is that necessary for a monolayer coverage. Parameter values: $\varepsilon_s = 1.78$, $\varepsilon = 4.84$,[20] $\lambda = 500$ nm, $\theta \leq 5 \times 10^4$ rad m^{-1}.[21]

sizeable rotation difference between the nanoparticle-covered and the free interface.

The voltage-dependent coverage of an ITIES with functionalized NPs as a function of the applied electrical potential has been calculated based on the free energy profile of a single NP at the ITIES.[4,5] Localised NPs were allowed to interact pairwise *via* screened electrostatic forces. Depending on the system parameters such as the NP size and charge, the contact angle with the two liquids, and the magnitude and polarity of the applied field, a stable or metastable bound state of the NP arises at the interface. Parameters for the free energy calculation used here are taken from ref. 5.† The increasing coverage of the ITIES with NPs for an increasingly positive applied potential is depicted in Fig. 3 for three NP sizes.

For a system with NP radius $R = 40$ nm and $\varepsilon = 4.84$, the rotation per unit length in the bulk, θ_b, and in the monolayer, θ_m, are 4×10^{-2} rad m^{-1} and 10^4 rad m^{-1} respectively. The net rotation when all particles are dispersed in the bulk liquid is

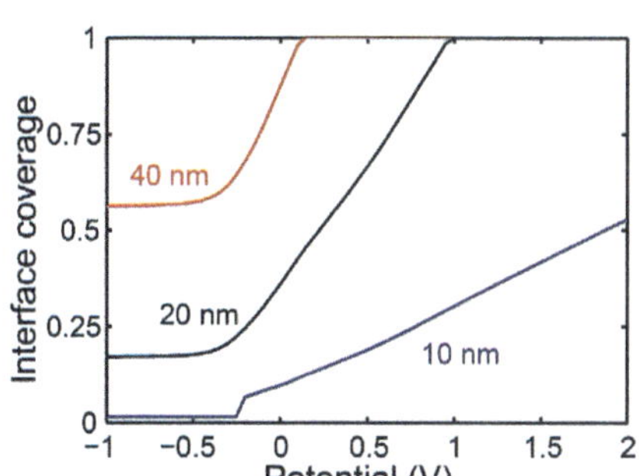

Fig. 3 A small variation in the applied potential drop between the oil phase and aqueous phase changes the interfacial coverage Γ dramatically. This is the combined result of the energy profile of a single NP and the electrostatic repulsion between NPs at the interface. Curves were calculated following ref. 5 and 22, as described in ESI.† Parameter values: $\varepsilon_1 = 78.8$ (water), $\varepsilon_2 = 10.7$ (1,2 dichloroethane), $z = -4000$, $\kappa_1 = 5 \times 10^7$ m^{-1} (0.2 mM), $\kappa_2 = 6 \times 10^8$ m^{-1} (5 mM), $\sigma_{12} = 0.03$ N m^{-1}, $\mu = 10^{-11}$, $\theta_{contact} = 0.55\pi$.

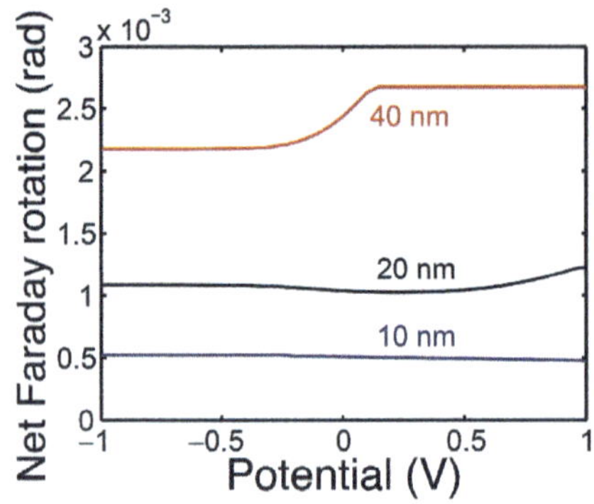

Fig. 4 The net Faraday rotation for light passing one time perpendicular to the interface varies with the applied potential. A change in the coverage as a function of the potential is assumed, as in Fig. 3. Significant effects appear only when a large enough potential is applied to lead to a monolayer coverage. Parameter values: $\varepsilon_s = 1.78$, $\varepsilon = 4.84$,[20] $\lambda = 500$ nm, $\theta \leq 5 \times 10^4$ rad m^{-1}.

$l\theta_b = 2.1 \times 10^{-3}$ rad, while from the monolayer it is $2R\theta_m = 2.7 \times 10^{-3}$ rad, leading to a difference in the polarization angle of 6×10^{-6} rad for a single pass (Fig. 4). The change in the Faraday rotation upon accumulation of NPs at the interface originates from the enhancement of the off-diagonal dielectric polarizability of a single NP due to the polarization fields of neighboring NPs. As more NPs populate the ITIES, the distance between them decreases and the polarizability enhancement becomes stronger, reaching a maximum for full coverage.

Discussion

Although the rotation of the polarization angle upon one pass through the interface is modest, the effect is amplified in a high quality factor cavity. The

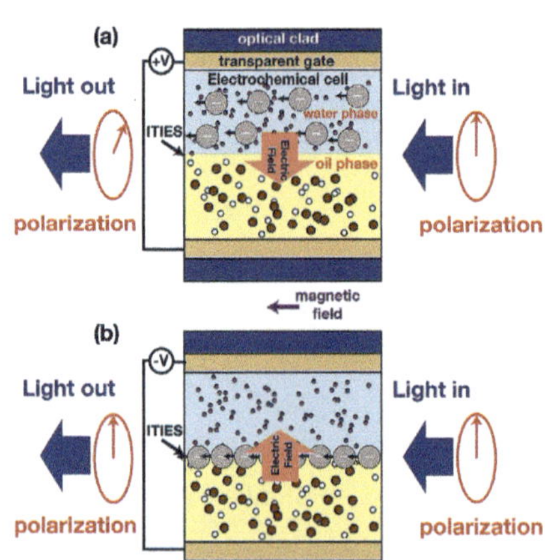

Fig. 5 Schematic of a waveguide version of Fig. 1. (a) The potential is such that the NPs remain in the bulk of the liquid, forming a dilute solution, and Faraday rotation occurs. (b) The potential is such that the NPs migrate to the liquid–liquid interface where, due to an interface-induced birefringence, the Faraday rotation is quenched. Not drawn to scale.

criteria for switching the cavity from completely transmissive to reflective is that the shift in resonant frequency due to the altered Faraday rotation would exceed the cavity linewidth ($\approx \omega/Q$, where ω is the resonant frequency of the cavity). This would occur for a cavity with a quality factor $Q > 10^4$.

Conclusions

We have presented a geometry for electrically-controllable Faraday rotation due to nanoparticle accumulation at a liquid–liquid interface. The enhanced polarization of the nanoparticles due to their accumulation at the interface leads to an enhancement of the normal-incidence Faraday rotation. This effect should be visible in a high quality factor ($Q > 10^4$) optical cavity. For a small electrochemical cell and voltages less than 1 V, electrical switching of a Faraday isolator can be achieved in this geometry.

As a concluding note, we emphasize that other nonlinear optical effects are possible that involve switching the Faraday rotation due to NP accumulation at the interface. As an example, in Fig. 5 we show a waveguide geometry. Here the highly anisotropic linear polarization of the nanoparticles when accumulated at the interface, seen from eqn (6), leads to an efficient quenching of the Faraday rotation for light propagating parallel to the interface. The linear birefringence that causes the quenching of the Faraday rotation can be compensated for with other linear birefringent elements in an optical device.

Acknowledgements

M. E. F. acknowledges an ARO MURI. A.A.K. acknowledges the European Science Foundation FP7-MPP-2011 grant "Nanodetector" and the grant of the Engineering and Physical Sciences Research Council UK, EP/L02098X/1.

References

1 H. H. J. Girault and D. J. Schiffrin, *Electroanal. Chem.*, 1989, **15**, 1–141.
2 Z. Samec, *Pure Appl. Chem.*, 2004, **76**, 2147–2180.
3 H. H. Girault, *Nat. Mater.*, 2006, **5**, 851–852.
4 M. E. Flatté, A. A. Kornyshev and M. Urbakh, *Proc. Natl. Acad. Sci. U. S. A.*, 2008, **105**, 18212–18214.
5 M. E. Flatté, A. A. Kornyshev and M. Urbakh, *J. Phys. Chem. C*, 2010, **114**, 1735–1747.
6 A. A. Kornyshev, M. Marinescu, J. Paget and M. Urbakh, *Phys. Chem. Chem. Phys.*, 2012, **14**, 1850–1859.
7 M. Hojeij, N. Younan, L. Ribeaucourt and H. H. Girault, *Nanoscale*, 2010, **2**, 1665–1669.
8 *Colloidal Magnetic Fluids*, ed. S. Odenbach, Springer, New York, 2010.
9 H. W. Davies and J. P. Llewellyn, *J. Phys. D: Appl. Phys.*, 1980, **13**, 2327–2336.
10 C. P. Pang, C. T. Hsieh and J. T. Lue, *J. Phys. D: Appl. Phys.*, 2003, **36**, 1764–1768.
11 P. M. Hui and D. Stroud, *Appl. Phys. Lett.*, 1987, **50**, 950–952.
12 D. A. Smith, Y. A. Barnakov, B. L. Scott, S. A. White and K. L. Stokes, *J. Appl. Phys.*, 2005, **97**, 10M504.

13 V. A. Turek, M. P. Cecchini, J. Paget, A. R. Kucernak, A. A. Kornyshev and J. B. Edel, *ACS Nano*, 2012, **6**, 7789–7799.
14 M. P. Cecchini, V. A. Turek, J. Paget, A. A. Kornyshev and J. B. Edel, *Nat. Mater.*, 2013, **12**, 165–171.
15 C. Maxwell-Garnett, *Philos. Trans. R. Soc. London, Ser. A*, 1904, **203**, 385–420.
16 B. N. J. Persson and A. Liebsch, *Phys. Rev. B: Condens. Matter Mater. Phys.*, 1983, **28**, 4247–4254.
17 M. Abe and M. Gomi, *Jpn. J. Appl. Phys., Part 1*, 1984, **23**, 1580–1585.
18 L. D. Landau, E. M. Lifshitz and L. P. Pitaevskii, *Electrodynamics of continuous media*, Pergamon Press, 2nd edn, 1984, vol. 8.
19 J. Topping, *Proc. R. Soc. London, Ser. A*, 1927, **114**, 67–72.
20 A. M. Hofmeister and K. R. Campbell, *J. Appl. Phys.*, 1992, **72**, 638–646.
21 W. Wettling, B. Andlauer, P. Koidl, J. Schneider and W. Tolksdorf, *Phys. Status Solidi B*, 1973, **59**, 63–70.
22 M. E. Flatté, A. A. Kornyshev and M. Urbakh, *J. Phys.: Condens. Matter*, 2008, **20**, 073102.

PAPER

Fluorescence axial nanotomography with plasmonics

Nicholas I. Cade,[ab] Gilbert O. Fruhwirth,[cd] Alexey V. Krasavin,[a] Tony Ng[c] and David Richards*[a]

Received 7th October 2014, Accepted 16th October 2014
DOI: 10.1039/c4fd00198b

We present a novel imaging technique with super-resolution axial sensitivity, exploiting the changes in fluorescence lifetime above a plasmonic substrate. Using conventional confocal fluorescence lifetime imaging, we show that it is possible to deliver down to 6 nm axial position sensitivity of fluorophores in whole biological cell imaging. We employ this technique to map the topography of the cellular membrane, and demonstrate its application in an investigation of receptor-mediated endocytosis in carcinoma cells.

1 Introduction

Plasmonic nanostructures have received considerable attention in recent years for their promise as new sensing platforms. This attention has focused principally on the strong confinement of electromagnetic fields, leading to exquisite sensitivity of localised surface plasmon resonances (LSPRs) to the local refractive index and to surface enhanced Raman spectroscopy, both approaches providing high sensitivity to surface-bound molecules. Here we illustrate the power of thin film plasmonics to additionally provide high sensitivity to variations in the spatial distribution of fluorescent molecules within a hundred nanometres of a metal surface, by exploiting modifications to the radiative and non-radiative decays experienced by the molecules.

A range of fluorescence imaging techniques have emerged in recent years which are able to provide spatial super-resolution in three dimensions,[1,2] enabling the nanoscale mapping of cellular structures. However, generally these techniques do not retain compatibility with live cell imaging, while there are inevitable compromises on imaging speed and area. On the other hand, in many

[a]*Department of Physics, King's College London, Strand, London WC2R 2LS, UK. E-mail: david.r.richards@kcl.ac.uk; Fax: +44 (0)20 7848 2420; Tel: +44 (0)20 7848 2753*
[b]*Cancer Research UK, London Research Institute, 44 Lincoln's Inn Fields, London WC2A 3LY, UK*
[c]*Richard Dimbleby Department of Cancer Research, Division of Cancer Studies, Randall Division of Cell & Molecular Biophysics, King's College London, London SE1 1UL, UK*
[d]*Department of Imaging Chemistry & Biology, Division of Imaging Sciences & Biomedical Engineering, King's College London, London SE1 7EH, UK*

cases the lateral resolution of >200 nm achievable with conventional confocal or multiphoton microscopy is in fact sufficient, and it is the diffraction-limited resolution of >500 nm along the optical axis (perpendicular to the substrate surface) which limit their effectiveness for the measurement of subtle spatial dynamics of molecules, such as receptors in a cell membrane. Although wide-field microscopies such as total internal reflection (TIRF) microscopy[3,4] and fluorescence interference contrast microscopy (FLIC)[5] do enable axial resolution <100 nm, the encoding of the axial position of fluorophores through fluorescence intensity is hindered by photo-bleaching and variations in labelling homogeneity, while these techniques are also only sensitive to fluorophores very close to the substrate surface.

We have previously demonstrated that confocal fluorescence lifetime imaging (FLIM) of cells adhered on both smooth[6] and nanostructured[7] noble metal films can be used to precisely determine the axial position of fluorescent molecules up to about 100 nm above the surface, with an axial position sensitivity of less than 10 nm in whole cell imaging, offering significant advantages over other methods for monitoring the redistribution of fluorophores above a surface. In particular, changes in fluorescence lifetime are insensitive to changes in fluorescence intensity, either from variations in expression levels or from photo-bleaching. We have demonstrated that this technique provides a novel bioassay for screening cell receptor internalisation,[6,7] enabling the sensitive measurement of the redistribution of proteins in a cell during endocytosis, even when averaging over a large area. For membrane-bound fluorophores, the measured lifetime variation can be converted into an absolute distance above the substrate for tomographic imaging with super-resolution accuracy, thereby providing information on changes in membrane morphology during dynamic processes.[7]

Cell imaging performed using a flat gold film,[6] which has also been demonstrated subsequently by Chizhik *et al.*,[8] exploits the large increase in non-radiative damping, and hence the reduction in the fluorescence lifetime, of molecules located at sub-wavelength separations above a smooth gold film.[9–11] However, in this approach, there is either a loss of signal from imaging through a semi-transparent film, or it is necessary to image through the sample from above. The latter case increases the background fluorescence in thick samples and limits the collection numerical aperture; furthermore, this geometry is not suitable for studying live biological processes when normally the sample would be excited from below using an inverted microscope.

Efficient excitation through the film is made possible with the use of a large area nanostructured plasmonic thin film.[7] This maintains the flexibility and lateral resolution of a standard confocal configuration suitable for use in live-cell imaging; in addition, the nanostructured film can be designed to optimise the spectral overlap of the localised surface plasmon resonances of the metal nanostructure with the molecular fluorescence, to achieve the greatest lifetime sensitivity and signal enhancement. In this case, the Purcell effect also leads to an enhancement in radiative decay,[12] resulting in an increase in fluorescence intensity alongside a reduction in fluorescence lifetime, for molecules in close proximity to the nanostructure surface.[13]

We report here on the application of nanostructured plasmonic thin films, in combination with fluorescence lifetime imaging, for axial nanotomography of

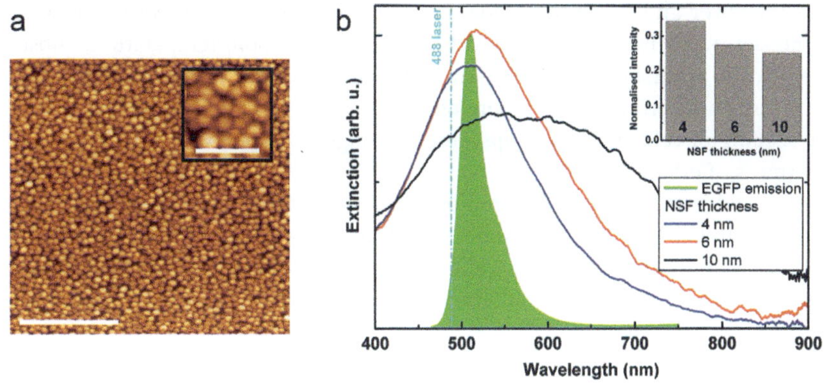

Fig. 1 (a) AFM image of a silver film with a 6 nm nominal Ag thickness; scale bar is 2 μm. Inset: enlarged region; scale bar is 500 nm. (b) Extinction spectra for large area nanostructured silver films (NSFs) of nominal thickness 4 nm (blue line), 6 nm (red) and 10 nm (black). Also shown is the emission spectrum of eGFP (filled green), and the 488 nm excitation laser. Inset: fluorescence intensity from cells on a NSF normalised to those on a nearby area of glass (a scratch in the silver film) for NSFs of different nominal thickness.

cells; in particular, we compare directly their use with that of flat solid metal films.

2 Experimental methods

2.1 Plasmonic substrates for cell culture and imaging

We have developed nanostructured silver films, comprising densely-packed self-organised distributions of silver nanoparticles, that can be easily and inexpensively manufactured over large continuous areas by thermal vacuum deposition.[7,14] A 1.3 nm layer of Cr was first thermally evaporated onto 0.2 mm glass coverslips (Agar Scientific, UK) to increase the adherence of Ag, followed by Ag with a nominal thickness of between 4 and 10 nm, in a vacuum chamber at 10^{-6} Torr (U300, Oerlikon Leybold, Germany). The evaporation rate (0.05 nm s^{-1}) and nominal metal thickness were measured during evaporation with a quartz crystal oscillator (E5550; Quorum Technologies, UK). The films were subsequently annealed for 30 min at 190 °C in an Ar atmosphere to create a nanostructured surface with a strong plasmon resonance; the nominal deposited thickness of the silver film and subsequent annealing allows the peak of the LSPR spectrum to be tuned to the required value.[15] Atomic force microscopy (AFM) images (see *e.g.* Fig. 1a) acquired in tapping mode (Caliber, Veeco Instruments, CA) and analyzed using Gwyddion 2.21 (http://gwyddion.net), indicate that the films consist of individual semi-spheroidal nanoparticles, closely packed with a density and diameter dependent on the nominal evaporated thickness of the film. The extinction spectra from the nanostructured silver films with different nominal thicknesses are shown in Fig. 1b, indicating a strong LSPR of these nanostructures, providing good spectral overlap with the emission spectrum of enhanced green fluorescence protein (eGFP).

For compatibility with cell culture, nanostructured silver films were coated with *N*-(2-mercaptopropionyl)glycine (MPG) which forms a protective monolayer over the silver and significantly lowers its cytotoxicity.[16] Coated coverslips were immersed in a 1 mM solution of MPG for 60 min at room temperature. Subsequently, the coverslips were twice washed with phosphate buffered saline (PBS, pH = 7.4) and sterilized by submerging into 70% (v/v) ethanol for 60 min. Before the seeding of cells, the coverslips were thoroughly washed with PBS to remove any traces of ethanol.

Flat gold surfaces were prepared by deposition of 30 nm-thick gold films on sixteen-well LabTek chamber slides (Thermo Fisher Scientific, Denmark) by thermal evaporation of gold (as above). The slides were disassembled before evaporation, and then reassembled and tested for integrity. No further treatment was required for cell culture.

2.2 Cell culture and treatment

The DNA of the human chemokine receptor CXCR4 was fused to the N-terminus of enhanced green fluorescent protein (GFP) and the resulting plasmid was used to generate a mammary adenocarcinoma cell line stably expressing CXCR4-GFP, as described elsewhere.[17] Cells were trypsinised for 5 min, seeded onto the metal-coated slides, and grown to the desired cell density. The cells exhibited normal growth on both MPG-treated nanostructured silver films and gold films, with the MPG coating preventing silver-induced cytotoxicity. Subsequently, cells were fixed in 4% (w/v) paraformaldehyde for 15 min, washed with PBS and rinsed with deionized water, before the samples were mounted in Mowiol 488 containing 2.5% (w/v) Dabco. All standard chemicals were either from Sigma-Aldrich (UK) or VWR (UK).

2.3 Fluorescence lifetime imaging

Fluorescence lifetime and intensity images of nanostructured silver film coated slides were obtained by imaging through the coverslip and silver film in a conventional inverted microscope geometry. A confocal microscope (TE2000E, Nikon Corporation, Japan) with custom built scanning stage was used with a 100× oil immersion lens (1.45 NA; Leica, Germany). Photon detection was carried out with an avalanche photodiode (Id-100; Id Quantique, Switzerland) and a synchronous time-correlated single photon counting (TCSPC) module (SPC-150; Becker & Hickl, Germany), with a band-pass filter (550/88, Semrock, IL) fitted before the detector. Excitation was at 488 nm and 20 MHz using a super-continuum laser (SC-450-2; Fianium, UK), with the output sent through a monochromator, laser line filter (488IN, Comar, UK), and beam expander.

Measurements of Au-coated slides were performed using a scanning confocal microscope (TCSP2; Leica, Germany) scanning at 400 Hz with a 63× oil immersion lens (1.4 NA; Leica), through a coverslip placed over the sample. Photon detection was with a photomultiplier tube (PCM-100; Becker & Hickl, Germany) and a synchronous TCSPC module (SPC-830; Becker & Hickl). A 464 nm long-pass filter (XE458; Omega Optical, USA) and EGFP band-pass filter (525/50M; Chroma Technology, USA) were fitted before the detector. Excitation was at 442 nm and 76 MHz using a Ti:Sapphire laser (Mira 900; Coherent Laser, USA) with frequency doubler.

All measurements were acquired in the focal plane of the metal surface, except where indicated. The laser power was adjusted to give average photon counting rates below the maximum rate of the photon counting electronics and far below the fluorescence saturation level. The photon arrival times were binned into 256 time windows over a period of 10–25 ns. Spatially resolved FLIM images of cells were acquired for 300–600 s at a resolution of 256 × 256 pixels. The data was typically binned with a radius of 2–5 pixels, and the decay transient for each pixel bin was fitted with a mono-exponential fit after correcting for background fluorescence and an instrument response time of ~180 ps (SPCImage, Becker & Hickl). The intensity of the resulting lifetime image was weighted with the total fluorescence intensity at each point.

3 Distance calibration

Fluorescent molecules in close proximity to a gold or silver surface (smooth or nanostructured) experience a modification of their fluorescence lifetime which is dependent on their separation from the surface. This provides a means for the determination of the axial position of a fluorophore from a noble metal surface, as we and others have shown,[7,8,18] provided there exists a calibration of the dependence of the modified fluorescence lifetime with the separation.

We have obtained an empirical calibration of the axial position using micro-beads surface labeled with fluorescein isothiocyanate (FITC) (M0162; Sigma, Gillingham, UK) (such that they closely mimic the geometry of membrane labelled cells, with photophysical properties similar to eGFP). The spherical geometry of the beads, of known diameter $2R$ (6–10 μm), permits the determination of the separation d of fluorophores from the plane of contact, using the radial distance r from the centre of symmetry (identified with the point of contact) in the FLIM images (see Fig. 2a).[6,7]

200 μL of the microspheres were centrifuged at 5000 rpm for 2 min, the supernatant was removed, 100 μL of Mowiol (ICN, USA) containing Dabco as an antifade was added and well mixed, and a few drops of the microspheres were placed on a slide and covered with a glass coverslip. Fluorescence lifetime images of the micro-beads were then acquired in the plane of the metal film surface (flat gold or nanostructured silver film), imaging only close to the point of contact, where the bead surface is close to the substrate surface and the surface-labelled fluorophores lie within the excitation focal volume. The lateral resolution of ~250 nm in our diffraction-limited confocal images means that our measurements average over a small range of axial separations. In the case of the nanostructured silver film substrate, this also results in FLIM measurements averaging typically over a number of silver nanoparticles at any given point (see Fig. 1a), such that it can be considered to be effectively a smooth surface.

For both types of surface, a reduction in the fluorescence lifetime is observed around the centre of the microsphere image, corresponding to locations where the fluorophores bound to the microsphere surface are close to the metal surface. For the flat gold film, there was an associated reduction in fluorescence intensity while, for the nanostructured silver film substrate, we observed a two-fold increase in fluorescence intensity, commensurate with the decrease in fluorescence lifetime. Lifetime analysis was performed with a constant bin radius of 3 pixels (~250 nm), and the fluorescence lifetime image was then radially averaged around the point of contact (constant r), to obtain a distance–lifetime profile. The average profiles for both nanostructured silver and flat gold film substrates are

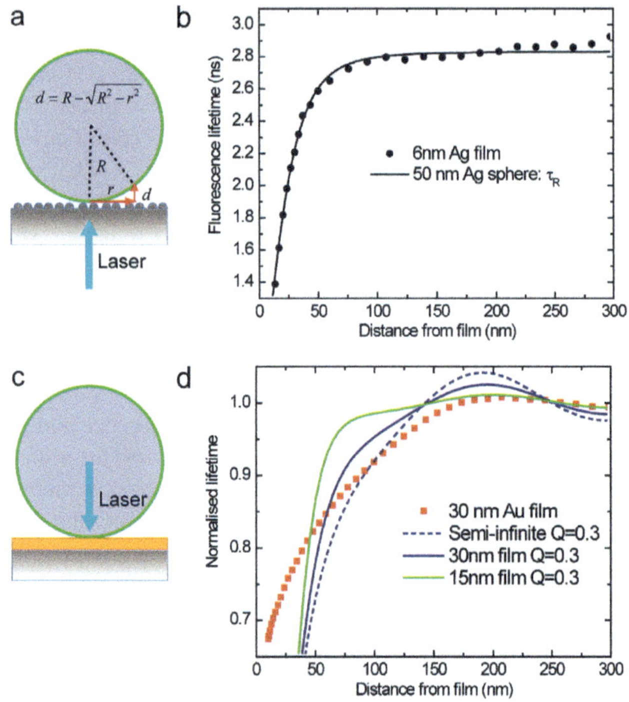

Fig. 2 Distance-dependence characterisation of fluorescence modification by a nano-structured silver film of nominal thickness 6 nm (a and b) and a flat 30 nm-thick gold film (c and d), using fluorescence lifetime imaging (FLIM) of FITC-surface-labelled micro-beads deposited on the surfaces. (a and c) Schematic illustrations of the measurements, indicating in (a) the method of determination of the fluorophore-surface separation, *d*, from the distance from the centre of the micro-bead FLIM image, and the known radius, *R*, of the micro-bead. (b and d) Fluorescence lifetime, normalised to the unmodified eGFP lifetime as a function of the axial distance above the surface. (b) For the nanostructured silver film, obtained by radially averaging the images from 3 beads of different diameter; the solid line is the polarization-averaged radiative lifetime τ_R of a dipole as a function of the distance away from a 50 nm Ag sphere in water, calculated at an emission wavelength of 510 nm. (d) Fluorescence lifetime dependence for the flat gold film obtained by radially averaging the FLIM image of an 8 μm diameter micro-bead. The continuous lines show the polarization-averaged radiative lifetime of a dipole, in water, as a function of the distance away from a Au film of thickness 15 nm (solid green line), 30 nm (solid blue line) and semi-infinite (dashed line), calculated at an emission wavelength of 510 nm and an unmodified quantum yield of 0.3.

shown in Fig. 2, with a rapid reduction in both cases in the fluorophore lifetime as the separation from the metal surface decreases. However, while the flat gold film indicates already a high degree of distance sensitivity of ~20 nm per 100 ps for separations less than 50 nm, the nanostructured silver film provides a much higher sensitivity of ~3 nm per 100 ps within ~50 nm of the surface.

The results of a theoretical analysis of the expected dependence of fluorescence lifetime on separation from the metal film are also shown in Fig. 2. The nanostructured silver film was modelled as an approximately homogeneous array of uncoupled spherical silver spheres in water, with the fluorescence lifetime of a FITC molecule calculated using a dipole–dipole model, as a function of the separation from the sphere;[19,20] the nanoparticle radius was taken to be 50 nm, corresponding to the average radius and height of the nanoparticles measured with AFM for the nanostructured silver film (of nominal thickness 6 nm). Decay rates were averaged over the tangential and radial orientations of the emitting dipole assuming an isotropic distribution, for an emission wavelength of 510 nm, with an excellent agreement with the experimental data (Fig. 2b); these calculations indicate that the reduction in the fluorescence lifetime is dominated by an enhancement in radiative decay. We note that the diffraction limited measurement, over a number of nanoparticles and a large number of molecules around the point of contact of the micro-bead, results in insensitivity to separations less than ~10 nm, such that our measurements do not probe the expected reduction of emission intensity at very small separations.[21,22]

On the other hand, in the case of a flat gold film (Fig. 2d), for separations less than 50 nm, the observed behaviour could not be reproduced from calculations of the expected fluorescence lifetime dependence (averaged over the tangential and radial orientations of the emitting dipole),[23,24] with the experiment demonstrating a slower change in lifetime than that predicted theoretically. Unlike a nanostructured film, for which an enhancement in fluorescence intensity and radiative decay rate occurs with the decreasing separation, fluorescence is quenched as a fluorophore approaches a flat gold surface, associated with an increase in only the non-radiative decay rate, from the non-radiative energy transfer to surface plasmons and the direct power dissipation in the gold film. The consequence of this is that, given the limited axial and lateral resolution of confocal imaging, few photons are detected from the molecules closest to the gold film: the fluorescence lifetime decay determined using TCSPC originates from fluorescence further away from the surface (although within the focal area), while the low signal levels also result in greater errors in the lifetime determination. The net effect of this is that, in reality, the first part of the empirical curve corresponds to larger distances, accounting for the departure from the theoretical prediction for separations less than 50 nm. The calculated fluorescence lifetime dependence for a flat gold film is also strongly dependent on both the fluorophore quantum yield and the thickness of the gold film, with a quantum yield of only 0.3 providing the best agreement for large separations.

The measured dependence of the fluorescence lifetime on the fluorophore separation from the metal film surface provides an empirical calibration of the separation as a function of the fluorescence lifetime in FLIM. There are, however, various sources of measurement error which limit the axial sensitivity of this approach. Firstly, the sensitivity to lifetime variations is limited by the instrumental response of the system of 180 ps, which is convolved with the fluorescence decay. Low photon counts can also limit significantly mono-exponential fits of the fluorescence decay and, as identified above, these can be dominant in the case of small separations in flat gold films. For nanostructured silver films, the accuracy in the relative sample height above the film is limited by the distribution in the nanoparticle heights within a binned area of the FLIM image (approximately the size of the confocal spot, ~250 nm); from the AFM height analysis, this is approximately ±5 nm. Propagating these uncertainties, we estimate for nanostructured silver substrates an error in the final height map of approximately ±6 nm when closest to the film and ±12 nm for distances ~65 nm.

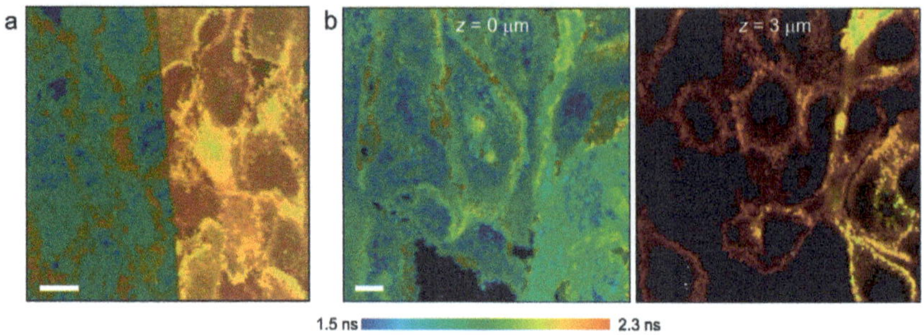

Fig. 3 Fluorescence lifetime images (intensity weighted) of carcinoma cells expressing eGFP on the CXCR4 membrane receptor: (a) at a NSF (left)–glass (right) boundary; scale bar 10 μm; (b) from a region of cells on a NSF, when imaging in the plane of the nanostructured silver film (left) and 3 μm above the NSF (right); scale bar 5 μm.

4 Nanotomography of biological cells

Metal-coated glass coverslips were then employed in a fluorescence imaging study of mammary carcinoma cells expressing the chemokine receptor CXCR4 C-terminally fused to eGFP at the plasma membrane (so eGFP is at the cytosolic side of the membrane). Fig. 3a shows a FLIM image of cells at the boundary between uncoated glass and a nanostructured silver film with a nominal film thickness of 6 nm. On the glass region, the cells have a very narrow lifetime distribution around that expected for eGFP; in contrast, similar cells on the nanostructured silver film show significant lifetime reduction, indicating that the eGFP is located within 50 nm of the surface. FLIM images acquired from a focal plane a few microns above the nanostructured silver film show no reduction in lifetime, compared to those acquired in the plane of the film (Fig. 3b), confirming that the lifetime modification occurs only over a small distance above the silver nanostructure. Note that such measurements >100 nm from the substrate would not be possible with TIRF.

FLIM of cells cultured on nanostructured silver-coated slides with different nominal film thicknesses, and hence nanoparticle sizes and LSPR extinction spectra (Fig. 1b),[14] all demonstrated the same behaviour. However, the fluorescence intensity was found to be greatest for a nominal thickness of 4 nm, and lower for nominal thicknesses of 6 nm and 10 nm (see inset Fig. 1b). This signal reduction may simply result from the increase in the coating material, although we note that the overlap of the extinction spectrum for the plasmon film with both the excitation wavelength and the eGFP emission is greatest for the 4 nm-film, such that the strongest fluorescence enhancement is expected through resonance with the localised surface plasmons of the silver nanostructure.

High resolution fluorescence intensity and associated fluorescence lifetime images of single cells are presented in Fig. 4, for both nanostructured silver film and flat gold substrates. Small changes in the membrane topography can be identified by the resulting variation in the fluorescence lifetime across the cells and so, height maps were obtained by applying the normalised empirical lifetime calibration (Fig. 2) to the FLIM image data, re-normalised to the unmodified cell fluorescence lifetime (the modal value for cells cultured on glass). We note that

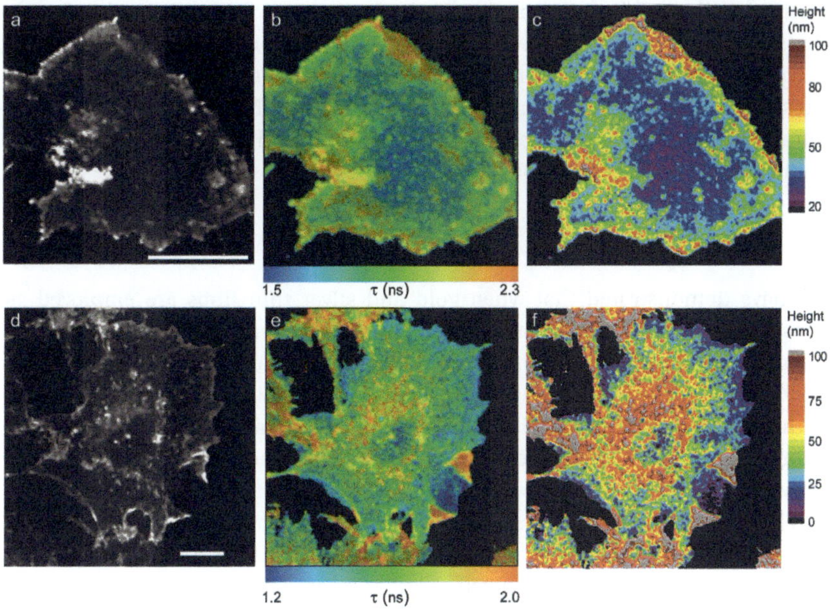

Fig. 4 FLIM and nanotomography of carcinoma cells expressing eGFP on the CXCR4 membrane receptor, cultured on (a–c) a nanostructured silver film (NSF) and (d–f) a flat gold film (scale bars 10 μm). (a and d) Fluorescence intensity and (b and e) fluorescence lifetime, demonstrating the plasmonic assisted reduction in the lifetime of fluorophores expressed in the ventral membrane of the cells. (c and f) Height maps derived from the fluorescence lifetime using calibration curves in Fig. 2; heights were binned in 3 nm intervals. Fluorescence quenching for smaller separations of fluorophores from the flat gold film results in an over-representation of longer lifetimes, such that the height map for the cell cultured on the flat gold film (f) is not considered to be so reliable.

since eGFP is fused to the C-terminus of the CXCR4 receptor, it is separated from the metal films by at least the width of the plasma membrane (~10 nm).

For a nanostructured silver film (Fig. 4c), large areas of the basal cell membrane are determined to be 30–50 nm from the substrate, noting that this averages over the diffraction-limited resolution of ~250 nm. In addition, the height map also identifies non-attached and vertical parts at the edge of the cell, which have been reported not to be attached to the substrate.[25] For the majority of the cell image, the lifetime data is very well fitted by a mono-exponential distribution, which suggests a negligible contribution to the signal from out-of-focus components.

The height map derived for the flat gold substrate (Fig. 4f) presents a very different picture, which we consider does not provide an accurate representation of the height variation of the basal cell membrane. The lower numerical aperture of the objective lens required for imaging from above, due to the opaque substrate, leads to a worse confocal axial resolution, hence resulting in the capture of additional fluorescence from eGFP elsewhere in the cell (*e.g.* other organelles containing CXCR4-eGFP receptors and also the dorsal cell membrane). This is accentuated in the case of a flat gold substrate because the reduction in lifetime is associated with quenching of the fluorescence intensity, such that fluorophores located >100 nm from the substrate with longer fluorescence lifetimes will provide a greater contribution to the total fluorescence signal. This leads to a distribution of lifetimes, and hence the corresponding height map becomes less reliable in these regions. This is less of an issue with nanostructured substrates, which provide both compatibility with high numerical aperture oil immersion lenses, and also lead to an enhancement (rather than reduction) of the fluorescence intensity for molecules close to the substrate.[7]

5 Conclusions

We have demonstrated that when gold and silver thin films are employed as substrates in biological cell imaging, the distance-dependent modification of the fluorescence lifetime allows the precise determination of the molecular location up to about 100 nm above the substrate surface, providing tomographic imaging of the basal cell membrane with axial super-resolution accuracy.[7] Changes in the fluorescence lifetime are insensitive to changes in the fluorescence intensity, either from variations in expression levels or from photo-bleaching.

Flat gold films appear at first sight to be attractive for such an application, and we have previously demonstrated the advantages of this approach in a novel FLIM-based assay for screening cell receptor internalisation, in which the fluorescence lifetime provided a very effective yes/no reporter of endocytosis.[6] However, when using flat metal films, it is the enhancement of non-radiative decay which results in the reduction of the lifetime with the separation from the film; hence, the intensity of the fluorescence signal near the metal is also reduced. This, together with limitations imposed on the collection numerical aperture for opaque substrates, can result in both low signal levels for small separations, making more difficult the reliable determination of the fluorescence lifetime, and in the simultaneous detection of contributions to the fluorescence signal from fluorescent labels both within close proximity of the substrate and also much deeper in the cell, hindering the reliable calibration of the distance from the substrate.

The use of nanostructured plasmonic thin films, on the other hand, provides an axial sensitivity of $\sim$6 nm over 100 nm above the nanostructure surface while maintaining the flexibility and lateral resolution of a standard confocal configuration suitable for use in live-cell imaging. For these nanoplasmonic thin films, radiative decay processes dominate, even at close proximity to the metal, leading to higher signal levels from fluorophores close to the substrate. The combination of this and the higher resolution offered by the high numerical aperture lenses available in this case, results in signals derived from well-defined locations within the cell, such as the basal membrane, providing a robust determination of the distance from the substrate from the measured fluorescence lifetime. We have shown elsewhere[7] that this approach is sensitive to the redistribution of proteins in a cell during endocytosis, even when averaging over a large area, offering significant advantages over other methods.

Acknowledgements

This work was supported by the UK EPSRC (EP/G029806/1). We would also like to acknowledge support from the Cancer Research UK/EPSRC/MRC/DoH-funded KCL and UCL Comprehensive Cancer Imaging Centre C1519/A10331.

References

1 X. Zhang, *Nat. Photonics*, 2009, **3**, 365.
2 S. W. Hell, R. Schmidt and A. Egner, *Nat. Photonics*, 2009, **3**, 381.
3 A. L. Mattheyses, S. M. Simon and J. Z. Rappoport, Imaging with total internal reflection fluorescence microscopy for the cell biologist, *J. Cell Sci.*, 2010, **123**, 3621–3628.
4 D. Axelrod, *Traffic*, 2001, **2**, 764.
5 D. Braun and P. Fromherz, *Appl. Phys. A: Mater. Sci. Process.*, 1997, **65**, 341.
6 N. I. Cade, G. Fruhwirth, S. J. Archibald, T. Ng and D. Richards, *Biophys. J.*, 2010, **98**, 2752.
7 N. I. Cade, G. Fruhwirth, T. Ng and D. Richards, *J. Phys. Chem. Lett.*, 2013, **4**, 3402.
8 A. I. Chizhik, J. Rother, I. Gregor, A. Janshoff and J. Enderlein, *Nat. Photonics*, 2014, **8**, 124.
9 R. R. Chance, A. H. Miller, A. Prock and R. Silbey, *J. Chem. Phys.*, 1975, **63**, 1589.
10 W. L. Barnes, *J. Mod. Opt.*, 1998, **45**, 661.
11 E. Fort and S. Grésillon, *J. Phys. D: Appl. Phys.*, 2008, **41**, 013001.
12 E. M. Purcell, *Phys. Rev.*, 1946, **69**, 681.
13 T. Ritman-Meer, N. I. Cade and D. Richards, *Appl. Phys. Lett.*, 2007, **91**, 123122.
14 N. I. Cade, T. Ritman-Meer and D. Richards, *Phys. Rev. B: Condens. Matter Mater. Phys.*, 2009, **79**, 241404.
15 K. Aslan, J. R. Lakowicz, H. Szmacinski and C. D. Geddes, *J. Fluoresc.*, 2005, **15**, 37.
16 J. Zhang, Y. Fu, D. Liang, R. Y. Zhao and J. R. Lakowicz, *Langmuir*, 2008, **24**, 12452.
17 L. S. Vermeer, G. O. Fruhwirth, P. Pandya, T. Ng and A. J. Mason, *J. Proteome Res.*, 2012, **11**, 2996.
18 M. Berndt, M. Lorenz, J. Enderlein and S. Diez, *Nano Lett.*, 2010, **10**, 1497.
19 R. Carminati, J.-J. Greffet, C. Henkel and J. Vigoureux, *Opt. Commun.*, 2006, **261**, 368.
20 A. Moroz, *Opt. Commun.*, 2010, **283**, 2277.
21 S. Kühn, U. Håkanson, L. Rogobete and V. Sandoghdar, *Phys. Rev. Lett.*, 2006, **97**, 017402.
22 P. Anger, P. Bharadwaj and L. Novotny, *Phys. Rev. Lett.*, 2006, **96**, 113002.
23 R. R. Chance, A. Prock and R. Silbey, *J. Chem. Phys.*, 1976, **65**, 2527.
24 G. W. Ford and W. H. Weber, *Phys. Rep.*, 1984, **113**, 195.
25 J. A. Broussard, W.-H. Lin, D. Majumdar, B. Anderson, B. Eason, C. M. Brown and D. J. Webb, *Mol. Biol. Cell*, 2012, **23**, 1486.

Faraday Discussions

PAPER

Surface plasmon enhanced energy transfer between gold nanorods and fluorophores: application to endocytosis study and RNA detection

Yinan Zhang,†‡[a] Guoke Wei,†[b] Jun Yu,[c] David J. S. Birch[a] and Yu Chen*[a]

Received 8th October 2014, Accepted 4th November 2014
DOI: 10.1039/c4fd00199k

Previously we have demonstrated surface plasmon enhanced energy transfer between fluorophores and gold nanorods under two-photon excitation using fluorescence lifetime imaging microscopy (FLIM) in both solution and intracellular phases. These studies demonstrated that gold nanoparticle–dye energy transfer combinations are appealing, not only in *Förster resonance energy transfer (FRET)* imaging, but also energy transfer-based fluorescence lifetime sensing of bio-analytes. Here, we apply this approach to study the internalization of gold nanorods (GNRs) in HeLa cells using the early endosome labeling marker GFP. The observed energy transfer between GFP and the GNRs indicates the involvement of endocytosis in GNR uptake. Moreover, a novel nanoprobe based on oligonucleotide functionalized gold nanorods for nucleic acid sensing *via* dye–GNRs energy transfer is demonstrated, potentially opening up new possibilities in cancer diagnosis and prognosis. The influence of oligonucleotide design on such nanoprobe performance was studied for the first time using time-resolved fluorescence spectroscopy, bringing new insights to the optimization of the nanoprobe.

I. Introduction

Förster resonance energy transfer (FRET) is frequently referred to as the 'spectroscopic ruler' and has been applied to a broad range of research, including distance distributions, metabolic sensing and protein and cell functions.[1,2] It occurs between donors and acceptors in close proximity, with an effective range of typically 1–8 nm for organic donors and acceptors, as limited by the nature of their dipole–dipole interaction. However, this constraint in the length scale of

[a]*Department of Physics, Strathclyde University, John Anderson Building, 107 Rottenrow, Glasgow G4 0NG, UK. E-mail: y.chen@strath.ac.uk*
[b]*Department of Physics, Beihang University, Beijing 100191, China*
[c]*Strathclyde Institute of Pharmacy and Biomedical Sciences, University of Strathclyde, Glasgow G4 0RE, UK*
† Both authors made the same level of contribution to this work.
‡ Current address: College of Physics, Jilin University, 2699 Qianjin Street, Changchun 130012, China.

detection can be extended dramatically when metallic particles act as acceptors since the distance dependence then scales at less than the 6th power.[3]

In addition to extending the upper range limit, gold nanoparticles are in many ways superior to organic dye molecules for energy transfer applications as they are photostable. This combines with their low toxicity, tunable absorption band and ability to conjugate to bio-molecules, making gold nanoparticles a versatile probe in biological imaging and sensing.[4–15] This is especially true for gold nanorods (GNRs), where their strong two-photon luminescence (TPL) makes them excellent fluorescence probes in biological imaging[16,17] as the intrinsic advantages of two-photon excitation (high spatial definition, increased specificity, reduced photobleaching and reduced excitation of endogenous fluorescence) can also be brought to bear. Previously we demonstrated that by incorporating GNRs into cells, the fast fluorescence decay time of the GNRs could be used to good effect to increase the contrast in Fluorescence Lifetime Imaging Microscopy (FLIM).[18]

Recent studies on energy transfer processes in Au nanoparticle assemblies have shown their applications in protein binding and conformational changes, RNA folding/unfolding, metal ion detection, *etc.*[19] While most studies used single photon excitation, we have reported the study of energy transfer between a dye and GNRs under two-photon excitation. Energy transfer between 4′-6-diamidino-2-phenylindole (DAPI), a commonly used DNA dye, and GNRs were observed under two-photon excitation using FLIM in both solution and intracellular phases.[15] It was also found that GNRs provided more efficient energy transfer than gold nanospheres with comparable size and concentration. This energy transfer enhancement was attributed to the longitudinal surface plasmon mode of GNRs overlapping with the incidence excitation wavelength. Surface plasmon effects in the energy transfer were also found in core/shell nanoparticles where gold nanorod cores were coated by dye (Rhodamine 800) doped polystyrene shells.[20] Enhanced energy transfer, indicated by a large decrease in the fluorescence lifetime of the fluorophore, was observed when the excitation wavelength matched the longitudinal surface plasmon band of the gold cores. Further energy transfer between Alexa Fluor 405 and gold nanorods under one- and two-photon excitation was demonstrated experimentally using FLIM and theoretically using a density matrix method.[21] The energy transfer *via* the dipole–dipole interaction was found to cause a decrease in the fluorescence lifetime. In this work, a new model for the dipole–plasmon interaction between the dye and metal nanoparticles for the two-photon process was developed where plasmon in the nanorods was considered as localized in all three dimensions (in stead of the one-dimensional treatment in the surface energy transfer (SET) model) because the size of the nanorod in all three dimensions is much smaller than the wavelength of the incident light. This description successfully accounts for the surface plasmon enhanced energy transfer under two-photon excitation.

Together these studies demonstrate that gold nanorod–dye energy transfer combinations are appealing, not only in FRET imaging, but also energy transfer-based fluorescence lifetime sensing of bio-analytes. Here, we describe a study of the internalization of GNRs in cells *via* energy transfer–based fluorescence lifetime imaging using an early endosome marker coupled with GFP. The observed energy transfer between the GNRs and GFP indicates the involvement of endocytosis in GNR uptake. Moreover, a novel nanoprobe based on functionalized gold nanorods for nucleic acid sensing through energy transfer fluorescence spectroscopy is demonstrated.

II. Experimental

A. Synthesis of Au nanoparticles

Gold nanorods were synthesized by the seeded growth method.[22] To synthesize the GNRs, 2.5 ml $HAuCl_4 \cdot 3H_2O$ (0.001 M) and 0.6 ml ice-cold $NaBH_4$ (0.01 M) were added into 7.5 ml cetyltrimethylammonium bromide (CTAB) (0.1 M) to prepare the seeds solution. The growth solution was synthesized by adding 0.15 M benzyldimethylhexadecylammonium chloride (BDAC), 25 ml $HAuCl_4 \cdot 3H_2O$ (0.001 M), 1 ml $AgNO_3$ (0.004 M) and 350 μl ascorbic acid (0.778 M) to 25 ml CTAB solution (0.1 M). Then 80 μl seed solution (2 hours after preparation) was injected into the growth solution to grow gold nanorods. The absorption spectrum shows a longitudinal plasmon mode centered at around 850 nm and a weak transverse plasmon mode at 550 nm.

B. Functionalization of Au nanorods with hairpin DNA (hpDNA)

The CTAB surfactant on the GNR surface was first replaced with mercaptohexanoic acid (MHA) using a previously reported round-trip phase transfer ligand exchange approach.[23] Thiolated oligonucleotides and the corresponding complementary oligonucleotides were purchased from Eurofins MWG Operon and Integrated DNA Technologies, respectively. The disulfide bonds of thiolated single strand DNA (ssDNA) were reduced by tris(2-carboxyethyl)phosphine hydrochloride (TCEP). The MHA–GNRs were conjugated with ssDNA through a salting aging process.[24,25]

C. Cell treatment with gold nanorods

Dispersions of gold nanorods were centrifuged to remove the excess CTAB and redispersed in deionized water twice (14 000 rpm, 5 min per cycle) with a final optical density about 1.0. Cells were treated with 100 μl of gold nanorods solution and incubated for 3 hours under standard cell culture conditions at 37 °C and 5% CO_2. The cells were washed thoroughly with phosphate buffered saline (PBS) to remove excess nanorods and fixed with 3.7% paraformaldehyde. After staining with 4′-6-diamidino-2-phenylindole (DAPI), the sample was dispersed on a glass slide and with a cover slit for imaging.

D. Lifetime measurements

Time-resolved fluorescence measurements were performed using the time-correlated single-photon counting (TCSPC) technique on an IBH Fluorocube fluorescence lifetime system (Horiba Jobin Yvon IBH Ltd., Glasgow, UK) equipped with an emission monochromator. A pulsed light-emitting diode (LED) of 649 nm operating at 1 MHz repetition rate was used as the excitation source. A longpass filter of 670 nm was used to minimize the detection of excitation light. Fluorescence decays were measured at the magic angle (54.7°) to eliminate polarization effects. Data analysis was performed using nonlinear least squares with the IBH iterative reconvolution software (DAS6 data analysis package). The fluorescence intensity decays were analyzed in terms of the multi-exponential model as the sum of individual single exponential decays:

$$I(t) = \sum_i B_i \exp\left(-\frac{t}{\tau_i}\right), \tag{1}$$

where τ_i are the decay times and B_i the associated amplitudes.

The fractional contribution of each lifetime component to the steady-state intensity is represented by

$$f_i = B_i\tau_i \Big/ \sum_k B_k\tau_k. \tag{2}$$

E. Fluorescence lifetime imaging microscopy (FLIM)

FLIM was performed using a confocal microscope (LSM 510, Carl Zeiss) equipped with a time-correlated single photon counting (TCSPC) module (SPC-830, Becker & Hickl GmbH). A femtosecond Ti:Sapphire laser (Chameleon, Coherent) was tuned at *850 nm* to generate TPL from gold nanorods. The laser pulse has a repetition rate of 80 MHz and a duration of less than 200 fs. Emission was collected with a 60× water-immersion objective (N.A. = 1.0) and a bandpass filter with a transmission window from 500 nm to 550 nm.

III. Results and discussion

A. Energy transfer based FLIM study on the internalization of GNRs in HeLa cells

Gold nanoparticles have great potential in biomedical application, including *in vitro* and *in vivo* imaging, sensing, drug delivery and cancer therapy because of their unique physical and chemical properties.[26–28] To exploit the potential, it is essential to understand the penetration of nanoparticles through biological barriers.[29] The study of intracellular pathways of gold nanoparticles and uptake mechanisms are critical for many of these applications, as they provide information on the uptake rate, route, final intracellular location, the effects of nanoparticles to cell organelles, *etc.* However this is not a trivial task. On one hand, the uptake process is a complex process influenced by both the properties of the cells, such as cell groups and types, and the characteristics of the nanoparticles, such as the size, shape, surface charge and coating conditions.[30–39] So far different nanoparticle pathways into cell membranes have been reported and mechanisms are still a subject of study.[40,41] On the other hand, conventional techniques used for tracking nanoparticles in cell cultures have both advantages and limitations. For example, traditional fluorescence microscopy can be used for *in vivo* or *in vitro* imaging of cell organelles *via* fluorescence labeling, but it is restricted by its diffraction-limited spatial resolution. It is difficult in some cases to conclude whether the overlapping pixel is real colocalization or just structures in approximation. Super resolution techniques can extend the limit of optical microscopy but only at a considerable increase in complexity. Similarly, transmission electron microscopy (TEM) has superior spatial resolution but the specimen has to be thinly sliced and imaged under vacuum, therefore is not suitable for *in situ* visualization of biological samples.

Here we introduce a two-photon energy transfer – based FLIM method that monitors the energy transfer between gold nanorods and fluorescent molecules in order to study cell uptake of GNRs. As the energy transfer process only occurs when two fluorophores are in close proximity (typically within 8 nm for Förster resonance energy transfer and double this distance for surface energy transfer), this FLIM-FRET method can directly indicate whether gold nanoparticles are enclosed in a specific cellular organelle, *e.g.*, endosomes, and can be applied both *in vivo* and *in vitro*. TPL and energy transfer processes have been combined to study the involvement of endocytosis in the intra-cellular trafficking of gold nanorods in HeLa cells. The Invitrogen CellLight® Early Endosomes-GFP, BacMam 2.0 kit was used in this work. The BacMam 2.0 kit offers a user-friendly technique which contains protein fusion of green fluorescent protein (GFP) and early endosome marker RAB5a. HeLa cells were treated with 5% GNRs (volume ratio) and the BacMam kit for 1 h then fixed for imaging experiments.

Fig. 1(a) shows a typical confocal image of HeLa cells labelled with Rab5a-GFP. Bright spots in the image are considered to be fluorescence from GFP, while no strong fluorescence signal but only weak cell autofluorescence can be seen from a control sample in Fig. 1(b). Fig.1(c) is a TPL intensity image taken from the same sample as in Fig. 1(a) with 850 nm femto-second laser excitation and 500–550 nm emission bandpass filter. Fig. 1(d) is the FLIM image of the same area where different colours in each image pixel varying from orange to blue represent fluorescence lifetimes ranging from 0 to 3 ns. The fluorescence lifetime was obtained by fitting the experimental decay curves to a single exponential decay model, as shown in Fig. 1(e). The fluorescence lifetime of GFP in cell culture is found to be around 2 ns with a symmetric distribution (Fig. 1(f)), consistent with previous reports.[43,44] Fig. 1 shows that Rab5a-GFP allowed the imaging of early endosomes in HeLa cells and emitted strong and stable fluorescence under both single and two-photon excitations.

To investigate the uptake process of gold nanorods by HeLa cells, we have incubated the cells with gold nanorods at 5% volume concentration at 37 °C for different time periods, from 30 minutes to 1 hour. FLIM images are shown in Fig. 2: from (a) to (c), the gold nanorods incubation time is 30 minutes, 45 minutes, and 1 hour respectively. Decay curves of all bright pixels in images have been fitted by a two-exponential model as the luminescence from gold nanorods has to be considered. The coded colour hereby represents the average lifetime τ_{ave}, which can be expressed by $\tau_{ave} = \sum_i^n B_i \tau_i \ (n = 4)$. In the FLIM images, green/blue pixels represent emission from GFP, while brighter orange/yellow spots in cells are considered to be signals from gold nanorods or overlapping areas where both emissions from GNRs and GFP were found. This is because the decay time scale of TPL from GNRs[18] is extremely small (<100 ps and comparable to the instrumental impulse response), while the intensity is much stronger than that from GFP (contribution over 80% of total signal). Therefore even though the lifetime of GFP is as long as 2 ns, the average lifetimes in these pixels are quite short, usually around several hundred picoseconds.

Fig. 2 shows that a few gold nanorods were found in cells after incubation for 30 min and 45 min, while a significant amount of nanorods has been found in cells after 1 h incubation, demonstrated by the increase in the number of the

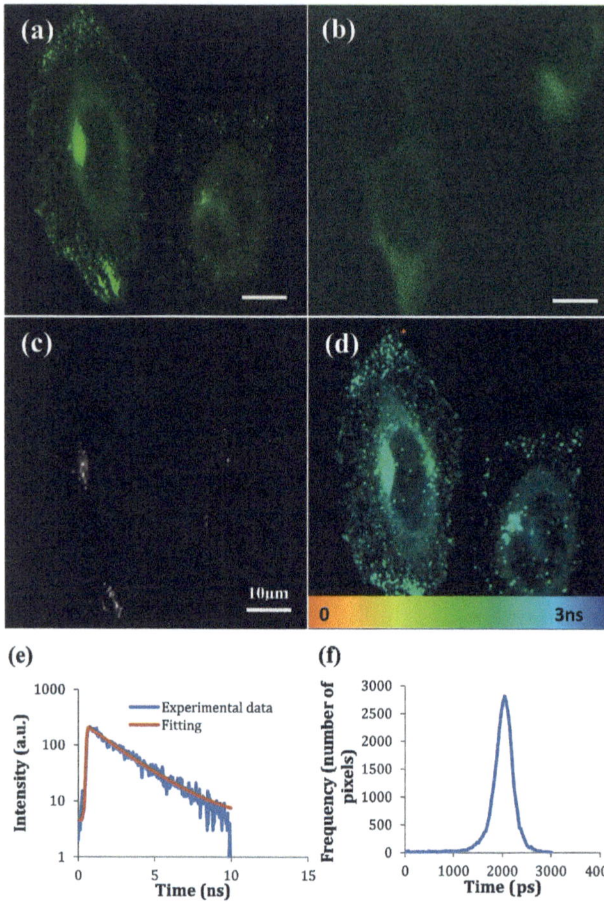

Fig. 1 (a) Confocal fluorescence image of HeLa cells treated with Rab5a-GFP, excited at 488 nm, emission collected with a longpass 505 nm filter; (b) HeLa cells without Rab5a-GFP; (c) and (d) intensity and FLIM image of two-photon excited HeLa cells treated with Rab5a-GFP, excited under 850 nm and emission collected with 500–550 nm bandpass filter. For the FLIM image, the lifetime ranges from 0 (orange) to 3 ns (blue); (e) typical decay curve of fluorescence from GFP in cell culture; (f) distribution of fluorescence lifetimes displayed in (d). The same scale bars apply for (a)–(d). The luminescence intensity in (a) and (b) are indicated by the brightness of the images from 0 (dark) to 255 (bright green spots in (a)).

pixels displaying shorter lifetimes, *i.e.*, an orange color. All of the bright spots in the FLIM images are found to be overlapping pixels, *i.e.*, the GNRs overlap with the Rab5a-GFP stained early endosomes and there are no isolated GNRs. A further analysis of the FLIM image has been carried out to examine whether there was energy transfer between GFPs and GNRs in order to confirm the involvement of early endosomes in the uptake process. To this end, the long lifetime component originating from GFP in the overlapping areas was analysed. As shown in Fig. 2(d), the long lifetime component subtracted from 58 overlapping pixels in different FLIM images have been sorted in 100 ps intervals from 1100 ps to 2400 ps. Compared to the symmetric lifetime distribution of GFP (Fig. 1(f)), lifetimes of

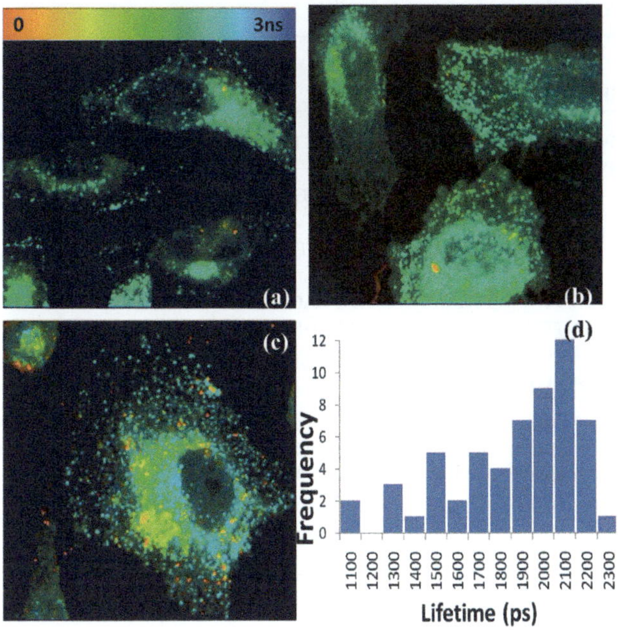

Fig. 2 FLIM images of Rab5a-GFP treated HeLa cells incubated with CTAB capped gold nanorods for (a) 30 minutes, (b) 45 minutes, and (c) 1 hour; (d) lifetime distribution of GFP in overlapping areas.

GFP in the vicinity of gold nanorods form an asymmetric distribution, showing a higher rate of reduced lifetimes with a long tail reaching 1.1 ns, which suggests energy transfer does exist between GFP and gold nanorods. This indicates that gold nanorods and Rab5a-GFP are in close proximity, implying the cell uptake of gold nanorods by endocytosis.

B. Gold nanorod-based energy transfer probe for RNA sensing

Messenger ribonucleic acid (mRNA) bridges genome information harboured within DNA to phenotypes primarily and collectively expressed by proteins. In most cases, the dynamics of mRNA in the cell directly reflects the levels of protein, hence the phonotypical expression. Because of its important roles in biology numerous methods have been developed to assess the mRNA expression in response to a variety of physiological and pathological cues.[42–44] Molecular beacons (MB) are one of the most promising probes for mRNA detection. MBs are usually designed with a hairpin-structured oligonucleotide with a fluorophore-quencher pair that undergoes a spontaneous fluorogenic conformational change upon hybridization with the complementary nucleic acid target.[45,46] MBs offer good opportunities in homogeneous assays of mRNA and real-time monitoring of the expression of mRNA inside living cells due to their good sensitivity and enhanced specificity.[47–50] However, traditional MBs suffer from the problem of lacking universal organic quenchers.[51] Quenching efficiency may vary from one fluorophore to another, limiting the signal/background ratio. Moreover, transfection reagents are required for cellular internalization of MBs. Recent studies

Table 1 DNA sequences used in this work

Name	Sequence (5′ to 3′)[a]	Hairpin T_m[b] (°C)	Hairpin ΔG[b] (kcal mol^{-1})
hpCy5_1	Cy5-CCGAG TTG GTG AAG CTA ACG TTG AGG CTCGG TTTTT-SH	48.2	−3.28
hpCy5_2	Cy5-CCGGTG GTG AAG CTA ACG TTG AG CACCGG TTTTT-SH	60.5	−5.66
hpAACy5	Cy5-CTGACTTG GTG AAG CTA ACG TTG AG CAAGTCAG AA-SH	58.9	−6.58
hpCy5f	Cy5-AATTT AAATTGAACTTG GTG AAG CTA ACG TTG AG CAAGTTCAATTT TTT TTT TTT T-SH	60.5	−10.12
cDNA_1[c]	CCT CAA CGT TAG CTT CAC CAA		
cDNA_2[d]	CTC AAC GTT AGC TTC AC		

[a] The underlined bases represent the stem sequence. [b] Predicted at 25 °C in 100 mM [Na$^+$] buffer by UNAFold software (http://www.idtdna.com). [c] Complementary DNA for hpCy5_1. [d] Complementary DNA for hpCy5_2, hpAACy5 and hpCy5f.

show that these limitations can be elegantly addressed by using functionalized gold nanoparticles (AuNPs). It has been proven that AuNPs are highly efficient quenchers for a range of organic fluorophores[48] and exhibit long-range fluorescence quenching capability.[15,21,52] Moreover, AuNPs functionalized with oligonucleotides show highly efficient cellular uptake without the need of transfection reagents and extraordinary intracellular stability against enzymatic degradation as well as enhanced binding capability of complementary nucleic acids.[53–55] Exploring the potential of gold nanoparticle based RNA probes, here we report a new nanoprobe based on functionalized gold nanorods and the influence of the hairpin structure on the quenching efficiency of this energy transfer pair.

Thiolated oligonucleotides and the corresponding complementary oligonucleotides used in this work are listed in Table 1. The underlined bases represent the stem sequences in their hairpin structure. Fig. 3 shows fluorescence decay

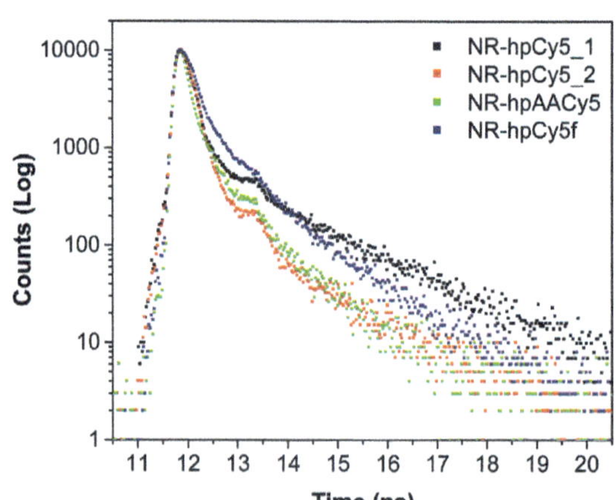

Fig. 3 Fluorescence decay curves taken from hpDNA functionalized GNRs in their initial quenching states.

Table 2 Fluorescence lifetimes from a three exponential model fitting

Sample	τ_1 (ns)	B_1 rel. ampl.	τ_2 (ns)	B_2 rel. ampl.	τ_3 (ns)	B_3 rel. ampl.	X^2
GNR–hpDNA–Cy5_1	1.803 ± 0.020	17.91%	0.326 ± 0.022	8.49%	0.034 ± 0.004	73.60%	1.14
GNR–hpDNA–Cy5_2	1.629 ± 0.052	5.22%	0.290 ± 0.011	15.38%	0.055 ± 0.002	79.39%	1.03
GNR–hpAACy5	0.942 ± 0.022	13.39%	0.238 ± 0.008	31.25%	0.030 ± 0.005	55.36%	1.56
GNR–hpDNA–Cy5f	1.225 ± 0.021	17.27%	0.408 ± 0.021	24.98%	0.057 ± 0.002	57.75%	1.26

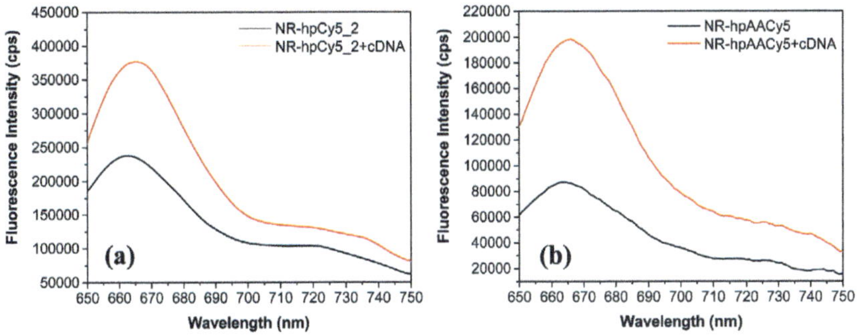

Fig. 4 Fluorescence spectra taken from (a) GNR–hpCy5_2 and (b) GNR–hpAACy5 before (black) and after (red) hybridization with their complementary oligonucleotides (500 nM).

curves taken from hairpin DNA (hpDNA) functionalized GNRs in their initial quenching states. Lifetime analysis using multiple exponentials reveals three lifetime components, as summarized in Table 2. Two lifetime components were found in the ns and sub ns range that can be attributed to Cy5, while the third short lifetime component in the tens of ps region is probably due to strong scattering from gold cores. Three probes, namely hpCy5_2, hpAACy5 and hpCy5f, have first lifetime components, 1.6 ns, 0.9 ns and 1.2 ns, shorter than the lifetime found from the Cy5–oligonucleotide in its free form (2.1 ns, 1.8 ns and 2.0 ns, respectively). In contrast, the first lifetime component of hpCy5_1 (1.8 ns) is comparable to its free form (1.96 ns). This suggests the existence of loose hairpins possibly due to a relatively low hairpin melting temperature (48.2 °C). The second lifetime component of all four samples are in the sub nanosecond range, indicating much reduced lifetimes arising from energy transfer between Cy5 and GNRs. A detailed comparison reveals that hpAACy5 surpasses other designs and has the highest energy transfer efficiency. This is the result of the short distance between Cy5 and the gold surface because hpAACy5 has only 2As while the other three designs have 5 or more Ts at the 3′ end. This improvement in the quenching efficiency results in a reduced background and thus an enhanced signal/background ratio. Fig. 4 compares the fluorescence spectra taken from GNR–hpCy5_2 and GNR–hpAACy5 before and after hybridization with their complementary oligonucleotides. An apparent increase in fluorescence intensity is observed in both cases, suggesting a successful switching from initial quenching states to final fluorescence states after hybridization. As expected, the hpAACy5 probe has a higher signal/background ratio than hpCy5_2. For the first time, this study shows that time-resolved fluorescence spectroscopy is a useful tool for investigating the quenching efficiency of hairpin DNA functionalized gold nanorods and reveals the significance of hairpin design on the performance of such a nanoprobe. It is anticipated that this study will bring insight to the optimization of such RNA nanoprobes.

IV. Conclusion

In summary, we have found that energy transfer provides more detailed information in biological studies when combined with the advantages of two-photon excitation microscopy of GNRs, as demonstrated here in the study of intra-cellular trafficking of GNRs in HeLa cells *via* GFP labelled early endosomes. It would seem that GNRs have further potential, not only in FRET imaging, but also in fluorescence lifetime-based intra-cellular sensing of bio-analytes. A new RNA probe based on hairpin DNA functionalized GNRs has been demonstrated. This method combines the strength of enhanced energy transfer in GNR–dye pairs and the TPL of GNRs, in addition to the other advantages of GNRs, such as photostability, self-uptake by cells, multiplexing and multifunctional capabilities, overcoming the technical limits suffered by other fluorogenic methods, *e.g.*, fluorescence protein labelling and hybrid fluorescence probes. This new mRNA probe thus potentially opens up new possibilities in cancer diagnosis and prognosis. We have also demonstrated that time-resolved fluorescence spectroscopy can be applied to investigate the influence of hairpin design on the initial quenching state and thus bring insights towards optimizing such nanoprobes.

Acknowledgements

The authors acknowledge W. Li and R. Y. M. M. Qotob for assistance during the experimental work. This work was supported by a BBSRC Award BB/K013416/1. G. Wei acknowledges financial support from China Scholarship Council (CSC).

References

1 O. J. Rolinski, D. J. S. Birch, L. J. McCartney and J. C. Puckup, *J. Photochem. Photobiol., B*, 2000, **54**, 26–34.
2 A. J. W. G. Visser, S. P. Laptenok, N. V. Visser, A. van Hoek, D. J. S. Birch, J.-C. Brochon and J. W. Borst, *Eur. Biophys. J.*, 2010, **39**, 241–253.
3 C. S. Yun, A. Javier, T. Jennings, M. Fisher, S. Hira, S. Peterson, B. Hopkins, N. O. Reich and G. F. Strouse, *J. Am. Chem. Soc.*, 2005, **127**, 3115.
4 X. Huang, S. Neretina and M. A. El-Sayed, *Adv. Mater.*, 2009, **21**, 4880–4910.
5 P. K. Jain, X. Huang, I. H. El-Sayed and M. A. El-Sayed, *Acc. Chem. Res.*, 2008, **41**, 1578–1586.
6 Y. Chen, Y. Zhang, D. J. S. Birch and A. S. Barnard, *Nanoscale*, 2012, **4**, 5017.
7 Y. Zhang, D. Xu, W. Li, J. Yu and Y. Chen, *J. Nanomater.*, 2012, 375496.
8 Y. Zhang, J. Yu, D. J. S. Birch and Y. Chen, *Proc. SPIE*, 2011, **7910**, 79101H.
9 S. Link and M. A. El-Sayed, *Int. Rev. Phys. Chem.*, 2000, **19**(3), 409–453.

10 M. B. Mohamed, V. Volkov, S. Link and M. A. El-Sayed, *Chem. Phys. Lett.*, 2000, **317**, 517–523.
11 R. A. Farrer, F. L. Butterfield, V. W. Chen and J. T. Fourkas, *Nano Lett.*, 2005, **5**, 1139–1142.
12 H. Wang, T. B. Huff, D. A. Zweifel, W. He, P. S. Low, A. Wei and J. Cheng, *Proc. Natl. Acad. Sci.*, 2005, **102**, 15752–15756.
13 K. Imura, T. Nagahara and H. m. Okamoto, *J. Phys. Chem. B*, 2005, **109**, 13214–13220.
14 X. H. Huang, I. H. El-Sayed, W. Qian and M. A. El-Sayed, *J. Am. Chem. Soc.*, 2006, **128**, 2115–2120.
15 Y. Zhang, D. J. S. Birch and Y. Chen, *Appl. Phys. Lett.*, 2011, **99**, 103701.
16 X. H. Huang, S. Neretina and M. A. EL-Sayed, *Adv. Mater.*, 2009, **21**, 4880.
17 Y. Chen, J. A. Preece and R. E. Palmer, *Ann. N. Y. Acad. Sci.*, 2008, **1130**, 201.
18 Y. Zhang, J. Yu, D. J. S. Birch and Y. Chen, *J. Biomed. Opt.*, 2010, **15**, 20504.
19 T. Sen and A. Patra, *J. Phys. Chem. C*, 2012, **116**, 17307–17317.
20 P. Gu, D. J. S. Birch and Y. Chen, *Methods Appl. Fluoresc.*, 2014, **2**, 024004.
21 C. Rachnor, M. R. Singh, Y. Zhang, D. J. S. Birch and Y. Chen, *Methods Appl. Fluoresc.*, 2014, **2**, 015002.
22 J. Murphy, T. K. Sau, A. M. Gole, C. J. Orendorff, J. Gao, L. Gou, S. E. Hunyadi and T. Li, *J. Phys. Chem. B*, 2005, **109**, 13875.
23 A. Wijaya and K. Hamad-Schifferli, *Langmuir*, 2008, **24**, 9966–9969.
24 S. J. Hurst, A. K. R. Lytton-Jean and C. A. Mirkin, *Anal. Chem.*, 2006, **78**, 8313–8318.
25 A. Wijaya, S. B. Schaffer, I. G. Pallares and K. Hamad-Schifferli, *ACS Nano*, 2009, **3**, 80–86.
26 V. Biju, *Chem. Soc. Rev.*, 2014, **43**, 744.
27 T. Bose, D. Latawiec, P. P. Mondal and S. Mandal, *J. Nanopart. Res.*, 2014, **16**, 2527.
28 L. A. Austin, M. A. Mackey, E. C. dreaden and M. A. Ei-Sayed, *Arch. Toxicol.*, 2014, **88**, 1391–1417.
29 S. Barua and S. Mitragotri, *Nano Today*, 2014, **9**, 223–243.
30 A. E. Nel, L. Mädler, F. Velegol, T. Xia, E. M. V. Hoek, P. Somasundaran, F. Klaessig, V. Castranova and M. Thompson, *Nat. Mater.*, 2009, **8**, 543–557.
31 S. J. H. Soenen and M. D. Cuyper, *Nanomedicine*, 2010, **5**, 1261.
32 S. D. Conner and S. L. Schmid, *Nature*, 2003, **422**, 37–44.
33 J. Davda and V. Labhasetwar, *Int. J. Pharm.*, 2002, **233**, 51–59.
34 T. Kato, T. Yashiro, Y. Murata, D. C. Herbert, K. Oshikawa, M. Bando, S. Ohn and Y. Sugiyama, *Cell Tissue Res.*, 2003, **311**, 47–51.
35 A. T. Florence and N. Hussain, *Adv. Drug Delivery Rev.*, 2001, **50**, S69–S89.
36 F. Zhao, Y. Zhao, Y. Liu, X. Chang, C. Chen and Y. Zhao, *Small*, 2011, **7**, 1322–1337.
37 A. Verma and F. Stellacci, *Small*, 2010, **6**, 12–21.
38 H. J. Parab, H. M. Chen, T. C. Lai, *et al.*, *J. Phys. Chem. C*, 2009, **113**, 7574–7578.
39 J. H. Huang, T. C. Lai, L. C. Cheng, *et al.*, *J. Mater. Chem.*, 2011, **21**, 14821–14829.
40 Z. Liu, W. Cai, L. He, N. Nakayama, K. Chen, X. Sun, X. Chen and H. Dai, *Nat. Nanotechnol.*, 2007, **2**, 47–52.
41 V. Mailänder and K. Landfester, *Biomacromolecules*, 2009, **10**, 2379–2400.

42 D. Sivaraman, P. Biswas, L. N. Cella, M. V. Yates and W. Chen, *Trends Biotechnol.*, 2011, **29**, 307.

43 B. A. Armitage, *Curr. Opin. Chem. Biol.*, 2011, **15**, 806.

44 S. R. Harry, D. J. Hicks, K. I. Amiri and D. W. Wright, *Chem. Commun.*, 2010, **46**, 5557.

45 S. Tyagi and F. R. Kramer, *Nat. Biotechnol.*, 1996, **14**, 303–308.

46 S. Tyagi, D. P. Bratu and F. R. Kramer, *Nat. Biotechnol.*, 1998, **16**, 49–53.

47 T. Matsuo, *Biochim. Biophys. Acta*, 1998, **1379**, 178–184.

48 D. L. Sokol, X. Zhang, P. Lu and A. M. Gewirtz, *Proc. Natl. Acad. Sci. U. S. A.*, 1998, **95**, 11538–11543.

49 N. Nitin, P. J. Santangelo, G. Kim, S. Nie and G. Bao, *Nucleic Acids Res.*, 2004, **32**, e58.

50 C. D. Medley, T. J. Drake, J. M. Tomasini, R. J. Rogers and W. Tan, *Anal. Chem.*, 2005, **77**, 4713–4718.

51 B. Dubertret, M. Calame and A. J. Libchaber, *Nat. Biotechnol.*, 2001, **19**, 365–370.

52 S. Mayilo, M. A. Kloster, M. Wunderlich, A. Lutich, T. A. Klar, A. Nichtl, K. Kürzinger, F. D. Stefani and J. Feldmann, *Nano Lett.*, 2009, **9**, 4558–4563.

53 N. L. Rosi, D. A. Giljohann, C. S. Thaxton, A. K. E. Lytton-Jean, M. S. Han and C. A. Mirkin, *Science*, 2006, **312**, 1027–1030.

54 D. S. Seferos, D. A. Giljohann, H. D. Hill, A. E. Prigodich and C. A. Mirkin, *J. Am. Chem. Soc.*, 2007, **129**, 15477–15479.

55 A. E. Prigodich, D. S. Seferos, M. D. Massich, D. A. Giljohann, B. C. Lane and C. A. Mirkin, *ACS Nano*, 2009, **3**, 2147–2152.

Faraday Discussions

PAPER

Control of Förster energy transfer in the vicinity of metallic surfaces and hyperbolic metamaterials

T. U. Tumkur,[a] J. K. Kitur,[a] C. E. Bonner,[a] A. N. Poddubny,[bc] E. E. Narimanov[d] and M. A. Noginov*[a]

Received 29th September 2014, Accepted 14th November 2014
DOI: 10.1039/c4fd00184b

Optical cavities, plasmonic structures, photonic band crystals and interfaces, as well as, generally speaking, any photonic media with homogeneous or spatially inhomogeneous dielectric permittivity (including metamaterials) have local densities of photonic states, which are different from that in vacuum. These modified density of states environments are known to control both the rate and the angular distribution of spontaneous emission. In the present study, we question whether the proximity to metallic and metamaterial surfaces can affect other physical phenomena of fundamental and practical importance. We show that the same substrates and the same nonlocal dielectric environments that boost spontaneous emission, also inhibit Förster energy transfer between donor and acceptor molecules doped into a thin polymeric film. This finding correlates with the fact that in dielectric media, the rate of spontaneous emission is proportional to the index of refraction n, while the rate of the donor–acceptor energy transfer (in solid solutions with a random distribution of acceptors) is proportional to $n^{-1.5}$. This heuristic correspondence suggests that other classical and quantum phenomena, which in regular dielectric media depend on n, can also be controlled with custom-tailored metamaterials, plasmonic structures, and cavities.

1. Introduction

Metamaterials – engineered composite materials containing subwavelength inclusions with tailored shapes, sizes, mutual arrangements and orientations[1,2] – fascinate scientists and engineers with their unparalleled responses to electromagnetic waves.[3–7] Thus, metamaterials with hyperbolic dispersion, whose dielectric permittivities in orthogonal directions have different signs[6,8–10]

[a]Center for Materials Research, Norfolk State University, Norfolk, VA 23504, USA. E-mail: mnoginov@nsu.edu
[b]ITMO University, St. Petersburg 197101, Russia
[c]Ioffe Physical-Technical Institute, St. Petersburg 194021, Russia
[d]Birck Nanotechnology Center, Department of Electrical and Computer Engineering, Purdue University, West Lafayette, IN 47907, USA

propagate waves with nearly unlimited wavevectors and have a broad-band singularity of the local density of photonic states.[11] The latter phenomenon, in spirit of the Fermi's golden rule, can control spontaneous emission[12–15] and reflectance of roughened metamaterials' surfaces.[16,17]

In this study, we have researched the effect of the dielectric environment, including the local density of photonic states ρ, on Förster energy transfer between donor and acceptor molecules.[18] The literature has several contradictory theoretical and experimental studies of the effect of the local density of photonic states on the Förster energy transfer. Thus, its rate was claimed to be dependent on[19–23] or independent[24–27] of the photonic environment. Experimentally, a high density of photonic states was reported to enhance the energy transfer rate in cavities,[20] modulate it[21,22] or leave it unaffected[24–27] in the vicinity of mirrors and planar interfaces, or inhibit it in plasmonic structures.[28] This broad range of claims and opinions makes our findings outlined below particularly important.

Förster energy transfer typically occurs between donors and acceptors situated in close proximity ($\ll\lambda$). This interaction of a Coulombic nature is usually mediated by a polarizable medium of the host matrix. At dipole–dipole character of the energy transfer, with fixed positions of donors and acceptors, and random spatial distributions of acceptors around donors, the emission kinetics of donors $I(t)$ (excited by short laser pulses) is given by[18]

$$I(t) = I_0 \exp\left(-(A + W)t - \gamma\sqrt{t}\right), \tag{1}$$

where I_0 is the initial emission intensity, A and W are the radiative and non-radiative emission decay rates of the donors, t is the time, $\gamma \approx 6.28{R_0}^3/\sqrt{\tau_0}$ is the energy transfer constant, N is the concentration of acceptors, and $\tau_0 = (A + W)^{-1}$ is the decay-time of donors in the absence of acceptors. Here

$$R_0 = \sqrt[6]{\frac{3}{2(2\pi)^5}\frac{\eta_0}{n^4}\int F(\bar{\nu})\sigma(\bar{\nu})\frac{d\bar{\nu}}{\bar{\nu}^4}} \tag{2}$$

is the characteristic distance at which the energy transfer rate in a pair of donor and acceptor is equal to the emission decay rate of the donor molecule, $\bar{\nu} = \omega/2\pi c$ is the frequency in cm^{-1}, ω is the angular frequency, c is the speed of light, $F(\bar{\nu})$ is the normalized radiation spectrum of donors, $\sigma(\bar{\nu})$ is the absorption cross section spectrum of acceptors, $\eta_0 = A/(A + W)$ is the quantum yield of spontaneous emission of donors (in the absence of acceptors), and n is the index of refraction.

2. Experimental results

In this work, thin polymeric films doped by donor and acceptor molecules were deposited on a variety of substrates including glass (control sample), 200 nm Ag film on glass, 200 nm Au film on glass, as well as hyperbolic metamaterials. Known metamaterials with hyperbolic dispersion include ordered arrays of metallic nanowires[29,30] and multi-layered metal/dielectric or semiconductor lamellar thin film structures.[13–15,31] The latter morphology was employed in our studies (see Section 5.1 for the fabrication of hyperbolic metamaterials and study of their dispersion properties).

In an optimal pair of matching dyes, the emission band of a *donor* should spectrally overlap with the absorption band of an *acceptor*. This condition is satisfied in a poly(methyl methacrylate) (PMMA) film co-doped with DCM dye molecules (donors), and DOTC dye molecules (acceptors), Fig. 1a (see Section 5.2). Trace 1 of Fig. 1b depicts the emission spectrum of such a film deposited on glass and pumped at $\lambda = 400$ nm into the absorption band of DCM. The emission band of the DOTC dye, clearly seen in this spectrum, manifests an efficient donor → acceptor energy transfer. The emission spectra in Fig. 1b, although useful for qualitative demonstrations, are not suitable for a quantitative analysis of the donor → acceptor energy transfer on top of metallic and metamaterials substrates. This is because emission and excitation spectra in a mixture of donor and acceptor molecules depend not only on the energy transfer rate γ, but also on the absorption strength of the donors, the radiative A and non-radiative W decay rates, as well as the directionality of the emission of both the donors and acceptors. As all of these parameters can be sensitive to local densities of photonic states, evaluating the efficiency of the donor → acceptor energy transfer from the analysis of emission spectra becomes an extremely difficult task.

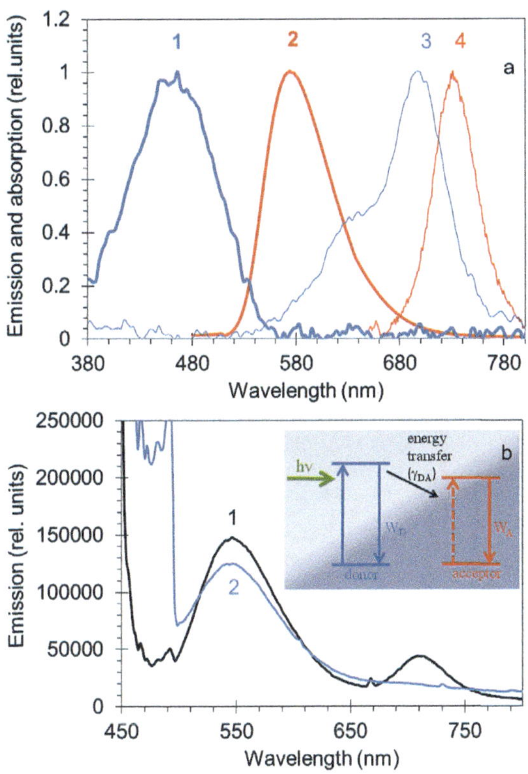

Fig. 1 Dye spectroscopy. (a) Absorption (1 and 3) and emission (2 and 4) spectra of DCM (1 and 2) and DOTC (3 and 4) dyes doped into PMMA. (b) Emission spectra of ≈80 nm thick PMMA films co-doped by DCM and DOTC molecules (concentrations ≈ 33 mM DCM and 20 mM DOTC) deposited onto glass (1) and silver film (2) substrates and pumped at $\lambda = 400$ nm into the absorption band of DCM (a similar effect was observed on top of a metamaterial with Ag as the outermost layer). Inset: schematic of Förster donor–acceptor energy transfer showing the absorption of light in the donor, a combination of radiative and non-radiative relaxation processes in the donor (W_D), donor–acceptor energy transfer (γ_{DA}), and the relaxation processes in the acceptor (W_A).

As proven by numerous studies,[32] a much better way to quantitatively characterize the efficiency of the Förster energy transfer is by recording and analyzing the emission kinetics of donors in presence of acceptors. Fig. 2 depicts the emission kinetics of DCM molecules embedded into a PMMA film, excited at 392 nm with the second harmonic of 150 fs Ti:sapphire laser, as well as the emission kinetics of the DCM molecules (donors) and DOTC molecules (acceptors) co-doped into PMMA (all three kinetics have been measured in dye doped PMMA films deposited on glass). The shortening of the emission kinetics of donors and delayed rise of the emission intensity of acceptors (Fig. 2), are the signature features of an efficient donor → acceptor energy transfer.[18]

Dividing the emission decay kinetics of donors measured in a co-doped PMMA film (described by eqn (1)) by that in the film doped with donor molecules only ($\propto \exp(-(A+W)t)$), we expected to single out the contribution of the Förster donor → acceptor energy transfer, $\propto \exp(-\gamma\sqrt{t})$. When we applied this procedure to emission kinetics 1 and 2 in Fig. 2, the resulting curve, plotted in the inset of Fig. 2 as $\{\ln(-\ln(I(t)))$ *vs.* $\ln(t)\}$, had slope 0.6, which is close to 1/2 (expected of the function $\propto \exp(-\gamma\sqrt{t})$). This validates our method and gives us a powerful tool, allowing us to separate the Förster contribution to the emission kinetics (which nominally has slope 1/2) from all other radiative and non-radiative decay processes (which have slope 1). The corresponding energy transfer constant γ can be determined by fitting the resultant experimental curve in the inset of Fig. 2 by $\propto \exp(-\gamma\sqrt{t})$.

We have found that in polymeric films co-doped by donors and acceptors on top of Ag and Au films as well as lamellar metamaterials with Ag as the outermost

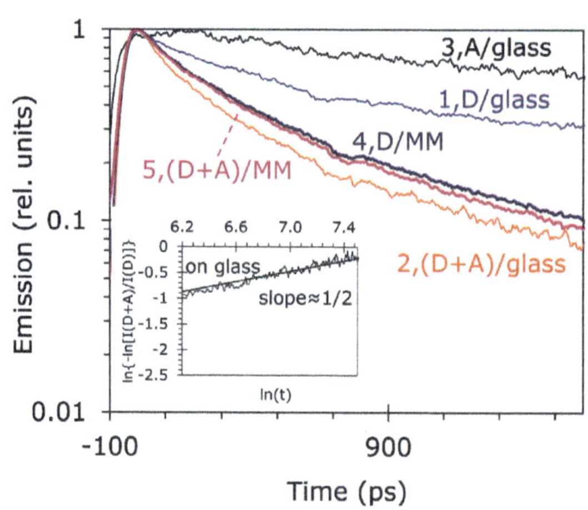

Fig. 2 Emission kinetics. Spontaneous emission kinetics in PMMA films deposited on glass: donors (DCM) in a singly doped PMMA film (1, D/glass), donors in the PMMA film co-doped with donors and acceptors (2, D + A/glass), and acceptors (DOTC) in the PMMA film co-doped with donors and acceptors (3, A/glass). Spontaneous emission kinetics in PMMA films deposited on metamaterial substrates with Ag as the top layer: donors (DCM) in a singly doped PMMA film (4, D/MM) and donors in the PMMA film co-doped with donors and acceptors (5, D + A/MM). All films were pumped into the absorption band of the acceptors at $\lambda = 392$ nm. Inset: the ratio of emission kinetics 2 and 1 from the main frame, showing slope ~1/2 when plotted as $\ln(-\ln(I(t)))$ *vs.* $\ln(t)$ (solid line indicates the slope equal to 1/2).

layer, the rate of the Förster energy transfer is strongly inhibited. This is evidenced by the emission spectra of donor–acceptor mixtures pumped into the absorption band of donors, which lack emission of acceptors at ~710 nm (Fig. 1b, trace 2), as well as emission decay kinetics of donors, which are almost the same in the absence (trace 4) and in the presence (trace 5) of acceptors, Fig. 2.

By comparing traces 1 and 4 of Fig. 2, one can also see that the emission kinetics of (single-doped) donors on top of a metamaterial is shorter than that on top of glass. Following ref. 11, 13, 15 and 25, we explain this phenomenon with a combination of spontaneous emission enhancement caused by the high local density of photonic states[11,33] (arguably a predominant effect) and nonradiative quenching of dye emission by the metallic surface.[34] As the lifetime shortening is

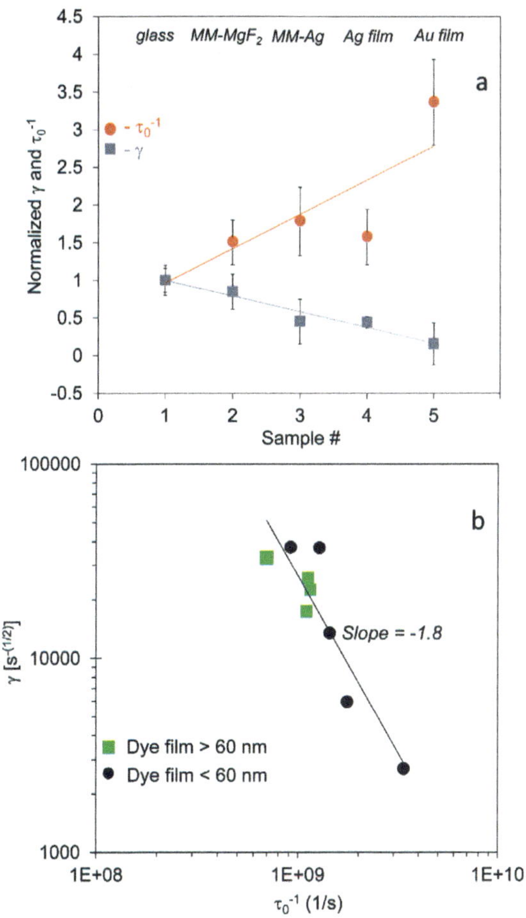

Fig. 3 Energy transfer summary. (a) Emission decay rates τ_0^{-1} and Förster energy transfer constants γ in dye-doped films deposited on top of glass (1), a metamaterial with MgF_2 as the outermost layer (2), a metamaterial with Ag as the outermost layer (3), a Ag film (4), and a Au film (5). All of the data points are averaged over several samples deposited on the same type of substrate and normalized to that on glass. (b) Values of τ_0^{-1} plotted against corresponding values of γ in the dye-doped PMMA films deposited on the top of samples 1, 2, 3 and 4 in (a).

not uniform through the films' thickness, the emission decay kinetics showed deviation from a pure exponential form. The respective emission decay rates have been calculated as $I_0 \Big/ \int_0^\infty I(t)\mathrm{d}t$.

At the same time, the emission kinetics of donors in the presence of acceptors is shorter on top of glass than on top of a metamaterial, (compare traces 2 and 5 in Fig. 2). This, combined with the nearly identical behavior of traces 4 and 5, is an unambiguous manifestation of the inhibition of the Förster energy transfer in the vicinity of lamellar metal/dielectric substrates (note, that the inhibition was weaker on top of a lamellar metamaterial, whose outermost layer was MgF_2, than on top a similar metamaterial, whose outmost layer was Ag).

The values γ obtained from polymeric films co-doped by donors and acceptors and corresponding emission decay rates τ_0^{-1} in films doped by donor molecules only have been measured in several tens of thin film samples (whose thickness ranged from 32 nm to 100 nm) deposited onto five types of substrates. The results of the measurements are summarized in Fig. 3a.

As one can see in Fig. 3a, the same metallic and hyperbolic metamaterial substrates and the same environments that enhance spontaneous emission decay (possibly with minor contributions from non-radiative decay), inhibit Förster energy transfer. This is the central experimental result of the present study. This observation correlates with the qualitative argument that, in regular dielectric media, the rate of spontaneous emission A (which we assume to be nearly $\propto \tau_0^{-1}$) is proportional to n, while $\gamma \propto n^{-1.5}$. The pairs of data points (γ, τ_0^{-1}) measured on top of multiple metamaterial, metallic, and glass substrates are plotted in logarithmic coordinates in Fig. 3b. The slope of the curve is equal to -1.8 ± 0.42, in a fair agreement with the heuristic argument ($A \propto n$, $\gamma \propto n^{-1.5} \Rightarrow \gamma \propto A^{-1.5}$). One can see that the emission lifetime shortening and the inhibition of the energy transfer are weaker in thick dye doped films (>60 nm) than in thin films (<60 nm), due to the larger distances from dye molecules to metal or metamaterial. Correspondingly, the spread of both τ_0^{-1} values and γ values is smaller in thicker films (>60 nm) than in thinner films (<60 nm).

3. Theoretical analysis and discussion

3.1 Comparison with the literature

Although our findings are consistent with those of ref. 28, in which the inhibition of the Förster energy transfer was observed in the vicinity of plasmonic nanoparticles, it seemingly disagrees with several other reports, including ref. 25, which claims an independence of the Förster energy transfer rate of the molecule-to-metal distance and the corresponding local density of photonic states. In the latter study, the distance between donor–acceptor pairs and the metallic mirror ranged between 60 nm and 270 nm, and the developed theoretical model predicted almost no dependence of the Förster energy transfer rate on the local density of states if the dielectric environment did not change on a scale smaller than the wavelength (~500 nm). In our work, the distances between donor–acceptor pairs and the metamaterial or metallic substrates were substantially smaller, ranging between 0 nm and 32 nm in thinner dye-doped films and between 0 nm and 95 nm in thicker films. This suggested that the effect of the substrate on the Förster energy transfer rate in our experiment could be larger than that in ref. 25. However, since the average molecule-to-substrate distances in our experiments were still large in comparison to the average donor–acceptor distance, $d = 4.4$ nm, the effect of the local density of photonic states on the rate of the Förster energy transfer had to be quantitatively evaluated as described below.

3.2 Theoretical formulation of the problem

The quasi-static result for the rate of the Förster energy transfer between donor and acceptor in an homogeneous medium reads

$$W_{\mathrm{DA}} = W_0 \left(\frac{R_0}{r_{\mathrm{DA}}} \right)^6 \tag{3}$$

where r_{DA} is the distance between donor and acceptor molecules, $W_0 = A + W = \tau_0^{-1}$ is the decay rate of donors in the absence of acceptors, and the Förster radius R_0 (at which the rate of energy transfer is equal to the spontaneous emission rate) depends on the intrinsic properties of the donor and acceptor as well as the dielectric permittivity of the medium (eqn (2)). In an arbitrary medium characterized by the position-dependent dielectric permittivity $\varepsilon(r)$, eqn (3) can be generalized to[35,36]

$$W_{\mathrm{DA}}(r_{\mathrm{A}},r_{\mathrm{B}}) = V\mathrm{Tr}[\hat{G}(r_{\mathrm{A}},r_{\mathrm{D}})\hat{G}^{+}(r_{\mathrm{A}},r_{\mathrm{D}})] \tag{4}$$

where $\hat{G}(\boldsymbol{r},\boldsymbol{r}')$ is the electromagnetic Green tensor satisfying the equation

$$\left[\Delta + \left(\frac{\omega}{c} \right)^2 \varepsilon(\boldsymbol{r}) \right] G_{\alpha\beta}(\boldsymbol{r},\boldsymbol{r}') = -4\pi \left(\frac{\omega}{c} \right)^2 \left(\delta_{\alpha\beta} + \frac{1}{(\omega/c)^2 \varepsilon} \frac{\partial^2}{\partial x_\alpha \partial x_\beta} \right) \delta(\boldsymbol{r} - \boldsymbol{r}') \tag{5}$$

and evaluated at the donor emission frequency ω; and α, β are the Cartesian indices. The trace in eqn (4) reflects averaging over the orientations of the donor and acceptor dipole matrix elements. The quantity V in eqn (4) depends only on the donor and acceptor properties, and all information about the electromagnetic environment is included into the Green function. The correspondence between eqn (3) and (4) can be recovered in the case of a homogeneous medium with the dielectric permittivity ε_{b}, for the donor and acceptor molecules separated by the distance much smaller than the wavelength, $r_{\mathrm{DA}} \ll c/\omega$. Under these assumptions, the Green function G depends only on the relative coordinate $\boldsymbol{r}_{\mathrm{DA}} = \boldsymbol{r}_{\mathrm{D}} - \boldsymbol{r}_{\mathrm{A}}$ and is reduced to the electrostatic dipole–dipole interaction term G_0

$$G_{0,\alpha\beta}(\boldsymbol{r}_{\mathrm{DA}}) = \frac{3r_{\mathrm{DA},\alpha}r_{\mathrm{DA},\beta} - {r_{\mathrm{DA}}}^2\delta_{\alpha,\beta}}{\varepsilon_{\mathrm{b}}{r_{\mathrm{DA}}}^5}. \tag{6}$$

Furthermore, the general result eqn (4) reduces to eqn (3), provided that $V = {\varepsilon_{\mathrm{b}}}^2 W_0 {R_0}^6/6$. Eqn (3) and (4) describe the rates of energy transfer between individual donor and acceptor molecules. For ensembles of donors and acceptors, there exists a distribution of the transfer rates depending on the intermolecular distance in the donor–acceptor pair. The decay kinetics of the given donor molecule is then determined by the transfer probabilities to different acceptors and strongly depends on the spatial distribution of the molecules. In the case when the acceptors are distributed homogeneously and independently in an infinitely large medium, the expression for the donor decay reads,[18]

$$I(t) = I_0 e^{-H(t)},$$
$$H(t) = \int_{r_{DA} > r_{min}} d^3 r_A n_A(r_A)\left(1 - e^{-W_{DA}(r_A, r_B)t}\right), \quad (7)$$

where the integration is performed over the spatial distribution of acceptors characterized with the concentration n_A. Here the radiative decay of the donors is neglected, *cf.* eqn (7) and (1). The small sphere with the radius r_{min} is excluded from the integration to account for the minimal distance between the molecules due to their finite size. For homogeneous medium, substituting eqn (3) into eqn (7) and neglecting r_{min}, we recover the energy transfer law eqn (1) with $H(t) \propto \sqrt{t}$.[18]

3.3 Effect of the silver substrate on the Förster energy transfer

The calculated effect of the silver substrate on the energy transfer kinetics is presented in Fig. 4. The donor and acceptor molecules are embedded in the semi-infinite dielectric layer lying on top of the silver substrate. The dielectric-silver interface corresponds to the plane $z = 0$. Acceptors are homogeneously distributed in a dielectric (polymer) film of thickness $d_{dye} = 30$ nm deposited on top of silver ($0 < z_A < d_{dye}$); the concentration of acceptors is equal to n_A. In order to simplify the treatment, in our model the dielectric film with dye is covered by the same un-doped polymer. In the experimental samples the dye film is bounded by air from the top. Since the permittivity contrast between polymer and air is smaller than that between polymer and silver, the effect of the polymer/air boundary is weaker than the effect of the polymer/silver boundary, and this

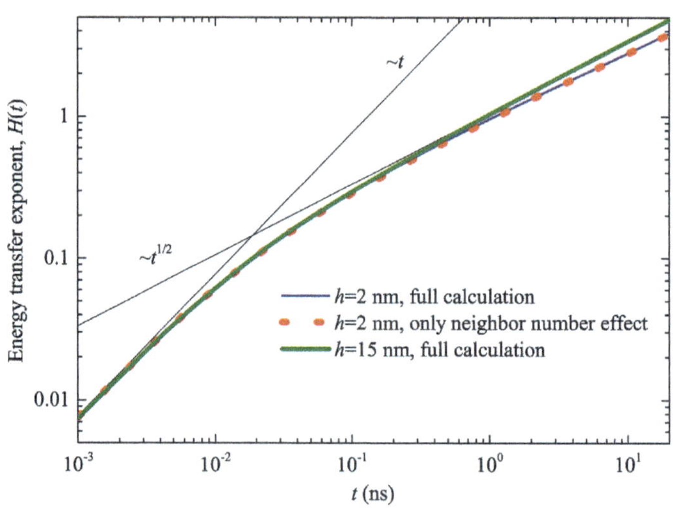

Fig. 4 Theoretical time dependence of the energy transfer function $H(t)$. The green and blue solid curves have been calculated for the donor height over the Ag substrate h equal to 15 nm and 2 nm, respectively. The red dotted curve has been calculated for $h = 2$ nm neglecting the modification of the Green function due to the substrate. The calculation has been performed for the following set of parameters: $W_0 = 10^9\ s^{-1}$, $\varepsilon_b = 2$, $n_A = 1.2 \times 10^{19}\ cm^{-3}$, $d_{dye} = 30$ nm, $\hbar\omega = 2.1$ eV, $r_{min} = 1$ nm (see Methods). Data from ref. 37 was used for the dielectric permittivity of Ag. The thin black lines are guides for the eye and have been obtained as linear fits with the slopes $H(t) \propto t$ and $H(t) \propto t^{1/2}$, respectively.

simplification should not significantly influence our results. The different curves in Fig. 4 correspond to two values of the height of the donor molecule $h \equiv z_D$ above the interface, $h = 2$ nm and $h = 15$ nm. The Green function has been evaluated according to ref. 38. At short times the decay law is exponential, with $H(t) \propto t$, which is due to the finite value of r_{min}. This part of the decay curves is independent of the distance from the substrate provided it is larger than r_{min}. At larger times, the decay kinetics of the donor in the bulk is described by the $H(t) \propto \sqrt{t}$ law (*cf.* green and thin black curves). For the donor located close to the substrate, the decay rate is slightly suppressed (blue curve). In principle, this suppression can be caused by a combination of two reasons: (i) the donor at the surface has a smaller number of acceptors in its vicinity than the donor in the bulk and (ii) the electromagnetic Green function near the surface is different from its bulk value, eqn (6).

The red dotted curve shows the contribution of the former effect only. It has been calculated using the bulk Green function that does not account for the presence of the substrate. Since this curve is very close to the blue solid curve, we conclude that the reduced number of available acceptors for the donors located near the surface of the polymeric film is the most important effect, while the Green function modification does not affect the energy transfer kinetics noticeably.

The analysis provided in Section 5.4 shows that, while the effect of the modification of the Green function on the energy transfer rate is weak, the spatial modification pattern is quite complex. Thus, depending on the position of the acceptor, the Förster energy transfer can be either increased or suppressed.

We further demonstrate (in Section 5.4) that in the framework of our model, assuming interaction of individual donors with individual acceptors, the effects of the dielectric environment on the (i) density of photonic states and (ii) Förster energy transfer practically do not correlate with each other, with the latter effect being much weaker. Hence, from only the value of the density of states, it is not possible to deduce whether the energy transfer rate for a given donor–acceptor distance r_{DA} will be enhanced or suppressed. This is explained by the fact that the energy transfer rate is determined by the electromagnetic modes with the wave vectors of the order of $1/r_{DA}$, while the density of states has contributions from all modes, in agreement with ref. 25.

3.4 Collective dye–plasmon interactions

In the consideration above, we have assumed that each donor or acceptor molecule interacts with the electromagnetic modes of the system independently. However, for dense enough molecular ensembles and small inhomogeneous broadening, the collective effects may play an important role and result in the formation of hybrid polaritonic excitations (recently, it has been theoretically predicted that collective effects are important for the energy transfer for an array of molecules coupled to a spherical nanoparticle[39]).

Moreover, when the interaction between the excitonic modes of the donor (acceptor) ensembles becomes stronger than their dampings, one can attain the strong coupling regime, characterized by the anticrossings in the dispersion curves based on the emission, absorption and reflection spectra and the formation of hybridized exciton and plasmon eigenmodes.[40,41]

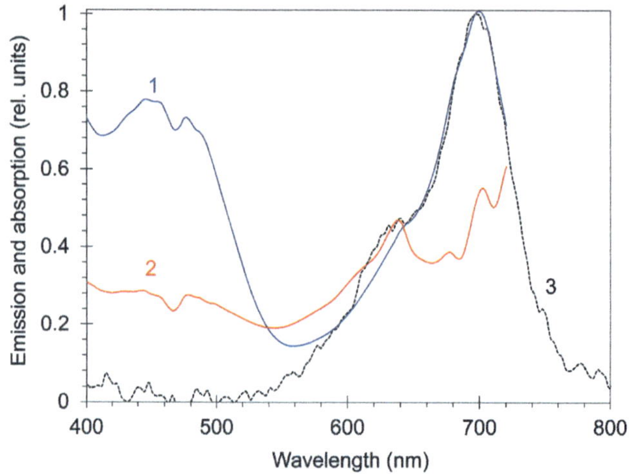

Fig. 5 The effect of a nonlocal dielectric environment on excitation spectra. Excitation spectra of the DOTC molecules in the PMMA film (co-doped by the DCM dye) deposited on glass (trace 1) and Ag/MgF_2 metamaterial (trace 2). Trace 3 (dashed line) shows the absorption spectrum of a DOTC doped PMMA film deposited on glass.

Although not studied by us systematically, the modification of absorption and/or excitation spectra of dye molecules in the vicinity of metallic films and multilayered metal/dielectric structures is routinely observed in a variety of systems. As an example, the excitation spectra of the DOTC molecules in the PMMA film (co-doped by the DCM dye) deposited on glass and the Ag/MgF_2 metamaterial are depicted in Fig. 5. The strong difference between the shapes of the two spectra suggests that the overlap integral (eqn (2)) and the rate of the Förster donor → acceptor energy transfer can be strongly modified by the nonlocal dielectric environments. Indeed, the energies, the oscillator strengths, and the lifetimes of the polaritonic modes are modified by the environment and are substantially different from those of the individual molecules. The detailed experimental investigation of this phenomenon as well as a comprehensive theoretical study of the Förster energy transfer in nonlocal dielectric environments is the subject of an ongoing research effort to be published elsewhere.

3.5 Potential applications

Note that while some phenomena, including photosynthesis and photovoltaics, benefit from donor → acceptor energy transfer, others suffer from it badly. The latter include migration of electronic excitation to defects, which quenches the luminescence of luminophosphors, and self-quenching of luminescence in Nd^{3+} doped laser crystals that limits the maximal concentration of active ions and gain. Correspondingly, the possibility of inhibiting the Förster energy transfer on demand can be very useful for a variety of applications.

4. Summary

To summarize, we show that the rate of the Förster energy transfer γ in an ensemble of donor and acceptor molecules is inhibited on top of metallic and hyperbolic metamaterial substrates – the environments with high local densities of photonic states, which enhance spontaneous emission rates A. Although our theoretical model, describing interactions between *individual* independent donors and acceptors, predicts a weak reduction of the energy transfer on top of the substrate, the calculated effect is orders of magnitude smaller than the experimental one. We infer that the observed effect can be due to strong collective interactions of the dye molecules and surface plasmon polaritons, which can affect the absorption and emission spectra and, correspondingly, change the rate of the Förster energy transfer.

5. Methods and extended data

5.1 Fabrication and characterization of metamaterial and metallic film substrates

The hyperbolic metamaterials samples, fabricated using the physical vapor deposition technique, consisted of seven 25 nm layers of Ag and six 35 nm layers of MgF_2 (with Ag on the top) or nine layers of Ag and nine layers of MgF_2 (with MgF_2 on the top). The thickness of the layers was monitored with a quartz crystal sensor and later double-checked with the DekTak 6M profilometer.

The dielectric permittivities in the directions parallel ($\varepsilon_{\|}$) and perpendicular ($\varepsilon\perp$) to the sample surface were calculated in the effective medium approximation[42] for a lamellar metal/dielectric structure,

$$\begin{aligned} \varepsilon_{\|} &= f\varepsilon_{\mathrm{m}} + (1-f)\varepsilon_{\mathrm{d}} \\ \frac{1}{\varepsilon_{\perp}} &= \frac{f}{\varepsilon_{\mathrm{m}}} + \frac{(1-f)}{\varepsilon_{\mathrm{d}}}, \end{aligned} \tag{8}$$

where ε_{m} and f are the permittivity and the filling factor of a metal ($0 < f < 1$), and ε_{d} is the permittivity of a dielectric. According to the spectra of the real parts of $\varepsilon_{\|}$ and $\varepsilon\perp$ (calculated at $f = 0.39$, corresponding to 25 nm of Ag and 35 nm of MgF_2), the material has a hyperbolic dispersion at $\lambda \geq 384$ nm, (left inset of Fig. 6).

Experimentally, $\varepsilon_{\|}$ and $\varepsilon\perp$ have been determined by measuring the angular reflectance profiles $R(\theta)$ in s and p polarizations (Fig. 6) and fitting them with the known formulas[30]

$$R = |r|^2 = \left\{ \begin{array}{lll} \left|\dfrac{\sin(\theta-\theta_t)}{\sin(\theta+\theta_t)}\right|^2, & \theta_t = \arcsin\left(\dfrac{\sin\theta}{\sqrt{\varepsilon_{\|}}}\right) & \text{s polarization.} \\ \left|\dfrac{\varepsilon_{\|}\tan\theta_t - \tan\theta}{\varepsilon_{\|}\tan\theta_t + \tan\theta}\right|^2, & \theta_t = \arctan\sqrt{\dfrac{\varepsilon_{\perp}\sin^2\theta}{\varepsilon_{\|}\varepsilon_{\perp} - \varepsilon_{\|}\sin^2\theta}} & \text{p polarization.} \end{array} \right\}. \tag{9}$$

The values $\varepsilon_{\|} = -2.57 + i0.56$ and $\varepsilon\perp = 5.56 + i0.64$ determined at $\lambda = 543$ nm (the wavelength corresponding to the Förster energy transfer range in our experiment, see the main text), are reasonably close to the theoretical prediction, (left inset of Fig. 6).

Furthermore, the calculated (normal incidence) reflectance spectrum is in good agreement with the experimental reflectance spectrum measured at a small

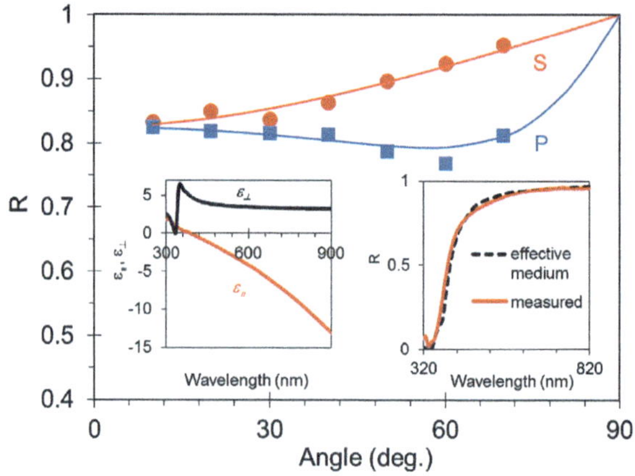

Fig. 6 Characterization of the metamaterial sample. Angular reflectance spectra of a typical Ag/MgF_2 multilayered sample (Ag as the top layer) in s and p polarizations, at $\lambda = 543$ nm. Markers – experiment, solid lines – fitting. Left inset: effective medium spectra of real parts of dielectric permittivities $\varepsilon_{\parallel}$ and $\varepsilon\perp$, calculated for a lamellar Ag/MgF_2 structure with Ag filling factor equal to 39%. Characters – experimentally measured values at $\lambda = 543$ nm. Right inset: reflectance spectrum of the lamellar Ag/MgF_2 sample. Solid line: experiment; dashed line: prediction of the effective medium theory, corresponding to a metal fill fraction of 39%.

incidence angle of 8 degrees, (right inset of Fig. 6). We thus conclude that (i) our samples have hyperbolic dispersion in the whole visible and near infrared range that is of interest to our studies, and (ii) the dispersion can be reasonably well described in the effective medium approximation.

5.2 Fabrication of dye-doped PMMA films

The DCM ([2-[2-[4-(dimethylamino)phenyl]ethenyl]-6-methyl-4*H*-pyran-4-ylidene]-propanedinitrile) and DOTC (3-ethyl-2-[7-(3-ethyl-2(3*H*)-benzoxazolylidene)-1,3,5-heptatrienyl]-benzoxazolium iodide) dyes, along with poly(methyl methacrylate) (PMMA, ~120 000 avg molecular weight) were dissolved in dichloromethane (in an ultrasonic bath at 25 °C) and spin-coated onto a variety of substrates as discussed in Section 2. In our experiments, the concentration of DCM molecules was equal to 2×10^{19} cm^{-3} (33 mM or 10 g l^{-1}) and the concentration of DOTC molecules was equal to 1.2×10^{19} cm^{-3} (20 mM or 10 g l^{-1}). The film thicknesses, measured with a Dektak 6M profilometer, ranged from 32 nm to 100 nm.

5.3 Emission kinetics measurements and analysis

The emission kinetics were detected with the Hamamatsu C5680 streak camera. The time resolution, determined by wide open slit of the streak camera and jitter of the laser, was ~80 ps. Appropriate combinations of color filters were used to selectively transmit the emissions of DCM and DOTC and block scattered pumping light.

In the analysis of the emission kinetics, we relied on formula (1) derived under the assumption that the physical size of the medium is much larger than the average donor–acceptor distance *d*.[18] We infer that the slight deviation of the experimentally measured slope in the inset of Fig. 2 from 1/2 could be due to a relatively small thickness of the dye-doped films, which exceeded $d = 4.4$ nm by approximately one order of magnitude.

5.4 Theoretical studies

Green function modification. Although, as shown in Section 3.3, the Green function modification practically does not affect the decay kinetics, it is still instructive to examine this effect in more detail. To this end, in Fig. 7 we analyze the dependence of the Förster energy transfer rate on the acceptor coordinate. The geometry of the problem is sketched in Fig. 7a.

Fig. 7b presents the spatial map of the transfer rate of the donor positioned above the Ag mirror to different acceptors. Fig. 7d shows the same map for donors in the bulk. Both panels look quite similar. To reveal the difference, in Fig. 7c we examine the ratio between the rates shown in panels b and d. In particular, Fig. 7c presents the spatial map of the energy transfer rates normalized to the corresponding values evaluated in the same points in a bulk dielectric; it is obtained by dividing the data in Fig. 7b by those in Fig. 7d and subtracting unity.

This figure demonstrates that while the relative modification of the energy transfer rate is weak, the modification pattern is quite complex. Depending on the position of the acceptor, the Förster energy transfer can be either increased or suppressed. Importantly, the energy transfer in the lateral direction, along the mirror surface, is suppressed for 10 nm < r_{DA} < 60 nm, (blue areas in Fig. 7c).

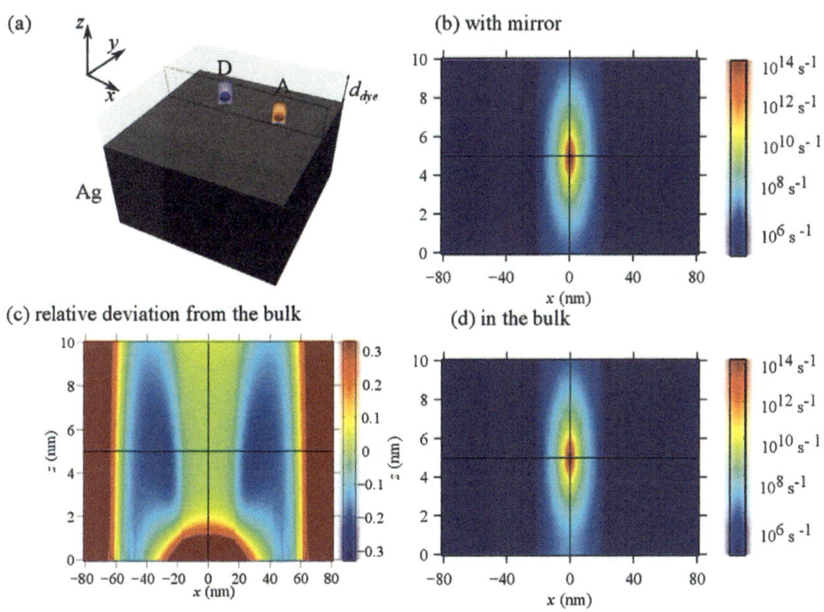

Fig. 7 Dependence of the Förster energy transfer rate on the acceptor coordinate (logarithmic scale). (a) Schematic illustration of the geometry of the problem. (b), (d) Transfer rate maps in the xz plane for the donor located above the mirror and in the bulk, respectively. (c) The energy transfer rate map, normalized to the bulk value as $W_{DA}/W_{DA}^{bulk} - 1$. The calculation has been performed for $r_D = h\hat{z}$ with $h = 5$ nm.

This suppression can be well described within the image charge approximation for the Green function.[43] We consider as an example the transfer between a copolarized donor and acceptor positioned at the same height from the surface. The relevant component of the Green tensor within the image charge approximation reads $G_{xx}(x_{\mathrm{DA}},0,0) = 4/[|x_{\mathrm{DA}}|^3(\varepsilon_{\mathrm{Ag}} + \varepsilon_{\mathrm{b}})]$, where $\varepsilon_{\mathrm{Ag}}$ is the dielectric permittivity of silver. This expression predicts an enhancement of the transfer at the surface plasmon resonance, corresponding to the condition $\varepsilon_{\mathrm{Ag}} + \varepsilon_{\mathrm{b}} = 0$. However, the frequency range of the spectral overlap of emission of donors and absorption of acceptors is below the resonance, so $|\varepsilon_{\mathrm{Ag}}| > |\varepsilon_{\mathrm{b}}|$. As a result, the value of G_{xx} is smaller than the bulk value $G_{0,xx} = 2/(|x_{\mathrm{DA}}|^3\varepsilon_{\mathrm{b}})$ and the transfer rate along the surface is suppressed, in agreement with Fig. 7c.

At the very large lateral distances, the energy transfer rate (normalized as in Fig. 7d) is increased compared to the bulk value (red areas on both the right hand and left hand sides of Fig. 7d). This is the surface plasmon-assisted transfer along the surface[44] – an effect, which is not captured by the quasi-static image charge model. The plasmon-assisted energy transfer becomes relatively large for lateral distances on the order of the inverse surface plasmon polariton (SPP) wave vector

$$k_{\mathrm{pl}} = \frac{\omega}{c}\sqrt{\frac{\varepsilon_{\mathrm{b}}\varepsilon_{\mathrm{Ag}}}{\varepsilon_{\mathrm{b}} + \varepsilon_{\mathrm{Ag}}}}, \tag{10}$$

for our parameters $\mathrm{Re}\{k_{\mathrm{pl}}{}^{-1}\} \sim 60$ nm, in agreement with Fig. 7d.

However, even though the relative value of the energy transfer rate with respect to the bulk value is increased, this practically does not affect the donor decay kinetics. The reason is that the absolute value of the Förster energy transfer rate for such large distances is quite small and the relative enhancement plays no role. In other words, the energy transfer rate is mostly determined by the short-range electromagnetic coupling between donor and acceptor, and the effect of the long-range plasmon assisted energy transfer is negligible. This effect can be more important at higher frequencies, which are closer to the surface plasmon resonance in silver. In fact, at higher frequencies, the surface plasmon wave vector k_{pl} is larger, so the plasmons can contribute more strongly to the energy transfer.

Fig. 8 presents the dependence of the energy transfer rate on the donor–acceptor distance. The rate has been determined by averaging the data in Fig. 7 over the angular position of the acceptor. The calculation demonstrates that the rate deviates from the bulk decay law $1/r_{\mathrm{DA}}{}^6$ only at very large distances, where the transfer is very weak.

Förster energy transfer and photonic local density of states. In this section, we analyze the effect of the silver substrate on the Förster energy transfer rate and on the photonic local density of states, which determines the rate of spontaneous emission (Purcell effect).[45–47] Fig. 9 compares the effect of the donor position above the substrate h on the Förster energy transfer with its effect on the Purcell enhancement factor f_{P}. Clearly, these two quantities are not correlated and the Purcell factor depends on h much more strongly than the energy transfer rate does. In fact, at h equal to several nanometers, f_{P} reaches the value of the order of a hundred, constituting strong quenching of the molecule emission due to the Joule heating of the metal.[18] The Purcell factor, which can be approximated as[48]

$$f_{\mathrm{P}}(h) = \frac{3}{8(\omega h/c)^3}\frac{\mathrm{Im}\{\varepsilon_{\mathrm{Ag}}\}}{|\varepsilon_{\mathrm{b}} + \varepsilon_{\mathrm{Ag}}|^2}, \tag{11}$$

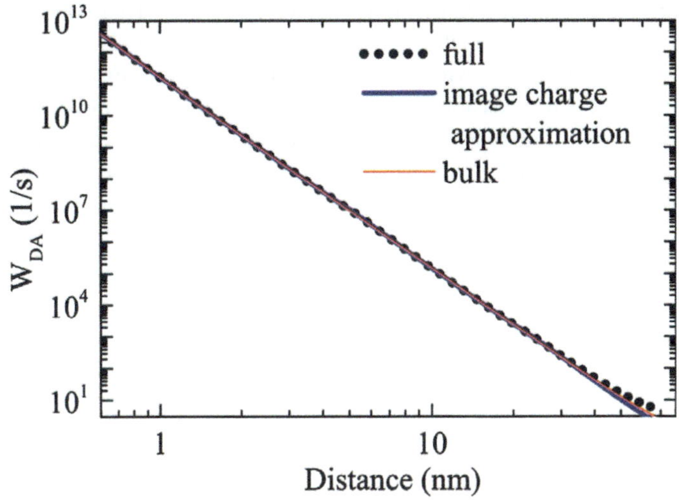

Fig. 8 Dependence of the angular-averaged energy transfer rate *versus* donor–acceptor distance r_{DA}. Full calculation – dotted line, calculation in the electrostatic image-charge approximation – thick solid blue line, and calculation with the bulk Green function that does not include the effect of the mirror into account – thin solid red line. The donor has been positioned at $r_D = h\hat{z}$ with $h = 5$ nm, other parameters are the same as in Fig. 4.

is very sensitive to the distance from the silver surface h. At the same time, the rate of the donor–acceptor energy transfer (calculated for fixed distances between donor and acceptor $r_{DA} = 4$ nm) depends on the height h weakly and

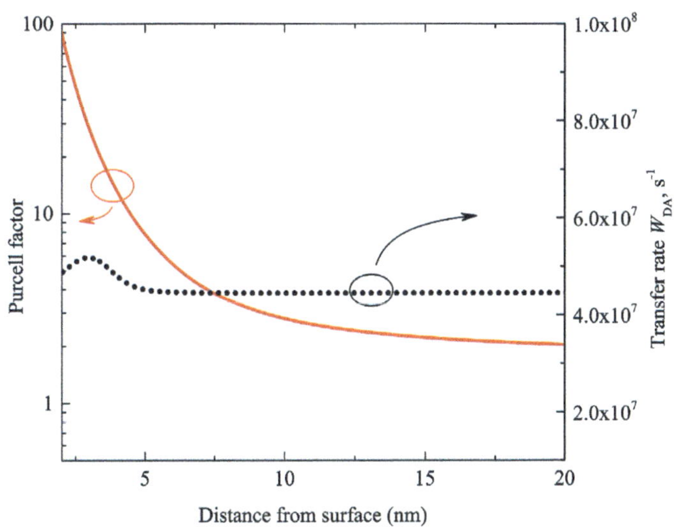

Fig. 9 Comparison of the effect of the silver substrate on the Purcell factor and on the energy transfer rate. The red solid curve shows the polarization-averaged Purcell factor as function of the height of the donor above the silver surface. The thick dotted curve shows the height dependence of the energy transfer rate. The transfer rates has been evaluated at donor–acceptor distances $r_{DA} = 4$ nm, and averaged over the angular distribution of acceptors. The calculation parameters are the same as in Fig. 4.

nonmonotonously (see the black dotted curve in Fig. 9). Therefore, the local dielectric environment affects the density of states and the Förster energy transfer rate in very different ways, with its effect on the energy transfer being much weaker. Hence, it is not possible to deduce whether the energy transfer rate for a given donor–acceptor distance r_{DA} will be enhanced or suppressed based on the density of states value only. This is explained by the fact that the energy transfer rate is determined by the (short-range) electromagnetic modes with the wave vectors of the order of $1/r_{DA}$, while the density of states has contributions from all modes, in agreement with the findings of ref. 25.

Acknowledgements

TUT, JKK, CEB, and MAN were supported by the NSF PREM grant DMR-1205457, NSF IGERT grant DGE-0966188, and AFOSR grant FA9550-09-1-0456. MAN and EEN acknowledge the ARO grant W911NF-14-1-0639. ANP acknowledges the support by the Government of the Russian Federation (Grant no. 074-U01) and the "Dynasty" Foundation. EEN was supported by ARO MURI, NSF Center for Photonics and Multiscale Nanomaterials, and Gordon and Betty Moore Foundation.

References

1 N. Engheta and R. W. Ziolkowski, *Electromagnetic Metamaterials: Physics and Engineering Explorations*, John Wiley & Sons, New York, 2006.
2 *Tutorials in Meta materials*, ed. M. A. Noginov and V. A. Podolskiy, Taylor & Francis, Florida, 2012.
3 J. Pendry, Negative refraction makes a perfect lens, *Phys. Rev. Lett.*, 2000, **85**, 3966–3968.
4 N. Fang, H. Lee, C. Sun and X. Zhang, Sub-diffraction-limited optical imaging with a silver superlens, *Science*, 2005, **308**, 534–537.
5 N. Engheta, Circuits with light at nanoscales: optical nanocircuits inspired by metamaterials, *Science*, 2007, **317**, 1698–1702.
6 Z. Jacob, L. V. Alekseyev and E. Narimanov, Optical Hyperlens: Far-field imaging beyond the diffraction limit, *Opt. Express*, 2006, **14**, 8247–8256.
7 H. Chen, C. T. Chan and P. Sheng, Transformation optics and metamaterials, *Nat. Mater.*, 2010, **9**, 387–396.
8 D. R. Smith and D. Schurig, Electromagnetic wave propagation in media with indefinite permittivity and permeability tensors, *Phys. Rev. Lett.*, 2003, **90**, 077405.
9 C. R. Simovski and P. A. Belov, Low-frequency spatial dispersion in wire media, *Phys. Rev. E: Stat., Nonlinear, Soft Matter Phys.*, 2004, **70**, 046616.
10 A. Salandrino and N. Engheta, Far-field subdiffraction optical microscopy using metamaterial crystals: Theory and simulations, *Phys. Rev. B: Condens. Matter Mater. Phys.*, 2006, **74**, 075103.
11 Z. Jacob, I. I. Smolyaninov and E. E. Narimanov, Broadband Purcell effect: Radiative decay engineering with metamaterials, *Appl. Phys. Lett.*, 2012, **100**, 181105.
12 M. A. Noginov, *et al.* Controlling spontaneous emission with metamaterials, *Opt. Lett.*, 2010, **35**, 1863–1865.

13 Z. Jacob, *et al.* Engineering photonic density of states using metamaterials, *Appl. Phys. B*, 2010, **100**, 215–218.
14 T. Tumkur, *et al.* Control of spontaneous emission in a volume of functionalized hyperbolic metamaterial, *Appl. Phys. Lett.*, 2011, **99**, 151115.
15 H. N. Krishnamoorthy, Z. Jacob, E. Narimanov, I. Kretzschmar and V. M. Menon, Topological transitions in metamaterials, *Science*, 2012, **336**, 205–209.
16 T. U. Tumkur, *et. al.* Control of reflectance and transmittance in scattering and curvilinear hyperbolic metamaterials, *Appl. Phys. Lett.*, 2012, **101**, 091105.
17 E. E. Narimanov, H. Li, Yu. A. Barnakov, T. U. Tumkur and M. A. Noginov, Reduced reflection from roughened hyperbolic metamaterial, *Opt. Express*, 2013, **21**, 14956–14961.
18 V. M. Agranovich and M. D. Galanin *Electronic Excitation Energy Transfer in Condensed Matter*, North Holland Publishing Company, Amsterdam, 1982.
19 H. T. Dung, L. Knöll and D. G. Welsch, Three-dimensional quantization of the electromagnetic field in dispersive and absorbing inhomogeneous dielectrics, *Phys. Rev. A*, 1998, **57**, 3931.
20 P. Andrew and W. L. Barnes, Förster energy transfer in an optical microcavity, *Science*, 2000, **290**, 785–788.
21 T. Nakamura, M. Fujii, K. Imakita and S. Hayashi, Modification of energy transfer from Si nanocrystals to Er^{3+} near a Au thin film, *Phys. Rev. B: Condens. Matter Mater. Phys.*, 2005, **72**, 235412.
22 T. Nakamura, M. Fujii, S. Miura, M. Inui and S. Hayashi, Enhancement and suppression of energy transfer from Si nanocrystals to Er ions through a control of the photonic mode density, *Phys. Rev. B: Condens. Matter Mater. Phys.*, 2006, **74**, 045302.
23 Q. C. Sun, *et al.* Plasmon-enhanced energy transfer for improved upconversion of infrared radiation in doped-lanthanide nanocrystals, *Nano Lett.*, 2013, **14**, 101–106.
24 M. de Dood, J. Knoester, A. Tip and A. Polman, Förster transfer and the local optical densityof states in erbium-doped silica, *Phys. Rev. B: Condens. Matter Mater. Phys.*, 2005, **71**, 115102.
25 C. Blum, *et al.* Nanophotonic control of the Förster resonance energy transfer efficiency, *Phys. Rev. Lett.*, 2012, **109**, 203601.
26 F. Schleifenbaum, *et al.* Dynamic control of Förster energy transfer in a photonic environment, *Phys. Chem. Chem. Phys.*, 2014, **16**, 12812–12817.
27 F. T. Rabouw, *et al.* Photonic effects on the Förster resonance energy transfer efficiency, *Nat. Commun.*, 2014, **5**, 3610.
28 K. S. Kim, *et al.* Switching off FRET in the hybrid assemblies of diblock copolymer micelles, quantum dots, and dyes by plasmonic nanoparticles, *ACS Nano*, 2012, **6**, 5051–5059.
29 J. Yao, *et al.* Optical negative refraction in bulk metamaterials of nanowires, *Science*, 2008, **321**, 930.
30 M. A. Noginov, *et al.* Bulk photonic metamaterial with hyperbolic dispersion, *Appl. Phys. Lett.*, 2009, **94**, 151105.
31 A. J. Hoffman, *et al.* Negative refraction in semiconductor metamaterials, *Nat. Mater.*, 2007, **6**, 946–950.

32 Yu. K. Voron'ko, *et al.* Nature of non radiative excitation-energy relaxation in condensed media with high activator concentrations, *Soviet Physics - JETP*, 1977, **44**, 251–261.

33 L. V. Alekseyev and E. E. Narimanov, *Tutorials in Metamaterials – Series in Nano-optics and Nano-photonics*, ed. M. A. Noginov and V. A. Podolskiy, CRC Press, Florida, 2012, pp. 209–224.

34 L. Novotny and B. Hecht, *Principles of Nano-Optics*, Cambridge Univ. Press, Cambridge, 2006.

35 G. Juzeliunas and D. L. Andrews, Quantum electrodynamics of resonant energy transfer in condensed matter, *Phys. Rev. B: Condens. Matter Mater. Phys.*, 1994, **49**, 8751.

36 H. T. Dung, L. Knöll and D. Welsch, Intermolecular energy transfer in the presence of dispersing and absorbing media, *Phys. Rev. A*, 2002, **65**, 043813.

37 P. B. Johnson and R. W. Christy, Optical Constants of the Noble Metals, *Phys. Rev. B: Solid State*, 1972, **6**, 4370.

38 M. S. Tomas, Green function for multilayers: Light scattering in planar cavities, *Phys. Rev. A*, 1995, **51**, 2545–2559.

39 V. N. Pustovit, A. M. Urbas and T. V. Shahbazyan, Cooperative amplification of energy transfer in plasmonic systems, *Phys. Rev. B: Condens. Matter Mater. Phys.*, 2013, **88**, 245427.

40 D. E. Gomez, K. C. Vernon, P. Mulvaney and T. J. Davis, Surface Plasmon Mediated Strong Exciton–Photon Coupling in Semiconductor Nanocrystals, *Nano Lett.*, 2009, **10**, 274–278.

41 P. Törmä and W. L. Barnes, Strong coupling between surface plasmon polaritons and emitters, Preprint at <arXiv preprint arXiv:1405.1661>, 2014.

42 J. E. Sipe and R. W. Boyd, *Optical Properties of Nanostructured Random Media – Topics in Applied Physics*, ed. V. M. Shalaev, Springer, New York, 2002, pp. 1–18.

43 L. D. Landau and E. M. Lifshitz, *Electrodynamics of Continuous Media*, Pergamon, Oxford, 1984.

44 V. N. Pustovit and T. V. Shahbazyan, Resonance energy transfer near metal nanostructures mediated by surface plasmons, *Phys. Rev. B: Condens. Matter Mater. Phys.*, 2011, **83**, 085427.

45 E. Snoeks, A. Lagendijk and A. Polman, Measuring and modifying the spontaneous emission rate of erbium near an interface, *Phys. Rev. Lett.*, 1995, **74**, 2459–2462.

46 M. S. Tomas and Z. Lenac, Spontaneous emission spectrum in an absorbing Fabry–Perot cavity, *Phys. Rev. A*, 1999, **60**, 2431–2437.

47 K. H. Drexhage, Influence of a dielectric interface on fluorescence decay time, *J. Lumin.*, 1970, **1**, 693–701.

48 V. M. Agranovich, Y. N. Gartstein and M. Litinskaya, Hybrid resonant organic–inorganic nanostructures for optoelectronic applications, *Chem. Rev.*, 2011, **111**, 5179–5214.

Faraday Discussions

PAPER

A surface plasmon enabled liquid-junction photovoltaic cell

Woo-ram Lee,[a] Syed Mubeen,[ab] Galen D. Stucky[a] and Martin Moskovits*[a]

Received 29th September 2014, Accepted 15th October 2014
DOI: 10.1039/c4fd00185k

Plasmonic nanosystems have recently been shown to be capable of functioning as photovoltaics and of carrying out redox photochemistry, purportedly using the energetic electrons and holes created following plasmonic decay as charge carriers. Although such devices currently have low efficiency, they already manifest a number of favorable characteristics, such as their tunability over the entire solar spectrum and a remarkable resistance to photocorrosion. Here, we report a plasmonic photovoltaic using a 25 μm thick electrolytic liquid junction which supports the iodide/triiodide (I^-/I_3^-) redox couple. The device produces photocurrent densities in excess of 40 μA cm^{-2}, an open circuit voltage (V_{oc}) of ~0.24 V and a fill factor of ~0.5 using AM 1.5 G solar radiation at 100 mW cm^{-2}. The photocurrent and the power conversion efficiency are primarily limited by the low light absorption in the 2-D gold nanoparticle arrays. The use of a liquid junction greatly reduces dielectric breakdown in the oxide layers utilized, which must be very thin for optimal performance, leading to a great improvement in the long-term stability of the cell's performance.

Main text

Metallic nanostructures capable of supporting intense localized surface plasmon (LSP) resonances at visible frequencies, which have been known for many decades, are recently gaining significant attention for solar energy conversion applications.[1-17] Among the promising plasmon enabled solar energy conversion strategies demonstrated are: (i) light trapping in which the optical properties of the plasmonic structure are exploited to improve the efficiencies of conventional semiconductor-based devices,[1,2] and (ii) producing devices which use directly the hot electrons and holes generated following LSP decay.[3,4] Based upon the results reported to date, it is not unreasonable to hope that plasmonic systems will one day prove to be competitive with, and perhaps superior to, traditional semiconductor-based devices as photovoltaics or, more likely, as artificial

[a]*Department of Chemistry and Biochemistry, University of California, Santa Barbara, CA 93106, USA. E-mail: moskovits@chem.ucsb.edu*
[b]*Department of Chemical and Biochemical Engineering, University of Iowa, Iowa 52242, USA*

photosynthesis devices, on account of the ease with which the light absorption properties of plasmonic systems can be tuned to encompass the entire intense portion of the solar spectrum, and the greater resistance of plasmonic systems to photocorrosion, which has limited the use of semiconductors in artificial photosynthesis.[5]

Recent reports by us and others[6–14] have convincingly demonstrated the fact that hot electrons and holes generated as a result of surface plasmon decay can be harvested to do useful work. One embodiment of such devices makes use of junctions between a plasmonic metal and an n-type semiconductor whose conduction band edge is more positive than the Fermi level of the metal. The Schottky barrier thus formed creates the necessary internal potential barrier which hot electrons with sufficient energies can either overcome or tunnel through to become conduction electrons in the conduction band of the semiconductor. The electrons can then either be forced through a load generating electrical power, or transferred to an appropriate catalyst to carry out chemical reduction. The positive charges left behind can then act as holes completing the circuit (in a photovoltaic device) or enabling chemical oxidation of an appropriate species through the intermediacy of an appropriate catalyst (in an artificial photosynthesis device).

While the plasmon-to-electron paradigm has been demonstrated in the context of photocatalysis, photosynthesis, and photodetection,[5–7,11–14] its potential as a photovoltaic strategy has been less widely demonstrated.[7,9–11,15–17] For example, we recently reported a solar cell device consisting of vertically oriented gold nanorods capped with a thin layer of TiO_2, in which all of the electrons and holes contributing to the solar-to-electric power generation were shown unequivocally to have a plasmonic origin.[7] Photoexcitation of the LSP resonance of the gold nanorods in that device ultimately resulted in the transfer of hot electrons to the conduction band of the TiO_2. A portion of the photon energy was converted to electron kinetic energy which appeared as open circuit photovoltage when collected by a low work function metal (titanium) electrode. A load connecting the two metals completed the circuit, maintaining electroneutrality and sustaining a measurable photocurrent. Despite its preliminary status, we were able to extract power with a respectable absorbed photon-to-electron conversion efficiency of 2.75%.

However, based on the observed current–voltage characteristics and quantum efficiency measurements, we concluded that high efficiency plasmonic solar cells could be realized if the quality of the semiconducting oxide layer can be substantially increased. In those devices, it was the thickness of the oxide layer what determined the cell efficiency. As the oxide layer thickness was reduced, the cell's measured photovoltage decreased rapidly even if the photocurrent increased slightly. Such performance degradation, likely due to dielectric breakdown, is a common property of thin layers of most semiconducting oxides, attributed to current leakage caused by traps and defects generated either during the growth process or by electric stress created during the operation of the cell.[18,19]

One way to minimize dielectric breakdown of thin oxides is by minimizing the damage at its interface. It is well established in electrolytic capacitors that the breakdown voltage of dielectric oxides can be substantially mitigated by using a liquid electrolyte.[20] The breakdown voltage follows a semi-log relationship with the electrolyte resistivity and is independent of the dielectric thickness.[21,22] By choosing an electrolyte with modest conductivity, it is possible to increase the breakdown voltage and decrease the leakage currents. The electrolyte must also be electrochemically stable in the operational potential window and compatible with the electrode materials and the oxide.

We show here that the above approach can be successfully applied in the development of plasmonic solar cells with stable photovoltages by fabricating a device based on Au nanoparticles coated with a thin layer of TiO_2, and investigating its photovoltaic behavior. Gold nanoparticles were chosen *in lieu* of gold nanorods to avoid 'lightening rod' effects at the metal–oxide interface.[23–25] The TiO_2 was decoupled from the counter electrode using the iodide/triiodide (I^-/I_3^-) couple in a thin electrolyte layer.

The fabrication protocol and the corresponding cell configuration of the plasmonic metal–semiconductor–liquid junction solar cell are depicted schematically in Fig. 1. Briefly, a 2 nm (mass thickness) gold film was e-beam deposited on a FTO-coated glass substrate. Rapid thermal annealing (RTA) of the gold coated FTO substrates at 400 °C in a N_2 atmosphere converted the thin gold film into a 2-D array of gold nanoparticles with an average diameter of ~17 nm and an inter-particle distance of ~15 nm. A 5 nm layer of TiO_2 was subsequently deposited by atomic layer deposition (ALD) on the gold coated FTO substrate, followed by thermal oxidation at 500 °C for 1 hour. Pt nanoparticles deposited on a second FTO coated glass served as the counter electrode. A 25 μm thick spacer constructed using electrically insulating thermal melt polymer was used to isolate the anode and the counter electrode and maintain the desired electrode-to-electrode spacing. The I^-/I_3^- electrolyte (0.1 M I_2, 0.1 M lithium iodide (LiI), 0.6 M tetrabutylammonium iodide in acetonitrile) was injected between the two electrodes through pre-drilled holes that were subsequently sealed (see Methods for details).

Fig. 2 shows the approximate energy positions of the plasmonic metal, the semiconductor and the electrolyte. Au ($E_{F,Au}$ = 5.1 eV *vs.* vacuum) and TiO_2 (E_{CB,TiO_2} = 4.2 eV *vs.* vacuum) form the plasmonic metal–semiconductor junction, creating a Schottky barrier of approximately 0.9 V at the metal–semiconductor interface.[26–28] The I^-/I_3^- electrolyte was chosen because of its proven performance in the field of dye sensitized solar cells. An important factor to consider is the location of the Fermi level (*i.e.* the electrochemical potential) of the I^-/I_3^- electrolyte with respect to the conduction band edge of TiO_2. Preferably, the acceptor level of the redox couple should be close to the conduction band edge of TiO_2 in order to extract conduction electrons efficiently. For the I^-/I_3^- electrolyte ($E_{F,I^-/I_3^-}$ = 4.8 eV *vs.* vacuum),[29] the redox potential is more positive (on the electrochemical scale) than the conduction band edge of TiO_2, creating a barrier between TiO_2 and I^-/I_3^-. To circumvent this, we deposited 2 nm of titanium nitride (TiN) on top of TiO_2 using ALD. The reported work function of the ALD deposited TiN is 4.2–4.4 eV *vs.* vacuum, which should make it a good material for the formation of an Ohmic contact with TiO_2.[30] Additionally, since TiN is a good conductor, the interface between TiN and I^-/I_3^- would not sustain any electrical fields.

The photon-to-electron conversion in this device occurs as follows. Visible light is absorbed by the gold nanoparticles producing LSP excitations, which decay to a large number of hot electrons. Hot electrons with sufficient energy overcome the Schottky barrier at the metal–semiconductor junction and enter the conduction band of TiO_2. Some energetic hot carriers might quantum

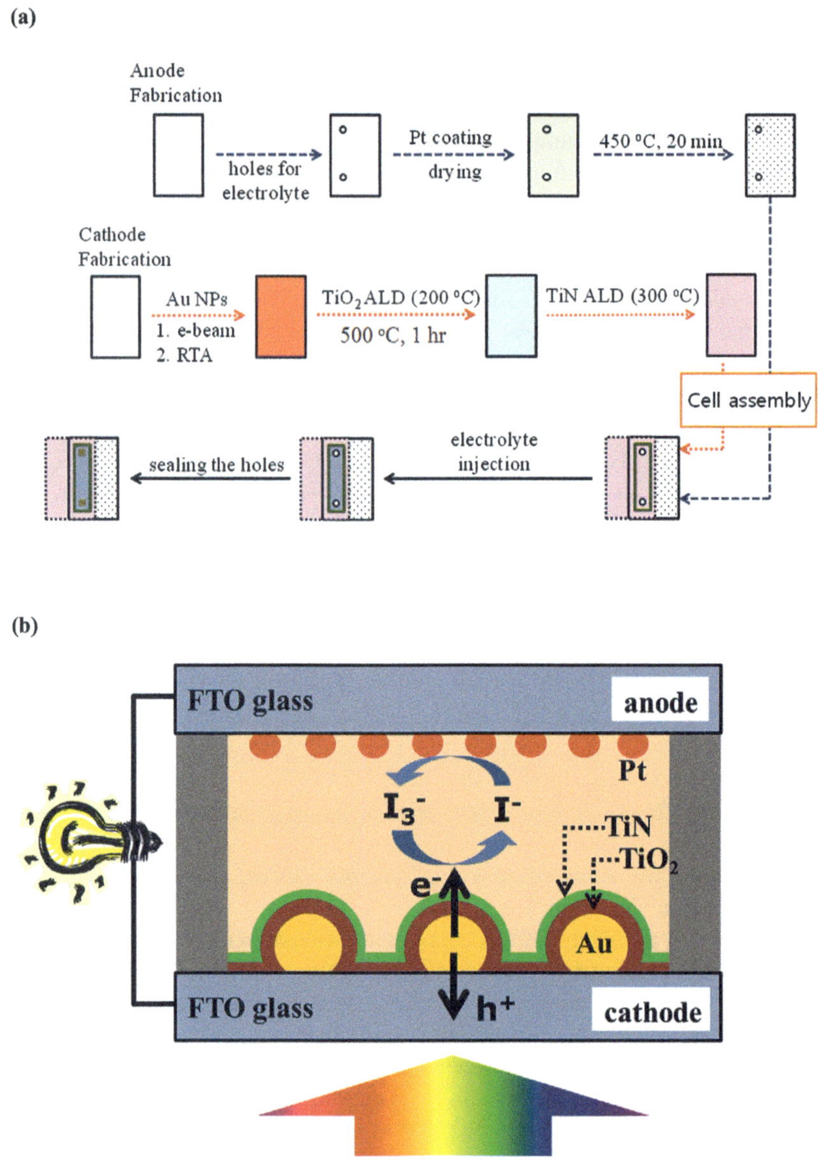

Fig. 1 (a) Schematic itemizing the steps in the fabrication process used to produce the plasmonic liquid-junction photovoltaic cell. (b) A schematic of the structure, components and materials comprising the cell.

mechanically tunnel through the barrier. The electrons conduct through the TiN layer to the surface where they carry out the I_3^- to I^- reduction. The positive charges left in the gold conduct through the load and carry out the oxidation step (I^- to I_3^-) on the Pt layer at the counter electrode, thereby restoring charge neutrality.

As in electrochemical dye-sensitized solar cells, this structure decouples the photon absorption process from charge separation and transport; however, it has

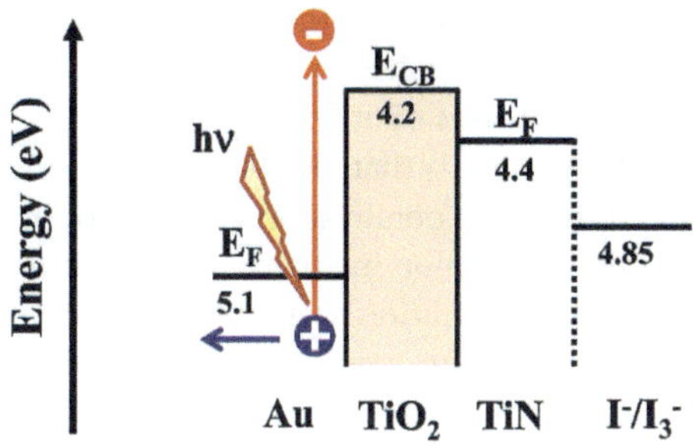

Fig. 2 Energy band diagram depicting the electron transfer pathway. Hot electrons generated by surface plasmon decay enter the conduction band of TiO_2 by overcoming the Schottky barrier at the Au–TiO_2 interface. Charge carriers conduct through the Ohmic junction and into the TiN layer and ultimately to the I^-/I_3^- species in the electrolyte.

the significant advantage that metals do not photobleach as dyes do.[31] A key feature of this device is that the cell can be kept sufficiently thin to eliminate the uncompensated resistance loses due to the electrolyte that arise in typical electrochemical systems. The presence of the electrolyte also increases the reliability of the cell by increasing the breakdown voltage and reducing the leakage current.

Fig. 3 displays the current density *vs.* voltage characteristics of a representative device when illuminated by simulated AM 1.5 G solar radiation (100 mW cm^{-2}). The active area of the device was 0.80 cm^2. Under AM 1.5 G conditions, we observed a short circuit current density of 46 μA cm^{-2}, a stable open circuit voltage (V_{oc}) of 0.24 V, a fill factor of 0.48 and a power conversion efficiency of ~0.005%. The short circuit current density and power conversion efficiency are limited by the low nanoparticle coverage and the consequently low photon absorption. Extinction measurements performed on a single 2-D array of gold nanoparticles yielded an extinction value of ~0.1 (the absorptance would be less than this since some of the light extinction results from scattering). Increasing the metal particle density or using cells in tandem would greatly improve the performance. Additionally, a significant improvement is expected by optimizing the reflection losses, the electrode–electrolyte configuration and the surface treatment, steps that are routinely used to improve the performance of conventional semiconductor devices. Incorporating such improvements could increase

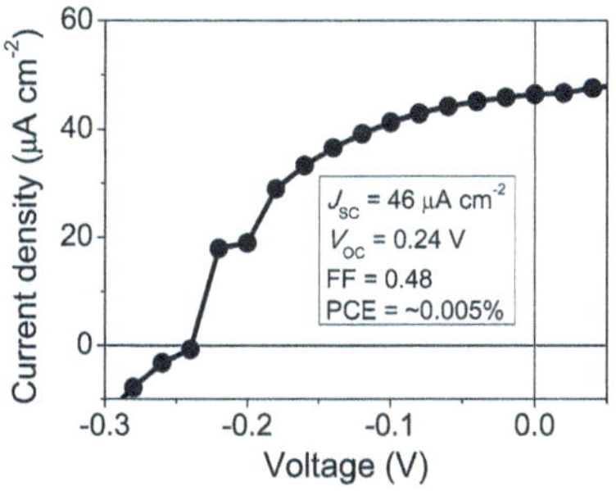

Fig. 3 Photocurrent density *vs.* voltage (*J*–*V*) measured under simulated AM 1.5 G full sun illumination (100 mW cm^{-2}) for a cell with an active area of 0.80 cm^2.

the short circuit current and consequently the cell efficiency by a factor of at least 20. Efforts to boost the open circuit voltage, such as those discussed below should produce further efficiency increases of at least a factor of 2.

In fact, a key result of this work is that, despite using less than a monolayer of gold nanoparticles, we were able to obtain short circuit current densities of ~0.05 mA cm^{-2}, wholly powered by plasmon-mediated charge carriers. The modest fill factor is due to losses from concentration overpotentials and can be minimized by optimizing the electrode–electrolyte configuration. By reducing the inter-electrode distance between the anode and the cathode, other constraints such as the absorption of light by the solution, the insufficient conductivity of the electrolyte, or uncompensated resistance losses can also be mitigated.

The semiconductor layer in our device does not take part in photon absorption, thus the semiconductor layer can be made very thin, essentially working uniquely as an electron filter. The maximum photovoltage depends on the Schottky barrier height between the metal and the semiconductor and on the redox potential of the electrolyte. Although what we report is a significant step in the achievement of photovoltages from plasmonic solar cell systems, the photovoltage obtained (0.24 V) is low. This is because the electrochemical potential of the selected redox couple is much more positive (on the electrochemical scale) than the conduction band of TiO_2. By selecting a redox couple whose Fermi level is close to the conduction band of TiO_2, one could significantly improve the photovoltage.

Conclusions

We report a plasmonic photovoltaic using a 25 μm thick electrolytic liquid junction which supports the iodide/triiodide (I^-/I_3^-) redox couple. The device produces photocurrent densities in excess of 40 μA cm^{-2}, an open circuit voltage (V_{oc}) of ~0.24 V and a fill factor of ~0.5 using AM 1.5 G solar radiation at 100 mW cm^{-2}. The use of a thin liquid layer with redox active molecules between the electrodes results in a cell which does not deteriorate on account of dielectric breakdown in the thin oxide layers, providing the basis for the development of durable and efficient plasmonic photovoltaics using a wide choice of both amorphous and polycrystalline oxides and inexpensive redox active couples, which can yield high photovoltages without the sacrifice in the overall power conversion efficiency that must often be made in solid-state cells, where material quality is more stringent. Indeed, even these very preliminary cells, meant to illustrate the principle, showed remarkable longevity: their current and voltage characteristics remained constant over multiple runs carried out over many weeks.

Methods

The solar cell was fabricated on two FTO-coated glass substrates (7 Ω sq^{-1}, 2.2 mm thickness, TCO22-7, Solaronix).

To prepare the cathode, the FTO glass was cleaned for 20 min. in a detergent solution (Deconex® 12 BASIC, Borer Chemie AG) using an ultrasonic bath, and rinsed with tap water, DI water and ethanol. A 2 nm (mass thickness) gold film was e-beam deposited on a FTO-coated glass substrate. Rapid thermal annealing

(RTA) of the gold coated FTO substrates at 400 °C in a N_2 atmosphere was carried out for 5 minutes to form a 2-D random arrangement of gold nanoparticles. TiO_2 and TiN layers were deposited in an Oxford FlexAl ALD tool. For TiO_2, tetrakis(dimethylamino)titanium (TDMAT) and water were used as the precursor and the reactant, respectively, at 200 °C, followed by thermal oxidation at 500 °C in an oven for 1 hour. Plasma-enhanced ALD of TiN was carried out at 300 °C using TDMAT and a mixture of H_2 and N_2 gas (3 : 1 ratio).

For the anode, two holes were drilled in the FTO glass, which had been washed as described previously. The FTO glass was coated with a thin layer of a 5 mM solution of $H_2PtCl_6 \cdot 6H_2O$ in isopropanol, dried at room temperature for 10 minutes, heated from room temperature to 450 °C at a rate of 25 °C min^{-1} and maintained at 450 °C for 20 minutes.

The two electrodes were sealed around their perimeter using a thermal melt polymer film (25 μm thickness, Meltonix 1170-25, Solaronix), so that a ~25 μm space was created between them. The liquid electrolyte (0.1 M I_2, 0.1 M lithium iodide (LiI), 0.6 M tetrabutylammonium iodide in acetonitrile) was injected in the space between the electrodes through one of the holes pre-drilled in the anode, and the holes were sealed with thermal melt polymer films and microscope cover slips. The active area of the device was 0.80 cm^2.

Photovoltaic measurements were carried out using a Newport Air Mass 1.5 Global (AM 1.5 G) full spectrum solar simulator as source at an irradiation intensity of 100 mW cm^{-2}. Current density–voltage (J–V) measurements were carried out using a Keithley 2400 Digital Source Measure Unit.

Acknowledgements

This work was supported by the Institute for Energy Efficiency, an Energy Frontier Research Center funded by the US Department of Energy, Office of Science, Office of Basic Energy Sciences (award no. DE-SC0001009). Two of us (GDS and WRL) wish to thank the National Science Foundation (DMR 0805148) for support. Support from the MRSEC Program of the National Science Foundation under Award no. DMR 1121053 is also gratefully acknowledged. We also made extensive use of the MRL Central Facilities at UCSB, a member of the NSF-funded Materials Research Facilities Network (http://www.mrfn.org).

References

1 K. R. Catchpole and A. Polman, *Opt. Express*, 2008, **16**, 21793.
2 H. A. Atwater and A. Polman, *Nat. Mater.*, 2010, **9**, 205.
3 C. Clavero, *Nat. Photonics*, 2014, **8**, 95.
4 A. Manjavacas, J. G. Liu, V. Kulkarni and P. Nordlander, *ACS Nano*, 2014, **8**, 7630.
5 S. Linic, P. Christopher and D. B. Ingram, *Nat. Mater.*, 2011, **10**, 911.
6 S. Mubeen, J. Lee, N. Singh, S. Krämer, G. D. Stucky and M. Moskovits, *Nat. Nanotechnol.*, 2013, **8**, 247.
7 S. Mubeen, J. Lee, W.-R. Lee, N. Singh, G. D. Stucky and M. Moskovits, *ACS Nano*, 2014, **8**, 6066.
8 J. Lee, S. Mubeen, X. Ji, G. D. Stucky and M. Moskovits, *Nano Lett.*, 2012, **12**, 5014.

9 S. Mubeen, G. Hernandez-Sosa, D. Moses, J. Lee and M. Moskovits, *Nano Lett.*, 2011, **11**, 5548.

10 F. Wang and N. A. Melosh, *Nano Lett.*, 2011, **11**, 5426.

11 F. P. G. de Arquer, A. Mihi, D. Kufer and G. Konstantatos, *ACS Nano*, 2013, **7**, 3581.

12 Y. Nishijima, K. Ueno, Y. Kotake, K. Murakoshi, H. Inoue and H. Misawa, *J. Phys. Chem. Lett.*, 2012, **3**, 1248.

13 M. W. Knight, Y. Wang, A. S. Urban, A. Sobhani, B. Y. Zheng, P. Nordlander and N. J. Halas, *Nano Lett.*, 2013, **13**, 1687.

14 V. Subramanian, E. Wolf and P. V. Kamat, *J. Phys. Chem. B*, 2001, **105**, 11439.

15 Y.-H. Su, Y.-F. Ke, S.-L. Cai and Q.-Y. Yao, *Light: Sci. Appl.*, 2012, **1**, e14.

16 Z. H. Chen, Y. B. Tang, C. P. Liu, Y. H. Leung, G. D. Yuan, L. M. Chen, Y. Q. Wang, I. Bello, J. A. Zapien, W. J. Zhang, C. S. Lee and S. T. Lee, *J. Phys. Chem. C*, 2009, **113**, 13433.

17 T. Bora, H. H. Kyaw, S. Sarkar, S. K. Pal and J. Dutta, *Beilstein J. Nanotechnol.*, 2011, **2**, 681.

18 J. J. O'Dwyer, *The theory of electrical conduction and breakdown of solids*, Clarendon Press, Oxford, 1973.

19 M. Fröhlich, *Theory of dielectrics*, Oxford University Press, 1949.

20 M. Jayalakshmi and K. Balasubramanian, *Int. J. Electrochem. Sci.*, 2008, **3**, 1196.

21 A. Guntherschulze and H. Betz, *Elektrolytkondensatoren*, Technischer Verlag Herbert Cram, Berlin, 1952.

22 F. J. Burger and J. C. Wu, *J. Electrochem. Soc.*, 1971, **118**, 2039.

23 P. F. Liao and A. Wokaun, *J. Chem. Phys.*, 1982, **76**, 751.

24 O. Benson, *Nature*, 2011, **480**, 193.

25 M. B. Mohamed, V. Volkov, S. Link and M. A. El-Sayed, *Chem. Phys. Lett.*, 2000, **317**, 517.

26 E. W. McFarland and J. Tang, *Nature*, 2003, **421**, 616.

27 J. Y. Park, H. Lee, J. R. Renzas, Y. Zhang and G. A. Somorjai, *Nano Lett.*, 2008, **8**, 2388.

28 J. Tang, M. White, G. D. Stucky and E. W. McFarland, *Electrochem. Commun.*, 2003, **5**, 497.

29 J. Datta, A. Bhattacharya and K. K. Kundu, *Bull. Chem. Soc. Jpn.*, 1988, **61**, 1735.

30 K. Choi, H.-C. Wen, H. Alshareefa, R. Harris, P. Lysaght, H. Luanc, P. Majhid and B. H. Lee, *Proceedings of Essderc 2005: 35th European Solid-State Device Research Conference, Book Series: Proceedings of the European Solid-State Device Research Conference*, 2005, pp. 101–104.

31 J. Kosar, *Light-sensitive systems*, Wiley, New York, 1965.

Faraday Discussions

PAPER

Optical forces in nanoplasmonic systems: how do they work, what can they be useful for?

T. V. Raziman, R. J. Wolke and O. J. F. Martin*

Received 17th November 2014, Accepted 24th November 2014
DOI: 10.1039/c4fd00224e

In this article, we share our vision for a future nanofactory, where plasmonic trapping is used to control the different manufacturing steps associated with the transformation of initial nanostructures to produce complex compounds. All the different functions existing in a traditional factory can be translated at the nanoscale using the optical forces produced by plasmonic nanostructures. A detailed knowledge of optical forces in plasmonic nanostructures is however essential to design such a nanofactory. To this end, we review the numerical techniques for computing optical forces on nanostructures immersed in a strong optical field and show under which conditions approximate solutions, like the dipole approximation, can be used in a satisfactory manner. Internal optical forces on realistic plasmonic antennas are investigated and the reconfiguration of a Fano-resonant plasmonic system using such internal forces is also studied in detail.

1 Introduction

Since the first observation by Ashkin in 1970 of the acceleration and trapping with optical radiation pressure of freely suspended micron-sized particles,[1] and the demonstration 15 years later by the same author of stable optical trapping in three dimensions using a single-beam gradient optical trap,[2] optical forces produced by a laser beam have been both extensively studied theoretically and applied for contactless manipulation of a broad variety of small objects, especially biological entities, for which optical tweezers have become routine. The following review articles provide an overview of this field of research.[3–6]

The mechanisms of optical trapping can be easily understood – even within the framework of ray optics – if one assumes that each photon forming the optical ray carries some momentum. This is illustrated in Fig. 1, inspired by one of the seminal papers of Ashkin.[2] A dielectric sphere placed in a homogeneous optical field will mainly feel a scattering force that pushes it along

Nanophotonics and Metrology Laboratory, Swiss Federal Institute of Technology Lausanne (EPFL), CH-1015 Lausanne, Switzerland. E-mail: olivier.martin@epfl.ch; Fax: +41 21 693 26 14; Tel: +41 21 693 26 07

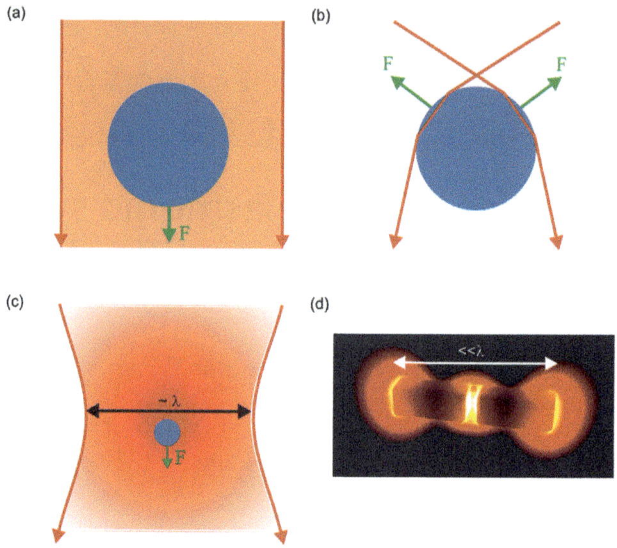

Fig. 1 Different optical trapping configurations. (a) In a homogeneous field, a particle is just pushed along by the scattering force. (b) In the focus of a strong lens, the gradient forces can laterally trap and catch the particle; these forces can be understood in the framework of ray optics, assuming that momentum is transferred from the incoming ray to the sphere at each reflexion or refraction. (c) In far-field optics, the focus produced by a lens is limited to about the wavelength λ; consequently, a very small particle does not "feel" the gradient forces and is merely pushed along by the scattering force. (d) A plasmonic nanostructure can produce extremely strong field gradients, which are not limited by the illumination wavelength, but simply determined by the geometrical features of the nanostructure. In that case, extremely small particles can be trapped near the gap of a plasmonic dipole antenna.

and is not much use for manipulating the sphere, Fig. 1(a). When the same sphere is placed in a strongly focused light beam, the lateral variations of the optical field now also give rise to lateral trapping forces – so called gradient forces – which can be used to trap and manipulate the sphere, Fig. 1(b). The dimensions of this strong focus produced by far-field optics are comparable to the wavelength of radiation used to illuminate the system. Hence, when the sphere becomes much smaller than the wavelength, it does not feel anymore the lateral field gradient and cannot be trapped, Fig. 1(c). This limitation can be overcome by using the very strong field gradient that is produced in the near-field of a plasmonic nanostructure, Fig. 1(d). Hence, in that case, it is possible to trap extremely small objects. It is actually a feature of the near-field that it can be confined to dimensions that are determined by the size of the object that produces the near-field, irrespective of the illumination wavelength.[7] This effect is dramatically intensified in plasmonic nanostructures, where the very strong near-field originates from the confinement of polarization charges at the surface of the metal.[8] We refer the reader to the recent review by the group of Romain Quidant for the development of this field of plasmonic trapping and the different geometries where it has been implemented.[9]

In this contribution, we focus on the optical forces produced by such a plasmonic system and enabled by the strongly localized near-field associated with the plasmon resonance. In Sec. 2, we review the physical mechanisms of optical forces and describe the theoretical framework required to analyse them; we also provide some insightful comparisons with the dipole approximation. We then investigate two facets of plasmonic forces: how they can act internally and externally to reconfigure a system. In Sec. 5 we present our vision for a nanofactory, where optical forces are used as the driving mechanism. The paper is summarized in Sec. 6.

2 Modelling optical forces in plasmonic systems

Maxwell's stress tensor

One must resort to Maxwell's equations to go beyond the simple ray optics picture sketched in Fig. 1(b) to describe the forces produced by light on an object.[10] Starting from the Lorentz force **F** caused by the electric field **E** and the magnetic induction **B** on a charged particle q,

$$\mathbf{F} = q(\mathbf{E} + \mathbf{v} \times \mathbf{B}), \tag{1}$$

and using Newton's second law to link the change of momentum d**P**/dt in the volume under study with the Lorentz force, one obtains Maxwell's stress tensor σ_{ij} in a medium with permittivity ε and permeability μ,

$$\sigma_{ij} = \varepsilon_0 \varepsilon E_i E_j + \mu_0 \mu H_i H_j - \frac{1}{2}(\varepsilon_0 \varepsilon E_k E_k + \mu_0 \mu H_k H_k)\delta_{ij} \tag{2}$$

which can be used to compute the optical force on a scatterer with boundary S,

$$\frac{\partial(\mathbf{P}_{\text{mech}} + \mathbf{P}_{\text{field}})_i}{\partial t} = \oint_S \sigma_{ij} n_j \mathrm{d}S, \tag{3}$$

where $\mathbf{P}_{\text{mech}}$ and $\mathbf{P}_{\text{field}}$ are the mechanical and electromagnetic momenta and n the outward normal to the surface S. Assuming harmonic electromagnetic fields, one can compute the time-averaged Maxwell stress tensor,

$$\langle \sigma_{ij} \rangle = \frac{1}{2}\Re\left\{ \varepsilon_0 \varepsilon E_i E_j^* + \mu_0 \mu H_i H_j^* - \frac{1}{2}\left(\varepsilon_0 \varepsilon E_k E_k^* + \mu_0 \mu H_k H_k^*\right)\delta_{ij} \right\}. \tag{4}$$

The time average of the electromagnetic momentum over a period is constant and its derivative in eqn (3) vanishes. Hence, the time averaged optical force acting on the scatterer is given by

$$\langle F_i \rangle = \oint_S \langle \sigma_{ij} \rangle n_j \mathrm{d}S. \tag{5}$$

Numerical implementation

Eqn (5) indicates that the knowledge of the electromagnetic field in the surroundings of a trapped object enables the calculation of the optical force on this object. In practice, the electromagnetic field is first computed on a mesh that follows a closed surface around the object, usually a sphere (Fig. 2(a)), and the optical force is computed using eqn (5).[11–15] Although this approach involves an additional overhead since a large number of surface points might need to be sampled to compute the optical force accurately, it is perfectly suitable for a variety of geometrical configurations, especially when the different scatterers are placed far apart. It often becomes cumbersome in the case of plasmonic trapping, where the trapped structure is very close to the plasmonic system producing the strong near-field, as illustrated in Fig. 2(b). To overcome this difficulty, we recently proposed an approached based on the surface integral equation (SIE), where the surface of the trapped particle is directly used to compute Maxwell's stress tensor.[16]

In short, we simulate the optical response of plasmonic structures with the SIE formulation, which utilises Green's dyadic function to solve Maxwell's equations in the frequency domain.[17–19] This method offers various advantages over formulations such as finite difference time domain and volume integral equations in terms of being able to model the scatterer surface very well and in the reduction of computational expenses by requiring the discretisation of only the scatterer surface. SIE simulations output the expansion of surface electric and magnetic currents (which correspond to the tangential magnetic and electric fields, respectively) over Rao, Wilton, and Glisson basis functions on the triangles the scatterer surface has been discretised into.[20] Relevant physical quantities such as fields and optical cross sections can be computed from these surface currents.

Maxwell's stress tensor is then directly evaluated over the triangles the scatterer surface has been discretised into, in terms of the surface electric $\mathbf{J}$ and magnetic $\mathbf{M}$ currents without computing the fields explicitly, Fig. 2(b). The expression for the force in terms of the currents over the triangles is

$$\mathbf{F} = \sum_T \int_T \mathrm{d}S_T \left\{ \frac{(\nabla\cdot\mathbf{M})(\nabla\cdot\mathbf{M}^*)}{2\omega^2\mu^*}\hat{\mathbf{n}} + \frac{(\nabla\cdot\mathbf{J})(\nabla\cdot\mathbf{J}^*)}{2\omega^2\varepsilon^*}\hat{\mathbf{n}} + \frac{i}{\omega}\left[\frac{\mu}{\mu^*}(\mathbf{J}\times\hat{\mathbf{n}})(\nabla\cdot\mathbf{M}^*) \right.\right.$$
$$\left.\left. + \frac{\varepsilon}{\varepsilon^*}(\hat{\mathbf{n}}\times\mathbf{M})(\nabla\cdot\mathbf{J}^*)\right] - \frac{1}{2}(\varepsilon\mathbf{M}\cdot\mathbf{M}^* + \mu\mathbf{J}\cdot\mathbf{J}^*)\hat{\mathbf{n}} \right\}. \quad (6)$$

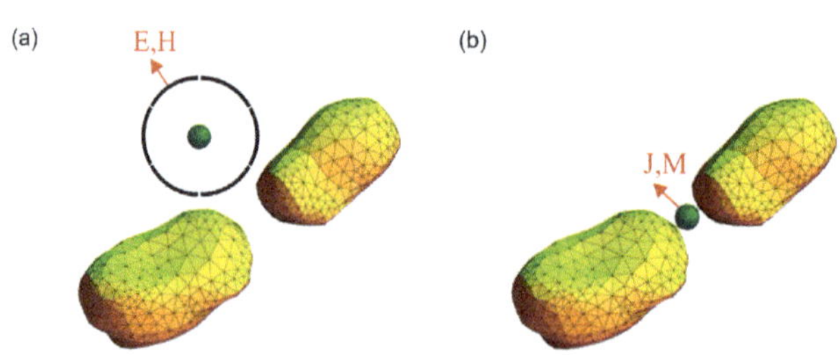

Fig. 2 Calculation of the optical forces on a nanostructure at the vicinity of a plasmonic dipole antenna using Maxwell's stress tensor. (a) In general, a virtual surface (*e.g.* the sphere in black) is defined around the nanostructure and the total electromagnetic field **E**, **H** is computed on that surface. The force on the object is then deduced using eqn (4). This approach is not well suited in the case of intricate trapping geometries, such as those encountered in plasmonic trapping. (b) Recently, we developed an alternative approach, where the optical force is computed with eqn (6) directly from the surface currents **J**, **M** on the trapped particle obtained from the SIE calculation for the complete system (plasmonic trapping antenna + trapped nanostructure).

Since the currents are expanded in terms of Rao, Wilton and Glisson basis functions which are linear over geometric coordinates, the integral over each triangle can be performed analytically, reducing further numerical errors during force computation.

Comparison with approximate solutions

The optical forces on small particles in a non-uniform electric field **E** can be approximated by the gradient force given by

$$\langle \mathbf{F} \rangle = \frac{1}{4} \Re\{\alpha\} \nabla \mathbf{E}^2, \tag{7}$$

where α is the dipolar polarisability of the particle under consideration. This approximate solution enables us to reveal another facet of plasmonic trapping, whereby the trapped particle supports plasmon resonances as well. Indeed, for a

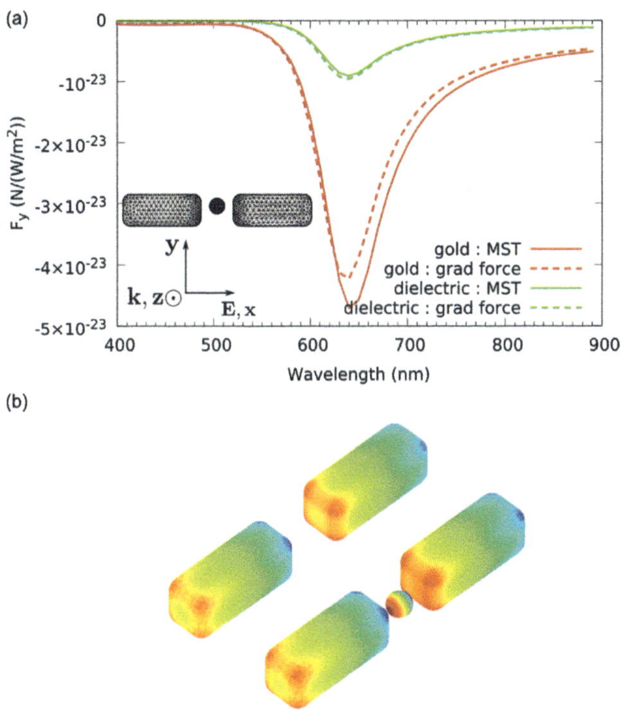

Fig. 3 Restoring forces on a nanoparticle displaced from the centre of a plasmonic gap antenna shown in the inset of (a). The antenna is made of gold with arms of dimensions 100 nm × 40 nm × 40 nm and a gap of 40 nm, and is placed with the antenna axis along *x*. A spherical particle of 10 nm radius is placed at the centre of the gap and displaced by 5 nm along *y*. The structure is illuminated by a plane wave propagating along *z* and polarised along *x*. (a) Comparison of restoring forces on a particle made of dielectric (glass) and metal (gold), computed using the Maxwell's stress tensor method and using the gradient force approximation. The gradient force method provides a good approximation for the force for the entire wavelength range. (b) Polarisation charges induced on the structure at the peak force wavelength both in the absence of the sphere and when the sphere is present. The presence of the sphere does not modify the charge distribution significantly.

given volume, a plasmonic particle can exhibit a much stronger polarizability than a dielectric particle.[21] This is advantageous in terms of detecting the trapping event,[22] and provides additional degrees of freedom for trapping a plasmonic nanostructure with a plasmonic trap, since the resonance of the coupled system can be fully tuned (from trapping to repelling) depending on the relative energy of both plasmon resonances.[23]

We next look at the applicability of this approximation by comparing it with the SIE results using Maxwell's stress tensor. The system under consideration is a gap antenna with arms of dimensions 100 nm $\times$ 40 nm $\times$ 40 nm made of gold. The antenna is illuminated by a plane wave propagating in the z-direction polarised along x. A sphere is placed at the centre of the gap between the antenna arms and displaced by 5 nm along y, and we look at the restoring force.

Fig. 3(a) shows the comparison between the forces calculated using the gradient force method and the Maxwell's stress tensor method for a gap of 40 nm and the displaced particle of radius 10 nm. We see that the gradient force approach is quite good for both the metallic (gold) particle and the dielectric (glass) particle, though the quantitative agreement is better for the latter. To understand better the reason behind this, we show the polarisation

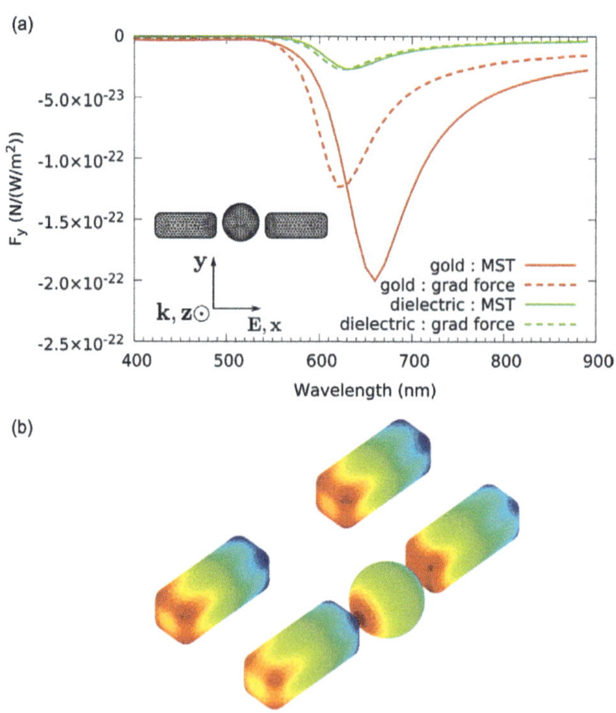

Fig. 4 Restoring forces on a larger nanoparticle of radius 30 nm, placed in a larger gap of 80 nm, as shown in the inset of (a), with other parameters the same as in Fig. 3. (a) Comparison of restoring forces computed using the two methods, showing that the gradient force approximation is now highly inaccurate for the case of the gold particle. (b) Polarisation charges induced on the structure are modified in the presence of the larger sphere, with a significant charge migration towards the gap faces in this case.

charges induced on the scatterer surfaces (calculated directly from the surface currents again) at the peak force wavelength in Fig. 3(b). It can be seen that the presence of the small sphere does not affect the charge distribution significantly, the result of which is that the gradient force calculation, which assumes that the presence of the sphere does not change the field distribution due to the other scatterers, works quite well.

We now repeat the calculation for a larger particle with 30 nm radius placed in an 80 nm gap, Fig. 4(a) shows the force computed using both methods. Though the results for the dielectric particles are quite reasonable, the gradient force shows a large mismatch with the Maxwell's stress tensor method for the metallic particle. The induced polarisation charges plotted in Fig. 4(b) explain why this happens. The inclusion of the large spherical particle affects the charge distribution on the antenna arms considerably, in particular, attracting the surface charges towards the gap faces. As a result, the electric field in the gap due to the arms changes a lot. Also, the electric field varies quite significantly over the dimensions of the particle unlike in the previous case. Consequently, the gradient force method is inadequate for force calculations in that case. Such behaviour even for particles quite smaller than the wavelength of light (at peak force wavelength, the gold sphere diameter is less than one-tenth the wavelength) shows the importance of using full Maxwell's stress tensor calculations for forces in plasmonic systems.

3 Internal forces in plasmonic systems

The vast majority of works reported on the transfer of momentum from an external field onto a nanostructure consider how the external field can modify the motion of the nanostructure. To the best of our knowledge, internal forces within the nanostructure induced by an external field have not yet been touched upon. Here, we look at the optical forces on the realistic gap antenna shown in the inset of Fig. 5(a), which discretization has been obtained from a fabricated structure.[24] The antenna is made out of gold, with arms of approximate dimensions 110 nm $\times$ 50 nm $\times$ 50 nm, and the gap between the antenna arms is approximately 20 nm. The gap antenna is illuminated by an x-polarised plane wave propagating in the z-direction. The scattering cross section for the structure as a function of the wavelength is shown in Fig. 5(a). The dipolar resonance of the gap antenna results in a strong scattering peak at 630 nm.

Fig. 5(b) shows the optical forces on the arms of the gap antenna. Clearly, the arms of the antenna attract each other throughout the wavelength range under consideration. Also, both arms are pushed in the z-direction due to the scattering force. Interestingly, we note that the pushing forces on the two arms are not identical due to the realistic and non-identical nature of the arms. Even the x-forces on the arms are not exactly equal in magnitude. In fact, there is also a small force on the arms in the y-direction (not shown in figure), again a consequence of the realistic and asymmetric nature of the structure.

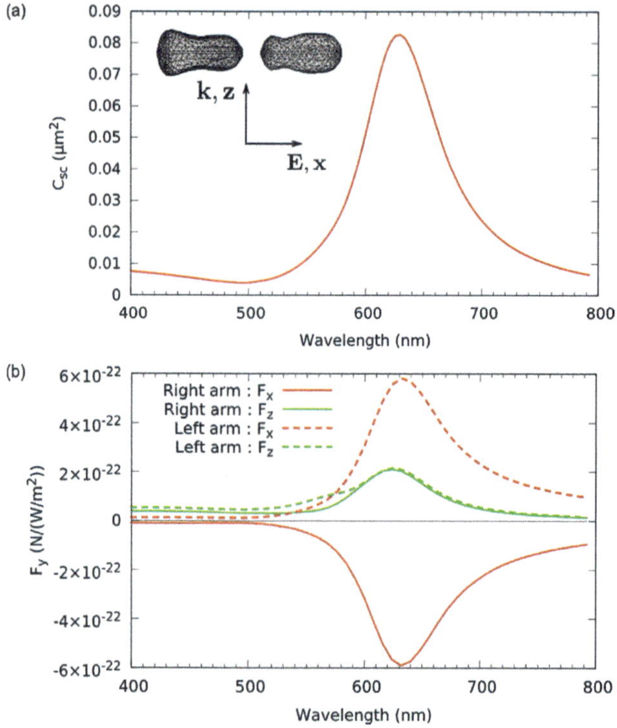

Fig. 5 Optical forces on a realistic gap antenna made of gold, shown in the inset of (a). The antenna is placed with the long axis along *x*, and illuminated by an incident plane wave propagating along *z* and polarised along *x*. (a) Scattering cross section of the antenna, showing the dipolar resonance peak near 630 nm. (b) Scattering force (along *z*) and internal force (along *x*) on the arms of the antenna. The realistic, asymmetric geometry results in unequal forces on the two arms.

4 Reconfiguring plasmonic systems using optical forces

The integration of nano-optical traps into a plasmonic system allows for a new type of reconfigurable plasmonic devices by utilizing the interaction between the trap and the trapped particle. Using specifically designed plasmonic nanostructures, the optical forces can be tailored to accommodate plasmonic colloidal particles at locations critical to the spectral response of the complete system.

Of particular interest are switchable Fano resonances, due to the large and broadband spectral variation that is not limited to spectral shifts or the creation of additional modes.[25–28] Furthermore, the strong field enhancements typical for the subradiant mode of a Fano resonance give rise to particularly strong optical forces. Under resonant laser excitation, the potential energies associated with these forces exceed those of Brownian motion, creating a potential well deep enough to trap particles at room temperature.

An example of such a system is heptamers of plasmonic discs. In the absence of the central particle, the ring-like hexamer exhibits a dipolar mode. Adding a

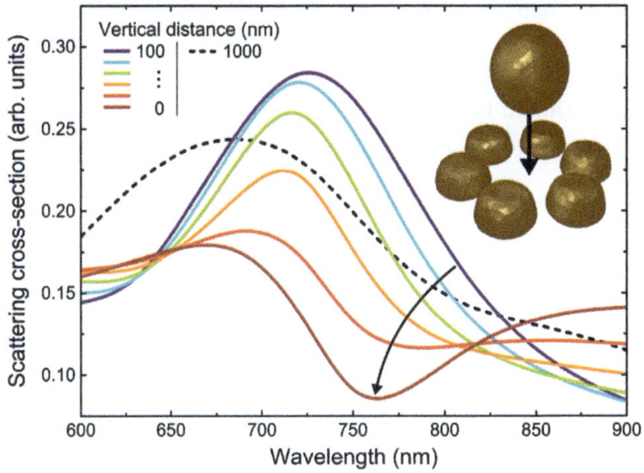

Fig. 6 Calculated scattering cross-section spectra of a hexamer of gold nanodiscs (d = 120 nm, h = 60 nm) in the presence of a gold sphere (r = 75 nm) located above the center of the structure, illuminated by a linearly polarized plane wave under normal incidence. Spectra are shown for different vertical distances of the sphere as measured from the bottom of the sphere to the bottom of the discs. As the sphere approaches the center of the hexamer (solid lines, top to bottom), a Fano resonance appears in the spectrum around 760 nm, due to the coupling between the hexamer's ring-like mode and the now supported linear mode. The dashed black line shows the uncoupled system, represented by placing the sphere 1 μm above the hexamer.

seventh particle in the center of the ring leads to the addition of linear modes to the system. More importantly, the coupling of ring-like and linear modes can lead to a Fano resonance. The spectral behavior of the plasmonic heptamer can be preserved when substituting the central disc by a spherical particle of similar size, making it an ideal candidate for a reconfigurable plasmonic system based on optical trapping of colloidal plasmonic nanospheres.

Fig. 6 shows SIE simulations of the scattering intensity of a gold hexamer that can be used for trapping. Included in the simulations is a gold sphere, positioned at different heights above the centre of the hexamer, ranging from 0 to 100 nm above the substrate (not included in the simulation), as indicated by the legend. As the sphere lowers into the trap, the spectrum changes and a distinct Fano resonance appears, with a subradiant mode around 760 nm for a distance of 0 (dark red line). The dotted line shows the combined spectrum of the hexamer and sphere without coupling, by placing the sphere at a height of 1 μm. The spectral responses for the two extreme cases are fundamentally different and easily distinguishable over a broad wavelength range.

5 Optical forces as drivers for a nanofactory

In this section, we would like to share our vision for realizing at the nanoscale the equivalent of a factory where structures are engineered through successive transformation steps. While, in the industrial revolution era, the different machineries operated in a factory would take their energy from a rather sophisticated line shaft, we propose that most of the operations required by such a nanofactory could be controlled with optical forces, as illustrated in Fig. 7.

As in a conventional factory, the first step is to move the new material to the location where it should be processed, Fig. 7(a). This can be achieved using the optical forces generated by surface plasmon polaritons propagating on a gold

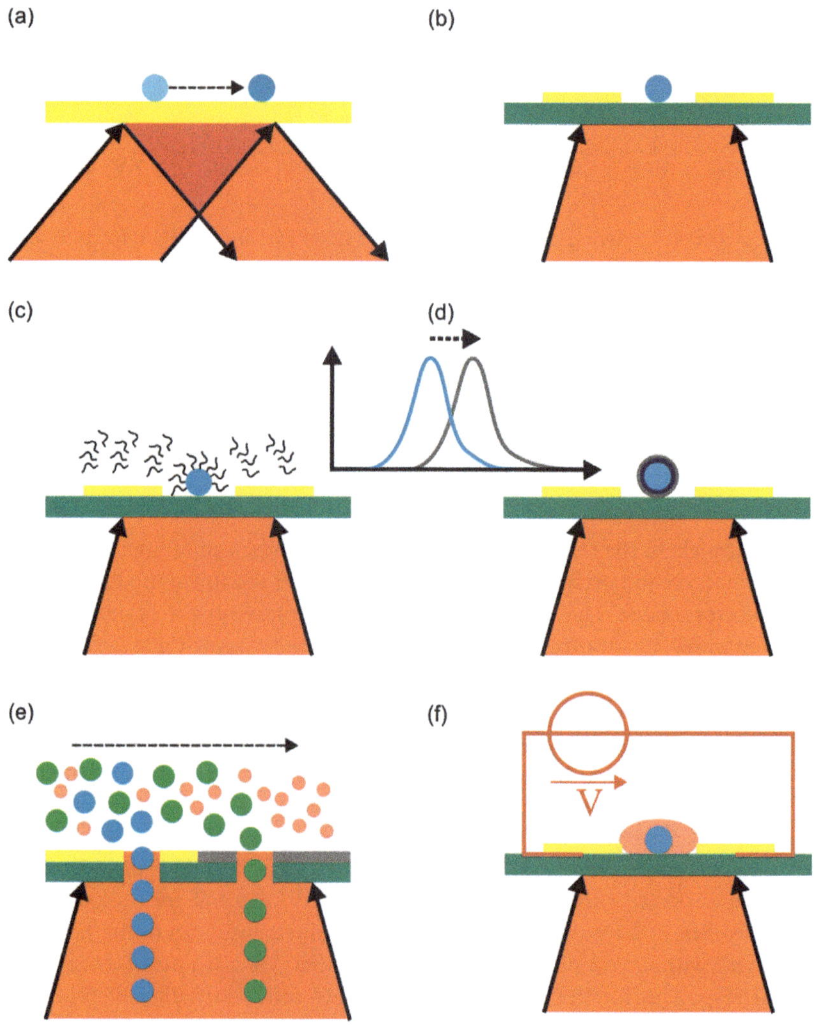

Fig. 7 The different manufacturing steps required for a nanofactory, where incoming nanostructures are modified and engineered into more complex products, can be controlled with plasmonic forces. (a) Plasmonic conveyor belt to move incoming nanostructures into position. (b) Immobilization of a nanostructure into its position. (c) Surface functionalization using chemistry to modify the nanostructure's surface. (d) Adjunction of new materials onto the nanostructure. Both processes can be monitored through the evolution of the spectral response of the nanostructure. (e) Sorting of final products or waste using a plasmonic sieve. (f) Additional manufacturing functions could be obtained by combining optical forces with electromagnetic fields applied directly onto the nanostructure under fabrication.

film.[29–31] Even in the simplest Kretschmann configuration, where a surface plasmon is excited by illumination through a prism, a standing plasmon can be produced on the surface of the metal. By controlling the phase of this standing wave, nanostructures can be moved along in a very controlled manner. For asymmetric trapped nanostructures, this approach can also be used to orient the nanostructure along a specific direction using the incident polarization.[32]

The nanostructure to be worked upon should then be immobilized. This can be performed using plasmonic trapping at the vicinity of a plasmonic antenna, Fig. 7(b). It could also be achieved using a reconfigurable structure, like that discussed in Sec. 4. Monitoring the optical response of the trapping structure provides means of following the immobilization of the trapped nanostructure, even when this structure is extremely small.[33] It should be mentioned that we assume that the entire nanofactory relies on microfluidics to bring in additional components or eliminate by-products.[34]

Once the nanostructure has been immobilized at a specific location, it can be worked upon. For example, specific surface chemistry can be used to functionalize its surface, Fig. 7(c). This surface functionalization can provide a broad variety of new functions,[35,36] that can be used for further nano-assembly[37] or to design a final product for specific applications, *e.g.* in biology.[38] Surface functionalization can be brought even further, by completely modifying the original nanostructure, Fig. 7(d). A typical example is that of the creation of multi-layered nanostructures, such as core–shell particles,[39] which broaden the possible optical responses for such nanostructures. Novel materials, like organic semiconductors, can also be combined with the carrying nanostructure to expand the range of possible applications, *e.g.* toward photovoltaics.[40]

For both surface functionalization and nanostructure modification, the evolution of the process can be followed in real time by monitoring the spectral changes in the optical response of the nanostructure (see the inset in Fig. 7). Especially for plasmonic nanostructures, minute variations of the geometry resulting from adsorption or aggregation, produce very noticeable modifications of the spectrum. This unique feature of a nanofactory driven by plasmonic forces enables extremely accurate control of the different steps required for the production of complex aggregates.

If the nanofactory relies on microfluidics to incorporate new components and eliminate waste or final products, a sorting step can be mandatory. This step can be required to sort between waste and final products, or to sort final products depending on their characteristics. This manufacturing step can easily be implemented with plasmonic trapping using a membrane and the extraordinary optical transmission phenomenon, Fig. 7(e).[9,41] Different degrees of freedom can be used here: the type of metal used – which determines the surface modes that can be excited and can dictate which particle will be selected, Fig. 7(e); the geometry of the apertures – which determines the resonances in the corresponding cavity; the surface chemistry – which can be used to selectively allow the passage of specific nanostructures.

So far, this foreseen nanofactory has used a combination of optical forces and surface chemistry to modify an original nanostructure and produce a more advanced system. It is very likely that in the future, the integration of electrical functions with such a system will provide further degrees of freedom, Fig. 7(f). By either directly contacting the nanostructure under work,[42] or immersing it in a very strong electromagnetic field, additional functionalities will be possible, including the specific growth of the structure by electrophoresis.[43] The combination of this approach with semiconductor nanomaterials will broaden even further the scope of possibilities for such a nanofactory.[44]

6 Summary

We have reviewed the numerical techniques for computing optical forces on nanostructures immersed in a strong optical field. In particular, we have shown that a recently introduced method that combines the surface integral equation with Maxwell's stress tensor is particularly well suited for computing optical forces produced by plasmonic structures on particles trapped at their ultimate vicinity. This fully electromagnetic approach has been compared with the dipole approximation for gradient forces calculations. While for specific geometries where the trapped object does not disturb much the illumination field, this approximation gives good results; however, it rapidly fails as soon as the interaction between the trapped particle and the structure producing the near-field becomes significant. Our data indicate that this is already the case for trapped particles as small as one tenth of the wavelength.

Studying a realistic optical antenna, we have shown that upon external illumination, both arms of the antenna undergo attractive optical forces. Due to the realistic shape of the structure, these forces are non-symmetrical, resulting in a torque on the overall structure. These internal forces are more complex for a nanostructure that includes several parts, as is the case for the heptamer that we have investigated. In that case, the seventh particle can be trapped in the centre of the structure, to complete its geometry. Upon trapping, we have observed very significant changes in the response of the system, especially the Fano resonance associated with such a structure can be fully controlled by the plasmonic trapping of one single nanoparticle.

Finally, we have shared our vision of a future nanofactory, where plasmonic trapping is used to control the different manufacturing steps associated with the transformation of initial nanostructures to produce complex compounds. All the different functions existing in a traditional factory can be translated at the nanoscale using the optical forces produced by plasmonic nanostructures.

Acknowledgements

Funding from the Swiss National Science Foundation (grant 200020_135452) is gratefully acknowledged.

References

1 A. Ashkin, *Phys. Rev. Lett.*, 1970, **24**, 156–159.
2 A. Ashkin, J. M. Dziedzic, J. E. Bjorkholm and S. Chu, *Opt. Lett.*, 1986, **11**, 288–290.
3 D. G. Grier, *Nature*, 2003, **424**, 810–816.
4 J. R. Moffitt, Y. R. Chemla, S. B. Smith and C. Bustamante, *Annu. Rev. Biochem.*, 2008, 77, 205–228.
5 K. C. Neuman and S. M. Block, *Rev. Sci. Instrum.*, 2004, **75**, 2787–2809.

6 M. Woerdemann, C. Alpmann, M. Esseling and C. Denz, *Laser Photonics Rev.*, 2013, **7**, 839–854.

7 O. J. F. Martin, C. Girard and A. Dereux, *Helv. Phys. Acta*, 1995, **68**, 195–196.

8 J. P. Kottmann, O. J. F. Martin, D. R. Smith and S. Schultz, *New J. Phys.*, 2000, **2**, 27.

9 M. L. Juan, M. Righini and R. Quidant, *Nat. Photonics*, 2011, **5**, 349–356.

10 J. D. Jackson, *Classical Electrodynamics*, Wiley, 3rd edn, 1998.

11 P. C. Chaumet and M. Nieto-Vesperinas, *Phys. Rev. B: Condens. Matter Mater. Phys.*, 2001, **64**, 035422.

12 D. Ganic, X. Gan and M. Gu, *Opt. Express*, 2004, **12**, 2670–2675.

13 M. L. Povinelli, M. Loncar, M. Ibanescu, E. J. Smythe, S. G. Johnson, F. Capasso and J. D. Joannopoulos, *Opt. Lett.*, 2005, **30**, 3042–3044.

14 E. Lamothe, G. Lévêque and O. J. F. Martin, *Opt. Express*, 2007, **15**, 9631–9644.

15 J. Chen, J. Ng, Z. F. Lin and C. T. Chan, *Nat. Photonics*, 2011, **5**, 531–534.

16 A. Ji, T. V. Raziman, J. Butet, R. P. Sharma and O. J. F. Martin, *Opt. Lett.*, 2014, **39**, 4699–4702.

17 M. Paulus and O. J. F. Martin, *Opt. Quantum Electron.*, 2001, **33**, 315–325.

18 A. M. Kern and O. J. F. Martin, *J. Opt. Soc. Am. A*, 2009, **26**, 732–740.

19 B. Gallinet, A. M. Kern and O. J. F. Martin, *J. Opt. Soc. Am. A*, 2010, **27**, 2261–2271.

20 S. Rao, D. Wilton and A. Glisson, *IEEE Trans. Antennas Propag.*, 1982, **30**, 409–418.

21 C. F. Bohren and D. R. Huffman, *Absorption and Scattering of Light by Small Particles* 1983.

22 A. Lovera and O. J. F. Martin, *Appl. Phys. Lett.*, 2011, **99**, 151104.

23 L. Huang and O. J. F. Martin, *Opt. Lett.*, 2008, **33**, 3001–3003.

24 A. M. Kern and O. J. F. Martin, *Nano Lett.*, 2011, **11**, 482–487.

25 B. Luk'yanchuk, N. I. Zheludev, S. A. Maier, N. J. Halas, P. Nordlander, H. Giessen and C. T. Chong, *Nat. Mater.*, 2010, **9**, 707–715.

26 B. Gallinet and O. J. F. Martin, *Phys. Rev. B: Condens. Matter Mater. Phys.*, 2011, **83**, 235427.

27 B. Gallinet and O. J. F. Martin, *Opt. Express*, 2011, **19**, 22167–22175.

28 A. Lovera, B. Gallinet, P. Nordlander and O. J. F. Martin, *ACS Nano*, 2013, **7**, 4527–4536.

29 G. Volpe, R. Quidant, G. Badenes and D. Petrov, *Phys. Rev. Lett.*, 2006, **96**, 238101.

30 D. B. Ruffner and D. G. Grier, *Phys. Rev. Lett.*, 2012, **109**, 163903.

31 O. Brzobohatý, V. Karásek, M. Šiler, L. Chvátal, T. Čižmár and P. Zemánek, *Nat. Photonics*, 2013, **7**, 123–127.

32 Y. Zhang, J. Wang, J. Shen, Z. Man, W. Shi, C. Min, G. Yuan, S. Zhu, H. P. Urbach and X. Yuan, *Nano Lett.*, 2014, **14**, 6430–6436.

33 W. Zhang, L. Huang, C. Santschi and O. J. F. Martin, *Nano Lett.*, 2010, **10**, 1006–1011.

34 L. Huang, S. Maerkl and O. J. F. Martin, *Opt. Express*, 2009, **17**, 6018–6024.

35 R. A. Sperling and W. J. Parak, *Philos. Trans. R. Soc. London, Ser. A*, 2010, **368**, 1333–1383.

36 R. Mout, D. F. Moyano, S. Rana and V. M. Rotello, *Chem. Soc. Rev.*, 2012, **41**, 2539–2544.

37 W. Li, P. H. C. Camargo, X. Lu and Y. Xia, *Nano Lett.*, 2009, **9**, 485–490.

38 A. Verma, O. Uzun, Y. Hu, Y. Hu, H.-S. Han, N. Watson, S. Chen, D. J. Irvine and F. Stellacci, *Nat. Mater.*, 2008, **7**, 588–595.

39 S. J. Oldenburg, R. D. Averitt, S. L. Westcott and N. J. Halas, *Chem. Phys. Lett.*, 1998, **288**, 243–247.

40 R. Plass, S. Pelet, J. Krueger, M. Grätzel and U. Bach, *J. Phys. Chem. B*, 2002, **106**, 7578–7580.

41 H. J. Lezec, A. Degiron, E. Devaux, R. A. Linke, L. Martin-Moreno, F. J. Garcia-Vidal and T. W. Ebbesen, *Science*, 2002, **297**, 820–822.

42 E. C. Garnett, W. Cai, J. J. Cha, F. Mahmood, S. T. Connor, M. Greyson Christoforo, Y. Cui, M. D. McGehee and M. L. Brongersma, *Nat. Mater.*, 2012, **11**, 241–249.

43 F. Iwata, M. Kaji, A. Suzuki, S. Ito and H. Nakao, *Nanotechnology*, 2009, **20**, 6.

44 M. Heiss, Y. Fontana, A. Gustafsson, G. Wüst, C. Magen, D. D. O'Regan, J. W. Luo, B. Ketterer, S. Conesa-Boj, A. V. Kuhlmann, J. Houel, E. Russo-Averchi, J. R. Morante, M. Cantoni, N. Marzari, J. Arbiol, A. Zunger, R. J. Warburton, A. Fontcuberta and I. Morral, *Nat. Mater.*, 2013, **12**, 439–444.

Faraday Discussions

DISCUSSIONS

Applications of plasmonics: general discussion

Vladimir Bochenkov, Jeremy Baumberg, Mikhail Noginov, Felix Benz, Hasan Aldewachi, Silvan Schmid, Viktor Podolskiy, Javier Aizpurua, Kaiqiang Lin, Thomas Ebbesen, Alexei A Kornyshev, James Hutchison, Katarzyna Matczyszyn, Samir Kumar, Bart de Nijs, Francisco Rodríguez Fortuño, James T. Hugall, Pablo de Roque, Niek van Hulst, Santhosh Kotni, Olivier Martin, F. Javier García de Abajo, Michael Flatté, Andrew Mount, Martin Moskovits, Pavel Ginzburg, David Zueco, Anatoly Zayats, Sang-Hyun Oh, Yu Chen, David Richards, Alessandro Belardini and Prineha Narang

DOI: 10.1039/c5fd90025e

F. Javier García de Abajo opened a general discussion of the paper by David Zueco: How do you interpret inelastic scattering in the ultrastrong coupling limit? Is part of the energy remaining in the system, while a photon with lower energy is emitted?

David Zueco responded: It is a Raman scattering process. The photon interacts with the hybrid (because they are ultrastrongly coupled) qubit-waveguide. After the interaction, the photon emitted has less energy. This energy stays in an excited state of the hybrid qubit-waveguide. We have checked this picture in our numerical calculations.

Jeremy Baumberg asked: Can you label clearly the curves of your energies shown, and show how they map onto P+ and P− at low *g*? Also, can you track how the excited final state that a single photon scattering off the system leaves changes as the coupling decreases from the ultrastrong regime? It is very unclear how to map your model of the inelastic scattering onto something involving the original strong coupling states. Can you trace out this in a figure, as it would be very useful. When one photon goes in, and comes out at a different energy, then there must be some energy left in the system, but where is it? Surely we must conserve energy in the system, but how does the lack of conservation of excitations work. Are you just saying that we have some sort of collective mode forming (like a plasmon)?

David Zueco replied: Thanks a lot for your interest. When a photon impinges on a qubit (or more) in the ultrastrong coupling regime we see that we have got inelastic scattering, provided that g and $\hbar\omega_{\text{in}}$ are large enough. In that case, a new photon emerges with a different energy. We show the probability of having this inelastic scattering in Fig. 1 as a function of input energy $\hbar\omega_{\text{in}}$ and coupling constant g. The output is a flying photon with different energy $\hbar\omega_{\text{out}}$ and the system does not come back to the ground state, but we find that instead it ends up in a dressed bound state $|E_2\rangle$. The energy of this new photon ($\hbar\omega_{\text{out}}$) will be that of the incident one ($\hbar\omega_{\text{in}}$) minus the gap between the excited and the ground state $|\text{GS}\rangle$:

$$\hbar\omega_{\text{out}} = \hbar\omega_{\text{in}} - (E_2 - E_{\text{GS}}). \tag{1}$$

Because of this condition, the input energy has got a lower bound:

$$\hbar\omega_{\text{in}} \geq \hbar\omega_{k=0} + (E_2 - E_{\text{GS}}) \tag{2}$$

This equation marks a boundary for existence of this process. The dotted white line in Fig. 1 renders this boundary using the computed $E_2(g)$, showing the involvement of $|E_2\rangle$ in the Raman process.

Let us now analyse the origin and properties of $|E_2\rangle$. Our Hamiltonian preserves parity $P = \text{e}^{i\pi N}$:

$$[P, H] = 0, \tag{3}$$

with N the number operator. As the ground state is even ($P|\text{GS}\rangle = |\text{GS}\rangle$), the input state is odd, so the output state is odd too. As this is a photon flying over $|E_2\rangle$, then $|E_2\rangle$ is even. The connection of $|E_2\rangle$ with known results is easier in the low g regime, where the Rotating Wave Approximation is valid and not only the parity but also the number of particles is well defined.

A known result in this case is that in the one-excitation sector, there are two bound states $|\Psi_\pm\rangle$, appearing below ($-$) and over (+) the band, with energies moving away from the band edges as g increases (see inset to Fig. 2). These states can be analytically computed (Longo *et al.*, ref. 2), having the expression:

$$|\Psi_\pm\rangle = a_\pm^\dagger|\text{GS}\rangle \qquad a_\pm^\dagger = \sum_x \phi_{x,\pm}a_x^\dagger + d_\pm\sigma^+ \tag{4}$$

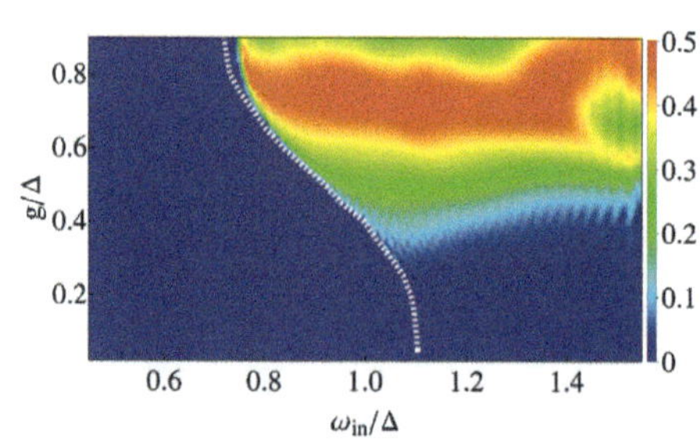

Fig. 1 Inelastic scattering. Probability of inelastic scattering in the full model as a function of both incident photon frequency ω_{in} and g. The white dotted line is the estimated boundary for the region where the photon frequency conversion occurs. Figure adapted from ref. 1.

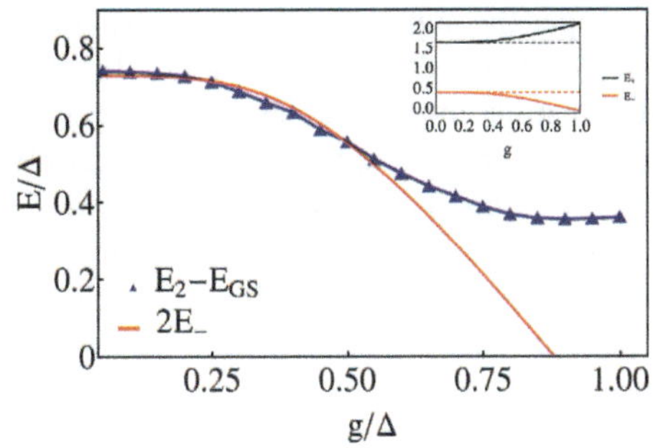

Fig. 2 $E_2 - E_{GS}$ *vs.* $2E_{\_}$. Gap in the ultrastrong coupling $E_2 - E_{GS}$ (blue triangles) and $2E_{\_}$ (solid red curve). As seen, they are almost equal for small g. We show the energies of the bound states in the Rotating Wave Approximation in the single-particle sector in the inset.

These states have got one excitation and odd parity. Then, they cannot be the state involved in the Raman scattering $|E_2\rangle$. However, we can consider states with two excitations, *e.g.* $(a_{\_}^{\dagger})^2|\mathrm{GS}\rangle$. It is still an eigenstate in the low coupling regime, because in such a case $d_{\_}$ goes to zero, so $|\Psi_{\_}\rangle$ is almost a bosonic excitation. Fig. 2 shows that the computed $E_2(g)$ is, at small g, in excellent agreement with $2E_{\_}$, showing that $|E_2\rangle \approx (a_{\_}^{\dagger})^2|\mathrm{GS}\rangle$, something that is further confirmed by the analysis of the populations of both qubit and surrounding cavities.

Summarising, we have traced back the state involved in that inelastic scattering to an eigenstate of a qubit coupled to an array of coupled cavities in the low coupling regime. It is the ground state in the two particle sector if g is small enough, or the first excited bound state with even parity for all g.

1 E. Sánchez-Burillo, D. Zueco, J. J. García-Ripoll and L. Martin-Moreno, *Phys. Rev. Lett.*, 2014, **113**, 263604.
2 P. Longo, P. Schmitteckert and K. Busch, *Phys. Rev. A*, 2011, **83**, 063828.

Javier Aizpurua asked: What happens if you insert losses in your description? Have you considered including the losses in your Hamiltonian? What would happen to the ultra-strong coupling if you started putting relevant losses in? Would it then be better to have weaker coupling with weak losses, or to have a modified ultra-strong coupling affected by the strong losses?

David Zueco answered: We did indeed include losses. We considered that the quantum emitter was radiating to other channels. In other words we considered the case of $\beta < 1$. The way we did it (beyond the master equation formalism) is described in the Supplementary Material of our recent article.[1] Essentially, our results are robust against reasonable losses, always such that the coupling emitter-waveguide is stronger than the other decays. It would be interesting to increase the losses, such that they are comparable to the coupling (while the latter is in the ultrastrong regime). We are currently investigating this.

1 E. Sánchez-Burillo, D. Zueco, J. J. García-Ripoll and L. Martin-Moreno, *Phys. Rev. Lett.*, 2014, **113**, 263604.

Pablo de Roque commented: Some years ago, J. Shapiro from MIT published a paper[1] entitled 'Single-photon Kerr nonlinearities do not help quantum computation'. In that paper the author gives a theoretical description of photon–photon interaction mediated by atomic nonlinearities. This theoretical description is within the framework of multi-mode treatment of single-photon nonlinearities, as opposed to common single-mode descriptions. He concludes that when taking into account the interaction of non-instantaneous nonlinearities with photon wave packets, the fidelities of the non-linear gates are not sufficient to implement any scalable quantum architecture, since the mentioned fidelities are not above the limits for applying quantum error correction protocols. My question is whether your formalism takes into account those finite-time responses. Conversely, I would like to have your opinion on whether the architecture proposed in your paper (*Faraday Discuss.*, 2015, DOI: 10.1039/C4FD00206G) would be satisfactory for a scalable quantum computer.

1 J. H. Shapiro, *Phys. Rev. A*, 2006, 73, 062305.

David Zueco responded: Unfortunately I am not an expert on quantum information processing. Thus, I do not know how to answer your question. If we consider finite-time responses, I would like to emphasize that we solve "exactly" the interaction of a waveguide with a quantum emitter. In our numerical simulations we do not make any Markovian or weak coupling approximations.

Pablo de Roque remarked: I would like to have your opinion on the feasibility of having g-factors as high as the ones normally present in theoretical predictions with atoms at optical frequencies, since coupling strength is proportional to the dipole moment, and those quantities are an order of magnitude smaller for atoms in the optical regime than for artificial superconducting qubits.

David Zueco replied: As you said, with artificial atoms (*e.g.* superconducting circuits) very high coupling strengths are possible. I recently heard that experimentalists are reaching 60% coupling (compared to the level splitting of the atom). In "natural" emitters, I do not know. I am just involved in some experimental collaboration with magnetic molecules coupled to superconducting cavities. They are trying to reach strong coupling (5-10% coupling) between the cavity and hundreds of molecules. Never with a single molecule.

F. Javier García de Abajo commented: Can dissipation play an important role in the system that you are studying? Can you use a traditional approach to include dissipation, such as density matrix or Monte Carlo? Also, how valid is the approximation of having Markovian processes in the ultrastrong coupling limit that you are considering? How can dissipation to a bath be realistically described?

David Zueco answered: In a previous paper,[1] we have considered dissipation. We studied the case where the qubit dissipates to other channels. Our calculations considered Ohmic spectral density and that the losses were smaller than the coupling to the line. Markovianity is a property of the bath. It is also related to a weak (system–bath) coupling limit, the so-called Davies limit. In principle, the Markovian character is independent of the system itself. Therefore, I do not see any point of considering if the system (qubit + cavity) is in the ultrastrong regime or not. Obviously the spectrum of the system (being in the ultra or not) will change and this could change the range of parameters where the system is weakly coupled to the bath. In other words, ultrastrong coupling is a property of the system. Whether this ultrastrong coupled system is coupled in a Markovian or non-Markovian way is another story. It is the way how it is coupled to the bath. If your question is about a dissipative theory of a system beyond weak (system–bath) coupling the answer is that it depends. In the literature, work already exists on this limit, by working within the path integral formalism for dissipative systems.

1 E. Sánchez-Burillo, D. Zueco, J. J. García-Ripoll and L. Martin-Moreno, *Phys. Rev. Lett.*, 2014, **113**, 263604.

Vladimir Bochenkov opened a general discussion of the paper by Alessandro Belardini: What is the nature of the optical activity in your system?

Alessandro Belardini answered: Our samples can be viewed as metasurfaces, so we cannot consider a propagation of the electromagnetic field through effective optically active media. Nevertheless a circular dichroism has been measured, thus indicating a geometrical extrinsic chiral response. This property originates from the relative orientation of a triad of vectors describing the direction of the wires, the direction of the impinging light and the direction of the surface normal.[1]

1 E. Plum, X.-X. Liu, V. A. Fedotov, Y. Chen, D. P. Tsai and N. I. Zheludev, *Phys. Rev. Lett.*, 2009, **102**, 113902.

Felix Benz commented: How well-defined is the tilt angle of the nanowires and how sensitive is the measurement to slight local variations? Is the homogeneity influenced by the growth temperature?

Alessandro Belardini replied: From analysing the SEM images of our samples, the tilt angle appears to be 60° with respect to the surface normal, with a symmetric distribution of approximately 10°. Measurements performed in different portions of the samples do not show substantial differences, revealing the good homogeneity of the samples with respect to the pump spot size (0.5 mm diameter). Unfortunately, we do not have samples with different tilt angles in order to investigate the behaviour as a function of the tilt angle. The substrate temperature does not affect the average wires orientation, but the spread of the angular distribution as shown by SEM images.

Niek van Hulst commented: The generated second harmonic will depend on plasmonic resonances with either fundamental or second harmonic wavelength. What is the typical length of the nanowires? Do you see any resonance effect as a function of the length?

Alessandro Belardini answered: From SEM images, the typical length of the wires is 400 nm for both samples. The linear transmission spectra do not display relevant resonant behaviour neither at 800 nm (pump) nor at 400 nm (SHG signal); this is also confirmed by the SHG signal intensity that results being only 2.5 times the signal generated by a flat Au layer.

Niek van Hulst remarked: The nanowires are made of bulk polycrystalline gold, with a zero χ^2 susceptibility. Thus what is the origin of the second harmonic? A surface effect, roughness or potentially a bulk component?

Alessandro Belardini answered: Due to the high surface *vs.* volume ratio we are pretty sure that the main contribution to our signal is related to the surface non-linear response. We can expect a weak contribution from the magnetic dipole non-linear response from the bulk. Anyway it is not possible to isolate bulk non-linear contributions in our experimental setup.

Viktor Podolskiy asked: It would be interesting to see the comparison of linear properties of your tilted wires to the predictions of effective medium theories (EMTs). Earlier EMTs have been developed by Mario Silveirinha; he has considered mostly high-conductivity wires, with emphasis on the low frequency regime. Interesting effects, such as inverted rainbows, have been predicted for tilted crossed wire metamaterials;[1] I'm not sure these will survive for tilted wires that are not crossed. More recently, our group has developed a nonlocal EMT with emphasis on the high-frequency, plasmonic response.[2]

1 M. G. Silveirinha, *Phys. Rev. Lett.*, 2009, **102**, 193903.
2 B. M. Wells, A. V. Zayats and V. A. Podolskiy, *Phys. Rev. B*, 2014, **89**, 035111.

Alessandro Belardini responded: We are completing the linear spectral characterisation, and then we will investigate the applicability of the EMT to our samples. It will be interesting to extend your results to our tilted nanowires.

Olivier Martin commented: SHG strongly depends on where you look at it; am I correct in understanding that you always measure SHG from the side of your sample? Have you ever considered looking at different angles and wouldn't this provide additional information on your sample, including on the orientation and homogeneity of the nanowires, since they can influence the SHG (see *e.g.* "Ultrasensitive optical shape characterization of gold nanoantennas using second harmonic generation"[1]).

1 J. Butet, K. Thyagarajan and O. J. F. Martin, *Nano Lett.*, 2013, **13**, 1787.

Alessandro Belardini responded: We measured SHG at the reflection angle for different incidence angles (ranging from 15° to 65°), but we did not check the SHG out of the reflection angle: this is an interesting suggestion.

Samir Kumar communicated: The authors have grown Au nanowires at two different temperatures (100 K and 300 K) by oblique angle deposition (OAD). From SEM images it seems that the nanowires grown at 300 K are more sharp and well separated. The effect of substrate temperature on the nanostructures grown by OAD has been studied very well by various groups. It is well known that at low substrate temperature the nanorods are sharper and longer than the nanorods grown at higher temperatures because of the limited surface diffusion. But your nanowires are the opposite of what has been observed. I want to know why your structures are contradicting the already established studies?

Alessandro Belardini communicated in reply: At first sight, we were surprised by our results as well. However, carefully analysing the SEM images, we noticed that the 100 K sample presents thinner wire diameters (10-20 nm) due to the low surface diffusion. Then, as the wires length increases during the growth process, groups of wires collapse forming pyramidal arrangements (bunches) of wires with a thicker base of 200 nm diameter (collecting more isolated wires) and with a tip of 80 nm diameter where the wires are connected. On the other hand, in the 300 K sample, the wires grow as larger columns from the base (30-40 nm) and are stable along their length (approx. 400 nm), thus remaining isolated wires.

Niek van Hulst opened a general discussion of the paper by Michael Flatté: You mention that YIG particles at the liquid–liquid interface cause the Faraday effect. Do these particles have to be oriented for the macroscopic effect to occur? Do you need an external field for orientation?

Michael Flatté replied: The particles themselves do not need to be oriented, but the magnetization of the particles does need to be oriented. This can be done with an external magnetic field. YIG has a very low coercive field, especially for near-spherical particles, so the magnetization can be oriented either parallel to the interface or perpendicular to it with a modest magnetic field.

Bart de Nijs asked: You show in your article (*Faraday Discuss.*, 2015, DOI: 10.1039/C4FD00210E) and presentation a very nice method of controlling the polarisation of light using nanoparticles attaching and detaching from a liquid–liquid interface. But you also show that this attachment then occurs at a certain potential. Does the detachment happen at the same potential or will there be some form of hysteresis? How reversible would the system be?

Michael Flatté replied: As we have found in previous studies (*e.g.* Fig. 2 of ref. 5 in our paper, *Faraday Discuss.*, 2015, DOI: 10.1039/C4FD00210E), it is possible at intermediate voltages to create a positive potential at the interface, but separated from the bulk by a large potential barrier. This situation can produce large hysteresis. Thus it would be most favorable for switching behavior to avoid the use of intermediate voltages for switching.

Bart de Nijs said: Is there any idea with what speed this switching can be done?

Michael Flatté answered: Yes, in previous studies (ref. 5 in our paper, *Faraday Discuss.*, 2015, DOI: 10.1039/C4FD00210E) we have estimated the speed with which a single layer of nanoparticles of similar size can be brought to, or away from an interface. Depending on system parameters, this time lies roughly between 0.1-1 s. However, using voltage pulsing techniques or stirring this can be sped up. For more details see ref. 1 and 2 below.

1 A. A. Kornyshev, A. R. Kucernak, M. Marinescu, C. W. Monroe, A. E. S. Sleightholme and M. Urbakh, *J. Phys. Chem. C*, 2010, **114**, 14885.
2 M. Marinescu, M. Urbakh, T. Barnea, A. R. Kucernak and A. A. Kornyshev, *J. Phys. Chem. C*, 2010, **114**, 22558.

Alexei A Kornyshev commented: Nanoparticles localise spontaneously in arrays at liquid interfaces, if the charge of their ligands and the concentration of electrolyte which screens the electrostatic repulsion but is not too strong to allow their fusion in the bulk are properly adjusted.[1] To move such particles from the interface, applying voltage across it, is a tricky task and has been achieved so far only for small particles,[2] because the capillary forces that keep them at the interface are very large. One of the ways to reach electrical control over the population of nanoparticles at the interface will be to start with such conditions of the electrolyte concentration and pH for which the nanoparticles do not yet self-assemble at the interface, but are very close to that situation. Since the electric field across the interface is localized within the back-to-back electrical double layers, the particles will reach the surface *via* random diffusion but not migration, and will be trapped at the interface by the electric field localized there. All this was discussed in detail in our feature article in the *Journal of Physical Chemistry C*.[3]

1 M. P. Cecchini, V. A. Turek, J. Paget, A. A. Kornyshev and J. B. Edel, *Nat. Mater.*, 2013, **12**, 165.
2 J. B. Edel, A. A. Kornyshev and M. Urbakh, *ACS Nano*, 2013, 7, 9526.
3 M. E. Flatté, A. A. Kornyshev and M. Urbakh, *J. Phys. Chem. C*, 2010, **114**, 1735.

Thomas Ebbesen asked: Can you clarify the size of the setup? Would it work faster if the dimensions were smaller?

Michael Flatté replied: The size of the setup in the lateral direction for light impinging perpendicular to the interface would be the spot size of the light beam, so a minimum size would be a few square microns. For the cell in the vertical direction the minimum thickness required is that to dilute the nanoparticles into the bulk, and bring them back to the interface, within ~0.1–1 s. That thickness is about 2 microns (ref. 5 in our paper, *Faraday Discuss.*, 2015, DOI: 10.1039/C4FD00210E). Similar sizes are appropriate for the waveguide geometry also presented.

Thomas Ebbesen remarked: What's the size of your particles and the concentration of the electrolyte? The electrolyte can of course screen the field enormously. Can you work with lower electrolyte concentrations?

Michael Flatté answered: Yes, the size of the particles cannot be too large or they will become trapped at the interface and cannot be removed, presumably not larger than 20–40 nm in diameter; however, manipulation with liquid–liquid interface tension (reducing it) may relax this constraint. The electrolyte concentration cannot be too high or the electric field becomes localized to the interface alone. Although the electric field gets larger by making the double layer at the liquid–liquid interface more compact, it will act at a smaller number of charges on the nanoparticles, and the effect of the field may eventually get weaker. We assume a concentration of 5×10^{-3} M, the same as in ref. 22 in our paper (*Faraday Discuss.*, 2015, DOI: 10.1039/C4FD00210E).

Alexei A Kornyshev added: Experiments with nanoparticles that can provide Faraday rotation have not yet been performed. But for metallic particles that we used to prove SERS at the liquid–liquid interface, the sizes were 18 nm in diameter, with presumably about 2 nm of surface-to-surface separation in the settled array, which warranted detection of analytes down to femtomolar amounts in 1 cm^3 . This was yet not an electrochemical cell, so there was no controllable electric field across the interface. The formation of the array was facilitated by mild centrifugation or just shaking, and took about 1 s.[1]

1 M. P. Cecchini, V. A. Turek, J. Paget, A. A. Kornyshev and J. B. Edel, *Nat. Mater.*, 2013, **12**, 165.

Andrew Mount asked: Is the radius, *R*, of the nanoparticle a hydrated radius, and is this likely to include *e.g.* the Stern layer of ions? Is the effective radius of the particle also likely to change with the ionic strength (controlled by the concentration of background electrolyte)? This one might expect as the optimum spacing between particles at monolayer coverage (their effective size) should increase with an increase in electrostatic repulsion between neighbouring nanoparticles; this could be caused by an increasing surface charge, which could be controlled by the contribution from the number of strongly electrostatically bound or adsorbed surface ions from the electrolyte, which could vary with electrolyte concentration. If so, what effect would this have on the calculated properties of the system, and to what extent has this been accounted for by *e.g.* incorporating the pairwise interaction of localised nanoparticles *via* screened electrostatic forces?

Michael Flatté responded: The radius of the nanoparticles is the radius of the YIG material. The layer of ligands around the YIG is anticipated to be very thin compared with the nanoparticle radius, as would any hydrated layer. Although the nanoparticle radius would not change due to a change in the ionic strength, there would be a change in the maximum allowed coverage of the interface. This has recently been demonstrated experimentally for gold nanoparticles.[1]

1 V. A. Turek, M. P. Cecchini, J. Paget, A. R. Kucernak, A. A. Kornyshev and J. B. Edel, *ACS Nano*, 2012, **9**, 7789.

Mikhail Noginov commented: Why do the nanoparticles assembled at the surface behave differently from the nanoparticles in the volume? Also, what is the surface filling factor for the nanoparticles?

Michael Flatté answered: For magnetization perpendicular to the interface, they behave differently than in the bulk because each nanoparticle also responds to the polarization field from the neighboring nanoparticles. This additional contribution from the polarization fields produces a larger dielectric response which yields a larger Faraday rotation. For magnetization parallel to the interface, these polarization fields produce a large birefringence that quenches the Faraday rotation. The surface filling factor is given by the coverage, Γ.

James T. Hugall opened a general discussion of the paper by David Richards: Using the lifetime imaging technique you describe, at any given lateral position in a sample, as soon as there are multiple fluorophores at multiple heights (axial positions), then the lifetime recorded will be an average over the many different fluorophores, and any precise height information/resolution will be lost. Therefore, this technique seems suited only to systems with a very well-defined, and singular, functionalised 2D surface within a 3D structure. In any real system, this requires that the fluorophores are highly localised to a 2D surface and thus requires prior knowledge of the system and a system in which there is no existence of similarly functionalised surfaces at different heights. Could you comment on this apparent limitation and the validity of this process for real biological systems? Are there ways to overcome it?

David Richards responded: Indeed, if fluorophores are located over a range of distances from the metal surface, the lifetime recorded will be an average resulting from the integrated fluorescence signal from the distribution of fluorophores. However, for studies of the cellular plasma membrane, the fluorescence lifetime-encoded axial position information provided by the technique can provide a valuable probe of the cellular plasma membrane. We have provided one example in this paper (*Faraday Discuss.*, 2015, DOI: 10.1039/C4FD00198B), in which we are able to map, through detection of the receptor CXCR4, the height of the plasma membrane above a nanostructured silver substrate. Elsewhere, we have demonstrated how this approach is able to provide a sensitive assay of receptor-mediated endocytosis,[1,2] both for nanostructured and flat metal films. For the nanostructured silver films the fluorescence intensity is also enhanced at the surface relative to distances >100 nm, commensurate with the lifetime modification,[2] providing an additional weighting of the lifetime to those fluorophores closest to the surface.

1 N. I. Cade, G. Fruhwirth, S. J. Archibald, T. Ng and D. Richards, *Biophys. J.*, 2010, **98**, 2752.
2 N. I. Cade, G. Fruhwirth, T. Ng and D. Richards, *J. Phys. Chem. Lett.*, 2013, **4**, 3402.

Niek van Hulst remarked: Lifetime is a very powerful observable as it depends on an enormous amount of parameters, *e.g.* it matters whether a molecule is parallel or perpendicular to the surface, differences in chemical environment have an effect, quenching to metal reduces the lifetime, and local density of states effects enhance or shorten the lifetime. Thus many parameters are influential and several assumptions are needed to use lifetime as an indicator of distance to a metal surface. Next, what is the range as a distance sensor? Quenching will dominate at the lowest few nanometers, yet at larger distance LDOS effects will take over. Thus surely you do observe distance-dependent effects, however how can this be made quantitative?

David Richards responded: We will first remark that measurements of fluorescence intensity, used to encode distance in TIRF microscopy, for example, are affected very significantly by variations in fluorophore and photobleaching, to both of which lifetime is relatively insensitive. Of those parameters which do affect lifetime, we are exploiting the distance-dependent changes in radiative and non-radiative lifetime, through local density of states effects and quenching, as a tool to specifically encode distance. From calculations of the distance-dependence of fluorescence lifetime, which provide strong agreement with our experimental observations, we have identified that in the case of the nanostructured silver substrate we present, enhancement of radiative decay rate dominates over the relevant distance range.[1] As we show in the present paper (*Faraday Discuss.*, 2015, DOI: 10.1039/C4FD00198B), the distance-dependence can be quantified using a calibration sample. Lifetime is indeed dependent on molecular orientation, but in our measurements we expect both a significant population of fluorophores within the focal area of excitation, and rapid fluorophore rotation, such that we are sensitive only to an average over all fluorophore orientations.

1 N. I. Cade, G. Fruhwirth, T. Ng and D. Richards, *J. Phys. Chem. Lett.*, 2013, **4**, 3402.

Niek van Hulst asked: The lifetime changes are particularly strong over the first few nanometers. I assume, looking at cellular dynamics, you will mainly observe intracellular membrane dynamics close to the metallic substrate, which could be useful for the study of membrane organisation?

David Richards responded: From the calibration measurements we have performed we find that the lifetime changes are particularly strong over the first 100 nm away from the plasmonic surface. In particular, we have demonstrated elsewhere that this technique provides sensitive detection of the internalisation of fluorescently labelled membrane receptors, providing a sensitive assay of receptor-mediated endocytosis.[1,2]

1 N. I. Cade, G. Fruhwirth, S. J. Archibald, T. Ng and D. Richards, *Biophys. J.*, 2010, **98**, 2752.
2 N. I. Cade, G. Fruhwirth, T. Ng and D. Richards, *J. Phys. Chem. Lett.*, 2013, **4**, 3402.

Martin Moskovits asked: This is a clever experiment in which you were able to obtain what is in essence tomographic structural information at the nanometer scale, by making use of the fact that the fluorescence lifetime decreases markedly as the fluorescence source approaches the cell–metal interface. This phenomenon should also occur for a flat metal surface rather than a nanostructured metal surface, whose geometry introduces both preparative complexities and uncertainty in position along the normal direction. Why is it necessary or beneficial to use a nanostructured surface for such experiments?

David Richards responded: We do indeed also demonstrate in our paper (*Faraday Discuss.*, 2015, DOI: 10.1039/C4FD00198B) the measurement of axial position from a flat gold surface with super-resolution sensitivity. However, because fluorescence lifetime is reduced only through an enhancement in non-radiative decay, leading to fluorescence quenching, we find that the resultant reduction in signal makes the determination of distance unreliable for separations very close to the metal surface. In contrast, a nanostructured silver film leads to an enhancement in fluorescence signal close to the surface, with fluorescence lifetime thereby providing a reliable measure of axial position with a sensitivity down to 6 nm. A thin film nanostructure also provides the additional advantage of being able to perform measurements on cells in a conventional inverted microscope geometry, whereas flat metal surfaces with poor optical transmission require illumination and detection from above the sample.

Martin Moskovits said: Benefitting from the fluorescence enhancement is helpful, but that would surely also restrict your observational territory to the lower portion of the cell. A great deal of biology occurs in the inaccessible portion of the cell.

David Richards replied: Because the technique we present uses conventional confocal microscopy, we are able to probe the whole body of the cell, with the measurement of fluorescence lifetime providing super-resolution distance discrimination within the first 100 nm. For example, as we demonstrate, this enables a straightforward assay of endocytosis, providing discrimination between e-GFP labelled proteins in the cell membrane and those that have been internalised. We note that, in contrast, in the case of total internal reflection fluorescence microscopy (TIRFM), widely used in studies of the cellular membrane, the observational territory is indeed restricted to the lower portion of the cell.

F. Javier García de Abajo enquired: Could you speculate on how much the combination of lateral and radial resolutions can be improved? The former could go down to ~10 nm by collecting many images made with one or a few emitting fluorophores each (*i.e.*, like in the stochastic optical reconstruction microscopy (STORM) technique), whereas the latter could rely on the impressive determination of the lifetime in your technique. Could these two perhaps be combined by going down to one or a few emitters per frame and looking not only at the lateral intensity profile for each individual emitter but also at how many photons are emanating from it, which should be connected with its lifetime, and this in turn with the presence of plasmonic decay channels (*i.e.*, its distance to the metal substrate)?

David Richards replied: We have estimated that the technique we present provides 6 nm axial distance sensitivity over a range of 100 nm from the surface (*i.e.* radial distance sensitivity, in the case of an isolated silver nanoparticle). This translates to precision for the determination of the height of a sheet of fluorophores, such as for the measurement of a cellular membrane presented in this paper (although the lateral resolution of the technique is still diffraction limited to ~200 nm). In this case the axial sensitivity is limited by the distance-dependence characteristics of the fluorescence lifetime and its determination; hence it could be improved further with tuning of the plasmonic substrate, reducing the instrumental response time, and using improved fluorescence decay fitting algorithms for low photon numbers (*e.g.* a Bayesian approach). On the other hand, the technique we present is not able to provide such axial resolution within a 3D volume of fluorophores, as is the case with localisation-based super-resolution microscopies, such as STORM and photoactivation localisation microscopy (PALM). However, as it is based around well-established fluorescence lifetime imaging (FLIM), it is able to offer acquisition times which are compatible with live cell imaging, generally not possible with PALM and STORM because of the need to measure large numbers of single-molecule images using those methods. To bring the two approaches together, the detection of single fluorophores presents first a challenge in the measurement of fluorescence lifetime decays using time-correlated single photon-counting, in particular within a full-frame wide-field modality required by PALM and STORM. Moreover, we would contend that the 3D resolution provided by these techniques would be superior to the plasmon-assisted technique we present, while the advantage of compatibility with conventional confocal microscopy in live cell imaging would be lost.

Niek van Hulst commented: As to combining your lifetime method with localisation microscopy, I think you should not be so pessimistic. Typically about ~1000 photocounts are enough for a single molecule to determine lifetime, or changes in lifetime. Thus, with regard to photon budget, one has sufficient signal to do both lifetime and STORM; the issue is more in single-point confocal *versus* wide-field microscopy. I do see an issue in using localisation microscopy with plasmonic structures to measure the position of emitters. The emission will couple to the plasmonic mode, thus one will localise the plasmonic mode instead of the molecule. A shift of many nanometers can occur due to the metal, and thus usually STORM on plasmonic-enhanced systems does go wrong due to the coupling to the plasmonic mode.

Olivier Martin addressed David Richards: The new axial tomography imaging technique you have established relies on the plasmonic substrate which serves as a reference for the signal measured for the cell that rests on that substrate, and the information recorded depends on the distance from that substrate. Recent experiments have used plasmonic nanoparticles to label the membrane of a cell;[1] couldn't one combine this labelling technique with your approach, to provide additional plasmonic references throughout the cell?

1 J. Wang, S. V. Boriskina, H. Wang and B. M. Reinhard, *ACS Nano*, 2011, **5**, 6619.

David Richards answered: With the approach discussed in the present paper (*Faraday Discuss.*, 2015, DOI: 10.1039/C4FD00198B) we assume there is a uniform plasmonic substrate over the whole sample (confirmed by atomic force microscopy and absorption spectroscopy), and hence do not need to consider any additional affinity factors due to introducing labelled nanoparticles that target specific sites. Nevertheless, we note that for the very high nanoparticle labelling densities reported by Wang *et al.*, a plasmonic sheet would be effectively created over the apical membrane; this would provide a means to discriminate fluorophores (through their lifetime) in close proximity to the membrane from those deeper within the body of the cell, in a similar manner to that which we have presented for the basal cell membrane, with application in, *e.g.*, assays of receptor-mediated endocytosis.[1]

1 N. I. Cade, G. Fruhwirth, S. J. Archibald, T. Ng and D. Richards, *Biophys. J.*, 2010, **98**, 2752.

Jeremy Baumberg asked: On the physics side there are strengths and weaknesses of using silver grains. But biologically the cells react very differently to smooth and rough surfaces, would you like to comment on that? Particularly thinking about roughness, length scales…

David Richards replied: We would first highlight that usually silver is considered to be toxic to cells. However, we have applied a protocol, identified in the present paper (*Faraday Discuss.*, 2015, DOI: 10.1039/C4FD00198B), whereby a coating of *N*-(2-mercaptopropionyl)glycine (MPG) over the silver plasmonic substrate permits normal cell growth. Due to its free –SH groups MPG binds well to silver, thereby inhibiting the toxic effect of silver. We do not have information about whether or how the MPG modifies the nanoscale morphology of the surface, but note that we did not observe qualitatively any obvious growth inhibition or change of the morphology of MTLn3E adenocarcinoma cells cultured on these surfaces.

Jeremy Baumberg responded: But particularly the conformation of a cell at the bottom which you can't normally measure, how does it adhere to the surface, and may it respond differently to different materials and structures?

David Richards replied: This could indeed be investigated. In principle other surface functionalisation techniques could be used to provide a thin barrier layer between the silver film and cellular membrane (*e.g.* fibronectin, PEGylation) without adversely affecting the plasmonic response. However, care is required to ensure there is no leakage of silver, prevented through the use of MPG as a barrier.

Katarzyna Matczyszyn opened a general discussion of the paper by Yu Chen: What are the dimensions of your nanorods? Do you play with various sizes or coatings? Also, how do you do the surface chemistry? For the functionality of the probes, the length of the DNA matters, were you dealing with the specific sequencing?

Yu Chen responded: The rods are about 40 nm long and 12 nm in diameter. We have studied rods of different aspect ratios and coatings (for details, please see ref. 7 in our paper, *Faraday Discuss.*, 2015, DOI: 10.1039/C4FD00199K). In this work, the hairpin was designed to target c-myc, with a loop containing 17–21 bases.

Katarzyna Matczyszyn said: How do you control the distance between the nanorods, though it depends on the sequence and persistence length of the DNA? Do you control the number of RNA molecules attached to a single nanorod?

Yu Chen answered: The nanorods are well-separated in the solution of low concentration, and the number of hairpins attached to a single nanorod can be tuned by changing the ratio of nanorods to hairpins in the salting aging process.

Olivier Martin asked: Endocytosis is a very complex process that depends on many conditions, including the charges on the particle, the pH, the hydrodynamic radius, *etc.* Do you think that the technique you have developed could improve understanding of the role of these different parameters and the monitoring of endocytosis as it happens? Also, is it likely that the energy transfer between the gold nanorods and the fluorophores is influenced by some of these parameters that influence endocytosis (charge, pH, *etc.*), which could then interfere with this type of measurements?

Yu Chen replied: Gold nanorods can be synthesized with different sizes/aspect ratios, and further functionalized with different surface chemistry/charge states, so it is possible to apply this method in living cells and study the factors that influence the endocytosis process. Fluorescence lifetime can be sensitive to local environment, *e.g.*, viscosity, pH *etc.*, in addition to energy transfer. Therefore, one should be careful about choosing a fluorophore, *e.g.*, using a fluorophore that is less sensitive to pH.

Jeremy Baumberg commented: Do we even know that the gold nanoparticles go in the same places as the gold nanorods? You're assuming you're tracking the same sequence of events—do we know that yet? Does the aspect ratio have some effect?

Yu Chen replied: Endocytosis is an important pathway for the uptake of gold nanoparticles, including gold nanorods. This work shows a potential way to study the intracellular pathway of gold nanorods and this method can also be applied to other internalisation processes. Regarding the effect of the aspect ratio, previous studies on the internalisation of gold nanorods with aspect ratios from 2 to 5 didn't find any obvious differences.

Mikhail Noginov asked: How do surface plasmons affect the energy transfer? What is the mechanism for this?

Yu Chen replied: Surface plasmons enhance the excitation of the molecules and also modify the non-radiative decay rate.

Hasan Aldewachi said: Have you studied the effects of different pathological conditions on the endocytosis process of the nanorod–dye combination, as these conditions might affect the mechanism by which endocytosis happens? Also, you mentioned that this combination is less toxic towards certain cells, does that mean that this combination is not applicable for all cell types for imaging and sensing?

Yu Chen responded: We haven't studied the effect of pathological conditions on the endocytosis process. The method discussed in this paper can be applied to this kind of study. The cytotoxicity of gold nanorods depends on the dose and on the surface coating (please see ref. 7 in our paper, *Faraday Discuss.*, 2015, DOI: 10.1039/C4FD00199K). So we can use this method in cells that are less tolerant to CTAB-capped gold nanorods, as shown in this paper.

Hasan Aldewachi commented: Why do you prefer to use the nanorods instead of nanoparticle spheres? The latter don't need to be prepared in CTAB, could it be more cell-friendly?

Yu Chen replied: Citrate-stabilized nanospheres are more cell-friendly than CTAB–gold nanorods are. However, gold nanorods have many advantages, including a tunable longitudinal surface plasmon mode and strong two-photon luminescence, *etc.*

Jeremy Baumberg asked: When you're looking at the hairpin system, the one thing you're going to be very sensitive to is heating of the system. Do you see any effects from that? Do you get enough photons out when the power is low enough to avoid heating?

Yu Chen responded: Thermal stability of the hairpin is an important issue and one of the factors to be considered in hairpin design. This paper (*Faraday Discuss.*, 2015, DOI: 10.1039/C4FD00199K) shows three stable structures and one less stable one. With an appropriate design, reliable signals can be obtained under normal measurement conditions.

Mikhail Noginov remarked: In equation 1 as presented in your slides ($k_T(r) = \frac{1}{\tau_D}\left(\frac{R_0}{r}\right)^6$), k has units of s^{-1}. In equation 2 ($k_T = 1 - \frac{F_{DA}}{F_D} = 1 - \frac{\tau_{DA}}{\tau_D}$), k is dimensionless. Two different quantities should not be denoted by the same letter.

Yu Chen replied: The second one should be ε ($\varepsilon = 1 - \frac{F_{DA}}{F_D} = 1 - \frac{\tau_{DA}}{\tau_D}$).

Olivier Martin commented: The imaging techniques presented in the papers from David Richards and Yu Chen (*Faraday Discuss.*, 2015, DOI: 10.1039/C4FD00198B and *Faraday Discuss.*, 2015, DOI: 10.1039/C4FD00199K) have great potential for uncovering some of the complex mechanistic processes taking place in living cells; in particular to shed light on the dynamics of these processes. Do you think that both techniques can be applied to living cells? If so, how fast can they perform? What kind of typical acquisition time can be reached, maybe by compromising on the spatial resolution?

Yu Chen replied: This technique can be applied in living cells. The typical acquisition time for a good quality image is a few minutes, though it can be reduced to half a minute by compromising on the spatial resolution. FLIM acquisition at 500 ms has recently been reported, by parallelizing the excitation and detection processes.[1]

1 S. P. Poland *et al.*, *Biomed. Opt. Express*, 2015, **6**, 277.

David Richards answered: FLIM imaging of living cells is relatively easily achievable at approximately video rates (see ref. 1 and other more recent papers by French and co-workers), with the acquisition time ultimately limited by the available photon budget. This results in a compromise between speed, the accuracy of determination of the fluorescence lifetime and the lateral image resolution, and the usual imaging restrictions such as photo-bleaching.

1 D. M. Grant *et al.*, *Optics Express*, 2007, **15**, 15656.

James Hutchison asked: I see you require three exponential components to fit your decays, and attribute them to energy transfer of various efficiencies or to scattering from the gold core. How confident are you in attributing each component of these decays to specific processes, given that (I assume) you float 6 variables to achieve a fit?

Yu Chen answered: Each rod has multiple dyes in different orientations and at different locations on the gold surface. Multiple exponential decays would be expected. The two lifetime components in the ns and sub ns range can be attributed to Cy5. The first lifetime component in the case of the less stable hairpin structure is close to the lifetime of free Cy5. Therefore the reduction in lifetime indicates the energy transfer and is used for comparison. The short lifetime component in the tens of ps region is likely due to scattering from gold cores.

Prineha Narang was invited to make some remarks to open a general discussion of the paper by Martin Moskovits, summarised as follows: To put these remarks into context, the decay of surface plasmon resonances is usually a detriment in the field of plasmonics, but the possibility to capture the energy normally lost to heat would open new opportunities in photon sensors, energy conversion devices and switching. In the context of hot-electron devices, the large extinction cross-section at a surface plasmon resonance enables nanostructures to absorb a significant fraction of the solar spectrum in very thin films. Despite the significant experimental work in this direction, a complete theoretical understanding of plasmon-driven hot carrier generation with electronic structure details has been elusive. Theoretical studies of plasmonic systems have traditionally focused on their optical response, including jellium models of nanostructured systems. While extremely valuable, these models do not capture the material dependence of this process and miss interband transitions in noble metals.[1] Recently we analyzed the quantum decay of surface plasmon polaritons and found that the prompt distribution of generated carriers is extremely sensitive to the energy band structure of the plasmonic material.[2] In this context, we view processes involving plasmons, electrons and phonons within a Feynman diagrammatic approach and treat plasmonic hot carrier generation fully from first principles. Built upon this general theoretical and computational framework, we will present results for Cu, Ag and Au on higher order processes such as multi-plasmon decays in metals that are critical for plasmon-driven upconversion and non-linear plasmonics.[3] The role of phonon-assisted intraband contributions in plasmonic hot carrier generation calculated from *ab initio* calculations will be shown. I discussed the implications of multi-plasmon decays on the design of hot carrier devices and plasmon-driven chemistry applications.

1 A. J. Leenheer, P. Narang, N. S. Lewis and H. A. Atwater, *J. Appl. Phys.*, 2014, **115**, 134301.
2 R. Sundararaman, P. Narang, A. S. Jermyn, W. A. Goddard III and H. A. Atwater, *Nat. Commun.*, 2014, **5**, 5788.
3 P. Narang, R. Sundararaman, A. S. Jermyn, A. M. Brown, W. A. Goddard III and H. A. Atwater, *in preparation*.

Alexei A Kornyshev opened a general discussion of the paper by Martin Moskovits†: In the effects that you have described, is there anything principally different from the photoemission from metal into electrolyte? This subject was very popular in the 1970s, due to the works of Brodsky, Gurevich, Rotenberg and Pleskov in the Soviet Union and Alan Bard in the USA (for a review see Ref. 1), but pioneered perhaps in the less-known work of Geoffrey Barker (the Electrochemistry Medal of the Royal Society of Chemistry is called the "Geoffrey Barker Medal").

1 A. M. Brodsky and Y. V. Pleskov, *Prog. Surf. Sci.*, 1972, **2**, 1.

Martin Moskovits answered: As you have correctly pointed out, such hot-carrier internal photoemission, from a metal to a semiconductor or to an electrolyte, has been known for a long time. For example, we have shown in the early 90′s that hot electrons can be extracted out of roughened metal surfaces directly as enhanced and directional photoemission into vacuum.[1] The directionality was

† The written version of the responses to the questions about Martin Moskovits' paper were completed by Syed Mubeen.

shown through a supporting theory based on light-induced drift, to indicate that the majority of the hot electrons arrived at the nanoparticle's surface ballistically. To use hot electrons in photovoltaic or photocatalytic systems, however, one needs to fabricate nanostructured devices that present interfaces appropriately chemically modified to extract and use these carriers before they lose their energy as heat. This is due to the short time length scales—approximately tens of nanometers—associated with rapid hot carrier dynamics. Further, when an appropriately nanostructured material (such as silver, gold or palladium as in this case) that can sustain high-quality localized surface plasmon resonances is illuminated with light of an appropriate wavelength, the optical fields radiated by the nanostructure and the incident field can coherently superimpose, redistributing the radiant energy so as to create regions of concentrated electromagnetic energy where the field amplitude can be enhanced by factors of tens or even hundreds, thereby enabling enhanced hot-carrier generation and photoemission. Also, the ability to tune localized surface plasmons in a prescriptive fashion by simply altering the architecture allows one to create panchromatic devices covering the entire solar spectrum.

1 V. M. Shalaev, C. Douketis and M. Moskovits, *Phys. Lett. A*, 1992, **169**, 205.

Thomas Ebbesen commented: I like your idea very much. I've been following a little bit the literature in Japan on this concept but the devices are much more macroscopic.[1,2] They produce hydrogen and oxygen in the right proportions, and the action spectra follow the surface plasmons resonance even in the IR. It looks ideal; the only problem is the very low quantum yield in their systems. What's the best quantum yield that you have obtained so far and what are the prospects for improving it?

1 Y. Nishijima, K. Ueno, Y. Kotake, K. Murakoshi, H. Inoue and H. Misawa, *J. Phys. Chem. Lett.*, 2012, 3, 1248.
2 Y. Zhong, K. Ueno, Y. Mori, X. Shi, T. Oshikiri, K. Murakoshi, H. Inoue and H. Misawa, *Angew. Chem. Int. Ed.*, 2014, **53**, 10350.

Martin Moskovits replied: We have achieved so far a maximum of 1.5% internal quantum efficiency for solid state photovoltaic devices and 1% internal quantum efficiency for photocatalytic devices.

Thomas Ebbesen continued: A 1% quantum yield is already very good. You have the advantage in your system that you store the energy in chemical products. If one converted hydrogen and oxygen using photovoltaics, the overall yields would be much lower. How did you get up to 1%, what did you change?

Martin Moskovits responded: Thank you for your positive comments. Internal quantum efficiency here is the absorbed photon-to-hot-electron injection and collection efficiency of the device, as measured by the currents in the case of PV devices and the amount of chemical products (H_2) in the case of photocatalytic devices. There are multiple steps that can decrease the quantum efficiency in the case of a plasmon-driven photocatalytic process (some of which can be applied to PV systems as well), such as: (i) the plasmon materials themselves—inefficient coupling of the plasmonic absorber material to the redox couple or to the catalytic system; (ii) the hot carriers themselves—inefficiencies in transporting the hot carriers to appropriate electrochemically-active sites to perform desired redox chemistries; (iii) the catalytic systems themselves—inefficiencies of the catalytic systems by themselves; and (iv), the interfacial charge transfer process—inefficiency of the 'electron filter' layer (mostly a semiconductor or insulator) in contact with the plasmonic metal. Consequently, the low efficiencies registered so far for plasmonic devices could be any or all of them. In this study (*Faraday Discuss.*, 2015, DOI: 10.1039/C4FD00185K) we tried to improve the quantum efficiencies by eliminating the intermediate 'electron filter' layer and 'co-catalyst' and used plasmonic metal as a standalone light absorber and catalytic unit. The above results also motivate us to believe that much improvement in efficiencies can be expected if one can better understand the overall chain of events, that begins with photoexcitation of the plasmon and terminates with the production of the desired product (fuel or electricity).

F. Javier García de Abajo commented: You have presented a very inspiring concept. I just have a question about the timescales involved. Let me consider for example a single 2 eV plasmon decaying into an electron–hole pair, with the hole near the Fermi energy, where the density of electron states is large, and the electron is then 2 eV above that level. It takes a few femtoseconds (~10 fs in gold) for the electron to release part of its energy, giving rise to a second electron–hole pair. In this cascade process, the number of hot charge carriers is increasing with time, while their average energy goes down, so the sum of their energies is the energy of the original plasmon minus the fraction that is dissipated by coupling to phonons. Although a ballistic regime is maintained during 100s of fs,[1] a diffusive regime is eventually reached. Therefore, we have one electron with 2 eV energy for $\sim$10 fs, then two electrons with lower energy for $\sim$10 fs, and eventually the energy is dissipated to phonons in a few picoseconds. I am then wondering what the efficiency of the hot electrons is for triggering chemical reactions at the surface of the metal. The penetration of the electron wave function towards the medium outside the metal consists of its evanescent tail, which extends more when the electron has higher energy relative to the Fermi level (*i.e.*, when it is closer to the confining potential barrier). Therefore, I would imagine that hotter electrons are present during the first tens of fs, while much lower-energy electrons are present during the rest of the time required for dissipation into phonons. What is, then, the timescale for efficiently triggering chemical reactions at the particle surface? The above picture could apply to a linear regime (*i.e.*, when the chemical reaction activated by hot electrons is produced at a rate that is linear with the light intensity). It is also possible that for larger light intensities a hot-electron plasma is maintained at a temperature of 100s of degrees due to the electron–phonon thermal barrier,[2] so a non-linear rate of chemical reaction is produced. I would appreciate to know your views on these issues.

1 S. D. Brorson, J. G. Fujimoto and E. P. Ippen, *Phys. Rev. Lett.*, 1987, **59**, 1962.
2 M. Perner, P. Bost, U. Lemmer, G. von Plessen, J. Feldmann, U. Becker, M. Mennig, M. Schmitt and H. Schmidt, *Phys. Rev. Lett.*, 1997, **78**, 2192.

Martin Moskovits answered: Those are very interesting questions. As pointed out, the dephasing times for localized surface plasmon (LSP) resonances are very short, in the order of 1–10 femtoseconds, the electron–phonon relaxations are in the order of 1–5 picoseconds and equilibration with its surroundings is achieved at even higher time scales. So the question is, what is the energy distribution of electrons in a plasmonic metal (here Pd) when it is CW illuminated at frequencies corresponding to the LSP of the Pd particle, and what will the energy distribution of the hot electrons be before a great deal of energy has been lost due to electron–phonon interactions. To our knowledge, no adequate theoretical treatment has yet been carried out to determine the energy distribution function of the electrons across different time scales—in all fairness though this is a rather formidable time-dependent calculation. However, since the LSP absorption in a plasmonic metal arises from the participation of thousands of virtual single-electron transitions (the transition dipole for the LSP in a single silver nanoparticle is ~20 000 D), we believe that at least one electron (maybe more than one) resulting from plasmon dephasing will occupy an energy level as high as $\hbar\omega_{SP}$ above the Fermi energy. Moreover, if the Pd particle is irradiated for a period of time (CW illumination), a new plasmon will be continuously excited; this will create hot electrons at rates much higher than electron–phonon relaxation rates, resulting in a steady state with many hot electrons occupying an energy level as high as $\hbar\omega_{SP}$ above the Fermi energy. So, if one could design plasmonic devices in such a way that a large number of plasmonic particles can be packed per cm^2, say for example 10^{11} particles per cm^2, one could potentially produce as many as 10^{24} electrons per second per cm^2 (based on the above time scale, one could create 10^{13} hot electrons per second, and assuming that there is at least one hot electron that has energies as high as $\hbar\omega_{SP}$ above the Fermi energy on plasmon dephasing). This is a lot of electrons, but the challenge is accessing these hot electrons for useful work, and with improvements in device engineering we hope that one could extract useful work with efficiencies rivalling conventional semiconductor-based devices. But currently, little is understood about how efficiencies depend on dynamics (the energy and momentum distribution of hot carriers), excitation intensity (and their linear and non-linear effects), pulse width, *etc.*

Mikhail Noginov said: In the cartoon scheme of the process, electrons are on one end and holes are on the other end. Reduction happens on one end and oxidation happens on the other end. However, at localized plasmon oscillations, positive and negative ends change with the frequency of light oscillation. Will the chemical reaction follow the change of polarity?

Martin Moskovits replied: The hot carriers produced as a result of plasmon decay are used to carry out the chemical reactions. As long as the laser is 'on', a new plasmon will be continuously excited incoherently with the previous plasmon, and hot carriers will be continually created with many hot electrons occupying energy states well beyond the Fermi energy, and it is these carriers that take part in the chemical reaction—the chemical reaction will not follow the change of polarity. But it is an intriguing question as to how the incident photon energy is distributed between the excited hot carriers and whether the relative oxidizing and reducing power of the plasmonic catalytic metal changes with time scale. We do not yet know the answer to the above.

Kaiqiang Lin asked: In your talk, I noticed that you have changed your hot electron harvesting system from gold and silver to palladium. For the palladium system, the special thing is that it has a strong photothermal effect, which means that the local temperature could be very high. So my question is whether the photothermal effect is helping the hot electron harvesting process, or are they just competing with each other for absorption energy? How can the contributions of these two parts be distinguished? Also, what kind of plasmonic material would be more promising?

Martin Moskovits responded: These are excellent points. Plasmon-mediated chemical processes may occur as a result of: (i) the hot charge carriers resulting from plasmon decay; (ii) intense local temperatures created as a result of concentrated electromagnetic fields that enhance the chemistry; and (iii), the plasmonic system behaving as an optical antenna, thereby improving light collection by concentrating or redirecting light. All of the above processes can either individually or collectively influence the photocatalysis process, and, it is important to understand the contribution from each process. We primarily believe that the increase in quantum efficiencies observed here (*Faraday Discuss.*, 2015, DOI: 10.1039/C4FD00185K) is primarily due to non-thermalized electrons, although one cannot completely eliminate thermal or local EM field enhancement effects. We place our argument based on the fact that most chemical reactions have energetics in the order of eV, and for the process to be a wholly thermally-driven process, extremely high temperatures need to be sustained with minimal heat losses. Currently we do not have a clear-cut solution as to how one could distinguish the contributions of these two parts, although we are aware of a few studies that use local temperature probes to study plasmon-induced thermal effects.

Felix Benz said: Is it possible to use alternative materials (*e.g.* TiN or ZrN) for emitting hot electrons? Would they be good candidates for large area applications like photocatalytic production of fuels?

Martin Moskovits responded: Certainly, in fact TiN nanoparticles have shown very good electrocatalytic activity towards the oxygen reduction reaction and have proved to be stable for the reduction of alkynes to alkenes.

Javier Aizpurua asked: In the dynamics of the electrons and holes in your antennas, there must be strong modifications to the lifetimes and to the mean free paths due to the particular configuration of the surfaces and interfaces forming the antenna. I am thinking of a different scale of distances, more connected with the roughness on the surfaces. These effects must have a tremendous impact as to how the electron makes it out from the antenna towards the catalyzer. Do you think this aspect is important in determining the efficiency? Have you checked how surface roughness, and other antenna surface effects affect the performance of your processes?

Martin Moskovits answered: Our initial experiments were performed on nanorough surfaces.[1] We were able to extract hot electrons directly from the rough surfaces and the majority of the hot electrons arrived at the nanoparticles surface ballistically. We haven't checked our quantum efficiency as a function of surface roughness, although we thoroughly agree that this will be a fascinating study to perform.

1 V. M. Shalaev, C. Douketis, J. T. Stuckless and M. Moskovits, *Phys. Rev. B*, 1996, **53**, 11388.

Pavel Ginzburg opened a general discussion of the paper by Mikhail Noginov by commenting: Thank you for presenting this very interesting contribution. My general question is on the theoretical interpretation of the experimentally obtained data. One of the strong assumptions, made for the fitting, is the single exponential nature of the decaying fluorophores. This assumption leads to the emergence of an additional square root of time component, if the acceptor molecule is present next to the donor. While the single exponential decay law holds for the majority of dyes in various solutions, the presence of the nearby nanostructure could make the decay process deviate from this single exponential behaviour. The decay dynamics will be dependent on the relative position and orientation of an emitting dipole relatively to a structure. My question is about the importance of those effects on the theoretical description of the FRET in the vicinity of hyperbolic nanostructured metamaterials.

Mikhail Noginov replied: This concern is, in principle, valid. However, the experimental emission kinetics of pure donors did not significantly deviate from a single exponential form. When we divided the kinetics of the donors in the presence of the acceptors by the kinetics of the donors only, we, in the first approximation, cancelled the contribution of radiative decay and non-radiative quenching (*e.g.* by metal) of the donors. In this way, we extracted the contribution of the donor-to-acceptor energy transfer to the emission kinetics.

Pablo de Roque asked: As far as I understood, the main claim of your paper (*Faraday Discuss.*, 2015, DOI: 10.1039/C4FD00184B) is that, given that you have an increase of the radiative decay rate, then the efficiency of the energy transfer decreases; did I understand correctly or am I wrong? Additionally, as pointed out by a different delegate previously, I presume that your metamaterial is quite inhomogeneous at the distance scales of intermolecular distances. This means that, for example, the local fields as a function of position in the metamaterial would look different in the plane of the metamaterial to out-of-plane of it. Given that your measurement averages over a relatively large volume of the metamaterial/donor–acceptor layer, I therefore think it is bold to adventure any direct link between changes in the decay rate of emitters and the rates of energy transfer.

Mikhail Noginov responded: We have experimentally observed that the same dielectric environments that enhance the emission decay rate of donors inhibit energy transfer to acceptors. At this time, we make a statement of this experimental fact and avoid any further speculation.

Michael Flatté said: As a theorist, looking at the experimental data showing the Foerster's energy transfer, the slope that you draw on the plot in the inset of Fig. 2 in your paper, *Faraday Discuss.*, 2015, DOI: 10.1039/C4FD00184B (slope ≈ 1/2), for the metamaterial, appears to be consistently above the line on the left and below the line on the right. Is that actually a fit or is it a guide to the eye? What is the actual power law? How confident are you that this is in the square root of time regime, as the actual slope doesn't appear to support that?

Mikhail Noginov answered: The slope of the curve in the inset of Fig. 2 in our paper (*Faraday Discuss.*, 2015, DOI: 10.1039/C4FD00184B) is ~0.6, much closer to 1/2 than to 1. Moreover, inhibition of the energy transfer on top of a metamaterial is evident from a comparison of curve 2 (emission of donors in the presence of acceptors on top of glass) and curve 5 (emission of donors in the presence of acceptors on top of a metamaterial) of Figure 2. The decay rate on top of a metamaterial is obviously slower. This conclusion does not depend on the $\sqrt{t}$ dependence of the energy transfer term.

Jeremy Baumberg enquired: In the emission–time plot, if you look at the acceptors alone on the metamaterial, you don't have that curve on there, so what do you see for that (would it be useful to add this to the graph)? If they are also very fast, we wouldn't expect to see that much difference with the metamaterial?

Mikhail Noginov responded: The behavior of acceptors in the presence of metal is qualitatively the same as that without metal.

Francisco Rodríguez Fortuño was invited to make some remarks to open a general discussion of the paper by Olivier Martin, saying: I would like to draw attention to the possibility for an interesting optical force. Typically, optical forces are associated with gradients in the amplitude of light, however the vector character of electromagnetic fields means that light polarization can also play an important role.

Works on optical torques induced by polarization are well known.[1] However, we ask whether polarization can be responsible for a mechanical net force which pushes the particle as a whole (unlike a torque). Prof. Anatoly Zayats and I, in collaboration with Prof. Nader Engheta and Prof. Alejandro Martínez, have studied the possibility of achieving a lateral force acting on a centrosymmetric particle near a surface: this force is perpendicular to the illumination direction and parallel to the surface, and it changes direction according to the polarization. The illumination can be a simple polarized plane wave with no incident field

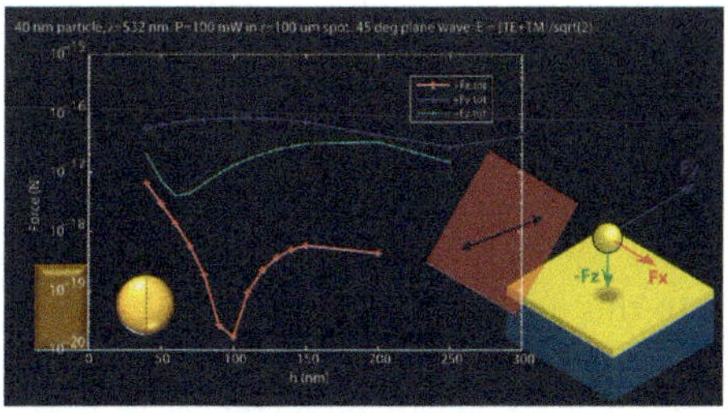

Fig. 3 The force acting on a 40 nm spherical gold nanoparticle suspended in water, at varying distances '*h*' over a 50 nm thick gold film on a glass substrate, and illuminated at 45 degrees incidence by a polarized plane wave at a wavelength of 532 nm, near the localized resonance of the particle.

gradients. The particle *need not be chiral, since the symmetry is broken by the polarization*. This suggests the fascinating possibility of a very simple and switchable "nano conveyor belt" that can act on several particles simultaneously, as part of the nanofactories envisioned by Prof. Olivier Martin.

What is the origin of this lateral force? Our inspiration came from recent works on the unidirectional excitation of electromagnetic modes. By illuminating at grazing angle a particle near a metallic surface, its dipolar scattering can couple to surface plasmons that are launched unidirectionally towards the left (or right) of the particle when right-handed (or left-handed) polarization is used in the illumination. We measured this experimentally in ref. 2. A simple explanation in terms of near field interference can be found in ref. 3. The phenomenon can be understood as an example of spin–orbit coupling of light, and is possible thanks to the transverse spin of guided modes.

The existence of a polarization-switchable lateral force in such a system follows from conservation of linear momentum. The net lateral momentum of the illumination is zero, and must be conserved, which means that *any lateral electromagnetic momentum carried by the asymmetric scattering of the particle (e.g. unidirectionally launched surface plasmons) must be accompanied by an equal and opposite mechanical momentum (recoil force) exerted on the particle*. I think that the ability to generate equal but opposite momenta of such different natures is a fascinating possibility: we are inducing electromagnetic momentum directed to one side, and harvesting the mechanical momentum directed to the other. Since, as described above, the unidirectional excitation of electromagnetic modes is controlled by the polarization of the incident plane wave, so must the direction of the lateral recoil force be, achieving the desired switchable "conveyor belt". The concept is sufficiently general to be applicable to many different scenarios. For example, a spin-dependent recoil force is expected in every instance in which spin–orbit coupling causes a directional excitation of fields.

1 R. A. Beth, *Phys. Rev.*, 1936, **50**, 115.
2 D. O'Connor, P. Ginzburg, F. J. Rodríguez-Fortuño, G. A. Wurtz and A. V. Zayats, *Nat. Commun.*, 2014, **5**, 5327.
3 F. J. Rodríguez-Fortuño, G. Marino, P. Ginzburg, D. O'Connor, A. Martínez, G. A. Wurtz and A. V. Zayats, *Science*, 2013, **340**, 328.

Silvan Schmid addressed Francisco Rodríguez Fortuño: The calculated force acting in the plasmonic particle seems to be very small. I am concerned that the random Brownian motion due to the photothermal heating of the particle will cancel the effect of this directional plasmonic force. As a comparison, the plasmonic forces acting on nanoparticle dimers are typically 4–5 orders of magnitude larger.

Francisco Rodríguez Fortuño responded: Thank you for bringing up this very important practical issue. My slides showed a recent simulation result obtained on the previous day. In this numerical calculation, we used a commercial electromagnetic solver to calculate the force acting on a 40 nm spherical gold nanoparticle suspended in water, at varying distances '*h*' over a 50 nm thick gold film on a glass substrate, and illuminated at 45 degrees incidence by a polarized plane wave at a wavelength of 532 nm, near the localized resonance of the particle. The force was computed using Maxwell's stress tensor. The power density of the plane wave was taken equivalent to that of a 100 mW beam focused on a 200 μm diameter spot. Directional excitation of surface plasmons accompanied by a polarization-dependent lateral force was confirmed. The specific slide is shown in the attached figure (Fig. 3). The order of magnitude of the force is around 10^{-19} to 10^{-17} N. First of all, I would like to acknowledge that this force is indeed orders of magnitude below typical plasmonic forces, and we are aware that such a small force would be 'drowned' by Brownian motion. For this observation I thank you. However, I would also like to stress that the simulation shown was simply intended to provide a confirmation of the mere *existence* of a lateral force, as we hoped from our previous theoretical study. Therefore, I think that it is premature to judge the practical applicability of the concept on the basis of this particular simulation, with very specific parameters. Clearly there is much room for improvement in the design of a practical experiment. Our last simulations show that increasing the size of the scatterer from 40 nm to 80 nm already results in a 1 order of magnitude increase in the amplitude of the force. Also notice that we assumed a beam diameter of 200 μm, which in practice can be reduced, increasing the incident power density, and therefore the force. Other considerations of materials, particle shapes and wavelengths can easily be explored. Also, an important result of the simulation presented lies not in the absolute value of the force, which depends greatly on the specific parameters, but instead in the relative amplitudes between the different force components. If we compare them, we see that the lateral force is rapidly approaching the same magnitude as the pressure force when the particle gets closer to the substrate. This means that we can achieve lateral polarization-dependent forces comparable to the pressure force, and the latter have been successfully measured experimentally (for various sizes of particles and materials). In an experiment we will coat the particles and the substrate to avoid sticking between them, and use nanofluidics to place the particles in close proximity to the substrate. In addition, if we find it difficult to increase the force significantly, we can resort to a statistical analysis of the Brownian motion in order to detect an average drift caused by the lateral force. The fact that the force is lateral (perpendicular to all other forces in the system) and switchable with the incident polarization provides a method of easy experimental discrimination. Opposite average drifts should be measured for opposite polarizations. This differential measurement allows any experimental systematic drift to be compensated for. As a final remark, unidirectional excitation has been clearly visible in various experiments related to optical spin–orbit coupling (see as an example ref. 1–4), and in all these cases there is an opportunity to measure a recoil force. A particularly promising case of force measurement would be that in which the emitters are single atoms, such as in ref. 1.

1 R. Mitsch, C. Sayrin, B. Albrecht, P. Schneeweiss and A. Rauschenbeutel, *Nat. Commun.*, 2014, **5**, 5713.
2 D. O'Connor, P. Ginzburg, F. J. Rodríguez-Fortuño, G. A. Wurtz and A. V. Zayats, *Nat. Commun.*, 2014, **5**, 5327.
3 F. J. Rodríguez-Fortuño, I. Barber-Sanz, D. Puerto, A. Griol and A. Martínez, *ACS Photonics*, 2014, **1**, 762.
4 J. Petersen, J. Volz and A. Rauschenbeutel, *Science*, 2014, **346**, 67.

Olivier Martin addressed Francisco Rodríguez Fortuño: I have one question and one remark to the invited comment you presented after my paper (*Faraday*

Discuss., 2015, DOI: 10.1039/C4FD00224E): firstly, it seems that the optical forces you calculate are orders of magnitude smaller than the typical optical forces for the type of illumination you use (usually these forces are in the tens of pN range), maybe because the lateral effect is small, but the scattering force should be in that order of magnitude as well. Secondly, I believe that the lateral force you observe on your particle is a mere transfer of the circular polarization made possible by the feedback produced by the surface. It is similar to a (rather difficult) carom billiard shot, where one gives the ball some transverse spin (*i.e.* the spin axis is along the ball motion). This results in a transverse motion of the ball, equivalent to the transverse force that you observe in your calculations.

Francisco Rodríguez Fortuño replied: Thank you very much for the comment. In regard to your question, I refer to my more lengthy reply to Silvan Schmid, who raised the same issue. The force I calculated was for a given specific scenario, in which a 40 nm diameter gold particle is illuminated by a plane wave whose power density is equivalent to 100 mW on a 200 μm diameter spot. The simulation was simply intended to confirm numerically the existence of the lateral force which we had predicted analytically. Is it possible that the small size of the particle and the relatively large area of illumination (resulting in a smaller power density) are the reasons for the low scattering force? In any case, it is exciting to see that the lateral F_x force comes very close in magnitude to the pressure F_y force (especially at small distances from the substrate), meaning that the lateral force is at least comparable with other well-known forces that have been successfully measured in the past. Regarding your insightful remark, certainly this is a good interpretation. The surface is absolutely required for this effect. The circular polarization together with the surface provide the necessary breaking of mirror symmetry that enables a lateral force. As a simple test of this, it is easy to prove using symmetry arguments that the lateral force is zero under normal illumination (incidence perpendicular to the surface). Building upon the spin discussion, in this case the illumination has a longitudinal spin (directed along the propagation direction, as is usual in circularly polarized light) which is transferred to the polarization of the particle. This rotating polarization then couples to a surface mode propagating laterally, but only in one direction, due to the transverse spin of any guided TM mode that exists in the nearby surface (with spin perpendicular to the propagation direction). This asymmetry in the excitation of modes provides the possibility of a lateral force. This 'spin interpretation' of the directional symmetry-breaking excitation of modes has been discussed in depth for experiments.[1,2]

1 J. Petersen, J. Volz and A. Rauschenbeutel, *Science*, 2014, **346**, 67.
2 D. O'Connor, P. Ginzburg, F. J. Rodríguez-Fortuño, G. A. Wurtz and A. V. Zayats, *Nat. Commun.*, 2014, **5**, 5327.

Niek van Hulst commented to Francisco Rodríguez Fortuño: You show a very interesting concept, to steer forces on particles by simply switching between left- and right-handed circular polarisation. As a comment, recently the Kuipers group at AMOLF-Amsterdam showed work on sub-wavelength apertures, acting both as electric dipolar sources and as magnetic sources.[1] Driving the aperture with circularly polarised light, in near field contact with a waveguide, they could show launching of light to the left or right direction depending on the direction of the inserted circular polarization. In that configuration, possibly forces for left–right steering might occur.

1 B. le Feber, N. Rotenberg and L. Kuipers, *Nat. Commun.*, 2015, **6**, 6695.

Francisco Rodríguez Fortuño answered: Absolutely, I agree that there is most probably a polarization-dependent force in that case. I think that, following conservation of momentum, we would expect to see a polarization-dependent "recoil force" in any experiment in which there is polarization-switchable directional emission of electromagnetic modes. Thank you for your comment and for bringing this exciting possibility to our attention.

F. Javier García de Abajo opened a general discussion of the paper by Olivier Martin: You are presenting a very inspiring concept, where optical forces are used to position nano-objects and make them interact. I am worried, however, that van der Waals forces might be comparatively larger than optical forces when the objects are in close proximity to a surface or to each other. Is this a source of concern?

Olivier Martin replied: It is true that when we started our first experiments, we were worried that van der Waals forces would be stronger than the optical forces, especially for extremely small distances between the trap and the trapped particle. Actually, the experiments showed otherwise.[1] In particular, Fig. 2d in ref. 1 indicated that the nanoparticle is released once the trapping laser is turned off. Later in time in that same figure, another nanoparticle makes an excursion through the antenna gap without being trapped (since the trapping laser is off), indicating that van der Waals forces will not be sufficient to trap the particle in the vicinity of the trap. One should keep in mind that all these experiments are performed in water, which screens the van der Waals interaction. Furthermore, calculations have indicated that the van der Waals force becomes significant only at extremely short distances from the surface.[2] In any case, this is definitely a very interesting question to address—and possibly to utilize—for producing additional degrees of freedom in those systems.

1 W. Zhang, L. Huang, C. Santschi and O. J. F. Martin, *Nano Lett.*, 2010, **10**, 1006.
2 L. Novotny and C. Henkel, *Opt. Lett.*, 2008, **33**, 1029.

F. Javier García de Abajo asked: Optical forces have been extensively used to trap and manipulate atoms, but doing this in close proximity to a surface remains a challenge, as van der Waals forces dominate at distances below ~100 nm for realistic light intensities. In the opposite extreme, large objects can feel much larger optical forces (*e.g.*, a gradient force roughly proportional to their volume or the pressure force that aims solar sails in spacecraft), which are extensively used in optical tweezers and have also proved useful to steer particles down to tens of nanometers in size.[1] Therefore, there seems to be a compromise between the size of the objects and their proximity to a surface. What are the size and distance ranges that you envision?

1 M. Geiselmann, R. Marty, J. Renger, F. J. García de Abajo and R. Quidant, *Nano Lett.*, 2014, **14**, 1520.

Olivier Martin responded: Actually, the positioning of nanoscopic polarizable entities in the hot spot of a plasmonic nanostructure was already pioneered in 2010 by W. Zhang *et al.*[1] Regarding the competition between van der Waals forces and optical forces, Novotny *et al.* conclude that it is only at very short distances (a few nanometres) that van der Waals can compete with optical forces.[2] Actually, in all the experiments we have performed, we did not observe any disturbance caused by van der Waals forces (see *e.g.* the releasing experiments reported in Fig. 2d in the aforementioned 2010 *Nano Letters* article[1]). One should also keep in mind that most optical trapping experiments are performed in water, which strongly screens the van der Waals potential.

1 W. Zhang, L. Huang, C. Santschi and O. J. F. Martin, *Nano Lett.*, 2010, **10**, 1006.
2 L. Novotny and C. Henkel, *Opt. Lett.*, 2008, **33**, 1029.

Jeremy Baumberg remarked: In the calculation of internal strain, can you convert the forces into an actual strain from the Young's modulus? It would be interesting to know, since it is likely to be extremely small, correct?

Olivier Martin answered: As a matter of fact, the electromagnetic forces on nanostructures are too small to result in any considerable strain. For typical illuminations, the forces on plasmonic nanostructures are of the order of piconewtons. Considering a nanostructure cross section of 1000 nm^2, this results in a stress of 1 kPa. Since the Young's moduli of plasmonic materials such as gold and silver are of the order of 100 GPa, the strain on the nanostructures is merely 10^{-8}. This value is too small to be of any consequence.

Santhosh Kotni commented: As the trapping of particles or molecules in the nanojunctions strongly depends on the intensity of the electromagnetic field in the hot spots, please could you explain the distribution of electromagnetic field

Fig. 4 Near field calculated for a distribution of ~1500 gold nanorods. Reprinted with permission from D. M. Solis, J. M. Taboada, F. Obelleiro, L. M. Liz-Marzán and F. J. García de Abajo, *ACS Nano*, 2014, **8**, 7559. Copyright 2014 American Chemical Society.

across the hot spots. Is the field uniform across the gap, and if not, how does the field intensity distribution affect the position of the particle in the hot spots?

Olivier Martin responded: This is an interesting question, the answer to which depends a lot on the type of plasmonic structure used for trapping. In the case of a dipole antenna as studied in Fig. 3 and 4 of our paper (*Faraday Discuss.*, 2015, DOI: 10.1039/C4FD00224E), we can deduct from the charge distributions that the field is quite homogeneous within the gap of the structure, with one single hot spot, where the particle will be trapped. So we can infer one trapping position in such a geometry. This is actually confirmed by earlier studies, where we computed the trapping potential—without the presence of the particle, though—and observed one single trapping well: see *e.g.* Fig. 3c in ref. 1. For more complex trapping structures, the potential will be more complicated and this is certainly an issue when designing such systems. For example, in the case of the trapping hexamer shown in Fig. 6 of our *Faraday Discussions* paper, the trapping potential has a 6-fold symmetry, with a well-defined minimum.

1 W. Zhang, L. Huang, C. Santschi and O. J. F. Martin, *Nano Lett.*, 2010, **10**, 1006.

Jeremy Baumberg asked: Please could you comment generally on how the optical forces compare to optical heating and convection flows. Do you see any way to get around all of that realistically?

Olivier Martin replied: Forces created by convection flows or even temperature gradients can indeed dramatically disturb optical trapping experiments on plasmonic nanostructures. This is a fairly unexplored field of research and it is difficult to find quantitative data on the corresponding forces. Braun *et al.* have performed experiments on thermophoretic trapping of small objects,[1] in a similar structure to the hexamer we discuss in our paper (*Faraday Discuss.*, 2015, DOI: 10.1039/C4FD00224E). They have observed trapping potentials caused by thermal gradients in the order of 10 KT, which is comparable to the trapping potential caused by a plasmonic nanostructure (see *e.g.* Fig. 3c in ref. 2). In experiments we performed, we have tried to minimize thermal effects using microfluidics: a relatively shallow microfluidic channel (10–15 microns) prevents the build-up of convection flows, while the water flow through the channel provides some additional cooling to the system. Still, this is definitely a very interesting question and—possibly—the combination of thermal effects with optical trapping could provide additional degrees of freedom for manipulations at the nanoscale.

1 M. Braun and F. Cichos, *ACS Nano*, 2013, 7, 11200.
2 W. Zhang, L. Huang, C. Santschi and O. J. F. Martin, *Nano Lett.*, 2010, **10**, 1006.

F. Javier García de Abajo opened a general discussion of the topics raised at the meeting by saying: Advances in nanofabrication and the ability of solving Maxwell's equations in the presence of nanostructured materials have been often argued to be responsible for the explosive growth of plasmonics. Additional understanding gathered from simple analytical models has been combined with simulations for increasingly complex systems. However, there are situations in which simulations of large systems are extremely useful, for example in the determination of the position, frequency of occurrence, and distribution of plasmonic hotspots in realistic, random structures. I would like to emphasise that such simulations are now mastered by a handful of groups that specialise in solving the electromagnetic problem. As an example, I attach a figure extracted from Ref. 1 (Fig. 4), in which a structure formed by over 1500 gold nanorods was considered, with a random arrangement of the particles (starting from a randomly oriented nanorod that approaches the system until its distance from any other surface is 1 nm, and repeating this process for each additional rod). The figure shows the near field, calculated with well-converged methods developed within the group led by Profs. J. M. Taboada and F. Obelleiro, and taking a few hours of simulation time in a computer cluster. In my opinion, the combination of suitable boundary element methods and acceleration procedures developed by the noted group sets a new standard for the complexity of plasmonic systems that we are currently capable of simulating within the plasmonics community.

1 D. M. Solis, J. M. Taboada, F. Obelleiro, L. M. Liz-Marzán and F. J. García de Abajo, *ACS Nano*, 2014, **8**, 7559.

Olivier Martin commented: The results presented in the aforementioned *ACS Nano* paper by D. M. Solis *et al.*[1] are truly impressive and illustrate the tremendous developments in the surface integral equation (SIE) method since the original work by A. M. Kern *et al.*,[2] its extension to periodic structures by B. Gallinet *et al.*[3] and to the calculation of non-linear effects by J. Butet *et al.*[4] There is a family of numerical methods, which bear similarities with the boundary equation method introduced by F. J. García de Abajo *et al.*[5] All these techniques take advantage of the integral form of Maxwell's equation and can handle complex boundary conditions, including surfaces and stratified backgrounds. They are also particularly well-suited to simulate realistic nanostructures and provide a guide for experiments, as shown by A. M. Kern *et al.* in the case of fluorescence.[6]

1 D. M. Solis, J. M. Taboada, F. Obelleiro, L. M. Liz-Marzán and F. J. García de Abajo, *ACS Nano*, 2014, **8**, 7559.
2 A. M. Kern and O. J. F. Martin, *J. Opt. Soc. Am. A*, 2009, **26**, 732.
3 B. Gallinet, A. M. Kern and O. J. F. Martin, *J. Opt. Soc. Am. A*, 2010, **27**, 2261.
4 J. Butet, K. Thyagarajan and O. J. F. Martin, *Nano Lett.*, 2013, **13**, 1787.
5 F. J. García de Abajo and A. Howie, *Phys. Rev. B*, 2002, **65**, 115418.
6 A. M. Kern and O. J. F. Martin, *Nano Lett.*, 2011, **11**, 482.

Sang-Hyun Oh said: With new techniques to produce nanogap structures, we also face significant challenges in computational modeling. For example, using atomic layer deposition, it is now possible to make 1–2 nm gaps that are as long as millimeters or even centimeters in perimeter.[1] From a modeling perspective, the extremely tight optical confinement in these nanogaps can give rise to an enormous mismatch in critical length scales (*e.g.* 1 nm gap *vs.* millimeter-wavelength incoming radiation). This mismatch can quickly overcome the capabilities of conventional FDTD or FEM methodologies. Similar problems are often encountered in aerospace engineering, and advanced techniques exist to tackle such multi-scale modeling problems. One such technique—known as Hybridizable Discontinuous Galerkin (HDG)—has been applied to terahertz nanogap modeling recently,[2] where the authors showed full 3D modeling of THz resonances in 2 nm gaps. I believe that such multi-scale modeling techniques will be useful in other problems such as sub-nanometer gaps, nonlocality, and atomic-scale materials.

1 X. Chen *et al.*, *Nat. Commun.*, 2013, **4**, 2361.
2 H.-R. Park, X. Chen, N.-C. Nguyen, J. Peraire and S.-H. Oh, *ACS Photonics*, 2015, **2**, 417.

Jeremy Baumberg commented: I'm interested in the comment about electron microscopy not entering here so much (EELS)—one reason is that we're in a chemistry institution. There's been some attempts to do chemistry in electron microscopy environments, but it is very difficult and you don't know that what you're seeing is useful in real systems. Thus the chemistry community are likely going to stay further away from it and work at ambient conditions. On the other side, we're also now looking at structures where atoms matter. Then electron microscopy is also not helpful really—it mostly looks at 3D arrangements of atoms down lines (or single-molecule-thick structures). It's hard to see in 3D, and especially things embedded inside structures. The really interesting thing about where the field has got in the optical domain, is that we're capable of doing better than electron microscopy now, if we only understood what the spectroscopic signatures we see tell us. We have to convince the electron microscopy community that this is really useful, that now optics tells you more then electron microscopy.

F. Javier García de Abajo remarked: Electron microscopies and spectroscopies were the techniques used in the early studies of plasmons (see, for example, ref. 1) and they have ever since remained as the tools of choice for achieving the best possible spatial resolution for mapping plasmons. I believe that they will play an equally important role in the future of the plasmonics community, as they combine sub-nanometer spatial resolution and the ability to probe optical excitations better than other existing techniques. In this respect, the advance in instrumentation is impressive, particularly with the arrival of aberration-corrected setups and improved monochromators in transmission electron microscopes (TEMs). There is really no other technique that combines the spatial and spectral resolution of electron microscopes.[2] As examples of applications to atomic-scale plasmonics, I find it really impressive that one can now map plasmons trapped by a single atomic defect in graphene,[3] or that the plasmonic response in metal gaps can be experimentally correlated with the morphology down to the atomic detail.[4] TEMs can thus study columns of atoms, but also atomically thin specimens, and they have a good vertical resolution when looking into the bulk with well-focused beams, which surpasses that of currently available optical setups. It is difficult to beat the picometer electron wavelength with light. Environmental microscopes are also offering a way of examining samples under realistic conditions, and in this context, there are many new directions, such as using graphene to wrap the specimen. But this is only part of the ongoing effort, as scanning tunnelling microscopes, low-energy and photoemission electron microscopes and secondary-electron microscopes combined with cathodoluminescence acquisition are providing further insight. Instead of showing which community does best, it is in our interest to join efforts. As a matter of fact, cathodoluminescence (CL) is a perfect example of that, combining the best of the two worlds: electron beams and optics, spatial resolution and spectral resolution. In my opinion, this explains why more groups are using CL, and there was even a representation during our *Faraday Discussion* from a manufacturer of a new line of CL setups (delmic.com), ready to be used for nanophotonics and plasmonics. Also, there is an increasing number of TEMs equipped with optical illumination, combining ultrafast capabilities and cold-field electron emission. In summary, the plasmonics community has been and surely will be benefiting from the impressive spatial resolution of electron microscopes, while extending their techniques to involve photonic technologies, including ultrafast optics. I believe that parts of the plasmonics community are engaged in extending this effort from complementary sides, so this is an exciting time for those interested in the interaction between electrons, photons and plasmons, with lots of win-win opportunities.

1 C. J. Powell and J. B. Swan, *Phys. Rev.*, 1959, **115**, 869.
2 F. J. García de Abajo, *Rev. Mod. Phys.*, 2010, **82**, 209.
3 W. Zhou, J. Lee, J. Nanda, S. T. Pantelides, S. J. Pennycook and J.-C. Idrobo, *Nat. Nanotechnol.*, 2012, 7, 161.
4 S. F. Tan, L. Wu, J. K. W. Yang, P. Bai, M. Bosman and C. A. Nijhuis, *Science*, 2014, **343**, 1496.

Faraday Discussions

PAPER

Nanoplasmonics: Concluding remarks

Niek F. van Hulst

Received 31st March 2015, Accepted 31st March 2015
DOI: 10.1039/c5fd90021b

The Faraday Connection

Faraday Discussions have a splendid reputation in stimulating scientific debates along the traditions set by Michael Faraday himself. The Faraday Discussion 178 on nanoplasmonics has been particularly appropriate. Nanoplasmonics is the science of harnessing the interaction of light and matter on the nanometer scale. Plasmonics is a highly interdisciplinary field bridging between physics, chemistry, modern nanotechnology and nanophotonics. Michael Faraday, firstly a chemist with original contributions in electrochemistry, paved fundamental roads in physics, establishing the basic concepts of the electromagnetic field in physics and the effect of magnetism on rays of light. He pointed out the peculiar effect metal particles have on light and the importance of the particle size. He mentioned gold as especially fitted for experiments, a tradition carried on until today.[1] As such, the spirit and scientific foundations of Michael Faraday have been ever-present at the Burlington house gathering in February 2015 (Fig. 1). Here, highlights of the presented papers and discussions are summarized with an eye on future perspectives.

Fig. 1 Michael Faraday, amidst the 4th century Lycurgus and 18th century Loving cup. Left, the glass Roman Lycurgus cup, with distinct colours in reflection and transmission due to embedded metal plasmonic nanoparticles (© Trustees of the British Museum). Right, the silver Loving cup used at the Faraday Discussions' Dinners to toast the memory of Mr. Marlow, Angela & Tony Fish, and the Faraday Division. (Image of Michael Faraday reproduced courtesy of the Library of the Royal Society of Chemistry.)

ICFO – Institut de Ciencies Fotoniques, Mediterranean Technology Park, Castelldefels – Barcelona, ICREA – Institució Catalana de Recerca i Estudis Avançats, Barcelona, Spain. E-mail: Niek.vanHulst@ICFO.eu; Web: http://www.ICFO.eu

Introduction

In the Burlington house corridors, asking around amongst the attendees as to the perspectives of nanoplasmonics I received quite disparate first-hand answers, varying from "metal nanoparticles are good for heating and non-linear optics" to "plasmonics is dead" and "hot electrons are the future". Clearly the field has developed some history already and is moving into new directions. A more balanced reflection on where the field stands today is needed.

Primarily, nanoplasmonics stands for metallic nanoparticles and nanostructures, which allow concentrating the light to dimensions far below the wavelength of light. As such, key strengths are: strong spatial confinement of the optical light field, intimately connected to the direction and polarization of both incident and emitted light; direct control of the radiative decay (engineering of LDOS); novel building blocks for light control, such as in flat optics applications; and the ideal combination of optical and electronic properties. Mark Brongersma nicely displayed these key strengths in his Introductory lecture on nanoplasmonics (DOI: 10.1039/C5FD90020D). Driven by the potential of nanoscale light control, plasmonics has developed fiercely over the last one or two decades. Strong field enhancement, nanoscale local confinement and emission control have all been shown. Pushing towards applications, many demonstrations of nanoplasmonic circuits, plasmonic assisted photovoltaics and even plasmonic nanolasers have been presented, placing the field at the top of the research agenda. Unfortunately, the use of metals at optical frequencies is associated with appreciable losses and heating, even more as the nanostructures are often non-crystalline. Only for far sub-micron sizes and distances the losses can be neglected. Many applications have been frustrated by this loss issue. Also, plasmonics is not compatible with complementary metal–oxide–semiconductors (CMOS), which complicates upscaling and the potential combination with silicon technology.

Many questions remain to be answered. Can novel plasmonic materials overcome the losses? What are the limits to spatial confinement? What is the role of non-local effects? Which plasmonic properties last in the quantum regime? Can one actively control the plasmonic response? What is the potential of plasmonics in strong coupling? Thus, to address such questions, current nanoplasmonics is moving into new directions. In recent years, many activities have been developed on the use of electron beams in plasmonics (electron energy loss spectroscopy (EELS), photoelectron emission microscopy (PEEM), cathode luminescence (CL)); the active manipulation of plasmonic structures; the expansion of the wavelength range towards both UV and MIR; the use of plasmonics in catalysis and photochemistry; new plasmonic (2D) materials, such as graphene, MoS_2 and BN; ultrafast atto- and femtosecond plasmonics; and more fundamental quantum plasmonics.

To capture these new developments, Faraday Discussion 178 focused on 5 themes:

- Plasmonic and new plasmonic materials.
- Surface plasmon enhanced spectroscopies.
- Quantum plasmonics, gain and spasers.
- Applications in nanophotonics.

- Sensing, imaging and chemistry applications of plasmonics.

Here, let's revisit and discuss particular highlights of the various themed sessions.

New materials

Clearly the major problem in nanoplasmonics is related to the material limitations, specifically the losses of gold and silver. The first day of the discussions was largely devoted to the quest for alternative superior materials. Vladimir Shalaev presented a new class of materials: transition metal nitrides (TiN, ZrN, HfN, TaN, AlN) as refractory optical materials, exhibiting promising plasmonic dispersion properties in the IR-Vis regime, along with high temperature durability (DOI: 10.1039/C4FD00208C). The dispersion of TiN is comparable to that of Au, yet TiN possesses superior hardness, a higher melting point, and CMOS compatibility and biocompatibility, all of which are important for applications. The novel material graphene is receiving great interest in the plasmonics community: graphene causes extreme localization of the optical field, together with fields increased over many orders of magnitude. Moreover, graphene, with a single Dirac point in the band-diagram, allows steep and broadband tuning through doping, *i.e.* a promising tunable plasmonic material. Inspired by the concept of atomically thin 2-dimensional materials, F. Javier García de Abajo presented 1–3 atom-thin gold nanodisks providing completely tunable optical absorption and potential for strong-coupling to quantum emitters (DOI: 10.1039/C4FD00216D). During the discussion, he extended the concept to graphene triangles, spanning 2 to 10 aromatic carbon hexagons, and "doped" with up to 6 charge carriers, showing the graphene plasmon tunability for absorption through the IR-Vis regime with a figure of merit better than those of Au and TiN. Interestingly, at the smallest scale of just 2 or 3 hexagons, such nanoscale graphene factually represents basic organic aromatic molecules such as anthracene, and the theory presented for plasmonic graphene also describes the optical properties of these charged molecules. Based on the intuitive graphene picture, García de Abajo coined the term "molecular plasmons", bridging the condensed-matter and molecular-physics schools, which generated an interesting debate. The concept for another new material was presented by Evgenii Narimanov by combining metamaterials and a photonic crystal in a so-called photonic hypercrystal, for which the plasmonic dispersion follows band gap zones, resulting in large wave-vectors and yet low losses (DOI: 10.1039/C4FD00207E). Viktor Podolskiy presented gold nanorods forming nonlocal plasmonic metamaterials for strong radiation coupling (DOI: 10.1039/C4FD00186A). Nader Engheta presented the concept of EMNZ (ε and μ near-zero) materials, with both relative permeability and permittivity near zero, which offer exceptional properties in light–matter interaction (DOI: 10.1039/C4FD00205A). With both ε and μ approaching zero, the curl of both electric and magnetic field approaches zero; that is, the light becomes static and the electric and magnetic fields decouple. Engheta showed the peculiar electromagnetic behaviour of optical circuit end-points filled with an EMNZ medium. EMNZ materials might actually exist at the Dirac point of artificial materials, topological insulators or in stacked layers.

Mind the gap: the ultimate limit of Ångstrom-gap plasmonics

Many nanoplasmonic structures can provide field enhancement, yet gap-nanoantennas composed by two metal particles separated by a narrow dielectric gap are particularly effective. The nanogaps underlie the spectacular enhancements observed in SERS. Classically, when the gap size approaches zero, one expects arbitrarily large charge densities at the opposite sides of a plasmonic gap, leading to extremely intense fields and strongly red-shifted resonances. Obviously the classical divergence in the gap cannot be correct as it disregards nonlocal and quantum effects. The classical local model fails to describe the smooth transition of the electron density at the interfaces, the strong nonlocal interactions, and the tunnelling regime metal nanoparticles. The proper understanding of such Ångstrom-gaps has been an important topic during this Faraday Discussion. Javier Aizpurua presented an overview of the aspects of the classical local regime, the nonlocal regime and the quantum regime of interaction (DOI: 10.1039/C4FD00196F). He presented his classical effective model, the Quantum Corrected Model (QCM), which correctly describes the main features of optical transport in plasmonic nanogaps, showing a non-local regime below the 2 nm gap and the onset of tunnelling below a threshold gap size of about 4 Ångstrom. Jeremy Baumberg, one of the first to identify the quantum regime in plasmonic gaps, presented his experimental work on plasmon resonance spectroscopy on sub-nm plasmonic gaps, coupling gold nanoparticles to a gold surface (DOI: 10.1039/C4FD00195H). The localised field of the coupled mode appeared extremely sensitive to morphological fluctuations and changes in the dielectric properties of the gap. Clear differences were observed in the spectral positions and intensity of the plasmonic modes on a variety of ultrathin molecular spacers such as cucurbiturils (both filled and empty) and graphene. Gap size, shape and facets, all matter, together with the effective "conductivity" of the molecule in the gap. Ångstrom sensitivity is highly promising for surface plasmon enhanced spectroscopies towards the single molecule limit. In an alternative approach to reach the Ångstrom regime in plasmonic gaps, Sang-Hyun Oh presented the use of atomic layer deposition (ALD) to grow controlled spacers with Ångstrom accuracy (DOI: 10.1039/C4FD00233D). He presented high-density metallic nanogap arrays for ultra-sensitive detection of thin films and biomolecules. Using ALD with Al_2O_3, the gaps could be varied up to 10 nm in order to tune the plasmonic spectra for dedicated SERS applications. Also, Masayuki Futamata presented a wide range of spectra using both nanogaps of silver dimers and silver nanoparticle-surface junctions (DOI: 10.1039/C4FD00188E). The critical role of the molecular junctions was nicely resolved in attenuated total reflection and SERS spectra, again a promising way to monitor molecular conformation. A radically different approach was put forward by Riccardo Sapienza with a plasmonic network consisting of a large cluster of nanoparticles with a multitude of nanogaps, fabricated with designed fractal topology (DOI: 10.1039/C4FD00187G). By changing the percolation of the network, the effects of both long-range interactions and local fields on the fluctuations in the local mode density were studied through the lifetime variation of embedded fluorophores. Local properties in the percolating plasmonic networks appeared to dominate the fluorescence variation.

Strong coupling and quantum plasmonics

Nanoplasmonic structures provide strongly enhanced and confined local fields, which establish ideal conditions to induce strong light-matter coupling in molecular materials. Sufficiently high concentrations of molecular materials in such strongly confined fields will form hybrid light–matter states: polaritonic states with a Rabi splitting which can range from over 100 meV up to almost 1 eV. Thomas Ebbesen presented strong plasmon induced light–matter coupling for various types of molecules (DOI: 10.1039/C4FD00197D). Remarkably, the static absorption of the red-shift polariton state did not lead to generation emission, nor to transient absorption of the same state. The excitation spectra appear very different from the static absorption of the coupled systems, the understanding of which needs further studies. Similarly striving for strong coupling, Sergey Bozhevolnyi presented a theoretical model of the relaxation dynamics of a quantum emitter resonantly coupled to a coherent state of localized surface plasmon (DOI: 10.1039/C4FD00165F). The model predicted relaxation dynamics with non-exponential decay and intriguing transient behaviour on the picosecond timescale. David Zueco went a step further in exploiting strong light–matter coupling for non-linear quantum optics (DOI: 10.1039/C4FD00206G). He calculated the propagation properties of multiple photons in 1-dimensional plasmonic waveguides coupled to multiple qubits, showing that the inter-qubit distance enhances the non-linear response in very strong coupling conditions and that inelastic processes are robust against the number of qubits.

Applications in nanoprobing, nanosensing, nanoruling and nanoimaging

The strong spatial confinement and local enhancement of optical fields in nanoplasmonic structures is bringing forward a wide variety of applications, largely in chemistry and biology. At Faraday Discussion 178 a variety of such applications was presented and discussed. The strong local fields are particularly advantageous for non-linear interactions with quadratic or even cubic response on the enhancement. Alessandro Belardini presented a second harmonic generation on self-assembled gold nanowires (DOI: 10.1039/C4FD00200H). By tilting and exciting the wires with circularly polarized pulsed light, he addressed the optical chiral response of the nanowire surface. Very much in the spirit of the early experiments by Michael Faraday himself, at FD178 Michael Flatté presented the electrical control of Faraday polarization rotation at a liquid/liquid interface using nanoparticles (DOI: 10.1039/C4FD00210E). Specifically, a monolayer of charged magnetic (yttrium iron garnet) nanoparticles was positioned at an electrified liquid/liquid interface, where the polarization fields of neighbouring nanoparticles enhance the Faraday rotation. Electrical control of the nanoparticle adsorption–desorption at the interface then provided electro-variable Faraday rotation.

The strong field enhancements at nanoplasmonic particles and surfaces, obviously imply a highly modified photonic density of states: radiative and non-radiative rates are changed, and even energy transfer rates can be affected. Particularly the fluorescence lifetime is the observable of choice to expose any rate changes. David Richards presented an interesting study using fluorescence lifetime changes, as a function of the distance to a metal surface, as a "nanoruler" to measure out-of-plane distances inside a biological cell; a kind of axial nanotomography exploiting the nanoscale range of the plasmonic interaction (DOI: 10.1039/C4FD00198B). Also using fluorescence lifetime imaging, Yu Chen showed the co-localisation and coupling of gold nanorods with fluorescent protein labels; moreover, by two photon luminescence the nanorods could individually be mapped (DOI: 10.1039/C4FD00199K). The intra-cellular movement of the nanorods was readily applied to monitor endocytosis. Whether the plasmonic mode density also affects the energy transfer rate is a topic of discussion. In practise, the influence depends mainly on the relative sizes of the plasmonic field extent and the molecular distance: a field gradient along the molecular spacing is required. Mikhail Noginov addressed the topic of Förster energy transfer by measuring lifetime changes of donor–acceptor pairs in the vicinity of metallic surfaces (DOI: 10.1039/C4FD00184B). Surprisingly, he concluded that the enhancement of spontaneous emission rates is accompanied by quite strong inhibition of the Förster energy transfer. The localized nanoplasmonic fields and field gradients have been shown to generate local forces on nanoparticles and are interesting as nano-optical tweezers. Olivier Martin presented a visionary picture to use optical forces in nanoplasmonic systems as drivers to control the different manufacturing steps in a nanofactory (DOI: 10.1039/C4FD00224E).

Hot electrons

Nanoplasmonic structures have definite promise in solar energy conversion applications. Firstly, plasmonic antennas are efficient in capturing light and have been explored for the improvement of semiconductor-based solar energy devices. Secondly, surface plasmons excited in Au nanoparticles decay into hot electrons with energies between the vacuum level and the work function of the metal, and the hot electrons and holes can be exploited directly, with the added advantage of their tunability over the solar spectrum. In this context, Martin Moskovits presented a plasmonic photocatalytic device: a 2D gold nanoparticle array combined with a thin electrolytic liquid junction supporting a redox couple (DOI: 10.1039/C4FD00185K). The device produced about 10^{23} hot electrons per cm^2 per second, resulting in photocurrent densities in excess of 40 $\mu A\ cm^{-2}$.

Ultrafast plasmonics

Finally it should be mentioned that ultrafast spectroscopy plays an increasing role in nanoplasmonics research. Femtosecond pulses allow coherent control of the spatio-temporal dynamics of electromagnetic excitations in nanoplasmonic structures and hybrid nanostructures. Control of the nanoscale field localization has potential applications for novel laser structures, the exploitation of optical non-linearities, in biochemistry, *etc.* No experimental work on ultrafast plasmonics was presented at FD178, yet Ortwin Hess presented a theoretical study of the spatio-temporal femtosecond dynamics of coherent amplification and lasing in a planar gain-enhanced nanoplasmonic structure, showing that a subwavelength localised vortex in the mode density can provide a stopped-light feedback allowing cavity-free surface-plasmon polariton nanolasing (DOI: 10.1039/C4FD00181H).

The collective set of papers presented at the Faraday Discussion reflects the wide range of advances in the field of nanoplasmonics in recent years. The field is gradually coming off age and, particularly, the growing diversity of the applications should be noted. From the fundamental point of view, it is shown again and again that, even at the nanometer scale and PHz frequencies, Maxwell's equations hold pretty well. It is mainly through the continuous advances in nanofabrication and new materials that Maxwell's equations unveil previously unimagined surprises. With a growing community active in nanoplasmonics, it can be anticipated that significant advances in the control of the nanoscale light–matter interaction will be achieved in the very near future. Surely Michael Faraday would have enjoyed participating in the inspiring discussion.

References

1 M. Faraday, *Philos. Trans. R. Soc. London*, 1857, **147**, 145–181.

Poster titles

Glass – metal composites, **A. Abdolvand**, *University of Dundee, UK*

Plasmonics for rapid laser micro-joining of dielectrics, **S. Zolotovskaya, G. Tang, Z. Wang and A. Abdolvand**, *University of Dundee, UK*

Electric field induced surface-enhanced Raman spectroscopy for multianalyte detection, **S. Sriram, S. Walia, A. Shah, P. Stoddart and M. Bhaskaran**, *RMIT University, Australia*

DNA origami based assembly of nanoparticle dimers for surface-enhanced Raman scattering detection, **V. Thacker, L. Herrmann, E. Hemmig, D. Sigle, T. Zhang, T. Liedl, J. Baumberg and U. Keyser**, *University of Cambridge, UK*

Control of emission direction of organic semiconductors by surface plasmon polaritons, **I. Samuel, G. Turnbull and S. Zhang**, *University of St Andrews, UK*

Biosynthesis and toxicity studies of the gold nanostructures of various shapes, **K. Matczyszyn, M. Klekotko, A. Podlesny, J. Olesiak-Banska, W. Krezel and M. Samoc**, *Wroclaw University of Technology, Poland*

Electrical excitation and imaging of optical nanoplasmons, **M. Cohen, Y. Abulafia and Z. Zalevsky**, *Bar Ilan University, Israel*

Optimal polarization conversion in coupled dimer plasmonic nanoantennas for metasurfaces, **L.-J. Black, Y. Wang, C. H. de Groot, A. Arbouet and O. L. Muskens**, *University of Southampton, UK*

Facile colorimetric measurement of dipeptidyl peptidase iv(dpp iv) activity *via* peptide capped gold nanoparticles, **H. Aldewachi, M. Hines, Z. Islam, N. Woodroofe and P. Gardiner**, *Sheffield Hallam University, UK*

Plasmonic nanoplate clusters for multimodal biomedical imaging, **A. Ray, J. Hwang, X. Wang and R. Kopelman**, *National Institute of Standards and Technology, USA*

Plasmonic interference effects in scattering of ultrashort electromagnetic pulses on silver nanospheres, **S. Svita and V. Astapenko**, *Moscow Institute of Physics and Technology, Russia*

High-throughput fabrication of nanometer-wide slit arrays with sacrificial layer etching method, **J. Jeong, J. Rhie, W. Jeon, C. S. Hwang and D.-S. Kim**, *Seoul National University, South Korea*

Transformation optics and hidden symmetries, **M. Kraft, J. B. Pendry, S. A. Maier and Y. Luo**, *Imperial College London, UK*

Coherent coupling of molecular resonators with a micro-cavity mode, **A. Shalabney, J. George, J. Hutchison, G. Pupillo, C. Genet and T. Ebbesen**, *ISIS & icFRC, University of Strasbourg and CNRS (UMR 7006), France*

Enhanced conductivity of organic semiconductors through strong light–matter interactions, **J. George, E. Orgiu, J. A. Hutchison, E. Devaux, J.-F. Dayen, B. Doudin, F. Stellacci, C. Genet, P. Samori and T. W. Ebbesen**, *ISIS, University of Strasbourg, France*

Detection of *P. Aeruginosa* bacteria by buckled PDMS silver nanorods arrays as active 3D SERS cage, **S. Kumar, P. Goel, N. Kalyani, P. Mishra and J. P. Singh**, *Indian Institute of Technology Delhi, India*

Tuning strong light-molecule coupling and modification of a phase transition, **S. Wang, T. Chervy, J. George, J. Hutchison, C. Genet and T. Ebbesen**, *ISIS & icFRC, University of Strasbourg, France*

Reciprocity on optical antennas, **A. Singh, G. Calbris, I. Hancu, J. Hugall and N. van Hulst**, *ICFO – The Institute of Photonic Sciences, Spain*

Ultra-strong coupling of molecular materials: spectroscopy and dynamics, **J. George, S. Wang, T. Chervy, A. Canaguier-Durand, G. Schaeffer, J.-M. Lehn, J. A. Hutchison, C. Genet and T. W. Ebbesen**, *Université de Strasbourg, France*

Quantum emitters near layered plasmonic nanostructures, **A. Pors and S. I. Bozhevolnyi**, *University of Southern Denmark, Denmark*

Investigating the SERS properties of DNA-nanoparticle conjugates functionalised with different Raman reporters, **D. Thompson, J. McClung, S. Mabbott, N. Sirimuthu, G. McNay, K. Faulds and D. Graham**, *University of Strathclyde, UK*

Developing tailored gold nanoarchitectures for biological imaging application, **Y.-C. Wang, F. Lesage and A. Kakkar**, *McGill University, Canada*

Light-induced field enhancement in polyacenes, **L. Bursi, A. Calzolari, S. Corni and E. Molinari**, *Università di Modena e Reggio Emilia and Istituto Nanoscienze CNR-NANO-S3, Italy*

Photodetection of broadband incoherent light with nanoridge plasmonics for on-chip photonic interconnection, **J.-H. Kim and J.-S. Yeo**, *Yonsei University, South Korea*

Probing the local potential in metallic nanopores by electrochemical active Surface Enhanced Raman Spectroscopy (EC-SERS), **Y. Li, C. Chen, K. Willems, S. Kerman, L. Lagae, G. Groeseneken, T. Stakenborg and P. Van Dorpe**, *imec & KU Leuven, Belgium*

A bridged rhombic nano-antenna with superimposing hotspots from visible, near infrared to mid-infrared region, **J. Yi, S.-Y. Ding and Z.-Q. Tian**, *Xiamen University, China*

Optical nanoscopy assisted by plasmonics, **E. Cortes and S. A. Maier**, *Imperial College London, UK*

Delocalization of nonlinear optical responses in plasmonic nanoantennas, **S. Viarbitskaya, O. Demichel, G. Colas des Francs and A. Bouhelier**, *Laboratoire Interdisciplinaire Carnot de Bourgogne, France*

A correlated study of LSPR, photoluminescence and SERS, **K. Q. Lin, J. Yi, S. Hu and B. Ren**, *Xiamen University, China*

The preparation of Letrozole loaded PLA-PEG nanoparticles as drug delivery systems, **S. Massadeh, M. Alaamery, D. Al Gudairi and O. Al Ahmad**, *King Abdulla International Medical Research Centre, Saudi Arabia*

Reversible compression of an optical piston through Kramers dynamics, **G. Schnoering and C. Genet**, *ISIS, Université de Strasbourg, France*

300 mm wafer-level, ultra-dense arrays of Au-capped nanopillars with sub-10 nm gaps as reliable SERS substrates, **J. Li, Y. Li, C. Chen, H. Jans, X. Xu, N. Verellen, I. Vos, Y. Okumura, L. Lagae and P. Van Dorpe**, *imec, Belgium*

Terahertz-radiation-induced ionization effect in a single gold nano island, **M. Seo, J.-H. Kang, H.-S. Kim, S. Lee and C. Kim**, *Korea Institute of Science and Technology, South Korea*

Enhanced antibacterial activity of nano-scale silver particles, **H. Tamiyakul, S. Dubas, S. Tanasupawat and W. Warisnoicharoen**, *Chulalongkorn University, Thailand*

Facile synthesis of magnetic SERS substrates *via* galvanic displacement of silver onto commercially available cobalt micro-powder, **S. Mabbott and D. Graham**, *University of Strathclyde, UK*

DNA base pair resolution for optimized enhancement of fluorescence by gold nanorods: biological applications, **C. Nogues, M. Gordel, R. Chenoufi, E. Henry, K. Matczyszyn, M. Samoc, E. Deprez and M. Buckle**, *CNRS, France*

Effect of non-locality on the Casimir forces between graphene sheets, **Y. Francescato and V. Giannini**, *Imperial College London, UK*

Exploiting Fano resonance in surface enhanced Raman scattering, **M. Hardy, R. McCarron, M. D. Doherty, K. M. Holsgrove, P. P. Mulholland, P. Dawson and R. J. Winfield**, *Queen's University Belfast, UK*

Fano-resonance plasmon-enhanced Raman scattering on metallic single-crystal surfaces coupled with nanoparticle clusters, **S.-Y. Ding, J. Yi and Z.-Q. Tian**, *Xiamen University, China*

Diffractive interface theory: A non-local susceptibility approach to metasurfaces, **C. Roberts, S. Inampudi and V. Podolskiy**, *University of Massachusetts - Lowell, USA*

SHG from metallic nano structures, **A. Salomon**, *Bar Ilan University, Israel*

Crumpled thin gold layer: Beating the rational SERS substrate design?, **L. Štolcová, J. Proška and M. Procházka**, *Czech Technical University in Prague, Czech Republic*

Ultrafast plasmonic switching *via* optical cavity modes, **C. McPolin, N. Olivier, J.-S. Bouillard, D. O'Connor, A. Krasavin, W. Dickson, G. Wurtz and A. Zayats**, *King's College London, UK*

Detection of organic pollutant by SERS using Diazonium salt-functionalized nanostructures, **N. Lidgi-Guigui, I. Tijunelyte, S. Betelu, A.-C. Schnepf, I. Ignatiadis, E. Guenin, J. Ibrahim, T. Toury, E. Rinnert and M. Lamy de la Chapelle**, *CSPBAT, Université Paris 13 - Sorbonne Paris Cité, France*

Nano-plasmonics tuned "click chemistry" monitored in situ by Surface Enhanced Raman Spectroscopy, **N. Lidgi-Guigui, I. Tijunelyte, E. Guenin, J. Ibrahim, T. Toury, E. Rinnert and M. Lamy de la Chapelle**, *CSPBAT, Université Paris 13 - Sorbonne Paris Cité, France*

Theoretical experimental and study of the SERS enhancement localization on gold nanodiscs, **N. Lidgi-Guigui, R. Gillibert, M. Sarkar, J.-F. Bryche, G. Barbillon, B. Bartenlian, M. Canva and M. Lamy de la Chapelle**, *CSPBAT, Université Paris 13 - Sorbonne Paris Cité, France*

Structure-function analysis of the Syk protein by Surface Enhanced Raman Spectroscopy and Western Blot, **N. Lidgi-Guigui, M. Cottat, N. Varin-Blank, C. Le Roy and M. Lamy de la Chapelle**, *CSPBAT, Université Paris 13 - Sorbonne Paris Cité, France*

Near field to far-field considerations of SERS measurements on nanocylinders in a periodic arrangement, **N. Lidgi-Guigui, M. Cottat, F. Colas, N. Guillot, A. Belhadj, R. Gillibert, A. Toma, E. DiFabrizio, P. Gucciardi and M. Lamy de la Chapelle**, *CSPBAT, Université Paris 13 - Sorbonne Paris Cité, France*

Multiplexed label-free detection of single proteins, **A. Sebesta, A. Weigel and P. Kukura**, *University of Oxford, UK*

Third-order nonlinearity of plasmonic hyperbolic metamaterials: Numerical design, **A. Neira, S. Peruch, M. Nasir, N. Olivier, W. Dickson, G. Wurtz and A. Zayats**, *King's College London, UK*

Confinement of hybrid plasmonic mode below 10-nm gap in a planar structure, **R. Mehfuz and P.Dawson**, *Queen's University Belfast, UK*

Ultrafast coherent nonlinear response in arrays of multipolar plasmonic resonators, **G. Sartorello, N. Olivier, J. Zhang, W. Yue, D. Gosztola, G. Wiederrecht, A. Krasavin, P. Ginzburg, G. Wurtz and A. Zayats**, *King's College London, UK*

Fabrication of 20 nm-gap in gold thin film using copper mask, **J. Rhie, Y.-M. Bahk, J. Jeong, J. Kim, G. Choi and D.-S. Kim**, *Seoul National University, Korea*

LDOS mapping of plasmonic nanoantennas using single molecules, **J. T. Hugall, A. Singh, P. M. de Roque, G. Calbris and N. F. van Hulst**, *ICFO -Institute of Photonic Science, Spain*

Measurement of electric near field enhancement on the nanometer sized metallic gaps in optical-near IR range, **T. Kang, J. Rhie, J. Kim, Y.-M. Bahk and D.-S. Kim**, *Seoul National University, Korea*

Metallic nanorings *versus* ring-shaped apertures in thin metal films: a comparative plasmon hybridization study, **V. E. Bochenkov and D. S. Sutherland**, *University of Aarhus, Denmark*

First principles calculations for surface plasmon decay dynamics, **P. Narang, R. Sundaraman, A. S. Jermyn, A. M. Brown, W. A. Goddard III and H. A. Atwater**, *California Institute of Technology, USA*

Ultrasensitive miRNA detection using a flexible nanoplasmonic bio-interface by transfer printing, **J. Lee and J.-S. Yeo**, *Yonsei University, Republic of Korea*

Nano-optics of molecular-shunted plasmonic nanojunctions, **F. Benz, C. Tserkezis, L. O. Herrmann, B. de Nijs, A. Sanders, D. O. Sigle, L. Pukenas, S. D. Evans, J. Aizpurua and J. J. Baumberg**, *University of Cambridge, UK*

The Skinner Prize for the best poster was jointly awarded to Ms Lucie Štolcová of Czech Technical University in Prague, Czech Republic, for her poster on beating the rational SERS substrate design with a crumpled thin gold layer and to Mr Mike Hardy of Queen's University Belfast, United Kingdom, for his poster on exploiting Fano resonance in surface enhanced Raman scattering.

List of participants

Professor Amin Abdolvand, *University of Dundee, United Kingdom*
Dr Patricia Abellan, *SuperSTEM, United Kingdom*
Dr Ali Adawi, *University of Hull, United Kingdom*
Professor Javier Aizpurua, *Spanish Council for Scientific Research CSIC-UPV/EHU, Spain*
Dr Pablo Albella, *Imperial College London, United Kingdom*
Mr Hasan Aldewachi, *Sheffield Hallam University Registry, United Kingdom*
Dr Muhammad Asif Bashir, *Seagate Technology, Northern Ireland*
Professor Jeremy Baumberg, *University of Cambridge, United Kingdom*
Dr Alessandro Belardini, *Sapienza Università di Roma - SBAI department, Italy*
Mr Felix Benz, *University of Cambridge, United Kingdom*
Mr Rodrigo Berte, *Imperial College London, United Kingdom*
Miss Julia Bieniewska, *King's College London, United Kingdom*
Mr Leo-Jay Black, *University Of Southampton, United Kingdom*
Dr Vladimir Bochenkov, *iNANO, Aarhus University, Denmark*
Dr Jean-Sebastien Bouillard, *University of Hull, United Kingdom*
Professor Dr Sergey Bozhevolnyi, *Centre for Nano Optics, University of Southern Denmark, Denmark*
Dr Avi Braun, *Imperial College London, United Kingdom*
Professor Dr Mark Brongersma, *Stanford University, U.S.A.*
Mr Luca Bursi, *Università di Modena e Reggio Emilia and Istituto Nanoscienze CNR-NANO-S3, Italy*
Professor Dr Irina Buyanova, *Linkoping, Sweden*
Dr Gaëtan Calbris, *ICFO, Spain*
Mr Javier Cambiasso, *Imperial College, United Kingdom*
Dr Monica Carreira Mendez, *Royal Society of Chemistry, United Kingdom*
Dr Marta Castro-Lopez, *King's College London, United Kingdom*
Dr Ruth Chantry, *SuperSTEM, United Kingdom*
Dr Yu Chen, *Strathclyde University, United Kingdom*
Professor Dr Weimin Chen, *Linkoping University, Sweden*
Mr Thibault Chervy, *ISIS, Université de Strasbourg and CNRS, France*
Dr Kim Chulki, *Korea Institute of Science and Technology, South Korea*
Mr Toon Coenen, *Delmic BV, Netherlands*
Mr Moshik Cohen, *Bar Ilan University, Israel*
Professor Lesley Cohen, *Imperial College, United Kingdom*
Dr Emiliano Cortes, *Imperial College London, United Kingdom*
Dr Paul Dawson, *Queens University Belfast, United Kingdom*
Mr Jacob Jan de Boer, *Delmic BV, Netherlands*
Dr Bart de Nijs, *University of Cambridge, United Kingdom*
Mr Pablo de Roque, *ICFO - The Institute of Photonic Sciences, Spain*
Mr Sander den Hoedt, *Delmic BV, Netherlands*
Mr Frederik Dieleman, *Imperial College London, United Kingdom*
Professor Thomas Ebbesen, *ISIS, University of Strasbourg, France*

Professor Nader Engheta, *University of Pennsylvania, U.S.A.*
Professor Michael Flatté, *University of Iowa, U.S.A.*
Dr Yan Francescato, *Imperial College London, United Kingdom*
Professor Dr Masayuki Futamata, *Saitama University, Japan*
Mr Michele Gaio, *King's College London, United Kingdom*
Dr Zhiqiang Gao, *National University of Singapore, Singapore*
Professor F. Javier García de Abajo, *ICFO - The Institute of Photonic Sciences, Spain*
Dr Sylwester Gawinkowski, *Institute of Physical Chemistry of the Polish Academy of Sciences, Poland*
Mr Sylvain Gennaro, *Imperial College London, United Kingdom*
Dr Jino George, *ISIS, University of Strasbourg, France*
Dr Vincenzo Giannini, *Imperial College London, United Kingdom*
Dr Pavel Ginzburg, *King's College London, United Kingdom*
Professor Duncan Graham, *University of Strathclyde, United Kingdom*
Dr Fredrik Hage, *SuperSTEM Laboratory, United Kingdom*
Mr Mike Hardy, *Queen's University Belfast, United Kingdom*
Ms Elisa Hemmig, *Department of Physics, University of Cambridge, United Kingdom*
Professor Dr Ortwin Hess, *Imperial College London, United Kingdom*
Dr James T. Hugall, *ICFO - Institute of Photonic Science, Spain*
Dr Paloma A, Huidobro, *Imperial College, Department of Physics, United Kingdom*
Dr James Hutchison, *ISIS, Université de Strasbourg and CNRS, France*
Mr Jeeyoon Jeong, *Center for Subwavelength Optics, Seoul National University, South Korea*
Mr Taehee Kang, *Seoul National University, South Korea*
Professor Sergei Kazarian, *Imperial College London, United Kingdom*
Professor Jacob Khurgin, *Johns Hopkins University, U.S.A.*
Ms Jeong-Hyeon Kim, *Yonsei University, South Korea*
Professor Alexei Kornyshev, *Imperial College London, United Kingdom*
Dr Santhosh Kotni, *Weizmann Institute of Science, Israel*
Mr Matthias Kraft, *Imperial College London, United Kingdom*
Dr Katrina Kramer, *Royal Society of Chemistry, United Kingdom*
Mr Samir Kumar, *Indian Institute of Technology Delhi, India*
Miss Sarah Latham, *Royal Society of Chemistry, United Kingdom*
Ms Jihye Lee, *Yonsei University, South Korea*
Mr Yi Li, *IMEC, Belgium*
Dr Nathalie Lidgi-Guigui, *CSPBAT, France*
Dr Katie Lim, *Royal Society of Chemistry, United Kingdom*
Mr Kaiqiang Lin, *Xiamen University, China*
Dr Samuel Mabbott, *University of Strathclyde, United Kingdom*
Professor Stefan Maier, *Imperial College London, United Kingdom*
Professor Dr Olivier Martin, *Swiss Federal Institute of Technology Lausanne (EPFL), Switzerland*
Dr Salam Massadeh, *King Abdulla International Medical Research Centre, Saudi Arabia*
Dr Katarzyna Matczyszyn, *Departament of Chemistry, Poland*
Dr Ryan McCarron, *Queen's University Belfast, United Kingdom*
Mr Cillian McPolin, *King's College London, United Kingdom*
Dr Reyad Mehfuz, *Queen's University Belfast, United Kingdom*
Dr Alberto Moscatelli, *Nature Nanotechnology, United Kingdom*
Professor Martin Moskovits, *University of California, Santa Barbara, U.S.A.*
Professor Andrew Mount, *The University of Edinburgh, United Kingdom*
Mr Jan Philipp Balthasar Mueller, *Harvard University, U.S.A.*
Ms Prineha Narang, *California Institute of Technology, U.S.A.*

Prof Evgeniy Narimanov, *Purdue University, U.S.A.*
Mr Andres Neira, *King's College London, United Kingdom*
Mr Michael Nielsen, *Imperial College London, United Kingdom*
Professor Mikhail Noginov, *Norfolk State University, U.S.A.*
Dr Claude Nogues, *CNRS, France*
Professor Sang-Hyun Oh, *University of Minnesota, Twin Cities, U.S.A.*
Dr Rupert Oulton, *Imperial College London, United Kingdom*
Sir John Pendry, *Imperial College London, United Kingdom*
Dr Anatoliy Pinchuk, *University of Colorado at Colorado Springs, U.S.A.*
Professor Viktor Podolskiy, *University of Massachusetts Lowell, U.S.A.*
Dr Anders Pors, *University of Southern Denmark, Denmark*
Ms Maria Pszona, *Institute of Physical Chemistry PAS, Poland*
Dr Mohsen Rahmani, *Imperial College London, United Kingdom*
Dr Aniruddha Ray, *National Institute of Standards and Technology, U.S.A.*
Ms Jiyeah Rhie, *Seoul National University, South Korea*
Professor David Richards, *King's College London, United Kingdom*
Mr Christopher Roberts, *University of Massachusetts - Lowell, U.S.A.*
Dr Francisco Rodríguez Fortuño, *King's College London, United Kingdom*
Miss Diane Roth, *King's College London, United Kingdom*
Dr Adi Salomon, *Bar Ilan University, Israel*
Professor Ifor Samuel, *University of St Andrews, United Kingdom*
Dr Riccardo Sapienza, *King's College London, United Kingdom*
Mrs Marta Sarnowska, *University of Warsaw, Poland*
Mr Giovanni Sartorello, *King's College London, United Kingdom*
Dr Silvan Schmid, *Technical University of Denmark, Denmark*
Mr Gabriel Schnoering, *ISIS, Université de Strasbourg, France*
Dr Zachary Schultz, *University of Notre Dame, U.S.A.*
Mr Aleksandar Sebesta, *University of Oxford, United Kingdom*
Dr Minah Seo, *Korea Institute of Science and Technology, South Korea*
Dr Atef Shalabney, *ISIS & icFRC, University of Strasbourg and CNRS, France*
Professor Vladimir Shalaev, *Purdue University, U.S.A.*
Miss Viktoryia Shautsova, *Imperial College London, United Kingdom*
Professor Dame Concita Sibilia, *Università di Roma La Sapienza, Italy*
Mr Anshuman Singh, *ICFO – The Institute of Photonic Sciences, Spain*
Ms Claire Springett, *Royal Society of Chemistry, United Kingdom*
Dr Sharath Sriram, *RMIT University, Australia*
Mr Sergey Svita, *Moscow Institute of Physics and Technology, Russia*
Miss Hathaichanok Tamiyakul, *Chulalongkorn University, Thailand*
Dr David Thompson, *University of Strathclyde, United Kingdom*
Professor Dr Niek van Hulst, *ICFO - The Institute of Photonics Sciences, Spain*
Dr Sviatlana Viarbitskaya, *Laboratoire Interdisciplinaire Carnot de Bourgogne, France*
Mr Shaojun Wang, *University of Strasbourg, France*
Miss Yu-Chen Wang, *McGill University, Canada*
Professor Jong-Souk Yeo, *Yonsei University, South Korea*
Mr Jun Yi, *State Key Laboratory of Physical Chemistry of Solid Surfaces, College of Chemistry and Chemical Engineering, Xiamen University, China*
Ms Ruth Zadik, *Royal Society of Chemistry, United Kingdom*
Professor Anatoly Zayats, *Kings College London, United Kingdom*
Mr Yu Zhou, *Imperial College London, United Kingdom*
Dr David Zueco, *Universidad de Zaragoza, Spain*
Ms Lucie Štolcová, *Czech Technical University in Prague, Czech Republic*